"It is known that solutions to probabilistic problems should be well-understood in the discrete-time models first before the same problems should be studied in the related continuous-time models. The authors fulfilled the outstanding gap very successfully by providing an extensive literature source on the most generally formulated multiple stopping problems in the discrete-time case."

Pavel V. Gapeev, *London School of Economics and Political Science, UK*

"The book presents an encyclopaedic approach to optimization problems with multiple stopping times as the strategies. The book is unique, there is no such position in the literature. Multiple stopping times appear in a natural way in many human activities and frequently people are not aware of possibility of optimization among such strategies. The authors show a number of motivating applications of the developed theory in sequential methods of statistics, selection problems, investment (management) problems and in behavioural ecology. Reading the book substantial mathematical skills are not required. The book is written to general audience of decision makers and for researchers in mathematics, statistics, economics, engineering, operation research and business administration. In many problems exact solutions are presented. Numerical methods are also shown."

Łukasz Stettner, *Institute of Mathematics Polish Acad. Sci.*

"I highly recommend this book for its comprehensive approach and meticulous attention to detail. It offers a concrete explanation and an extensive bibliography that will help clarify motivations and support the development of effective solutions. Featuring over 450 carefully selected papers and books, it is an invaluable resource for anyone working on the optimal stopping problem, stochastic optimization with sequential procedures, and more. The content spans multiple stopping models, stopping games, financial applications, and numerical asymptotic solutions, providing readers with the insights they need to identify the most suitable subjects for their work. Don't miss the opportunity to add this essential title to your collection."

Masami Yasuda, *the emeritus professor at Chiba University.*

"This book offers a comprehensive exploration of the fascinating field, bridging theoretical foundations with practical examples. A valuable resource, it caters to statisticians, financial mathematicians, and stochastic control experts. I wholeheartedly encourage fellow researchers and practitioners to dive into its insights. Optimal stopping strategies, discussed within, benefit not only statisticians but also professionals in finance and banking."

Philippe J. S. De Brouwer, *HSBC and AGH University of Krakow*

"Writing a comprehensive and coherent review of a broad scientific field is always a challenge, and this is certainly true for the theory and applications of optimal stopping. From the Foreword (Introduction) and the detailed table of contents, it is evident that the authors G. Sofronov and K. Szajowski have invested significant effort into this endeavour, and their work deserves sincere commendation. The book covers a vast array of optimal stopping methods and includes numerous application examples, ensuring that even seasoned experts will find new and intriguing material. The authors' expertise and dedication are apparent, making this book an invaluable resource for researchers in optimal stopping theory and those interested in its practical applications."

F. Thomas Bruss, *Professor emeritus and Invited Professor at the department of Mathematics, Université Libre de Bruxelles*

"This is a must-read for any scholar wishing to understand both the theory and applications of multiple stopping problems in the discrete-time setting. Sofronov and Szajowski should be applauded for providing the reader with one of the most comprehensive and accessible resources on this subject to date."

Philip Ernst, *Chair in Statistics and Royal Society Wolfson Fellow*

Multiple Stopping Problems

This book presents the theory of rational decisions involving the selection of stopping times in observed discrete-time stochastic processes, both by single and multiple decision-makers. Readers will become acquainted with the models, strategies, and applications of these models.

It begins with an examination of selected models framed as stochastic optimization challenges, emphasizing the critical role of optimal stopping times in sequential statistical procedures. The authors go on to explore models featuring multiple stopping and shares on leading applications, particularly focusing on change point detection, selection problems, and the nuances of behavioral ecology. In the following chapters, an array of perspectives on model strategies is presented, elucidating their interpretation and the methodologies underpinning their genesis. Essential notations and definitions are introduced, examining general theorems about solution existence and structure, with an intricate analysis of optimal stopping predicaments and addressing crucial multilateral models. The reader is presented with the practical application of models based on multiple stopping within stochastic processes. The coverage includes a diverse array of domains, including sequential statistics, finance, economics, and the broader generalization of the best-choice problem. Additionally, it delves into numerical and asymptotic solutions, offering a comprehensive exploration of optimal stopping quandaries.

The book will be of interest to researchers and practitioners in fields such as economics, finance, and engineering. It could also be used by graduate students doing a research degree in insurance, economics or business analytics or an advanced undergraduate course in mathematical sciences.

Georgy Sofronov received his PhD degree in Probability Theory and Mathematical Statistics from Moscow State University in 2002. He has held academic positions at several universities including the University of Queensland and the University of Wollongong. Currently, he is an Associate Professor in Statistics at Macquarie University. He serves on the editorial boards of Statistical Papers and Methodology and Computing in Applied Probability. His research interests include Markov chain Monte Carlo simulation, the Cross-Entropy method, change-point problems and optimal stopping rules.

Krzysztof Szajowski received his PhD degree and habilitation in Mathematical Sciences from the Technical University of Wrocław in 1980 and 1996, respectively. Since 1973, he has held academic and visiting research positions at Wroclaw University of Technology, Delft University of Technology, Purdue University and the Institute of Mathematics of the Polish Academy of Sciences. Currently, he is an Emeritus Professor at Wroclaw University of Science and Technology. He is a member of the editorial board of *Mathematica Applicanda* and former its editor-in-chief, as well as a member of the editorial board of the journal *Scientiae Mathematicae Japonicae* and *Annals of Dynamic Games*. His current research interests lie in probability theory and mathematical statistics, applied mathematics, change-point detection, optimal stopping problems and game theory models.

Multiple Stopping Problems

Unilateral and Multilateral Approaches

Georgy Sofronov and Krzysztof Szajowski

CRC Press
Taylor & Francis Group
Boca Raton London New York

CRC Press is an imprint of the
Taylor & Francis Group, an **informa** business
A CHAPMAN & HALL BOOK

Designed cover image: *The cover figure/picture is the author's own responsibility and construction. The tools which the author has used to do that is Wolfram Mathematica©*

First edition published 2025
by CRC Press
2385 NW Executive Center Drive, Suite 320, Boca Raton FL 33431

and by CRC Press
4 Park Square, Milton Park, Abingdon, Oxon, OX14 4RN

CRC Press is an imprint of Taylor & Francis Group, LLC

ISBN: 978-1-032-52543-3 (hbk)
ISBN: 978-1-032-52544-0 (pbk)
ISBN: 978-1-003-40710-2 (ebk)

DOI: 10.1201/ 9781003407102

Typeset in Latin Modern font
by KnowledgeWorks Global Ltd.

Publisher's note: This book has been prepared from camera-ready copy provided by the authors.

Contents

Contributors

Georgy Sofronov
Macquarie University
Sydney, NSW 2109, Australia

Krzysztof Szajowski
Wrocław University of Sci. & Tech.
Wrocław, Dolnoslaskie, Poland

Foreword

THIS BOOK contains selected topics in stochastic control: theory of discrete-time multiple stopping and stopping games. Part I with Chapter 1 only serves as a presentation of the selected models that were or could be formulated as the stochastic optimization problem with multiple stopping times as the strategies. It is a guide to the topics covered in the book. The objective of modeling is to comprehend and manage the system or environment. Section 1.1 recalls both optimal stopping models and problems that lead to optimization in more complex structures of random fields. Section 1.2 reminds us that optimally chosen stopping times are a key component of sequential statistical procedures. It is precisely sequential statistical methods that have shown that selecting the appropriate stopping time can be a fundamental problem in many stochastic modeling issues. However, numerous issues modeled using stochastic processes and problems that boil down to selecting the optimal stopping time(s) for these models have necessitated a broader view of this type of modeling. Examples of these issues are discussed in Section 1.3. The next Sections 1.4 specify models with multiple stopping by one decision-maker, and 1.5 multilateral stopping of stochastic sequences. The last Section 1.6 of the chapter is devoted to leading applications: multiple stopping in sequential methods of statistics (estimation problems, adaptive sampling, hypothesis testing, clinical trials, disorders detection problem), selection problems (unilateral or multilateral) with multiple attempts, buying–selling problems, behavioral ecology problems. The initial sections provide reasons and motivations for why optimal multiple-stopping tasks are discussed separately in another monograph. The expansion of the material presented in the sections of Chapter 1 is contained in Parts II and III.

Chapter 2 of Part II offers various perspectives on the set of model strategies, their interpretation, and methods of creation. In Section 2.1, the basic notations and definitions necessary for formulating models of multiple stopping of stochastic processes are introduced, while in Section 2.2, general theorems regarding the existence and structure of solutions are discussed. More details on the analysis of optimal stopping problems are found in Sections 2.3–2.5. Multiple stopping as an optimization problem warrants a separate study, particularly due to the potential for conflicting goals among decision-makers (v. Chapter 3 of Part II). Part III of the book focuses on applications of models based on multiple stopping of stochastic processes with discrete time. Separate chapters cover sequential methods of statistics (Chapter 4), the applications in finance and economics (Chapter 5), the generalization of the best-choice problem (Chapter 6), and numerical and asymptotic solutions for optimal stopping problems in Chapter 7.

Additional chapters (Appendices IV) of the book draw upon concepts from probability theory, risk theory, and optimization theory. Offers definitions and simplified explanations of certain advanced tools in these domains. The book focuses on students in various disciplines, including economics, engineering, operations research, finance, business, and mathematics. It is designed to be accessible to graduate students in economics and business administration, while also being suitable for advanced undergraduates in mathematics and science who have completed calculus and elementary probability courses. Although some sections may require a slightly stronger mathematical background, readers are encouraged not to be discouraged by technical content beyond their current understanding. Such sections can be skipped without impeding comprehension of the main ideas and results. In particular, the presentation avoids delving into measure theory, ensuring accessibility to a wider audience with diverse backgrounds.

List of Figures

List of Tables

Notations

The following notations and abbreviations are used in this manuscript:

$\mathbb{N}, \widetilde{\mathbb{N}}, \overline{\mathbb{N}}$	the set of natural numbers (p. 8), $\widetilde{\mathbb{N}} = \mathbb{N} \cup \{0\}$ (p. 8), $\overline{\mathbb{N}} = \widetilde{\mathbb{N}} \cup \{+\infty\}$ (p. 10)
$(\Omega, \mathcal{F}, \boldsymbol{P})$	a probability space (p. 50)
$\boldsymbol{P}_X(\cdot)$	the probability distribution of the random variable X (p. 293)
$\boldsymbol{E}(\cdot)$	the expected value of a random variable (p. 294)
	the expectation operator with respect to $\boldsymbol{P}$ (v. (1.1) on p. 10).
$\binom{n}{k}$, $\boldsymbol{C}_n^k = \frac{n!}{k!(n-k)!}$	the binomial coefficient (v. 197)
$Z_{m_1,\ldots,m_k}$, $m_j \in \mathbb{N}$, $j = 1, 2, \ldots, k$	a random field (2.1) (p. 49)
$\boldsymbol{\mathcal{F}} = \{\mathcal{F}_{m_1,\ldots,m_k}\}_{m_j \in \mathbb{N},\ j=1,2,\ldots,k}$	a filtration (see p. 12 and 50)
$\{\delta_n\}_{n\in\mathbb{N}}$, $\tau(\omega)$	the Markov time (stopping time) related to the binary decision sequences (v. p. 12)
$\boldsymbol{\tau}_j = (\tau_1, \ldots, \tau_j)$, $j = 1, \ldots, k$	j-multiple stopping rule (j-MSR) (Definition 2.1 on p. 50)
$a^- = -\min\{a, 0\}$	the absolute value of the negative part of the number a (v. 61)
$a^+ = \max\{a, 0\}$	the positive part of the number a (v. 51)
$\boldsymbol{x} = (x_1, x_2, \ldots, x_n)$	the vector (v. 11)
$\boldsymbol{x}_{-j} = (x_1, \ldots, x_{j-1}, x_{j+1}, \ldots, x_n)$	the vector without j-th component (v. 94).
$(a, \boldsymbol{x}_{-j}) = (\boldsymbol{x}_{-j}, a)$	the vector $(x_1, \ldots, x_{j-1}, a, x_{j+1}, \ldots, x_n)$ (v. 94)
$\sum_{i=1}^n p_i = p_1 + p_2 + \ldots + p_n$	the sum of $p_1, p_2, \ldots, p_n$ (v. p. 10)
$\prod_{i=1}^n p_i = p_1 \cdot p_2 \cdot \ldots \cdot p_n$	product of $p_1, p_2, \ldots, p_n$ (p. 11)
$\mathbb{I}_A(\cdot)$	the indicator function of the set A (v. 27)

Part I

Preface on Multiple Stopping Models

THE DEVELOPMENT of research on stochastic processes has led to optimization tasks understood in various ways. Mathematical models of system control under various random factors, with clearly defined requirements as to the behavior of the system, led to questions about the control of system parameters in a way that guaranteed its optimal or expected behavior. Stochastic optimal control is a subfield of control theory that deals with the existence of uncertainty, either in observations or in the noise that drives the evolution of the system. The system designer assumes, in a Bayesian probability-driven fashion, that random noise with known probability distribution affects the evolution and observation of the state variables. Stochastic control aims to design the time path of the controlled variables that perform the desired control task with minimum cost, although defined, despite the presence of this noise. The control theory of stochastic systems has distinguished tasks in which acceptable strategies are moments of stopping observation. An example is the sequential estimation of the parameters of the observed process, the testing of statistical hypotheses, or the selection of the starting and ending time of taking products in the production process. Due to the importance of this type of model, they were selected for in-depth research. We distinguish optimization tasks when the goal is to determine the extreme value of the problem (minimum or maximum), and, whenever possible, to determine a strategy that realizes this value, and problems in which the quality of the process is described using many criteria (multi-criteria) (v. Belton and Stewart (2002)). We can rationalize goals in multi-criteria problems, but the book is restricted to the case where the different criteria are related to different decision-makers. We are looking for the rational behavior of the parties involved in the decision process. One of the important and frequently used approaches to the definition of rationality is the assumption that the process is of interest to many decision-makers. Each of them has a specific share in the observed process and a strategy for its modification. Methods of game theory can be used to determine rational behavior.

In the first part, we will present a discrete-time stochastic model in which the moments of stopping constitute the strategies of one or more decision-makers. The one-decision-maker model of multiple stopping was first formulated by Haggstrom (1967), and the two-decision-maker model by Dynkin (1969). Later work expanded the classes of model-defining processes (v. Nikolaev (1977, 1998)[1], Sakaguchi (1978), Móri (1984), Nikolaev and Sofronov (2007), Sofronov (2013, 2016, 2018)), and rationalization methods with more decision-makers (v. Yasuda et al. (1980, 1982), Stadje (1985), Nakai (1997), Sakaguchi (1978, 1995), Szajowski (1993)).

It is worth mentioning here that in the early history of research on optimal stopping (optimal control, Markov decision processes) until the end of the 1990s, practically all approaches to solving such problems consisted of searching for analytical solutions or, in case of difficulties, adopting near-optimal solutions. Since the optimal strategy defines the area of states of the observed process, after reaching which we obtain the desired optimal state, the tasks were reduced to searching for the edge of the optimal stopping area. It was noticed that numerical construction of the boundary is a good way, which triggered both deterministic and stochastic methods of constructing these

[1]The tally of Nikolaev's paper is listed on Math-Net.ru.

boundaries (v. Belomestny and Schoenmakers (2018)). Some crucial techniques in such an approach to multiple stopping problems are the subject of the Chapter 7.

The history of optimal stopping problems, a subfield of probability theory, also begins with gambling. One of the first discoveries is credited to the eminent English mathematician Arthur Cayley (1875) of the University of Cambridge. In 1875, he found an optimal stopping strategy (stopping rule) for purchasing lottery tickets (v. Ferguson (1989), Hill (2009)).

1

Motivation for the Multiple Stopping and Selected Application

> ***The Trial of the Pyx*** is a judicial ceremony in the United Kingdom to ensure that newly minted coins from the Royal Mint conform to their required dimensional and fineness specifications. ··· Coins to be tested are drawn from the regular production of the Royal Mint. The Deputy Master of the Mint must, throughout the year, randomly select several thousand sample coins and place them aside for the Trial. These must be in a certain fixed proportion to the number of coins produced. For example, for every 5,000 bimetallic coins issued, one must be set aside, whereas for silver Maundy money the proportion is one in 150. ··· If the coinage is found to be substandard, the trial carries a punishment for the Master of the Mint of a fine, removal from office, or imprisonment. The last master of the mint to be punished was Isaac Newton in 1696. v. The Royal Mint sends largest coins in its history to 700-year-old Trial of the Pyx
>
> Royal Mint description. D. Thomas (2023)

In THIS CHAPTER, we describe the theoretical and applied motivation for multiple stopping in general as a model of one or multiple (multiple multilateral) decision-makers. Since decision-makers operate in conditions of random parameters of the surrounding environment, we assume that for modeling purposes, there is a fixed probability space $(\Omega, \mathcal{F}, \boldsymbol{P})$. We also outline the book's structure and introduce several typical application examples.

There are different versions of rational stopping of stochastic sequences by one decision-maker.

1. *Optimal Stopping Theory* which involves finding the optimal time to stop a stochastic process to maximize the expected value of a given objective function. The decision-maker observes a sequence of random variables and must decide when to stop observing them. The optimal stopping rule depends on the distribution of the random variables and the objective function.

DOI: 10.1201/9781003407102-1

2. *Sequential Analysis* is an approach which involves making sequential decisions based on the observations of a stochastic process. The decision-maker must choose whether to continue to observe the process or stop and take action based on the information obtained so far. The decision to continue or stop is based on a predetermined stopping rule.
3. *Bayesian Stopping Rules* approach is based on Bayesian methods. In such models, we determine when to stop observing a stochastic process. The decision-maker updates their beliefs about the underlying distribution of the process based on the observations and determines when to stop based on a prespecified utility function.
4. *Monte Carlo Methods* (MCM) involves generating random samples from a stochastic process to estimate the underlying distribution. The decision-maker can use these estimates to determine when to stop observing the process based on a prespecified stopping rule.

There are many other variations of rational stopping of stochastic sequences by a decision-maker, but these are a few examples. The choice of approach depends on the nature of the problem and the preferences of the decision-maker (v. Samuel (1961)).

1.1 Stochastic, Dynamic Phenomena Modeling

Motto:
From Concept to Control: Navigating Stochastic Phenomena through Mathematical Modeling and Optimal Decision-Making

The original aim of the models of stochastic phenomena was to accurately represent them. A significant milestone is achieving a model whose parameters can be adjusted to closely replicate the observed phenomenon. The mathematical modeling process typically involves several stages. Initially, a general mathematical framework is conceived for the variables involved in the phenomenon. Subsequently, efforts focus on refining the model to achieve the best fit, often involving parameter input and statistical methods. This refinement process, known as calibration, is closely tied to data collection. Sequential analysis, a statistical technique that allows staged study execution, allows researchers to stop analysis once preliminary results suffice to confirm the hypothesis of the investigation. The crux of our interest lies in determining the optimal juncture for incremental decisions, which is the central theme of this book.

The next step was to use the model to modify the actual process. The motivation was control tasks, model fit, and various activity-related questions, making mathematical modeling able to understand and control the situation. Optimal control is an extension of the calculus of variations and is a mathematical optimization method to derive control policies (v. Sargent (2000)). The method is largely due to the work of Lev Pontryagin and Richard Bellman achieved in the mid-20th century, after contributions to the calculus of variations by McShane (1938) (cf. Bryson (1996)). Optimal control can be seen as a control strategy in control theory. It is commonly accepted to define the purpose or criteria of an activity that modifies a phenomenon.

Of course, it is important to start here by discussing and defining what we mean by criteria and how we model them. The key moment is to define the criterion. One of the dictionary definitions of "criterion" (The Chambers Dictionary) is "measure or norm for a judgment". From a decision-making point of view, this would mean some kind of norm whereby one particular choice or course of action may be considered more desirable than another. The consideration of different choices or modus operandi becomes a multi-criteria decision-making (MCDM) problem when many such standards are largely contradictory to each other. For example, even with simple personal choices, such as choosing a new home or flat, the price, availability of public transport, and personal safety may be relevant criteria. Management decisions at the corporate level, both in the public and private sectors, typically involve the consideration of a much wider range of criteria, especially when seeking consensus among very different interest groups. In the following, we will focus on tasks in which we have one decision-maker with one goal (criterion), and we will also increase the number of decision-makers, but each of them has one goal defined by its criterion. Problem-solving methodology, where each decision-maker considers several criteria at the same time, is the main topic, for example, of the monograph by Belton and Stewart (2002).

When calibrating probabilistic models, we encounter problems of sequential analysis, consisting of the independent repetition of a set experiment in a Bayesian scheme. This can be described in terms of stochastic control problems, with the strategy being limited to a set of stopping moments. We then have control problems and sequence problems in which, after each observation, there is a choice between experiments. At the turn of the 1950s and 1960s, many excellent mathematicians and specialists, using mathematical models in various fields, mainly economics and technology, created models in this truncated version of optimal control tasks.

During the discussion of the motivation for this approach, it is impossible to ignore these considerations. Siegmund (1967) discussed some of them, which we will return to later in this book since the basic formulation of these models has been generalized to multiple optimal stop (MOS) tasks. Example 6 from Siegmund (1967) refines the formulation and solution from Elfving (1967).[1] Let a sequence of random variables $\eta_1, \eta_2, \ldots$ appear at the moment of jumps in a homogeneous Poisson process (**HPP**).[2] random variables ξ_n with the exponential distribution having parameter λ and let the moments of jumps be $\tau_0 = 0$, $\tau_n = \sum_{j=0}^{n} \xi_j$. Define a nonincreasing function $r : \Re^+ \to \Re$ that should be interpreted as discounting of the rewards. The problem states to find the stopping moment σ^* that is optimal for the functional $\boldsymbol{E}(\eta_n r(\tau_n))$. Under the general assumption $\int_0^\infty r(t)dt < \infty$, using the continuous aspect of the problem, Elfving (1967) showed that he can derive a differential equation

[1]Elfving was looking for a model related to the market problem. Imagine a man owning a commodity, e.g., a house, which is for sale. Offers in varying amounts come in now and then. The longer you postpone selling, the more you lose due to deterioration, interest losses, or the like. At each offer, he must decide whether to accept it or wait for a better one. (A more romantic example would be that of a girl scrutinizing successive suitors.) The biographical note can be found of Gustav Elfing in Wikipedia or on MT MacTutor biogram Erik Gustav Elfing.

[2]Independent and identically distributed.

(ODE) for the boundary $y(\cdot)$ of the optimal stopping region. The stopping rule σ^* is defined by the boundary function $y(\cdot)$ if $\sigma^* = \inf\{n \in \bar{\mathbb{N}} : \eta_n \geq y(\tau_n)\}$. This model shows that for an important class of processes with continuous time, optimization tasks come down to the optimal stopping of processes with discrete time, and the stopping region is determined by the borderline, which can be determined analytically.

The above examples show that there are many important applications in the class of control models in which the payment is dependent on the state and the moment associated with the emergence of this state. If the stoppage combined with the state selection does not modify the process, then further observation with the combined state selection is a natural generalization of the basic optimal stop task. These premises mean that it is worth focusing efforts on a thorough examination of this generalization and showing its usefulness (cf. Stadje (1985), Krasnosielska-Kobos (2009, 2010, 2015)). This formulation was initiated by Haggstrom (1966) by proposing a generalization of the problems of optimal stopping. To sum up, from the stochastic control models, we choose a specific type of task in which the strategy sets work well to select the moment in which "we get the state" based on a criterion that depends on this state but is also possible from the very moment (the moment of action). The basic task boils down to selecting one moment and the state of the process at this moment (cf. Snell's (1952), Chow et al. (1971), Peskir and Shiryaev (2006)). In Haggstrom (1966), the author showed how to generalize the results from the optimal stopping problem to the multiple optimal stopping.

The same ideas apply to the formulation of the control problem. Illustrative was Bayes' strategy of testing a simple hypothesis versus a simple alternative with m available experiments, each with its own fixed cost. This problem involves the selection of a stopping time defined by posterior probability. A sufficient sample size, and thus the cessation of sampling, occurs when the posterior probability falls below one threshold or exceeds the other.

From these tasks, the formulation of the optimal multiple-stopping (MS in short) naturally appeared in Haggstrom's (1967) article. The article by Haggstrom (1966) discussed optimal stopping for processes indexed by elements if $\mathbf{J}$, a locally finite partially ordered set, such as $\widetilde{\mathbb{N}}^d = (\{0\} \cup \mathbb{N})^d$, where $\mathbb{N} = \{1, 2, \ldots\}$ is the set of natural numbers, or a tree with a minimal element 0. In each $\boldsymbol{t} \in \mathbf{J}$ the family of events, the σ-field, is determined: $\boldsymbol{\mathcal{F}} = \{\mathcal{F}_{\boldsymbol{t}}\}_{\boldsymbol{t}\in\mathbf{J}}$. The elements of the family $\boldsymbol{\mathcal{F}}$ are increasing according to the order in $\mathbf{J}$ defined by relation $\prec$, i.e. for every $\boldsymbol{s}, \boldsymbol{t}$ in $\mathbf{J}$ such that $\prec$ we have $\mathcal{F}_{\boldsymbol{s}} \subset \mathcal{F}_{\boldsymbol{t}}$. His basic notion was *a control variable*, a random variable τ defined on Ω with values in $\mathbf{J}$ such that $\{\omega : \tau \preceq \boldsymbol{t}\} \in \mathcal{F}_{\boldsymbol{t}}$ holds for each $\boldsymbol{t}$. In the general partially ordered case, this notion is no longer suitable. Some of the first results and questions about stopping points on a plane are due to Cairoli (1971).

Krengel and Sucheston (1980) introduced the notion of tactic and the notion of a stopping rule τ given by a tactic. The latter notion is easily seen to be equivalent to what Cairoli and Dalang (1996) renames accessible stopping points. This is often equivalent to what Mandelbaum and Vanderbei (1981) called *predictable stopping points*.

Roughly speaking, let S be a countable partially ordered set which is locally finite, and let $\{\mathcal{F}_{\boldsymbol{s}}\}$ be an increasing family of sub-σ-algebras of a probability space

$(\Omega, \mathcal{F}, \boldsymbol{P})$. A strategy $\{\sigma_n\}$ is an increasing sequence of S-valued stopping points such that σ_{n+1} is $\mathcal{F}_{\sigma_n}$-measurable for every $n \in \mathbb{N}$ (cf. the idea of tactic in Krengel and Sucheston (1981) and an optional increasing path from Walsh (1981)). Let a policy π be a pair (σ_n, τ), where $\{\sigma_n\}$ is a strategy, and $\tau : \Omega \to \mathbb{N}$ be an $\mathbb{N}$-valued stopping time with respect to $\{\mathcal{F}_{\sigma_n}\}_{n \geq 0}$. Denote $\alpha(\pi) = \sigma_\tau$. Assume that $\{X_{\boldsymbol{s}}\}_{\boldsymbol{s} \in S}$ is the Snell[3] envelope of the process $\{Z_{\boldsymbol{s}}\}_{\boldsymbol{s} \in S}$. Suppose that there exists an optimal policy π^* such that $\boldsymbol{E}[Z_{\alpha(\pi^*)}] = \sup_\pi \boldsymbol{E}[Z_{\alpha(\pi)}] = \boldsymbol{E}[X_0]$. Then there exists an optimal policy $\pi = (\sigma_n, \tau)$ such that $\tau = \inf\{n \in \mathbb{N} : X_{\sigma_n} = Z_{\sigma_n}\}$, and $\boldsymbol{E}\left[X_{\sigma_{n+1}} \mid \mathcal{F}_{\sigma_n}\right] = X_{\sigma_n}$ on $\{\omega : n < \tau(\omega)\}$. In other words, a set $A \subset S$ is a stop set of π if $\tau \leq \inf\{n \in \mathbb{N} : \sigma_n \in A\}$, and a set B is a go set of π if $\alpha(\pi) \notin B$.

In the following considerations, we will focus on multiple stops, which can be considered a continuation of the idea from Haggstrom's (1967) article. The choice of particular moments of stopping can be linked to different goals and even different decision-makers.

Partially ordered indices.

1. Let S be a countable partially ordered set with the smallest element 0 such that the set $U(s)$ of direct successors of each s is finite and each set $\{t \in S : t \leq s\}$ is finite. Let $\mathcal{G}_s$ be a σ-algebra corresponding to the *present* s. The σ-algebra $\mathcal{F}_s[\mathcal{H}_s]$ generated by all G_t with $t \leq s$ [with $t \geq s$] is the s-past [s-future]. $\{G_t\}$ is said to be Markov if $\mathcal{F}_s$ is conditionally independent of $\mathcal{H}_s$ given G_s for all s. Lawler and Vanderbei (1983) considered the optimal control problem for processes adapted to $\{\mathcal{F}_s\}$. They define Markov strategies and show that $\{G_s\}$ is Markov if and only if, for all processes adapted to $\{\mathcal{F}_s\}$, the strategy with maximal expected reward is a Markov strategy (cf. The results of Irle (1981) for the case $S = \mathbb{N}$).

1.2 Classical Sequential Analysis Problems

Motto:
Charting New Territories: Navigating Classical Sequential Analysis Problems.

Unlike unplanned "data peeping", which is a methodological lapse and increases the risk that a false hypothesis is incorrectly considered true (a type 1 error), sequential analysis includes corrections for multiple data testing, thus controlling the type 1 error at a nominal significance level. Data download is stopped according to the predefined

[3]James Laurie Snell (January 15, 1925 in Wheaton, Illinois – March 19, 2011 in Hanover, New Hampshire) was an American mathematician and educator. The Snell envelope, used in stochastics and mathematical finance, is the smallest supermartingale that dominates the price process. The Snell envelope refers to results in Snell (1951, 1952) paper.

observation-stopping rule. As a result, a conclusion can often be drawn earlier than in the classical approach of testing statistical hypotheses or estimation, which reduces the cost of the study. It is especially useful in areas where the acquisition of each observation is extremely expensive, dangerous, or otherwise difficult, for example, in many years of medical clinical trials (see Section 1.6.5).

The initiation of systematic research on sample size control is attributed to Wald (1945, 1947), as well as Wallis (1980), and Milton Friedman (v. Berger (2017), also Wald and Wolfowitz (1949)), who developed tools for more efficient quality control of industrial production during World War II. The value of these procedures in keeping military production at an appropriate level of quality was immediately recognized and led to the classification of information about sequential method research being classified. At the same time, George Barnard (1946) was working on similar issues in Great Britain with a group of experts. Another early contribution to the method was made by K.J. Arrow with D. Blackwell and MA Girshick (1949) and Shiryaev (1963a, 1963b) (v. Dobrushin et al. (1963)).

Peter Armitage introduced the use of sequential analysis in medical research, especially in the area of clinical research. Sequential methods have become increasingly popular in medicine; see the work of Stuart J. Pocock (2013) (v. Tang et al. (1993)), which provided clear recommendations to control error rates of type 1 in sequence designs (v. Jennison and Turnbull (2000)).

Other needs resulted in the introduction of models of controlled stochastic processes. Before the presentation of the examples, let us look at the theory for a general class of discrete-time stochastic control problems. When creating a model of a controlled random process, we assume that there is a sufficiently rich probabilistic space $(\Omega, \mathcal{F}, \boldsymbol{P})$ that allows for the implementation of the following structure. In a standard discrete-time stochastic optimal control problem, the objective is to maximize (or minimize) a functional of the form

$$\boldsymbol{E}\Big[\sum_{n=0}^{T} C(n, X_n, u_n) + F(X_T)\Big], \tag{1.1}$$

where $T \in \overline{\mathbb{N}} = \{0, 1, \ldots\} \cup \{\infty\}$, X is some *controlled* process, u_n is the control applied at time n, and F and C are given real-valued functions. A typical example is when X follows a controlled scalar stochastic equation of the form

$$X_{n+1} = \mu(X_n, u_n, Y_{n+1}), \tag{1.2}$$

where Y is the stochastic noise process, and we have some initial condition $X_0 = x_0$. Later in the book, we will consider specific dynamics and cost functions, but in this informal section, we restrict ourselves to this case and, for simplicity, assume that there are no constraints on the scalar control u_n. Let us also mention that the existence of probabilistic measures on the trajectories of the controlled process is not a trivial task and requires a deep analysis of various aspects of the model. Further considerations require clarifying the details.

Define state spaces $\mathbb{E}_n \subset \Re$, endowed with a σ-algebra $\mathcal{E}_n$, and state history $\mathbb{H}_n = \bigtimes_{j=0}^{n} \mathbb{E}_j = \mathbb{E}_1 \times \ldots \times \mathbb{E}_n$, endowed with the product σ-algebra

$\mathcal{H}_n = \prod_{j=0}^{n} \mathcal{E}_j$[4]. Analogously, define action spaces $\mathbb{A}_n$ with σ-field $\mathcal{A}_n$ and the decision history $\mathbb{U}_n = \prod_{j=1} \mathbb{A}_j$ with σ-filed $\mathcal{U}_n$, a product σ-algebra.

Let $D_n \subset \mathbb{H}_n \times \mathbb{U}_n$ be a measurable subset of $\mathbb{H}_n \times \mathbb{U}_n$ and denote the set of admissible state-action pairs of path histories at moment n. To have a well-defined problem, we assume that D_n contains the graph of measurable mappings

$$f_n : \mathbb{H}_n \to \mathbb{U}_n \text{ for every } n \in \overline{\mathbb{N}},\ \boldsymbol{f}_n = (f_1, \ldots, f_n), \tag{1.3}$$

that is, $(\boldsymbol{x}_j, \boldsymbol{f}_j(\boldsymbol{x}_j)) \in D_j$ for all $x_j \in \mathbb{H}_j$, $j = 1, 2, \ldots, n$. For $\boldsymbol{x}_j \in \mathbb{H}$, the set $D_n(\boldsymbol{x}_n) = \{\boldsymbol{a}_n \in \mathbb{U}_n : (\boldsymbol{x}_n, \boldsymbol{a}_n) \in D_n\}$ is the set of admissible actions in states $\boldsymbol{x}_n$ at moments $0, 1, \ldots, n$.

Next, we introduce the notion of a strategy $u_n : \mathbb{H}_n \times \mathbb{U}_{n-1} \to \mathbb{A}_n$, the sequence of decision functions that determine the n-th decision based on the history of the states up to the moment n, and the list of previous decisions at moments $0, 1, \ldots, n-1$. Since the system is stochastic, a strategy must determine actions for every possible state of history $\boldsymbol{x}_n \in \mathbb{H}_n$ of the system and the previous actions $\boldsymbol{a}_{n-1} \in \mathbb{U}_{n-1}$ for every time point $n \in \overline{\mathbb{N}}$. A measurable mapping $u_n : \mathbb{H}_n \times \mathbb{U}_{n-1} \to \mathbb{A}_n$ with the property $u_n(\boldsymbol{x}_n, \boldsymbol{a}_{n-1}) \in D_n(\boldsymbol{x}_n)$ for all $\boldsymbol{x}_n \in \mathbb{H}_n$, is called a decision rule at time n. We denote by $\overrightarrow{U}_n$ the set of all decision rules at time n. A sequence of decision rules $\pi = (\boldsymbol{u}_0, \boldsymbol{u}_1, \ldots, \boldsymbol{u}_{N-1})$ with $\boldsymbol{u}_n \in \overrightarrow{U}_n$ is called the N-stage policy or the N-stage strategy. If the decision-maker (**DM**) follows a policy $\pi = (\boldsymbol{u}_0, \boldsymbol{u}_1, \ldots, \boldsymbol{u}_{N-1})$ and observes at time n the history of states $\boldsymbol{x}_n$ of the system, then the action that he or she chooses is $u_n(\boldsymbol{x})$. This means, in particular, that the decision at time n depends on the state of the system at times $0, 1, \ldots, n$. The decision-maker bases her decision on the whole history $(x_0, a_0, x_1, \ldots, a_{n-1}, x_n)$, but the important application allows simplifying the model.

The optimal stopping problem as well as the multiple-stopping problem[5] one can get as the special case of the controlled stochastic processes model. Let us assume that the sets of decision rules consist of binary functions, namely $\epsilon_n : \mathbb{H}_n \times \mathbb{U}_{n-1} \to \mathbb{A}_n = \{0, 1\}$. If the rewards are $C(n, X_n, \delta_n) = \delta_n \psi(n, X_n)$, where $\delta_n = \prod_{j=0}^{n-1}(1-\epsilon_n)\epsilon_n$. In this way, the strategy δ_n selects the profit from the payment determined by the state of the process at the moment n. This means that there is no influence on the return of the states before and after the first moment n with $\epsilon_n = 1$.

In probability theory, in particular, in the study of stochastic processes, a stopping time (also Markov time, Markov moment, optional stopping time, or optional time) is a specific type of "random time": a random variable whose value is interpreted as the time at which a given stochastic process exhibits a certain behavior of interest. A stopping time is often defined by a *stopping rule*, a mechanism for deciding whether to *continue* or *stop* a process based on the present position and past events, and which will almost always lead to a decision to stop at some finite time. Although in this paragraph we list concepts such as Markov time and stopping time as equivalent concepts, we will distinguish them after Shiryaev (1978) (v. Sect. 1.2, p. 5).

---

[4] $\prod_{i=1}^{n} p_i = p_1 \cdot p_2 \cdot \ldots \cdot p_n$

[5] The optimization problem related to the multiple-stopping model (MSM).

The history of optimal stopping problems, a subfield of probability theory, also begins with gambling. One of the first discoveries is credited to the eminent English mathematician Arthur Cayley (1875) of the University of Cambridge. In 1875, he found an optimal stopping strategy (stopping rule) for purchasing lottery tickets (v. Ferguson (1989), Hill (2009)). In the notation of this section, the stopping time τ is the random variable defined in the same probability space as the underlining process $\{X_n\}_{n=0}^{T}$ that supports binary decision sequences δ_n:

$$\tau(\omega) = \inf\{n \in \overline{\mathbb{N}} : \delta_n = 1\}. \tag{1.4}$$

Let us define the filtration $\{\mathcal{F}_n\}$ related to the underlining process X_n as

$$\mathcal{F}_n = \sigma\{X_0, X_1, \dots, X_n\}. \tag{1.5}$$

Definition 1.1. *Let filtration* $\{\boldsymbol{\mathcal{F}} = \mathcal{F}_n\}_{n=0}^{T}$ *be defined in a measurable space* $(\Omega, \mathcal{F})$. *The random variable* $\tau : \Omega \to \overline{\mathbb{N}}$ *is said to be a* Markov time with respect to the filtration $\boldsymbol{\mathcal{F}}$ *iff we have* $\{\omega : \tau(\omega) = n\} \in \mathcal{F}_n$ *for every* $n \in \mathbb{N}$.

In other words, this means that based on the information about $X_0, X_1, \dots, X_n$, we can be sure whether an event $\{\omega : \tau(\omega) = n\}$ has occurred or not. That is, if $X_0, X_1, \dots, X_n$ have been performed (if we are at time n), then we can be sure if event $\{\omega : \tau(\omega) = n\}$ applies or not.

Definition 1.2. *The Markov time* $\tau = \tau(\omega)$ *defined in the probability space* $(\Omega, \mathcal{F}, \boldsymbol{P})$ *is said to be a* stopping time *or a* finite Markov time *if* $\boldsymbol{P}(\tau < \infty) = 1$.

When we restrict ourselves to stopping strategies, the objective (1.1) is

$$\boldsymbol{E}[\psi(\tau, X_\tau)]. \tag{1.6}$$

The dynamics mentioned in the relationship (1.2), does not have to be a functional form, and the payment for the final state can be included in the function $\psi(\cdot, \cdot)$.

Theorem 1.1 (Optional Stopping Theorem). *Let* τ *be a stopping time for the martingale* $\{X_n\}_{n=0}^{\infty}$. *Then* $\mathbf{E}[X_\tau] = \mathbf{E}[X_0]$ *if at least one of the following conditions holds:*

a. τ *is finite (that is,* a stopping time*) and there exists a finite constant* C_1 *such that* $|X_n| \le C_1$ *for all* $n \le \tau$.
b. τ *is bounded, that is, there exists a finite constant* C_2 *such that* $\boldsymbol{P}(\tau \le C_2) = 1$.
c. $\mathbf{E}[\tau]$ *is finite and there exists a finite constant* C_3 *such that* $\mathbf{E}[|X_{n+1} - X_n||\mathcal{F}_n] < C_3$ *for* $n = 0, 1, \dots$.

This theorem is also called the **stopping time theorem**. So for the functional (1.6), we have an interesting conclusion.

Corollary 1.7. *If the process $Y_n = \psi(n, X_n)$, a sequence of payments related to the observation of the process $\{X_n\}_{n=0}^{\infty}$, which forms a martingale, then, with the assumptions of Theorem 1.1,* $\sup_\tau \boldsymbol{E}[Y_\tau] = \inf_\tau \boldsymbol{E}[Y_\tau] = \boldsymbol{E}[X_0]$. *Each stopping moment realizes an extreme value.*

It shows that variational problems for functional (1.1) can have a solution: There is a finite extreme value of this functional and strategies exist to obtain this value. Searching for extreme values of (1.1) is an important issue. As part of this task, we want to not only state that the functional reaches an extreme value but also construct a method of calculating these values and indicate a strategy, here it is the stopping time, that realizes this extreme.

1.3 Classical Optimal Stopping Problems

Motto:

Exploring Optimal Decision-Making: Navigating Visible Objectives and Decision-Maker Capabilities

In this section, we want to present illustrative models from the theory of optimal stopping, emphasizing the visible objectives of the decision-maker and their capabilities. Objectives are typically quantified using numerical criteria and take into account the states of the modeled process, as well as the costs of observation and decision-making. Capabilities, on the other hand, are linked to the resources of the decision-maker and sometimes to the natural constraints that limit access to states of the modeled process. At this point, we are not attempting to formulate a general model encompassing all possible conditions. Instead, we will focus on the role of information about the observed process (its states and trajectories) and the consequences of this knowledge, its nature, and permissible decision-making strategies.

We will utilize the classic secretary problem for this purpose, assuming that the candidates in the selection procedure are significantly different and their suitability for the position is measurable as a percentage of the required knowledge, skills, and experience. Candidate's characteristics can be evaluated during the recruitment procedure but they are unknown before the interview with the candidate. Although knowledge acquisition can be gradual, recruiters often analyze complex resumes provided by candidates. In this illustration, we assume that the analysis of all the material about each candidate occurs during the interview.

In developing a mathematical model for the aforementioned problem, we proceed as follows. Let $X_1, X_2, \ldots, X_n$ be independent and identically distributed (i.i.d.) random variables uniformly distributed on the interval $[0,1]$. Define R_k, $k = 1, 2, \ldots, n$, as the absolute rank of X_k, where the ranks are determined in ascending order, that is,

$$R_k(\omega) = \sum_{j=1}^{n} \mathbb{I}_{\{\omega: X_j(\omega) \le X_k(\omega)\}}(\omega). \tag{1.7}$$

This means that the smallest observation has the rank 1. Furthermore, let $\mathfrak{T}_n$ be the set of all finite stopping rules concerning filtration $\{\mathcal{F}_k\}_{k=1}^n$, where $\mathcal{F}_k = \sigma\{X_1, X_2, \ldots, X_k\}$, that is, $\mathfrak{T}_n = \{\tau : \{\omega : \tau(\omega) = k\} \in \mathcal{F}_k,\ 1 \le k \le n,$ and $\sum_{j=1}^n \boldsymbol{P}(\{\omega : \tau(\omega) = j\}) = 1\}$. For a more comprehensive overview of the selection problems, we encourage the readers to refer to Section 6.1.

Problem 1.1 (Robbins' problem).[a] *The problem is to find the value*

$$V(n) = \inf_{\tau \in \mathfrak{T}_n} \boldsymbol{E}[R_\tau],$$

where R *is given by* (1.7), *including its asymptotic behavior as* n *tends to infinity, and also the stopping rule* $\tau^* \in \mathfrak{T}_n$ *that achieves this value, namely*

$$\tau^* := \tau_n^* = \arg\max \boldsymbol{E}[R_\tau].$$

[a]The story of the formulation of the problem can be found in the paper by Bruss (2005). Herbert Robbins' biogram was written by Lai and Siegmund (2016) and the Robbins' publication list juxtaposed by Editorial Staff (2003).

Remark 1. *Let us examine the problem presented. The objective is to create a selection procedure in which statistically, across multiple proceedings, we will achieve an outcome that is, on average, the best. At the moment of decision, we have complete knowledge about the candidates analyzed previously and those currently being considered. The problem presented in this way is challenging and the authors do not know its solution. How can we modify the requirements (expectations) of the selector, hoping that, for the altered problem, a solution can be found? Another modification of this problem may involve the knowledge of the selector. Excessive knowledge can be problematic for making decisions. Preliminary bounds for* $V(n)$ *can be found in a straightforward manner straightforwardly by looking at related problems. We can limit knowledge gathering to the current ranks of candidates at the moment they are being evaluated. If we further constrain ourselves to search for the best candidate, the problem becomes simpler and allows for estimating the payout in the Robbins' problem (v. Ferguson (1989), Samuels (1991), Freeman (1983), Rose (1982)).*

The collection of modifications of the problem is presented in Chapter 6. It is astonishing to see the phenomenon of extraordinary interest in relatively straightforward problems. Undoubtedly, one such category includes the secretary problem, which, under numerous modifications, has been revisited, adjusted, generalized, and analyzed for at least half of the 20th century (see, for example, Б. А. Березовский, А. В. Гнедин (1984)). We will leverage this attention-grabbing characteristic of the problem and utilize it to demonstrate selected technical phenomena that arise in problems involving multiple stopping of stochastic sequences.

The original formulations referred to guessing the position of a fixed object in a permutation of n objects revealed sequentially, leading to selection. Almost equivalent is the formulation we will employ here (and in many other places in this book). We will assume that these objects form a sequence of observations of independent continuous random variables with a fixed distribution (for simple models, the distribution is irrelevant except that it is continuous). The original formulation did not speak of the limiting origin of the objects, so it was "free of the distribution of object characteristics" (cf. Ferguson (1989), Gardner (1960)). Fairly precise formulations in this weakened spirit are presented by Gilbert and Mosteller (1966).[6]

Problem 1.2 (The classical secretary problem (**CSP**)). *A decision-maker sequentially observes the* relative ranks, $\rho_1, \rho_2, \ldots, \rho_n$ *of the* $\{X_k\}_{k=1}^n$, *where*

$$\rho_k(\omega) = \sum_{j=1}^{k} \mathbb{I}_{\{\omega: X_j(\omega) \le X_k(\omega)\}}(\omega), \tag{1.8}$$

and wants to maximize $\boldsymbol{E}\mathbb{I}_{\{R_\tau=1\}} = \boldsymbol{P}(R_\tau = 1)$, *where R is the absolute rank of the k-th observation defined by* (1.7).

The decision-maker's strategies belong to the set $\mathfrak{S}_n$ of all finite stopping rules with respect to filtration $\{\mathcal{G}_k\}_{k=1}^n$, where $\mathcal{G}_k = \sigma\{\rho_1, \rho_2, \ldots, \rho_k\}$, i.e. $\mathfrak{S}_n = \big\{\tau : \{\omega : \tau(\omega) = k\} \in \mathcal{G}_k, 1 \le k \le n$, and $\sum_{j=1}^n \boldsymbol{P}(\{\omega : \tau(\omega) = j\}) = 1\big\}$. The problem is the well-known secretary problem. Lindley (1961) published the solution first, showing that the limiting optimal probability to choose the rank 1 is $1/e$.

Problem 1.3 (The full information secretary problem). *The problem is to find the value*

$$V(n) = \inf_{\tau \in \mathfrak{T}_n} \boldsymbol{E}[\xi_\tau], \tag{1.9}$$

where $\{\xi_k\}_{k=1}^n$ *is given by*

$$\xi_k = \mathbb{I}_{\{\omega: X_k(\omega) = \max\{X_1(\omega), X_2(\omega), \ldots, X_n(\omega)\}\}}(\omega), \tag{1.10}$$

including its asymptotic behavior as n tends to infinity, and also the stopping rule $\tau* \in \mathfrak{T}_n$ *that achieves this value, namely*

$$\tau^* := \tau_n^* = \arg\max \boldsymbol{E}[\xi_\tau]. \tag{1.11}$$

[6] Due to the steadily increasing interest in the best-choice model, which will be discussed at appropriate junctures, it is worthwhile to acquaint ourselves with the history of the problem, early works on the subject, and their chronology. The review articles by Rose (1982), Freeman (1983), Ferguson (1989) and Samuels (1991) will be helpful in this regard.

The situation is as in the Problem 1.2, where the decision-maker wants to maximize the probability of selecting rank 1, except that the distribution of the X_k is known to the decision-maker and he/she can use the observations $X_1, X_2, \ldots, X_n$ (not only their ranks), i.e. the decision-maker's strategies are stopping times with respect of the filtration $\{\mathcal{F}_k\}_{k=1}^n$ (belonging to $\mathfrak{T}_n$). It is known as full information (FI) BCP. The optimal probability of winning is now 0.5801 (v. Gilbert and Mosteller (1966), and the precise construction of the solution can be seen in the paper by Bojdecki (1977/78)). The random variable ξ_k, defined by (1.10), is equal 1 on the event ω iff $X_k(\omega)$ is equal the global maximum of the sequences $X_1, X_2, \ldots, X_n$. The popular formulation of the problem is to select the best option as it appears, based on the online observation, without recall. The mathematical model of the problem is the OSP given by (1.9) and its solution is determining the value function $V(n)$ and the optimal strategy given by (1.11).

Problem 1.4 (Minimizing expected rank in the case without information). *Let the decision-maker base her decision only on sequential observation of relative ranks, but aim to minimize $\mathbf{E}[R_\tau]$ on $\mathfrak{S}_n$.*

This problem is more difficult than those in Problems 1.2 and 1.3. A heuristic approach was given by Lindley (1961), but the resulting equations were too crude to derive the existence and value of the limit. This was achieved by Chow et al. (1964). The limiting optimal value (minimal expected loss) is

$$\lim_{n\to\infty} V(n) = \prod_{j=1}^{\infty}\left(\frac{j+2}{j}\right)^{\frac{1}{j+1}} \cong 3.695$$

(v. also Dubuc (1973)).

Problem 1.5 (The full-information expected-rank problem). *Natural complement of the Problems 1.2-1.4 is the minimization of $\mathbf{E}[R_\tau]$ in $\mathfrak{T}_n$.*

In this case, the decision-maker can base her decision on the sequential observation of i.i.d. random variables with a fully known distribution and wants to minimize the expected rank. Hence, this is Robbins' problem as in Problem 1.1 (v. Bruss (2005)). Only partial results are obtained. For example, Assaf and Samuel-Cahn (1996) demonstrated that for this problem, the limiting optimal expected rank is at least 1.85. Furthermore, for a specific subclass of threshold rules, this expected rank falls within the range of 2.295 to 2.327.

Robbins (1989) discusses the game variant of BCP.

Consider two cases of the secretary problem: in Problem 1.2, the payoff is 1 if we choose the best of the n applicants, 0 otherwise, and we want to maximize the expected payoff, and in Problem 1.4, the loss is the absolute rank of the person selected (1 for the best,. . ., n for the worst), and we want to minimize the expected loss. When all n possible orders of the applicants are equally likely, the solutions to both problems have been known since the late 1960s. And when the probabilities of the various permutations are controlled by an antagonist, so that Problems 1.2 and 1.4 become game-theoretical (minimax) problems, their solutions are also in the literature (v. problem 7 on page 60 of Chow et al. (1971), and page 89 of Chow et al. (1964)).

1.3.1 The best choice problem (BCP) with 2 stops

It is quite intuitive to extend the simple best-choice selection problem (BCP), where the decision-maker selects a single object, and for which an optimal stopping model with stopping times as strategies can be relatively easily formulated, to a scenario where the decision-maker is allowed to make two selections. The optimization task of sequentially choosing two objects, with the aim that at least one of the selections is correct, naturally leads to the formulation of an optimal two-stop model. This model involves selecting two stopping times as part of a single optimization process. Let us say σ and τ are the stopping times so that $\boldsymbol{P}(\omega : 1 \in \{R_\sigma, R_\tau\})$ is maximized. In Section 2b[7] of Gilbert and Mosteller (1966), we read the following analysis on this matter.

> In this section, we allow the player to have two choices, retaining the other conditions of the dowry problem. If either of his choices is the tag with the largest number, the player wins.
>
> . . .
>
> An argument similar to that used in Section 2a shows that the optimum strategy belongs to the class of strategies indexed by a pair of starting numbers (r, s), $r < s$. The first choice is to be used on the first candidate starting to withdraw r, and once the first choice is used, the second choice is to be used on the first candidate starting to withdraw s. Once the first choice is made, the problem is reduced to the one-choice problem of Section 2a, and the second choice will be used on the first candidate starting with s^*, as defined by inequalities (2a[8]-4) . . .
>
> To compute the probability of winning with strategy (r, s), the ways of winning are broken into three mutually exclusive events:
>
> a. win with first choice (never use second choice),
> b. win with the second choice and no choice is used before s,
> c. win with the second choice and the first choice is used in one of the positions $r, r+1, \ldots, s-1$.

---

[7] *The dowry problem with two choices: Exact and asymptotic theories.*
[8] The section *The dowry problem with one choice.*

By extensions of the methods used to derive equation (2a-2), it is easy to show that

$$
\begin{aligned}
&\boldsymbol{P}(a) = \frac{r-1}{n}\Big(\frac{1}{r-1} + \frac{1}{r-2} + \ldots + \frac{1}{n-1}\Big) && r > 1,\\
&\boldsymbol{P}(a) = \frac{1}{n} && r = 1,\\
\text{(2b-1)}\quad &\boldsymbol{P}(b) = \frac{r-1}{n}\sum_{v=s+1}^{n}\sum_{u=s}^{v-1}\frac{1}{(u-1)(v-1)}, && s > r \geq 1,\\
&\boldsymbol{P}(c) = \frac{s-r}{n}\sum_{s}^{n}\frac{1}{v-1}, && s > r \geq 1.
\end{aligned}
$$

$$\pi(r, s, n) = \boldsymbol{P}[\text{win with } (r, s)] = \boldsymbol{P}(a) + \boldsymbol{P}(b) + \boldsymbol{P}(c).$$

Gilbert and Mosteller (1966) do not provide a precise derivation of the above formulas. The authors leave this to the reader, assuming that the classical model of the problem of selecting the best object, based on the assumption that the permutations of sequentially available candidates are equally probable, allows them to be obtained. The problem of optimal selection (r, s) is solved (v. Table 3 of Gilbert and Mosteller (1966)). Haggstrom (1967) on p. 1623 presented the problem[9] with the detailed scheme of the solution algorithm. Further research on solving the problem of determining the parameters of the optimal strategy can be found in Sakaguchi (1978). He reduced the task to solving a sequence of auxiliary optimization problems and showed that each of them is *monotonic problem*, which implies that the optimal strategies for each of them are of one-step look-a-head (**OLA**) type.[10] It is also possible to apply the *odds algorithm* (v. Ano et al. (2010)) to the best choice problem (**BCP**) with multiple selection. Further combinatorial intricacies within the model are explored in the recent paper by Liu et al. (2023b, 2023a).

Based on the strategic analysis of the problem, we can determine the probability of success for a pair of stopping times defined by thresholds r (for the first stop) and s, $s > r$, for the subsequent second stop. Using the model embedded in a Markov chain (see Section 6.2), we reduce the task to a double-stopping of the Markov chain defined in Section 6.2.3. Using formulas for transition probabilities (6.8), payout functions (cf. (6.2), (6.50)), and Bellman's principle, we can construct the value functions of the problem truncated to the set of stopping strategies allowing stops from a fixed moment. The detailed solution of the problem is presented in Section 6.6.1.

The presented **BCP** features an exceptionally convenient payout function structure, which can be expressed as a sum of payouts from successive decisions. This facilitates the application of the Bellman principle and the sequential search for optimal stopping times. Additionally, we have the opportunity to transform the problem into an optimal two-stop Markov chain problem when the payout is the sum of payouts from states at stopping times – the payouts at individual stops depend only on the

---

[9] Example *b. Dowry problem with two choices*.

[10] The one step look a head (OLA) was introduced by Ross (1970).

states of the chain at those times. As a result, the task of optimal selection of two successive stopping times is reduced to two optimal stopping problems.

Another example will demonstrate that solving the optimal double-stopping problem can also be effectively addressed when the payout structure from selected states does not allow for a straightforward separation of decision effects. Among simple generalizations of the best choice problem is the task of optimal selection between the two main options with a single stop (v. Gilbert and Mosteller (1966) p. 49, Section 2d). With two choice decisions and two stopping times at hand, one can attempt to select the two best objects. A solution to the problem was provided by Nikolaev (1977) using Haggstrom's model, and by Tamaki (1979) utilizing a suitable **MDP** construction presented in detail in Section 6.7. Independently, Glasser (1978) worked on such generalizations. The results concerning the problems formulated in his doctoral thesis, among others (m, m, m)-SP (when it is possible to make m choices leading to selection of the m best objects), are the subject of publications by Glasser et al. (1983). The embedded Markov chain approach, which is analytically the most convenient approach for some of these models, is described in Section 6.7.1.

The **CSP** was later generalized and modified in many interesting ways (see the mentioned reviews), including the case of an unknown n (see, e.g. Presman and Sonin (1972)).

1.3.2 Variation on the number of potential options

The basic formulations of the decision-making problem assume that the decision-maker knows the number of options available and must make a decision based on the knowledge of these options assessed up to the decision point. A refinement of this task involves removing this constraint. The random-horizon problem was a challenge, with errors also in some approaches. Formulated and investigated first by Presman and Sonin (1972) was tackled by others. Irle (1980) proposed a method of obtaining optional stopping rules, which differs from the usual backward reduction method, for such an extension and applied it to the best choice problem of Presman and Sonin (1972). Although essentially no new results are obtained, the method (credited to an unpublished report by Rasche (1975)) should be of interest to us. Due to this research, the author gives a counterexample to erroneous Theorem 3.1 of Rasmussen (1975), which was also used by Rasmussen and Robbins (1975).

An example where the reasoning behind not needing to know the number of available options was examined is discussed by Cowan and Zabczyk (1976, 1979) (also see Section 1.1, and Elfving (1967)). Similarly, as in **BCP**, the object is to choose (at the instant of its arrival) the best of a sequence of candidates arriving in random order, knowing only whether or not each successive candidate is "a leader" (that is, it is better than all previous ones). The successive arrival times form a Poisson process on $[0, T]$ with known intensity λ. T is known, and the total number of candidates is random, with a Poisson distribution, where if k candidates have arrived in time t, the distribution of the number of additional candidates depends only on $T - t$ (cf. the model considered by Presman and Sonin (1972) where if k candidates have arrived, the distribution of the number of additional candidates depends only on k). The monograph on the random horizon selection problems was written by Tamaki (2023).

From the standard theory of optimal stopping of a Markov chain, the optimal rule is to select the i-th candidate, where i is the smallest k for which the k-th candidate has relative rank 1 and $\lambda(T-t) \leq y_k$, where $\{y_k\}$ is the increasing sequence of numbers such that if $\lambda(T-t) = y_k$, then the success probabilities for the rules *stop now* and *stop at the next relatively best* are equal. The $\{y_k\}_{k=1}^{\infty}$ are the solutions to the equations

$$\sum_{n=0}^{\infty} \frac{y_n}{n!(n+k)} = \sum_{n=1}^{\infty} \frac{y_n}{n!(n+k)} \sum_{j=0}^{n-1} \frac{1}{k+j}.$$

The above problem is based on natural assumptions. However, most of them can be relaxed to obtain a model that is better suited to real-life conditions. The Poisson process can be replaced by a less restrictive renewal process. This innocent generalization necessitates the application of more universal methods for finding the solution. Although the horizon T is fixed, it is worthwhile to allow it to be random. This also requires the development of optimization methods. Detailed considerations that led to the solution of these problems were independently provided by Bojdecki (1977) and Zabczyk (1977). It has been proven that a solution to our problem exists, although its form (unfortunately) may not be very clear. The general method presented sometimes allows one to obtain the solution in a simpler form. An example of this is when the horizon follows an exponential distribution, where everything can be calculated explicitly.

The study of multiple-stopping problems with random horizons is a rich and active area of research, with numerous results and techniques developed to address various aspects of the problem. Return to paper by Gilbert and Mosteller (1966) and the section of the paper "The dowry problem with r choices". The problem was generalized to the case where the number of applicants N is random by Móri (1984) with an explanation of the contribution to the problem of others. Under the assumption of no information, **BCP** up to r of the applicants can be sequentially selected to maximize the probability that one of these r is the best of all N. Consider a scenario in which the random variable N can follow any distribution as long as it satisfies the condition: $\frac{1}{p_i}\sum_{j=i+1}^{\infty}\frac{p_j}{j}$ is nonincreasing, where $p_i = \boldsymbol{P}(N=i)$. Furthermore, let us investigate asymptotic results for sequences of indexed families N_λ, where λ approaches infinity. In this scenario, the optimal selection strategy resembles that of fixed applicant sizes. Specifically, there exists a sequence of values $v_1, v_2, \ldots$ such that, for any given r, asymptotically, the best strategy is to select the candidate who is relatively the best after a certain proportion of applicants have arrived. This proportion is given by $\exp(-v_{r+1-k})$. Furthermore, in the limit, the optimal probability of selecting the best candidate is simply the sum of exponential terms up to r. This is evidently a multiple stopping problem as formulated in Section 1.4 and the references therein. Regarding the calculation of the sequence $\{v_i\}_{i=1}^{r}$ Gilbert and Mosteller listed the first eight of the v_i's (called "$-\log u_r$" by them) in Table 4 of their paper, but provided precious few clues as to where those numbers came from. Móri (1984) precisely provided an answer,

$$-v_k = 1 + \sum_{j=1}^{k-1} \frac{1}{(k+1-j)!} v_j^{k+1-j},$$

as well as the suggestion that this formula evidently appeared first in a paper by Sakaguchi (1978) using the general method for the optimal stopping problem created by Dynkin (1963) (v. Samuels (1987)).

We also mention the paper by Kawai and Tamaki (2003a, 2003b), where the problem stated and solved by Guseĭn-Zade (1966) for the fix horizon was extended assuming that the random number N of applicants is bounded. They obtained explicit results, and for uniformly distributed N they prove that the optimal policy is of the form of first waiting some time, then for some time choosing the next best one, and after this, hoping for success with the next best or second-best candidate.

1.4 Multiple Stopping by One Decision-Maker

The martingale treatment of stochastic control problems is based on the idea that the correct formulation of Bellman's principle of optimality for stochastic minimization problems is in terms of a submartingale inequality. The value function of dynamic programming is always a submartingale and is a martingale under a particular control strategy if and only if that strategy is optimal. Local conditions for optimality in the form of a minimum principle can be obtained by applying Meyer's submartingale decomposition along with martingale representation theorems, conditions for the existence of an optimal strategy can also be stated (v. Davis (1993)). The martingale approach is used for optimal stop and impulse control.

1.4.1 Dual representation of a multiple stopping problem

The main approaches to Dual Representations of Multiple-Stopping Problems (DRMSP for short) are the marginal and pure martingale approaches of Meinshausen and Hambly (2004) and Schoenmakers (2012), respectively. It is shown by Chandramouli and Haugh (2012) that these dual representations, as well as their more recent extensions to problems with volume constraints, can be derived in a simple unified manner using the general duality theory based on information relaxations (v. Appendix C.1). It is also possible to derive pure martingale representations for other multiple-stopping problem (MSP), including problems with refractive index constraints.

The notation used by Schoenmakers (2012) the multiple optimal stopping problems is defined as follows. There are T time periods and a probability space $(\Omega, \mathcal{F}, \boldsymbol{P})$ with associated filtration, $\boldsymbol{\mathcal{F}} = \{\mathcal{F}_t\}_{t=0,\dots,T}$, where $\mathcal{F}_T = \mathcal{F}$. We have a positive adapted reward process g_t, which satisfies $\sum_{t=1}^{T} \boldsymbol{E}[\| g_t \|] < \infty$. This reward process may be performed *exercised* $L \leq T$ times, subject to the restriction that it may not be performed more than once on any given date. For notational convenience, below we define $\mathcal{F}_t := \mathcal{F}_T$ and $g_t := 0$ for all $t > T$. For the same reason, we will also suppress the dependence of all stochastic processes on any state variables that may be driving the dynamics of the system. The goal is then to maximize the expected reward by optimally choosing the L exercise times. We have the following definition.

Definition 1.3. *For $0 \leq t \leq T$ and each L, we define $\mathcal{S}_t(L)$ to be the set of $\mathcal{F}$-stopping vectors $\tau_t := \{\tau^1, \ldots, \tau^L\}$ such that $t \leq \tau^1 < \tau^2 < \ldots < \tau^L \leq T$.*

1.4.2 Prophet inequalities when a mortal has multiple stops.

In the theory of *prophet's inequalities* one compares the expected outcomes of a sequential game between a prophet, who knows the outcomes of the problem beforehand, and a mortal who sees the outcomes sequentially, i.e. when they arrive. In the classical setting, the expectations $\boldsymbol{E}(\max_{1\leq i\leq n} X_i)$ and $\sup_\tau \boldsymbol{E}X_\tau$, where τ runs through the class of finite stopping times, are compared. The sequence $X_1, \ldots, X_n$ can be a sequence of arbitrarily dependent, independent, or i.i.d. random variables (v. Harten et al. (1997), Schmitz (2000) for the review of early results on the subject).

Consider an optimal stopping problem where a gambler observes a sequence of bounded random variables $X_1, X_2, \ldots$, and attempts to select the largest observation, using only the information provided by previous observed values. The problem of finding the gambler's optimal stopping time of the gambler and his expected optimal gain has been the subject of a large body of research. A prophet inequality gives a bound that binds the best possible expected gain of the gambler to the expected maximum of the random variable sequence (the best possible expected gain of the prophet). Krengel and Sucheston (1977) proved interesting results for generalizations of martingales (so-called "semiamarts" v. Krengel and Sucheston (1981)). In this rather theoretical context, they showed that the expectation $\boldsymbol{E}(\sup_n X_n^i)$ for independent random variables Xn and for "arithmetic" means can be bounded by a certain multiple of the stopping time value $\sup_\tau \boldsymbol{E}(X_\tau)$. They interpreted these results as bounds for the disadvantage of a gambler (who is restricted to using stopping rules) compared to an omniscient (but not omnipotent) opponent. A short time later, Krengel and Sucheston used for this opponent, who is endowed with complete foresight, the much more catchy notion of prophet – and this notion was adopted by all subsequent authors. In a short period, several mathematicians researching stochastic analysis, in particular Robert Kertz and Ted Hill[11] – proved further prophet inequalities and characterized the sets of all possible pairs $(\sup_\tau \boldsymbol{E}(X_\tau), \boldsymbol{E}(\sup_n X_n))$ (*prophet regions*) for interesting classes of stochastic processes; a new branch of probability theory was developed, called prophet theory (the phonetic identity with "*profit theory*" may have been a nice addition)

[11] v. hill.math.gatech.edu/

Krengel and Sucheston (1977)[12,13] found an inequality for the ratio of the expected gain of the prophet to the expected gain of the gambler when the random variables $X_1, X_2, \ldots$ are independent, and Hill and Kertz (1981) published inequalities for the difference of expected gains in the case of independent random variables and of dependent random variables published in Trans. Am. Math. Soc. two years later (v. in Hill and Kertz (1983)). All these inequalities are sharp. In another direction, there has been some research on finding the optimal stopping time and the optimal expected gain for a problem where the gambler incurs a cost for each value observed. MacQueen and Miller (1960) and Chow and Robbins (1967, 1963, 1961) considered this problem. Jones (1990) gave sharp inequalities for the difference between the expected gains of the gambler and the prophet in the case of independent random variables when there is a search cost.

Assaf and Samuel-Cahn (2000) analyzed an extension of the classical problem when the mortal has more than one chance to stop the sequence $X_1, \ldots, X_n$. Their main result is that the prophet can never do better than $\frac{k+1}{k}$ times the mortal. The stopping times that the authors used to do that are of the pure threshold type with some additional randomization.

For $k = 1$ the classical prophet inequality $\boldsymbol{E}(\max_{1 \le i \le n} X_i) \le 2 \sup_\tau \boldsymbol{E} X_\tau$ follows. For $k > 1$, the inequalities proved are not sharp in the sense that $\frac{k+1}{k}$ is not the best constant possible.

Hill and Kennedy (1989) generalized prophet inequalities for single sequences to compare for optimal stopping of several parallel sequences of independent random variables. For example, if $X_{i,j}$, $1 \le i \le n$, $1 \le j < \infty$, are independent nonnegative random variables, then $\boldsymbol{E}(\sup_{i,j} X_{i,j}) \le (n+1) \max_i \sup\{E(X_{i,\tau}) : t\mathcal{T}$ is a stop rule for $X_{i,1}, X_{i,2}, \ldots\}$, and this bound is the best possible. Applications are made to comparisons of the optimal expected returns of various alternative methods of stopping parallel processes. The result was extended by Boshuizen (1990).[14] The problem can be described as follows. Suppose that a player plays sequences of games in different rooms of a gambling house. Let us assume that his expected reward is the maximum of the optimal stopping values of these sequences. The research here is devoted to comparing values with possible outcomes for memoryless (threshold) stopping rules. The authors compare the expected supremum of all games (as obtainable by a prophet) with the expected reward under optimal rules (without foreseeing abilities), provided that the games can be represented as sequences of uniformly bounded independent variables.

1.4.3 Applications of multiple optimal stopping rules

Here we will introduce several applications of multiple optimal stopping rules. These applications will be considered in more detail in Part III.

---

[12] Short biography see `https://de.wikipedia.org/wiki/Ulrich_Krengel`

[13] Louis Sucheston Obituary (v. https://en.everybodywiki.com/Louis_Sucheston)

[14] This is a part of his PhD thesis (v. Boshuizen (1991)).

TABLE 1.1
The thresholds k_N^* and the probabilities of obtaining the best object P_N^* in the classical **BCP**.

N	1	2	3	4	5	6	7	8
k_N^*	1	1	2	2	3	3	3	4
P_N^*	1.0000	0.5000	0.5000	0.4583	0.4333	0.4278	0.4143	0.4098

Problem 1.6. *[Formulation of the Multiple Best Choice problems.] The Best Choice Problem or Secretary Problem is an important class of sequential decision problems, which arise in a wide variety of fields, including psychological, economic, and ecological applications. In the case of a single choice, the problem can be formulated as follows. We have a known number N of objects numbered $1, 2, \ldots, N$, so that, say, an object numbered 1 is classified as "the best", ..., and an object numbered N is classified as "the worst". It is assumed that the objects arrive one by one in random order, i.e. all $N!$ permutations are equiprobable. It is clear from comparing any two of these objects which one is better, although their actual number still remains unknown. After having known each sequential object, we either accept this object (and then the choice is made), or reject it and continue observation (it is impossible to return to the rejected object).*

An optimal selection rule implies that one observes and lets go an $k_N^* - 1$ object and continues observing till the time τ^*, at which time the best object from all the preceding objects makes its first appearance (if the best object does not appear by the moment N, we have to stop at this moment in any case). Table 1.1 can easily be constructed.

For large N, it is approximately optimal to ignore a proportion $e^{-1} = 36.8\%$ of the objects and then accept the next relatively best applicant, if any. The probability of obtaining the best object P_N^* is then approximately e^{-1}.

There is an extensive literature on **BCP**. For its history, previous works on it and formulation of some extensions, see papers by Б. А. Березовский, А. В. Гнедин (1984), Berezovskij et al. (1986), Ferguson (1989), Samuels (1991), and Gilbert and Mosteller (1966).

The best choice problem can arise in sequential decision problems in psychological and economic applications such as selling a house, hiring a secretary or seeking a job. It can be employed to analyze some behavioral ecology problems such as sequential mate choice or the optimal choice of the place of foraging.

All of these works considered the case where a single object was chosen. However, one can hire several secretaries or sell more than one house. Besides, in some species females sequentially mate with different males within a single mating period. Note also that an individual can sequentially choose more than one place to forage. Preater (1988), in his technical report, specifically adopted (m, k)-**BCP** notation for a subclass of problems where we allow for choice actions m to select the k-best items, $1 \leq k \leq m$. The $(1, 1)$-**BCP** is the classical **BCP**. Slightly generalizing this

TABLE 1.2
Absolute (a_1, a_2, a_3) and relative (ξ_1, ξ_2, ξ_3) ranks (the case $N = 3$).

Absolute ranks	Relative ranks
$(1, 2, 3)$	$(1, 2, 3)$
$(1, 3, 2)$	$(1, 2, 2)$
$(2, 1, 3)$	$(1, 1, 3)$
$(2, 3, 1)$	$(1, 2, 1)$
$(3, 1, 2)$	$(1, 1, 2)$
$(3, 2, 1)$	$(1, 1, 1)$

notation, we extend the class of such problems to the (m, k, r)-Secretary Problem, where with a maximum of m selections, we aim to obtain at least r out of the k best, $1 \le r \le \min\{k, m\}$.

Here, we consider a generalization of **BCP** to (k, k, k)-**SP** when it is possible to make k choices leading to the selection of the k best objects. The problem can be formulated in a similar way as Problem 1.6, but our aim now is to find stopping rules that maximize the probability of choosing k best objects.

Denote by $(a_1, a_2, \dots, r_N)$ any permutation of numbers $(1, 2, \dots, N)$. (1 corresponds to the best object, and N corresponds to the worst one) If a_i is the m-th object in the quality order of $(a_1, a_2, \dots, a_i)$, we write $\xi_i = m$ for all $i = 1, 2, \dots, N$. a_i is called the absolute rank, and ξ_i is called relative rank. Table 1.2 shows the absolute and relative ranks for all permutations for the case $N = 3$.

Remark 2. *Put $\eta_i = 1+$ number of objects from $(a_1, a_2, \dots, a_i) < a_i$. If our objective is to find a procedure such that the expected gain $\boldsymbol{E}(\eta_{\tau_1} + \dots + \eta_{\tau_k})$, $k \ge 2$ is minimal, then the problem is called the MBC problem with minimal summarized rank. In this case, $Z_{\boldsymbol{m}_k} = -\boldsymbol{E}(\eta_{\tau_1} + \dots + \eta_{\tau_k})$.*

Let $(i_1, \dots, i_k)$ be any permutation of numbers $1, 2, \dots, k$. A rule $\boldsymbol{\tau}_k^* = (\tau_1^*, \dots, \tau_k^*)$, $1 \le \tau_1^* < \tau_2^* < \dots < \tau_k^* \le N$ is an optimal rule if

$$\begin{aligned} \boldsymbol{P}_N^* &= \boldsymbol{P}\Big\{ \bigcup_{(i_1,\dots,i_k)} \{a_{\tau_1^*} = i_1, \dots, a_{\tau_k^*} = i_k\} \Big\} \\ &= \sup_{\tau} \boldsymbol{P}\Big\{ \bigcup_{(i_1,\dots,i_k)} \{a_{\tau_1} = i_1, \dots, a_{\tau_k} = i_k\} \Big\} \end{aligned} \tag{1.12}$$

where $\boldsymbol{\tau}_k = (\tau_1, \dots, \tau_k)$. We are interested in finding the optimal rule $\boldsymbol{\tau}_k^* = (\tau_1^*, \dots, \tau_k^*)$.

By $Z_{\boldsymbol{m}_k}^{(i)_k} = Z_{m_1,\dots,m_k}^{i_1,\dots,i_k}$ denote a conditional probability of event $\{a_{m_1} = i_1, \dots, a_{m_k} = i_k\}$ with respect to σ-algebra $\mathcal{F}_{\boldsymbol{m}_k}$, generated by observations $(y_1, \dots, y_{m_k})$, and put

$$Z_{\boldsymbol{m}_k} = \sum_{(i_1,\dots,i_k)} Z_{\boldsymbol{m}_k}^{(i)_k}. \tag{1.13}$$

Using (1.12), we get the value of the game v

$$\boldsymbol{P}_N^* = \boldsymbol{E}Z_{\tau^*} = \sup_{\tau\in\mathfrak{S}_1} \boldsymbol{E}Z_\tau = v. \tag{1.14}$$

Thus we reduce the **BC** problem of k objects to the problem of multiple stopping of the random sequence Z_{m_k}.

Remark 3. *In the context of multiple choice problems, it is worth paying attention to single choice problems, when the decision-maker sets different goals. Although the natural formulation is to select a candidate with the minimum expected rank (v. Chow et al. (1964)), the original problem was to select the best candidate (v. Gardner (1960)). A simple generalization of this last issue is to choose one of the best* k *(v. Guseĭn-Zade (1966)). In the above tasks, the structure of the optimal strategy was based on thresholds, as shown by the qualitative analysis of the problem. Therefore, the solution boils down to maximizing the probability of success, which, given the known form of the optimal strategy, comes down to maximizing a deterministic function of many variables. A statistician has a more difficult task when he wants to select a candidate with a fixed absolute rank or, more generally, with a rank from a fixed set of candidate ranks. The known results in the direction are:*

- *Selection of the candidate with a given rank* a *(v. Rose (1982), Szajowski (1982). Lin et al. (2019),*
- *Various extension are included in the paper by Goldenshluger et al. (2020). The selection of a candidate with the rank of a given set is treated in Suchwalko and Szajowski (2002).*
- *The crucial was observation by Preater (1993) the difficulties of selecting two items with given properties of the candidates.*
- *A version of the selection process was proposed by Vanderbei (1980). The decision-maker checks a ranked population in random order. At each step, the observer accepts or rejects the observed sample. As a result, he gets a partition of the population into two parts: accepted and rejected. The decision-maker is interested only in the case when all accepted samples are better than all rejected ones. Under this condition, the observer maximizes the probability of obtaining a fixed size* k *of the accepted part. It is also considered an arbitrary gain function* $\Phi(k)$*. These problems are some kind of MBC.*
- *Let us also recall papers by Wilson (1991), Majumdar (1991), and Rukavicka (2022).*
- *We can generalize the MBC problem to the case where the gain is the probability of choosing k objects with given ranks* $r_1, r_1, \ldots, r_k$*,* $1 \le r_1 < \ldots < r_k \le N$*. Let* $i_1, \ldots, i_k$ *be a permutation of the integer numbers* $r_1, r_2, \ldots, r_k$*. Then the optimal rule* $\boldsymbol{\tau}_k^* = (\tau_1^*, \ldots, \tau_k^*)$*, sequence* Z_{m_k}*, and the value of the game* v *are defined in a similar way as in (1.12), (1.13), and (1.14).*

Multiple optimal stopping in Bernoulli trails.

Let $\xi_1, \ldots, \xi_N$ be independent and identically distributed Bernoulli random variables, $\boldsymbol{P}(\xi_i = 1) = p = 1 - \boldsymbol{P}(\xi_i = 0) = 1 - q, 0 < p < 1$. Let R_j^i denote the length of an interval of successful trials, that is, the number of successive ones counting backward from i-th observation,

$$R_j^i = \sum_{n=j}^{i} \mathbb{I}(\xi_n = \cdots = \xi_i = 1). \tag{1.15}$$

Define the total gain after k, $k \geq 2$, stops the following way

$$Z_{\boldsymbol{m}_k} = R_1^{m_1} + R_{m_1+1}^{m_2} + \cdots + R_{m_{k-1}+1}^{m_k} - cm_k, \quad 1 \leq m_1 < m_2 < \cdots < m_k,$$

where c is a cost paid for each observation, $c < 1$. If $c \geq 1$, then $Z_{\boldsymbol{m}_k} \leq 0$ and the optimal strategy is trivial: stop at the beginning of the sequence.

Multiple selling problem.

Assume that we sequentially observe identically distributed independent random variables $\xi_1, \xi_2, \ldots$, where the value ξ_n can be interpreted as the value of an asset (for example, a house) at time n. We have k identical objects, which we want to sell, at each time n we receive one offer, and we are not allowed to recall past offers. If we decide to sell at times $m_1, m_2, \ldots, m_k$, $1 \leq m_1 < m_2 < \cdots < m_k$, we get the gain

$$Z_{\boldsymbol{m}_k} = \xi_{m_1} + \xi_{m_2} + \cdots + \xi_{m_k}.$$

The problem is related to the Moser's (1956) consideration of the well-known Cayley problem (see, for example, Ferguson (1989)).

Problem 1.7. *[Multiple change-point detection.] Let $\xi_1, \xi_2, \ldots$ be a sequence of random variables that is observed sequentially. At unknown moments $\theta_1, \theta_2, \ldots, \theta_k$, a distribution will change to some other distribution, and we want to detect the change points as soon as possible. Assume that after k stops at times $m_1, m_2, \ldots, m_k, 1 \leq m_1 < m_2 < \cdots < m_k$, we get a gain*

$$Z_{\boldsymbol{m}_k} = -\sum_{i=1}^{k} c_i(1 - \pi_{m_i}^{(i)}) - \sum_{j=0}^{m_1-1} (m_1 - j)p_{j,m_1}^{(1)} - \sum_{i=2}^{k} \sum_{j=m_{i-1}}^{m_i-1} (m_i - j)p_{j,m_i}^{(i)},$$

$$p_{j,m_1}^{(1)} = \boldsymbol{P}\{\theta_1 = j \mid \mathcal{F}_{m_1}\}, \quad j \geq 0,$$
$$p_{j,m_i}^{(i)} = \boldsymbol{P}\{\theta_i = j \mid \mathcal{F}_{m_1,\ldots,m_i}\} = \boldsymbol{P}\{\theta_i = j \mid \mathcal{F}_{(\boldsymbol{m})_i}\}, \quad j > m_{i-1},$$
$$\pi_{m_1}^{(1)} = \boldsymbol{P}\{\theta_1 \leq m_1 \mid \mathcal{F}_{m_1}\},$$
$$\pi_{m_i}^{(i)} = \boldsymbol{P}\{\theta_i \leq m_i \mid \mathcal{F}_{m_1,\ldots,m_i}\} = \boldsymbol{P}\{\theta_i \leq m_i \mid \mathcal{F}_{(\boldsymbol{m})_i}\},$$

where $p_{\infty,\infty}^{(1)}$ is the posterior probability of no change, $p_{\infty,\infty}^{(i)}$ is the posterior probability of $(i-1)$-multiple change, $c_1, c_2, \ldots, c_k$ are some positive constants representing the costs of false alarms, that is, of stops before the changes have actually occurred.

1.4.4 Multiple stopping strategies in inventory control

Consider a situation in which a commodity is stocked to satisfy a continuing demand. We assume that the replenishment of stock takes place at the end of periods labeled $n = 0, 1, 2, \ldots$ and we assume that the total aggregate demand for the commodity during period n is a random variable ξ_n, independent, identically distributed, whose distribution function is a discrete type

$$\boldsymbol{P}(\xi_n = k) = a_k \text{ for } k = 0, 1, 2, \ldots,$$

where $a_k \geq 0$ and $\sum_{k=0}^{\infty} a_k = 1$. The stock level is examined at the end of each period.

A replenishment policy is prescribed by specifying two nonnegative critical numbers s and $S > s$. They are interpreted as follows. If the amount of stock at the end of the period is not greater than s, then an amount sufficient to increase the quantity of stock available up to the level S is immediately obtained. However, if the available stock is over s, then no replenishment of the stock is carried out. Let X_n denote the quantity on hand at the end of the period n just before restocking. The states $\mathbb{E}$ of the process $\{X_n\}$ consist of the possible values of the stock size, that is, $\mathbb{E} = \{S, S-1, \ldots, +1, 0, -1, -2, \ldots\}$, where a negative value is interpreted as an unfilled demand that will be satisfied immediately upon restocking.

1.5 Multilateral Stopping of Stochastic Sequences

Optimal stopping games were introduced by Dynkin (1969) as an extension of the optimal stopping problem which has been actively studied since the middle of XX century.[15] The special structure of the sets of players' strategies, or rather the rules of the game, made it possible to prove the existence of an equilibrium point and to construct strategies in equilibrium. Let the state space $\mathbb{E}$ and the payoff function $g : \mathbb{E} \to \Re$ be given. The two decision-makers have the right to stop the process ξ_n: the first **DM** on subset E_1 and the second **DM** on E_2 of $\mathbb{E} = E_1 \cup E_2$. It is assumed that $E_1 \cap E_2 = \emptyset$. The strategies of the players in the game are the Markov moments τ, σ, subject $\xi_\tau \in E_1$, and $\xi_\sigma \in E_2$. If the process is stopped at the state, $x \in \mathbb{E}$ then the first player pays to the second $g(x)$. Let the process ξ_n have the Markov property and start from $x \in \mathbb{E}$. The mean payoff for the second player is equal to $g(x, \tau, \sigma) = \boldsymbol{E}_x g(\xi_{\tau \wedge \sigma})$. The solution to the problem for the Markov chain was given by Frid (1969).

The early investigations of the stopping game focused on the model as follows. There is a probability space $(\Omega, \mathcal{F}, \boldsymbol{P})$, a filtration of σ-algebras $\{\mathcal{F}_t\}$, $\mathcal{F}_t \subset \mathcal{F}$ with either $t \in \mathbb{N}$ (discrete time case) or $t \in \Re^+$ (continuous time case), $\{\mathcal{F}_t\}$-adapted

[15] Optimal stopping and, in particular, its game version was often discussed on Dynkin's undergraduate seminar at Moscow State University in the end of 1960 which resulted in seminal paper Frid (1969), Kifer (1971) and others (cf. Kifer (2013)).

payoff process $\{\xi_t\}$, and a pair of $\{\mathcal{F}_t\}$-adapted 0-1 valued "permission" processes $\varphi_t^{(i)}$, $i = 1, 2$ such that the player i is allowed to stop the game at time t if and only if $\varphi_t^{(i)} = 1$. If the game is stopped at the time t, then the first player pays the second player the sum ξ_t. Clearly, if $\varphi_t^{(1)} = 1$ and $\varphi_t^{(2)} = 0$, we arrive at the usual optimal stopping problem. Observe that in the one-player optimal stopping problem, the goal is the maximization of the payoff, and the corresponding supremum always exists (maybe infinite), so only optimal or almost optimal stopping times remain to be found. In the game version, the existence of the game value is the question of which should be resolved first, and only then can we look for optimal (saddle point) or almost optimal stopping times of the players.

It seems that the term "Dynkin's game" appeared first in Alario-Nazaret et al. (1982), where the version of the game was investigated for continuous time processes.[16]

1.6 Leading Applications

The model we present in this monograph is related to making decisions based on observations. Acquiring knowledge is costly, but wrong decisions are equally costly. We limit ourselves to the analysis of the moments of choosing to interrupt observations to make a decision, although we are interested in issues in which the problem is divided into the selection of several moments and moments of stopping to accept (stop) several states.

We will formulate sample statistics tasks in the Introduction in Section 1.6.1. We will take into account the problem of detecting change points as a special task of estimating and testing statistical hypotheses, which may naturally use multiple optimal stopping tools and algorithms.

The theory of optimal stopping sequences of random variables perhaps overuses the task of selecting the best object and its natural modifications. We will not give up these examples because if we are looking for a larger number of selected objects, the multiple-stop model is the most natural tool. An introduction to multiple selections is presented in Section 1.6.8.

1.6.1 Sequential methods of statistics

In statistical investigations, the *sequential method* and the *size of the sample* are two crucial aspects that affect the reliability and efficiency of the analysis.

Sequential method of statistics involves collecting data sequentially, analyzing it as it becomes available, and making decisions or drawing conclusions based on intermediate results. This approach is particularly useful in situations where resources

[16]Further details can be found in the paper by Kifer (2013). The tally of papers by Y. Kifer is on Math-net.ru.

are limited or where time is of the essence, such as clinical trials (see Sect. 1.6.5), quality control in manufacturing, or online data analysis. One common application of the sequential method is sequential hypothesis testing, where hypotheses are tested as data become available, allowing for early termination of experiments if significant results are observed.

The advantages of the sequential method include increased flexibility, the ability to adapt to changing conditions, and potentially faster decision-making compared to traditional fixed-sample methods.

The **size of the sample** refers to the number of observations or data points collected for statistical analysis. Larger sample sizes generally give more reliable estimates and reduce variability. Increasing the sample size improves the precision and test power, making it easier to detect true effects. However, larger samples need more resources.

Sequential methods interact with the sample size. They can be used regardless of sample size, but their effectiveness varies. With small samples, sequential methods are advantageous as they allow adaptive strategies when initial results are unclear. In fields like clinical trials or quality control, sequential methods monitor processes and adjust based on interim results. *Intermediate results are obtained at successive moments, and choosing optimal stopping points is often a challenging task. Hence, there is a need to develop methods for the optimal selection of multiple stopping moments.*

In each case, decisions are made at different data collection stages and sequential methods allow flexible adjustments based on the results. Even with large samples, sequential methods can be valuable, especially for quality control or real-time data analysis.

In conclusion, sequential statistical methods offer flexibility and efficiency, especially when resources are limited. The size of the sample affects the precision and power, and larger samples generally give more reliable results. The interaction between sequential methods and the sample size depends on the specific context and goals of the analysis.

Optimal stopping is a decision-making strategy used in sequential statistical analysis, particularly in situations where data are collected sequentially and decisions must be made on when to stop the data collection process. It involves determining the optimal point at which to halt data collection based on observed data, to achieve a specific objective such as maximizing expected utility or minimizing expected costs.

Let us list the key points about optimal stopping as a tool in sequential statistical analysis:

Objective: The primary objective of optimal stopping is to decide on when to stop collecting data to achieve a specific goal. This could involve maximizing the probability of making a correct decision, minimizing expected costs, or maximizing expected utility.

Sequential decision-making: Optimal stopping involves making sequential decisions based on observed data as it becomes available. At each decision point, the decision-maker must choose whether to continue collecting data or to stop and make a final decision based on the available information.

Utility and cost functions: To determine the optimal stopping rule, it is necessary to define the utility or the cost functions that quantify the consequences of different decisions. These functions reflect the preferences or objectives of the decision-maker and are used to evaluate the outcomes associated with different stopping points.

Bayesian and frequentist approaches: Optimal stopping can be approached from both Bayesian and frequentist perspectives. In the Bayesian framework, prior beliefs are updated based on observed data to compute posterior probabilities, which are then used to make optimal decisions. In the frequentist framework, optimal stopping rules are derived based on statistical properties such as likelihood ratios or p-values.

Applications: Optimal stopping is commonly used in various fields such as clinical trials, quality control, finance, and operations research. For example, in clinical trials, optimal stopping rules can be used to determine the optimal sample size or to decide when to stop a trial early if significant results are observed.

Challenges: Implementing optimal stopping strategies can be challenging due to uncertainty in model parameters, computational complexity, and the need to balance statistical properties with practical considerations such as resource constraints.

In general, optimal stopping provides a rigorous framework for making sequential decisions in statistical analysis, allowing decision-makers to achieve their objectives while efficiently using resources.

1.6.2 Multiple stopping in sequential statistical analysis

It is a concept that arises in sequential statistical analysis when there are two sequential decisions to be made, each with its own stopping rule. This approach is used in various fields to optimize decision-making processes, also based on statistical data, in situations where multiple sequential actions are required. Let us present some classes of such statistical problems.

Clinical trials: In clinical trials (refer to Section 1.6.5), double optimal stopping can be employed to identify the best timing for interim analyses and the final analysis. The first stopping rule may involve deciding when to conduct interim analyses to assess the trial's progress and potential early efficacy or futility.

The second stopping rule is used to determine when to stop the trial altogether based on the results of the final analysis, typically involving considerations of statistical significance, clinical relevance, and ethical considerations.

Quality Control: In quality control processes, double optimal stopping may be used to determine when to take corrective action during production and when to stop production altogether. This type of problem has many forms of formulation. The detection of change points or the disorder problem leads to double, or multiple optimal stopping problems defined in the classical manner (see Haggstrom (1966)). It is discussed in Section 1.6.6 and the motivation is as follows.

The first stopping rule might involve monitoring process parameters and deciding when to intervene to correct deviations from the desired performance. The second stopping rule may include deciding when to halt production entirely if the process is consistently out of control or if the product quality does not meet specified standards.

Supply management involves multiple stops in a sophisticated approach to capacity planning in a two-level supply chain, incorporating advanced forecasting models, collaboration with partners, and strategic decision-making regarding capacity reservation contracts to optimize performance and meet customer needs efficiently. Capacity planning involves determining the production capacity needed by an organization to meet its demand for products or services. In a supply chain context, it is about ensuring that each link in the chain, from suppliers to manufacturers to distributors, has the appropriate capacity to meet demand while minimizing costs and maximizing efficiency.

Capacity reservation contracts allow companies to secure access to a certain level of capacity in advance, ensuring that they have sufficient production or distribution capabilities when needed. Determining the optimal time to offer these contracts involves analyzing demand forecasts, capacity constraints, market conditions, and other factors to minimize costs while meeting customer demand effectively (v. Li et al. (2022) and the references given there).

Sequential Hypothesis Testing: Double optimal stopping can also arise in sequential hypothesis testing, where there are two decisions to be made: whether to continue sampling data and whether to reject the null hypothesis. The first stopping rule determines when to perform interim analyzes to assess the accumulated evidence and potentially make an early decision about the null hypothesis. The second stopping rule determines when to stop collecting data altogether and make a final decision based on the accumulated evidence, typically involving considerations of statistical significance and practical significance.

Financial Decision-Making: In financial decision-making, double (multiple) optimal stopping may be applied in scenarios such as portfolio management or option trading. The first stopping rule could involve deciding when to rebalance a portfolio based on market conditions or changes in asset prices. The second stopping rule may involve deciding when to exit a position or liquidate a portfolio based on predefined criteria such as risk tolerance, profit targets, or market conditions.

In each of these applications, double (or multiple) optimal stopping allows decision-makers to optimize their actions by making two (many) sequential decisions with appropriate stopping rules, balancing the need for timely intervention with the risk of premature decisions. This approach helps to ensure efficient resource allocation, effective risk management, and informed decision-making in complex sequential processes. However, we focus our attention on models of special type.

The focus of some sequential statistical methods is on conducting multiple hypothesis tests while controlling the FWER and efficiently allocating resources through adaptive sampling strategies. While this may involve sequential decision-making, the primary objective is not to optimize the stopping time, but rather to control the error rate and allocate resources efficiently. This philosophy of investigation leads to the conclusion that sequential analysis methods must be married with multiple testing error control methods to prevent false discoveries.

The result is a multistage step-down procedure that adaptively tests multiple hypotheses while preserving the family-wise error rate and extends Holm's (1979) step-down procedure to the sequential setting, yielding substantial savings in sample size with a small loss in power (v. Hochberg and Tamhane (1987)). This philosophy of investigation leads to the conclusion that sequential analysis methods must be married with multiple testing error control methods to prevent false discoveries. The thesis by Song (2013) presents sequential methods that control the multiple testing error rate, of which we consider the two most commonly used: familywise error rate (FWER) and false discovery rate (FDR). Conventional multiple hypothesis tests use step-up, step-down, or closed testing methods to control the overall error rates. Combining these methods with adaptive multistage sampling rules and stopping rules to perform efficient multiple hypothesis testing in sequential experimental designs, one can find in the paper by Bartroff and Lai (2010). For further reading on sequential multiple hypothesis tests, one can take dissertations by De (2012), Chen (2019), and references given therein.

Optimal multiple-stopping problems typically arise in situations such as clinical trials, where researchers must decide when to stop a trial based on interim results to minimize costs or maximize the probability of making a correct decision.

1.6.3 Multiple optimal-stopping vs. adaptive multistage sampling

Adaptive multistage sampling and optimal multiple-stopping are related concepts, but they are not exactly the same. Both involve making sequential decisions during the data collection process, but they have different objectives and applications.

Adaptive Multistage Sampling

is a method used in survey sampling or experimental design, where the sampling process is adjusted or adapted based on the information gathered at earlier stages. The main objective of adaptive multistage sampling is to improve the efficiency of the sampling process by allocating resources more effectively and reducing sampling variability. This method allows for flexibility in the sampling design and can lead to more accurate estimates with fewer resources compared to traditional fixed sampling designs.

Optimal Multiple Stopping

involves making sequential decisions about when to stop data collection or experimentation to optimize a specific objective, such as minimizing costs or maximizing utility. This concept is commonly encountered in situations like clinical trials, where researchers must decide when to stop a trial based on interim results to minimize costs or maximize the probability of making a correct decision.

While both *adaptive multistage sampling* (**AMS**) and *optimal multiple stopping* (**OMS**) involve sequential decision-making, they differ in their objectives and applications. Adaptive multistage sampling focuses on improving sampling efficiency and reducing variability, while optimal multiple-stopping focuses on optimizing a specific objective related to the stopping time of the data collection process.

1.6.4 Multiple optimal stopping vs. multiple hypothesis testing

Multiple-stopping algorithms in sequential multiple-hypothesis testing are not the same as the optimal multiple-stopping problem (**OMSP**). Both algorithms are based on the selection of operating moments, but there are significant differences between them.

Objective Differences: The primary objective of sequential multiple hypothesis testing is to control the family-wise error rate (FWER) or false discovery rate (FDR) while conducting multiple hypothesis tests sequentially. In contrast, the optimal multiple-stopping problem typically involves making sequential decisions about when to stop sampling data to optimize a specific objective, such as minimizing costs or maximizing utility.

Decision Timing: In sequential multiple-hypothesis testing, decisions are made at predefined interim analysis points based on accumulating data. These decisions are focused on controlling error rates and managing resources effectively. On the other hand, the optimal multiple-stopping problem involves making decisions about when to stop sampling data based on predefined criteria or thresholds related to the specific objective being optimized.

Resource Allocation vs. Stopping Time: Sequential multiple hypothesis testing algorithms primarily focus on efficient resource allocation and error control, rather than optimizing the stopping time of the experiment. In contrast, the optimal multiple-stopping problem is concerned with finding the optimal stopping time to achieve a specific objective, such as minimizing costs or maximizing utility.

Decision Context: The context in which decisions are made differs between sequential multiple-hypothesis testing and the optimal multiple-stopping problem. In sequential multiple-hypothesis testing, decisions are made in the context of conducting multiple-hypothesis tests while controlling error rates. In contrast, the optimal multiple-stopping problem involves making decisions in the context of optimizing a specific objective, such as making correct decisions with minimal costs.

Methodological Approach: Sequential multiple hypothesis testing algorithms employ statistical methods and procedures designed specifically for controlling error rates and managing resources in the context of conducting multiple hypothesis tests sequentially. In contrast, the optimal multiple-stopping problem requires the development of decision rules and stopping criteria tailored to the specific objective being optimized, which may involve different methodological approaches.

From the above, concluding, while both sequential multiple-hypothesis testing and the optimal multiple-stopping problem involve making sequential decisions, they differ in their objectives, decision timing, context, and methodological approaches. Therefore, it is essential to distinguish between them when discussing sequential decision-making in statistical analysis.

1.6.5 Optimal multiple-stopping vs. clinical trials

Clinical trials are very specific research on treatment methods for patients under care. The effects of treatment are not immediate. The approach taken to a patient's illness may be effective or require modification. In particular, errors are difficult and costly when the illness affects multiple patients simultaneously, easily spreads to others, and variants of the infection emerge. With large-scale occurrences of this phenomenon, we refer to it as an epidemic. We can assist doctors in making informed decisions by modeling both the progress of treatment for individual patients and the phenomenon of infection spreading. To help researchers make decisions about when to stop the trial based on interim results, to optimize a specific objective, such as minimizing costs or maximizing the probability of making a correct decision. Based on relevant models, determining optimal moments for implementing selected strategies is a crucial element. Let us examine the significance of optimal stopping points (for verifying the applied treatment method, changing the direction of research, etc.). It is how optimal multiple-stopping rules are typically applied in clinical trials:

Minimizing Costs: One common objective in clinical trials is to minimize costs while still achieving statistically significant results. Optimal multiple-stopping rules help researchers determine the optimal time to stop the trial to achieve this objective. By stopping the trial early if the interim results show clear efficacy or lack of efficacy, researchers can save resources that would otherwise be spent on continuing the trial.

Maximizing Efficiency: Optimal multiple-stopping rules also help maximize the efficiency of clinical trials by reducing the time and resources required to conclude. By stopping the trial early if the interim results meet predetermined criteria for efficacy or futility, researchers can expedite the process of determining the effectiveness of the treatment being tested.

Balancing Risks and Benefits: Optimal multiple-stopping rules allow researchers to balance the risks and benefits of continuing the trial. If interim results show clear efficacy with minimal risks, researchers can choose to stop the trial early to provide treatment to patients sooner. In contrast, if interim results show potential harm or lack of efficacy, researchers may choose to stop the trial early to minimize harm to participants.

Adaptive Design: Optimal multiple-stopping rules are often incorporated into adaptive trial designs, where the trial is modified based on accumulating data. These designs allow researchers to adapt the trial in response to interim results, such as modifying sample size, treatment arms, or endpoints, while still maintaining statistical validity.

In conclusion, the optimal multiple-stopping rules are a valuable tool in clinical trial design, allowing researchers to make informed decisions about when to stop the trial to achieve specific objectives while balancing risks and benefits for participants.

In the next part, we will demonstrate how multiple optimal stopping is applied in tasks that make extensive use of a foundational model, once we've defined the objective function and the observed process model. One classic problem involves identifying instances when the observed process undergoes changes in its characteristics, which we refer to as its disruption (see Section 1.6.6). If we have any prior knowledge about these moments of change, we can use Bayesian methods to transform the task into a model of multiple optimal stopping.

1.6.6 The problem of disorder,

known in the literature as the "change point problem" or the "disorder problem", is one of the most important issues of statistical inference (v. Shiryaev (2019), Tartakovsky (2020)). We deal with it wherever we want to answer the questions:

1. Are the observed random quantities homogeneous?
2. If the answer to this question is negative, are there homogeneous segments, and how do we locate them?

In scientific research, the comparison of the results of experiments is an important methodology. There is no single procedure for properly carrying out such a task with the use of statistical methods. Therefore, it must be clearly stated that we are talking here about a certain group of tasks that have common features, and because of this, it is possible to propose a uniform concept for conducting the analysis. This common feature is the dynamic nature of the observation. This is because the order of arrival of the data is essential. The methodology for detecting a change point uses

- stochastic control theory;
- theory of estimation and hypothesis testing (e.g. SPRT mentioned on p. 37);
- classic and Bayesian approach;
- finite sample inference and sequential methods.

With these issues, we naturally ask whether the observations are homogeneous (consistent with one-model calibration) or whether a parameter specifying the moment or moments of disorder should be introduced. Therefore, even in the formulation of the problem, it can be predicted that the goal is to determine many parameters and moments of disorder. This applies both to processes analyzed in real time and to the study of completed observations (historical data). In this section, we present an example model for such issues. A parameterization of them is shown in Problem 1.7. The issues are developed and discussed in more detail in Section 4.3.

The demand for this approach came from various statistical issues. The greatest demand for saving effort in collecting data appeared in cryptology (cipher-breaking tasks during World War II) and quality research, and more precisely in the quality control of industrial production. The first works devoted to the issue of production quality control were the construction of the Cumulative Sums Method (CUSUM in short) test[17]: Shewhart (1931), Shewhart and Deming (1939) and works Page,[18] Girshick and Rubin (1952). The mathematical formulation of the detection problem in radar systems was of great importance when it was necessary to optimally detect changes in the distribution during successive observations of a sequence of random variables. The mathematical formulation of this task was made by A.N. Kolmogorov in the late 1950s (v. Shiryaev (2006, 2019), and the first report on this topic is the work of Shiryaev (1961a, 1961b). The approach used in this work uses methods of optimal stopping theory. The practical significance of the "disorder" model resulted in considerable interest in research on this issue, which is visible in the monograph Brodsky and Darkhovsky (1993, 2000), Tartakovsky et al. (2015), and the collections edited by Carlstein et al. (1994) and Basseville and Benveniste (1986).

1.6.7 Detection of two disorders

To present this group of problems, we will discuss, for example, the task of determining homogeneous segments in a Markovian sequence. Homogeneity is understood here as a constant mechanism of transition probabilities, and a "disorder" is a point of change in transition probabilities. It extends the results on the problem of the detection of change points for Markov processes generalizing the results contained in the publications Sarnowski and Szajowski (2011), Szajowski (2011) and Ochman-Gozdek and Szajowski (2013). The short description is as follows. A random sequence having segments that are homogeneous Markov processes is registered. Each segment has its transition probability law, and the length of the segment is unknown and random. The transition probabilities of each process are known. The joint *a priori* distribution of the disorder moments is given. The detection of the disorder is rarely precise. The decision-maker accepts some deviation in the estimation of the disorder moments. The note aims to indicate a complete segment of the homogeneous process or a given length segment between disorders with maximal probabilities. The case with various precisions for overestimation and underestimation of the middle point is analyzed, including situations where the disorders do not appear with positive probability. The observed sequence, when the change point is known, has Markov properties. The results explain the structure of an optimal detector under various circumstances, show new details of the solution construction, as well as insignificantly extend the range of application. The motivation for this investigation is the modeling of the selection of suspicious observations in the experiments. Such observations can be treated as outliers or disturbed. The objective is to detect such an inaccuracy immediately or in a very short time before or after its appearance with the highest probability. The problem is reformulated to ensure optimal stopping of the observed sequences. A detailed analysis of the problem is presented to show the form of the optimal decision function. The application of the results to the analysis of piecewise

[17]In statistical Quality Control (QC in short), the CUSUM (or cumulative sum control chart) is a sequential analysis technique developed by E. S. Page of the University of Cambridge. It is typically used to monitor change detection. CUsUM was announced in Biometrika, in Page (1954), a few years after the publication of Wald's sequential probability ratio test (SPRT in short).

[18]Ewan Stafford Page. In Wikipedia. Ewan_Stafford_Page

deterministic processes with change points appearing at the moment of jumps is shown (see Herberts and Jensen (2004), Poor and Hadjiliadis (2009), Ferenstein and Pasternak-Winiarski (2011)).

The task of detecting disorders also has other formulations derived from systems in which a well-described system changes its dynamics in such a way that it can be assumed that we are dealing with switching between models. Each approach needs a detailed description of the model. Sometimes, this means changing one of the parameters. In continuous-time models, this can sometimes be done by changing one of the parameters (Mao and Yuan 2006)). Statistical analysis of data from a system whose model is switched processes means, among other things, the task of detecting moments of switching (*disorder*) (v. Shiryaev (2006, 1961b, 1961a), Tartakovsky (2010), Nikolaev (1998).

Formalization of the problem.

Suppose that the process $X = \{X_n, n \in \mathbb{N}\}$, $\mathbb{N} = \{0, 1, 2, \ldots\}$, is observed sequentially. It is obtained from Markov processes by switching between them at random moments θ_1 and θ_2 in such a way that the process after θ_1 starts from the state $X_{\theta_1 - 1}$ and after θ_2 starts from the state $X_{\theta_2 - 1}$. It means that the state at moment $n \in \mathbb{N}$ has a conditional distribution given the state at moment $n - 1$, where the formulae that describe these distributions have different forms: one for $n < \theta_1$, the second for $\theta_1 \leq n < \theta_2$, and another for $n \geq \theta_2$. Our objective is to indicate the segment of a given length between disorders with the maximum probability based on the observation of X. This data model appears in many practical problems of quality control (see Shewhart (1931), Shiryaev (1961a) and in the collection of the papers Basseville and Benveniste (1986), traffic anomalies in networks (see Tartakovsky et al. (2006)), epidemiology models (see Baron (2004)). The model considered here generalizes the basic problem stated in Szajowski (1992) (see also Bojdecki (1979), Yoshida (1983), Szajowski (2011)).

The classical disorder problem is limited to the case of switching between sequences of independent random variables (see Bojdecki (1979)). Some developments of the basic model can be found in Yakir (1994) where the optimal detection rule of the switching moment has been obtained when the finite state-space Markov chains are disordered. Moustakides (1998) formulates conditions that help reduce the problem of quickest detection for dependent sequences before and after the change in the case of independent random variables. Our result admits a Markovian dependence structure for switched sequences. We obtain an optimal rule under the probability-maximizing criterion.

The formulation of the problem can be found in Problem 1.8. The main result is presented in Section 4.4.3.

Problem 1.8. *[Detailed description of the double disorder model.] Let* $(\Omega, \mathcal{F}, \boldsymbol{P})$ *be a probability space that supports a sequence of observable random variables* $\{X_n\}_{n \in \mathcal{N}}$ *that generate filtration* $\mathcal{F}_n = \sigma(X_0, X_1, \ldots, X_n)$. *The random variables*

X_n take values in $(\mathcal{E}, \mathcal{B})$, where $\mathcal{E}$ is a subset of $\Re$. Space $(\Omega, \mathcal{F}, \boldsymbol{P})$ also supports unobservable random variables θ_1, θ_2 with values in $\mathbb{N}$ and the following distributions:

$$\boldsymbol{P}(\theta_1 = j) = \mathbb{I}_{\{j=0\}}(j)\pi + \mathbb{I}_{\{j>0\}}(j)\bar{\pi}p_1^{j-1}q_1, \quad (1.16)$$

$$\boldsymbol{P}(\theta_2 = k \mid \theta_1 = j) = \mathbb{I}_{\{k=j\}}(k)\rho + \mathbb{I}_{\{k>j\}}(k)\bar{\rho}p_2^{k-j-1}q_2 \quad (1.17)$$

where $j = 0, 1, 2, ...,$ $k = j, j+1, j+2, ...,$ $\bar{\pi} = 1 - \pi$, $\bar{\rho} = 1 - \rho$.

Additionally, we consider Markov processes $(X_n^i, \mathcal{G}_n^i, \boldsymbol{P}_x^i)$ on $(\Omega, \mathcal{F}, \boldsymbol{P})$, $i = 0, 1, 2$, where the σ-fields $\mathcal{G}_n^i$ are the smallest σ-fields for which $(X_n^i)_{n=0}^{\infty}$, $i = 0, 1, 2$, are adapted, respectively. Let us define process $(X_n)_{n\in\mathbb{N}}$ in the following way:

$$X_n = X_n^0 \mathbb{I}_{\{\theta_1 > n\}} + X_{n-\theta_1+1}^1 \mathbb{I}_{\{X_0^1 = x_{\theta_1-1}^0, \theta_1 \le n < \theta_2\}} + X_{n-\theta_2+1}^2 \mathbb{I}_{\{X_0^2 = x_{\theta_2-\theta_1}^1, \theta_2 \le n\}}. \quad (1.18)$$

We infer on θ_1 and θ_2 from the observable sequence $(X_n, n \in \mathbb{N})$ only.

DP 1: *Detection of the first change before the second one.* We aim to stop the observed sequence X_n described by (1.18) between the two disorders without direct observation of the component used in its mathematical model. This can be interpreted as a strategy for protecting against a second failure when the first has already occurred. The mathematical model of this is to control the probability $\boldsymbol{P}_x(\tau < \infty, \theta_1 + d_1 \le \tau < \theta_2 - d_2)$ by choosing the stopping time $\tau^* \in \mathcal{S}$ for which

$$\boldsymbol{P}_x(\theta_1 + d_1 \le \tau^* < \theta_2 - d_2) = \sup_{\tau \in \mathcal{T}} \boldsymbol{P}_x(\tau < \infty, \theta_1 + d_1 \le \tau < \theta_2 - d_2). \quad (1.19)$$

DP 2: *Detection the moment of the first and the second change.* we aim to indicate the moments of switching with a given precision d_1, d_2 (Problem $\mathrm{D}_{d_1 d_2}$). We want to determine a pair of stopping times $(\tau^*, \sigma^*) \in \mathcal{T}$ such that for every $x \in \mathbb{E}$

$$\boldsymbol{P}_x(\mid \tau^* - \theta_1 \mid \le d_1, \mid \sigma^* - \theta_2 \mid \le d_2) = \sup_{\substack{(\tau,\sigma)\in\mathcal{T} \\ 0 \le \tau \le \sigma < \infty}} \boldsymbol{P}_x(\mid \tau - \theta_1 \mid < d_1, \mid \sigma - \theta_2 \mid \le d_2). \quad (1.20)$$

The problem has been considered in Szajowski (1996) under natural simplification that there are three segments of data (i.e. there is $0 < \theta_1 < \theta_2$). In Section 4.4.3, the problem D_{00} is analyzed.

1.6.8 Multilateral selection of the secretary

Multilateral optimal stopping.

Decision models with many decision-makers are the natural extension of optimization problems. The case of decision problems with optimal stopping as strategies were extended to the game model by Dynkin (1969).

In many decentralized labor markets, job candidates are offered positions at very early stages in the hiring process. It has been argued that these early offers are an effect of the competition between employers for the best candidate. The timing of offers based on theoretical models on the classical secretary problem was analyzed by various researchers (v. Immorlica et al. (2006)). The systematic review was presented by Sakaguchi(1995)). We consider a secretary problem with multiple employers and study the equilibria of the induced game. Our results confirm the observation of early offers in labor markets: for several classes of strategies based on optimal stopping theory, as the number of employers grows, the timing of the earliest offer decreases.

Bilateral selection as stopping game.

Sources of questions about the selection modeling or the selection of candidates for a vacant position by employers can be found in the works of Presman and Sonin (1975) (see also Sonin (1976) where some issues are cleared). It is worth recalling the assumptions of the model from their work here because the further development of research in this area consisted of determining the meaning of these assumptions. The conditions in question come down to the following.

1. a stream of a certain number of candidates, let's say n, among which no two are alike, are reported to the candidates evaluating in a random order;
2. During their work as a selector, players do not obtain any information about the behavior of other selectors, and the order in which the candidates apply to each of them is independent. This means that we allow the possibility of simultaneous interviews with the candidate by different selectors. There are many ways to do this physically. For example, applicants undergo an agreed survey and a presentation on a fixed topic that is recorded. Both the survey and the recording are on the server, with the possibility of parallel access to them by a team of selectors. It can therefore be said that there are many identical copies of a set of candidates and that each player-selector operates under the conditions of the secretary problem on his copy. He checks his copy and decides to detain himself independently of the other selectors.
3. The goal of each player is to maximize the likelihood that they will stop first at the best object.

Such a model means competition between selectors and does not have to be associated with actual selection. In practice, we deal with a similar decision-making process when many experts randomly search for the closest pattern (e.g. a fingerprint) or use several different algorithms. The goal is to identify the right candidate without delay. The criterion is the highest chance of hitting (proper selection).

In the future, we will refer to this model as the basic model. One of the goals of further considerations will be to identify subtle modifications to assumptions and the impact of these changes on the necessary modifications to the mathematical model.

Consider a modification by Cownden and Steinsaltz (2014) which, according to the authors, is to make the task more realistic or in line with practical methods of proceeding with the selection of employees from a fixed pool of candidates. N

candidates are scrutinized by the K selectors. During the procedure, candidates are randomly assigned to selectors who can accept or reject the candidate immediately after the interview. Rejected candidates do not return to the pot. The activity of the selector ends when the candidate is selected or when the pool of available candidates is running out (is empty).

Within the conventional framework of the so-called secretary problem, Rose (1982) introduce two decision-makers, one of whom is dominant (has priority in selection decision and assignment). The objectives are to select the two best objects and immediately assign them to the decision-makers so that the dominant one obtains the best object. He obtained optimal strategies for selection and assignment and the maximum probability of achieving the objectives. An important suboptimal strategy was investigated: the dominant decision-maker acts to maximize the probability of obtaining the very best object while, independently, the other decision-maker acts to maximize the probability of selecting the second best was researched. Although it has intuitive appeal, this suboptimal strategy is rather poor. The two decision-makers may trade with each other the objects they select without affecting their chances of obtaining the desired objects. An interesting by-product of this study is that a simple closed-form solution is obtained for the problem of selecting the second-best object, which is an interesting topic for further investigation (v. Szajowski (1982), Vanderbei (2021), Suchwalko and Szajowski (2002), Bayón et al. (2019)).

1.6.9 Buying-selling problems

An agent-based modeling has been used to model the individual behavior and decision-making process of people in the market. This model is a representation of the decision-making processes of humans in the housing market as for example. Every agent is identified through his position, the time interval he exists in the system, his individual and social properties, and his behavior. Different features make different mentalities in different agents to process uncertain information they gain from the environment and the other agents to make two different types of events, i.e., supplies and demands. The model tries to forecast and analyze the dynamics of the housing market in a given region (city, county) by modeling the behavior of individuals who are living in this city and constructing the model on the interdependency of these behaviors and the information that flows in the system.

These models are described in the aspect of techniques applied in the mentioned problem in Section 5.1. The modeling approach we have used and the conceptual framework of the model will be presented there. For a better understanding of the limitations of mathematical modeling and its role in practice, let us think about the housing market. The mathematical modeling problem of the housing market involves the utilization of agent-based modeling to simulate the individual behaviors and decision-making processes of participants within the market. This model serves as a representation of the decision-making processes observed in the real-world housing market. Each agent in the model is characterized by various attributes, including their position, duration within the system, individual and social properties, and behavioral tendencies. These agents exhibit diverse mentalities that influence how they process uncertain information obtained from the environment and interactions with other agents, ultimately leading to the generation of supply and demand events.

The primary objective of the model is to forecast and analyze the dynamics of the housing market within a specified region, such as a city or county. This is achieved by capturing the behaviors of individuals residing in the region and constructing the model to account for the interdependence among these behaviors and the flow of information within the system. This consideration aims to present the modeling approach employed and the conceptual framework of the model developed to understand the intricacies of the housing market dynamics and the expectations of the house owners (sellers) and agents buying a house. The natural part of the preparing transaction is to collect information about various aspects. The natural strategy is the choice of the offer by the buyer. In economics literature the model of buying house is also interpreting as "job search problem".[19]

Let us focus on a *buying-selling problem* when two (or more) stops (contracts) are required. The aim is to construct optimal stopping rules in the mathematical abstraction of the trade, and the value of a contract obtained in such a way. The first models of house selling with strategies for the buyer was given by Sakaguchi (1961) as an application of the general optimal stopping problem (v. also Bellman et al. (1961), Saario (1994)). These are the first works on utilizing such an approach to construct rational agent behaviors in the real estate market. Because the model is abstract, the subject of trade is not significant; what matters are parameters such as supply and demand, the method of price set by the seller and buyer, and negotiation possibilities.

Bruss and Ferguson (1997) considered a generalization of the house-selling problem to selling k houses. The early formulation of house-selling models can be found in books by Kaufman (1963), and papers by Saario and Sakaguchi (1990), Mazalov and Falko (2008), Faller and Rüschendorf (2013). Let the offers X_i, $i = 1, 2, \ldots, n$, be *i.i.d.* random k-vectors having a common k-dimensional *cdf* with finite second moments. The decision-maker is asked to choose k stopping times $N_1, \ldots, N_k$, one for each component, where the payoff is $\sum_{j=1}^{k}$ (the jth component of X_{N_j}) minus the total observation cost until all houses are sold. The optimality equation is

$$V(K) = -c + \boldsymbol{E} \max_{\emptyset \subset S \subset K} \{\sum_{i \in S} X_i + V(K \setminus S)\},$$

with $V(\emptyset) = 0$, where $X = (X_1, \ldots, X_k)$, $K = \{1, 2, \ldots, k\}$, and $V(K)$ is the value obtained by using the optimal stopping rule. A simple description of the optimal rules is found. An interesting convexity property, $V(A)+V(B) \leq V(A \cup B)+V(A \cap B)$, is proven. Solutions for the case where the distribution of X is independent uniform in $[0, 1]^k$, for $k = 2$ and 3 (identical houses), are derived as functions of $c > 0$. Extensions are made to problems with recall (when $k \geq 2$) of past offers, and also to problems with a discount.

[19] How many searches should an unemployed person looking for a job undertake, considering each search costs him (v. Lippman and McCall (1976a, 1976b, 1981), Hall et al. (1979)).

1.6.10 Valuation and trading of options

The Chicago Board Options Exchange was established in 1973, which set up a regime using standardized forms and terms and trading through a guaranteed clearing house. Financial options from a mathematical modeling perspective are characterized by their mathematical rigor, reliance on stochastic processes, and the need to account for various factors such as volatility, risk, and market efficiency to accurately *price options* and *inform investment decisions.*

Multiple stops in this modeling take various parts. In the context of optimal stopping theory, it plays a significant role in financial option modeling, especially in scenarios where decision-making occurs over time (e.g. American options, game options) or in sequential steps. In modeling, we distinguish two primary objectives:

Valuation of Options This involves using mathematical models and analysis to determine the theoretical value of options based on factors such as the current price of the underlying asset, the option's strike price, time to expiration, volatility, and interest rates. Common models for option valuation include the Black-Scholes model and its variations, as well as more complex models that account for factors such as stochastic volatility or jumps in asset prices.

Trading of options Once options are valued, investors can engage in trading activities to buy or sell these contracts on exchanges or over-the-counter markets. Trading options allow investors to speculate on the future direction of asset prices, hedge against risks, or generate income through premiums received from selling options. Options trading strategies can range from simple directional bets to more complex combinations of options contracts designed to achieve specific risk-return objectives.

Let us mention an explicit application of multiple-stopping in the mentioned objectives:

Exotic Options Beyond standard European and American options, mathematical modeling extends to exotic options with non-standard features, such as barrier options, Asian options, and look-back options. These options often require more complex models and numerical techniques to price accurately. The techniques applied are the optimal multiple-stopping problem and the stopping games.

Early Exercise of American Options One common application of multiple-stopping is in the modeling of American options, which can be exercised at any time before expiration. Optimal stopping theory helps determine the optimal time to exercise American options, considering factors such as the current market price, volatility, and interest rates. This decision involves comparing the immediate payoff from exercising the option early to the expected future payoff from continuing to hold the option.

Real Options Analysis Real options analysis applies option pricing principles to evaluate investment decisions in situations where managers have the flexibility to delay, expand, or abandon projects over time. Multiple stopping is used to determine the optimal timing of investment decisions, considering factors such as project value, uncertainty, and delay cost.

Sequential Decision-Making in Portfolio Management In portfolio management, investors often face multiple decision points over time, such as when to buy, sell, or hold assets. Optimal stopping theory can help investors make these decisions by analyzing the expected returns and risks associated with different investment strategies. This includes deciding when to rebalance a portfolio, when to exercise options, or when to take profits or cut losses.

Dynamic Hedging Strategies For financial institutions and traders who hold options as part of their portfolio, multiple-stopping is used to develop dynamic hedging strategies that adjust positions over time in response to changing market conditions. These strategies aim to minimize risk exposure and maximize returns by continuously rebalancing the portfolio based on option pricing models and market signals.

Multiple-stopping techniques, rooted in optimal stopping theory and stopping games, provide a framework for decision-making in financial option modeling, helping investors and institutions optimize their strategies in dynamic and uncertain environments (v. Section 3.1).

1.6.11 Behavioral ecology problems

The multiple-stopping model can be used to analyze some behavioral ecology problems such as sequential mate choice or optimal choice of the place of foraging; see, for example, Cheng et al. (2014), Dombrovsky and Perrin (1994), Hutchinson and Halupka (2004), Real (1990, 1991), Wiegmann and Angeloni (2007). Analysis of polygyny and polyandry, in particular sequential polyandry (see, for example, Ligon (1999)), is another compelling reason to use a multiple-stopping model (MSM). In fact, in some species, active individuals (generally, females) sequentially mate with different passive individuals (usually males) within a single mating period (see, for example, Gabor and Halliday (1997) and Pitcher et al. (2003)). Note also that an individual can sequentially choose more than one place to forage. Therefore, we can consider a random variable ξ_n as the quality of the item (potential partner or foraging site) that appears at time n. An active individual can either accept the item (in this case one sample has been selected), or reject it and continue the observation.

Part II

Multiple-stopping and stopping games

> The d selectors all observe the same sequence of n objects and do so simultaneously. Each selector ranks each successive object relative to its predecessors. These relative ranks form the basis for jointly selecting only one object. The objective studied here is to maximize the minimum of the selectors' probabilities of choosing the best of all n objects.
>
> Samuels (2000)

THE DEVELOPMENT of research on stochastic processes has led to optimization tasks understood in various ways. Mathematical models of system control under various random factors, with clearly defined requirements as to the behavior of the system, led to questions about the control of system parameters in a way that guaranteed its optimal or expected behavior. Stochastic optimal control is a sub-field of control theory that deals with the existence of uncertainty either in observations or in the noise that drives the evolution of the system. The system designer assumes, in a Bayesian probability-driven fashion, that random noise with known probability distribution affects the evolution and observation of the state variables. Stochastic control aims to design the time path of the controlled variables that perform the desired control task with minimum cost and is somehow defined despite the presence of this noise. The control theory of stochastic systems has distinguished tasks in which acceptable strategies are moments of stopping observation. An example is the sequential estimation of the parameters of the observed process, testing of statistical hypotheses, or the selection of the starting and ending time of taking products in the production process. Due to the importance of these types of models, they were selected for in-depth research. We distinguish optimization tasks when the goal is to determine the extreme value of the problem (minimum or maximum), and, whenever possible, to determine a strategy that realizes this value, and problems in which the quality of the process is described using many criteria (multicriteria) (v. Belton and Stewart (2002)). We can rationalize goals in multi-criteria issues, but the book is restricted to the case where the different criteria are related to different decision-makers. We are looking for the rational behavior of the parties involved in the decision process. One of the important and frequently used approaches to the definition of rationality is the assumption that the process is of interest to many decision-makers. Each of them has a specific share in the observed process and a strategy for its modification. The methods of game theory can be used to determine rational behavior.

While there is extensive literature on optimal stopping rules when only one stop is required, multiple optimal stopping rules have been far less developed. Haggstrom (1967), Nikolaev (1980, 1981, 1998) and Stadje (1985) generalized the theory of optimal stopping rules to a case where more than two stops are required.

The martingale treatment of stochastic control problems is based on the idea that the correct formulation of Bellman's principle of optimality for stochastic minimization problems is in terms of a submartingale inequality. The value function of dynamic programming is always a submartingale and is a martingale under a particular control strategy if and only if that strategy is optimal. Local conditions for optimality in the form of a minimum principle can be obtained by applying Meyer's submartingale decomposition along with martingale representation theorems, conditions for the existence of an optimal strategy can also be stated (v. Davis (1993)). The martingale approach is used for optimal stopping and impulse control.

Game theory is a general term for the study of rational decision-making. Decisions are made by agents (parties, players), and the mathematical model allows defining acceptable decisions and their combined impact on the level of satisfaction of the parties (referred to as payouts) (v. Myerson (1991), Laraki et al. (2019)). So, as part of this research, we build models to represent situations where several actors or players make choices and where this set of individual behaviors produces an outcome that affects them all. Since agents may evaluate these outcomes differently, game theory is also the study of rational behavior within this framework. The adopted way of seeing reality in game theory, the game theory paradigm, is an approach to the behavior of participants in the decision-making problem between conflict and cooperation.

In this part, we will present a discrete-time stochastic model in which the moments of stopping constitute the strategies of one or more decision-makers. The one-decision-maker model of multiple-stopping was first formulated by Haggstrom (1967), and the two-decision-maker model by Dynkin (1969). Later work expanded the classes of model-defining processes (v. Nikolaev (1977, 1998), Nikolaev and Sofronov (2007), Sofronov (2013, 2016, 2018)), and rationalization methods with more decision-makers (v. Yasuda et al. (1980, 1982), Stadje (1985), Nakai (1997), Sakaguchi (1995), Szajowski (1993)).

2

Multiple Optimal Stopping Rules

Motto:
Unlocking opportunities through strategic pauses: The Power of optimal multiple-stopping.

THE MOTIVATION to abstract the optimal stop tasks from the stochastic control models was explained in Section 1.1. We are thinking about stochastic control models in which the strategy sets work well for selecting the moment at which "we get the process state" based on a criterion that depends on this state, but it is also possible from the very moment of action. The basic task is to choose one moment and the state of the process at this moment (v. Chapter 1). For similar reasons, we extend the class of optimal stop models, which we study with selected methods, allowing multiple choices of moments and states. Let $\xi_1, \xi_2, \ldots$ be a sequence of random variables with a known joint distribution. A decision-maker observes the ξ_n sequentially and can stop at any time. If the decision-maker stops on m_1 after observing $(\xi_1, \ldots, \xi_{m_1})$, then another sequence $\xi_{m_1,m_1+1}, \xi_{m_1,m_1+2}, \ldots$, which depends on $(\xi_1, \ldots, \xi_{m_1})$, appears and a new optimal stopping task for this new sequence to be solved. If i stops were made at times $m_1, m_2, \ldots, m_i$, $1 \leq i \leq k-1$, then the decision-maker observes a sequence of random variables $\xi_{m_1,\ldots,m_i,m_i+1}, \xi_{m_1,\ldots,m_i,m_i+2}, \ldots$ whose distribution depends on $(\xi_1, \ldots, \xi_{m_1}, \xi_{m_1,m_1+1}, \ldots, \xi_{m_1,m_2}, \ldots, \xi_{m_1,\ldots,m_i})$. The decision to stop at times m_i, $i = 1, 2, \ldots, k$, depends solely on *the values* of the basic random sequence already observed and not on future values. After k is stopped, $k \geq 2$, the decision-maker receives *a gain*

$$Z_{m_1,\ldots,m_k} = g_{m_1,\ldots,m_k}(\xi_1, \ldots, \xi_{m_1,m_1+1}, \ldots, \xi_{m_1,\ldots,m_k}), \tag{2.1}$$

where $g_{m_1,\ldots,m_k}(\cdot)$ is a known function, which shows the dependence of the gain $Z_{m_1,\ldots,m_k}$ on the values of the sequence[1]. In particular, $g_{m_1,\ldots,m_k}$ can be linear, that is, $Z_{m_1,\ldots,m_k} = \xi_{m_1} + \cdots + \xi_{m_k}$ (e.g., see Nikolaev and Sofronov (2007)). An important case is where $\{\xi_i\}_{i=0}^N$, where N can be finite or infinite, is a Markov sequence (for further details, see Nikolaev (1998)). Other examples supporting the multiple-stopping approach and gain functions can be found in Sofronov (2013) and Targino et al. (2017), and the multiple disorder detection considered by Yoshida (1983) and Szajowski (1996). The objective is to find stopping rules that maximize the expected gain and the value of the game.

[1]A family of random variables indexed with elements from a partially ordered set, especially when elements of the subset $\mathbb{N}^k$ or $\Re^{+k}$, are called in the literature *a random field*. The monograph Edgar and Sucheston (1992) is devoted to the detailed definitions of such processes, as well as generalizations of the optimal stopping tasks for them. Although our considerations narrow down the sets of indexes, it is worth bearing in mind that this is a subclass of issues formulated for random fields.

DOI: 10.1201/9781003407102-2

2.1 Definitions Needed to Formulate Multiple-Stopping Problem

The mathematical framework of the multiple optimal stopping problem consists of the following:

(a) a probability space $(\Omega, \mathcal{F}, \boldsymbol{P})$;

(b) a nondecreasing sequence of σ-subalgebras $\{\mathcal{F}_{m_1,\dots,m_i}\}_{m_i > m_{i-1}}$ of σ-algebra $\mathcal{F}$ such that

$$\mathcal{F}_{m_1,\dots,m_{i-1}} \subseteq \mathcal{F}_{m_1,\dots,m_i} \subseteq \mathcal{F}_{m_1,\dots,m_{i-1},m_i+1}$$

for all $i = 1, 2, \dots, k$, $0 \equiv m_0 < m_1 < \cdots < m_{i-1}$. This family is called *a filtration*;

(c) a random process compatible with filtration

$$\{Z_{m_1,\dots,m_k}, \mathcal{F}_{m_1,\dots,m_k}\},$$

for any fixed integers $m_1, \dots, m_{k-1}$, $1 \leq m_1 < m_2 < \cdots < m_{k-1}$.

Within this framework, the decision to stop at times n depends only on the observed values of $\xi_1, \dots, \xi_n$ and not on any future values. We also assume that it is not possible to stop twice at the same discrete time, that is, $1 \leq m_1 < m_2 < \cdots < m_k$.

Following Haggstrom (1967), Nikolaev (1980, 1981, 1998, 1998), Sofronov (2013) we will introduce necessary definitions, denotations and present a precise model of the optimal multiple-stopping problems.

Assumption 2.1 (MSR). *Let us assume that the collection of integer-valued random variables τ_j, $j = 1, \dots, i$, have the following properties:*

(a) $1 \leq \tau_1 < \tau_2 \cdots < \tau_i < \infty$ *($\boldsymbol{P}$-almost surely (a.s. for short)),*

(b_j) $\{\omega : \tau_1 = m_1, \dots, \tau_j = m_j\} \in \mathcal{F}_{m_1,\dots,m_j}$ *for all* $m_j > m_{j-1} > \dots > m_1 \geq 1$; $j = 1, 2, \dots, i$.

Definition 2.1. *A collection of integer-valued random variables* $\boldsymbol{\tau}_i = (\tau_1, \dots, \tau_i)$ *having properties **MSR** is called an* i-multiple-stopping rule, *where* $1 \leq i \leq k$. *A* k-multiple-stopping rule *with* $k > 1$ *is called* a multiple-stopping rule.

The following notation is used[2]:

$$\boldsymbol{\lambda}_i = (\lambda_1, \lambda_2, \dots, \lambda_i),\ \boldsymbol{\lambda}_1 = \lambda_1,\ \boldsymbol{E}_{\boldsymbol{m}_i}\xi = \boldsymbol{E}(\xi \mid \mathcal{F}_{\boldsymbol{m}_i}),$$

$$\boldsymbol{A}_{\boldsymbol{m}_i}\xi = \boldsymbol{E}_{\boldsymbol{m}_i}\left(\sup_{m_{i+1}} \boldsymbol{E}_{\boldsymbol{m}_{i+1}}\left(\cdots\left(\sup_{m_{k-1}} \boldsymbol{E}_{\boldsymbol{m}_{k-1}}\xi\right)\cdots\right)\right),$$

[2]In the narrative of this chapter, we consider a problem of maximizing the expected utility. The formulation of the minimization problem is analogous to the natural, obvious changes.

where ξ is an arbitrary random variable. In particular, the conditional expectations relative to the σ-subalgebras $\mathcal{F}_{m_1}$ will be denoted by $\mathbf{E}_{m_1}$.

The following conditions must be met for all expectations considered. There are several other conditions sufficient to ensure that the expectations exist, e.g.

$$(A^+): \ \boldsymbol{E}\left(\sup_{m_1} \boldsymbol{A}_{m_1}\left(\sup_{m_k} Z_{m_k}\right)\right) < \infty, \tag{2.2}$$

assumption (13) in Theorem 2 of the paper by Stadje (1985) and several conditions discussed in Kösters (2004). For further consideration, we assume that condition (2.2) is fulfilled for Z_{m_k}.

For each multiple-stopping rule $\boldsymbol{\tau} = (\tau_1, \ldots, \tau_k)$, we define $Z_{\tau} \equiv Z_{\tau_1,\ldots,\tau_k}$ as

$$Z_{\tau}(\omega) = \begin{cases} Z_{m_1,\ldots,m_k}, & \tau_1(\omega) = m_1, \ldots, \tau_k(\omega) = m_k,\ m_k > \cdots > m_1 \geq 1, \\ -\infty, & \text{if either } \tau_i(\omega) \geq \tau_{i+1}(\omega) \text{ or } \tau_k(\omega) = \infty. \end{cases}$$

Let $\mathfrak{S}_m$ be a class of multiple-stopping rules $\boldsymbol{\tau} = (\tau_1, \ldots \tau_k)$ such that $\tau_1 \geq m$ ($\boldsymbol{P}$-a.s.).

Definition 2.2. *The function $v_m = \sup_{\boldsymbol{\tau} \in \mathfrak{S}_m} \boldsymbol{E} Z_{\boldsymbol{\tau}}$ is called the m-value of the game.*

In particular, if $m = 1$, then $v = v_1$ is called the *value of the game*.

Definition 2.3. *A multiple-stopping rule $\boldsymbol{\tau}^* \in \mathfrak{S}_m$ is called an optimal multiple-stopping rule (OMSR) in $\mathfrak{S}_m$ if $\boldsymbol{E} Z_{\boldsymbol{\tau}^*}$ exists and $\boldsymbol{E} Z_{\boldsymbol{\tau}^*} = v_m$.*

The aforementioned condition (2.2) ensures the existence of $\boldsymbol{E} Z_{\boldsymbol{\tau}}$ for all $\boldsymbol{\tau} \in \mathfrak{S}_m$ and the finiteness of the value v_m.

2.2 The Structure of Solutions and Their Values

Our problem is to find the optimal rule for the problem of multiple stops and its value v_m, as stated in Definition 2.3. To this end, let us introduce the sequences $\{V_{\boldsymbol{m}_i}\}$ and $\{X_{\boldsymbol{m}_i}\}$, $i = 1, 2, \ldots, k$, which allows us to find the structure of the optimal rule and the value of the game. After properly defining the sequences, we will determine multiple-stopping rules $\boldsymbol{\tau}^*$. Let $\mathfrak{S}_{\boldsymbol{m}_i}$ be a class of i-multiple-stopping rules $\boldsymbol{\tau}_i = (\tau_1, \ldots, \tau_i)$, $i = 1, 2, \ldots, k$, with $\tau_1 = m_1, \ldots, \tau_{i-1} = m_{i-1}$, $\tau_i \geq m_i$ ($\mathbf{P}$-a.s.). In particular, we have $\mathfrak{S}_{\boldsymbol{m}_1} \equiv \mathfrak{S}_{m_1}$ the class of all stopping times τ_1 such that $\tau_1 \geq m_1$ ($\boldsymbol{P}$-a.s.). We put $X_{\boldsymbol{m}_k} = Z_{\boldsymbol{m}_k}$ and define by backward induction on i from $i = k$:

$$V_{\boldsymbol{m}_i} = \operatorname*{ess\,sup}_{\boldsymbol{\tau}_i \in \mathfrak{S}_{\boldsymbol{m}_i}} \boldsymbol{E}_{\boldsymbol{m}_i} X_{\boldsymbol{\tau}_i}, \tag{2.3}$$

$$X_{\boldsymbol{m}_{i-1}} = \boldsymbol{E}_{\boldsymbol{m}_{i-1}} V_{\boldsymbol{m}_{i-1}, m_{i-1}+1}, \qquad i = k, k-1, \ldots, 1, \tag{2.4}$$

where $X_0 \equiv 0$.

Remark 4. We emphasize that most of the statements in this section are almost surely valid. We shall not mention this in what follows.

It follows from the results of the general theory of optimal stopping (see, e.g., Chow et al. (1971) and Haggstrom (1967)) that the sequence $V_{\boldsymbol{m}_i}$ satisfies the recurrent equation (Bellman equation)

$$V_{\boldsymbol{m}_i} = \max\{X_{\boldsymbol{m}_i}, \boldsymbol{E}_{\boldsymbol{m}_i} V_{\boldsymbol{m}_{i-1}, m_i+1}\}. \tag{2.5}$$

Moreover, if the strategy

$$\tau_{\boldsymbol{m}_i} = \inf\{l \geq m_i : X_{\boldsymbol{m}_{i-1}, l} = V_{\boldsymbol{m}_{i-1}, l}\} \tag{2.6}$$

is finite (**P** -as), then the i-multiple-stopping rule $(\boldsymbol{m}_{i-1}, \tau_{\boldsymbol{m}_i})$ is optimal in $\mathfrak{S}_{\boldsymbol{m}_i}$ and $V_{\boldsymbol{m}_i} = \boldsymbol{E}_{\boldsymbol{m}_i} X_{\boldsymbol{m}_{i-1}, \tau_{\boldsymbol{m}_i}}$.

If the optimal strategy in $\mathfrak{S}_{\boldsymbol{m}_i}$ exists, then $\boldsymbol{P}(\tau_{\boldsymbol{m}_i} < \infty) = 1$. A sufficient condition for the existence of an optimal strategy in $\mathfrak{S}_{\boldsymbol{m}_i}$ is Snell's (1952) condition $\lim_{m_i \to \infty} X_{\boldsymbol{m}_i} = -\infty$.

From (2.3) and (2.4), we obtain

$$X_{\boldsymbol{m}_{i-1}} = \boldsymbol{E}_{\boldsymbol{m}_{i-1}} X_{\boldsymbol{m}_{i-1}, \tau_{\boldsymbol{m}_{i-1}, m_{i-1}+1}},$$

for $\tau_{\boldsymbol{m}_{i-1}, m_{i-1}+1} < \infty$.

It can be shown that the random variables $X_{\boldsymbol{m}_i}, i = 1, 2, \ldots, k$, are bounded from above by integrable random variables. Indeed, using the definition of $X_{\boldsymbol{m}_i}$ and the notations introduced before, we have

$$\begin{aligned} X_{\boldsymbol{m}_i} &= \boldsymbol{E}_{\boldsymbol{m}_i} \left(\operatorname*{ess\,sup}_{\tau_{i+1} \in \mathfrak{S}_{\boldsymbol{m}_i, m_i+1}} \boldsymbol{E}_{\boldsymbol{m}_i, m_i+1} X_{\boldsymbol{m}_i, \tau_{i+1}} \right) \\ &\leq \boldsymbol{E}_{\boldsymbol{m}_i} \left(\sup_{m_{i+1}} X_{\boldsymbol{m}_{i+1}} \right) \leq \boldsymbol{A}_{\boldsymbol{m}_i} \left(\sup_{\boldsymbol{m}_k} Z^+_{\boldsymbol{m}_k} \right) \equiv B_{\boldsymbol{m}_i}. \end{aligned} \tag{2.7}$$

From condition (2.2), it follows that $B_{\boldsymbol{m}_k} = Z^+_{\boldsymbol{m}_k}$ and $B_{\boldsymbol{m}_i}$ are integrable. We have

$$\boldsymbol{E} X_{\boldsymbol{m}_{i-1}, \tau_i} \leq \boldsymbol{E} \left(\sup_{m_i} X_{\boldsymbol{m}_i} \right) \leq \boldsymbol{E} \left(\sup_{m_i} B_{\boldsymbol{m}_i} \right) < \infty.$$

Thus the expectations in (2.3) and (2.4) are well defined.

Let us now consider the sequence $\{X_{m_1}, \mathcal{F}_{m_1}\}_{m_1 \geq 1}$. It follows from (2.5) that the sequence $\{V_{m_1}\}_{m_1 \geq 1}$ satisfies the recurrent equation

$$V_{m_1} = \max\{X_{m_1}, \boldsymbol{E}_{m_1} V_{m_1+1}\}.$$

Combining condition (2.2) and (2.7), we get $\boldsymbol{E}\left(\sup_{m_1} X_{m_1}\right) < \infty$. Therefore, from the general theory of optimal stopping, we obtain the following formula

$$\boldsymbol{E} V_{m_1} = \sup_{\sigma \in \mathfrak{S}_{m_1}} \boldsymbol{E} X_{\sigma}. \tag{2.8}$$

Furthermore, if there exists an optimal stopping time in $\mathfrak{S}_m$ for the sequence $\{X_{m_1}, \mathcal{F}_{m_1}\}_{m_1 \geq m}$, then the stopping time $\tilde{\tau}_{1,m} = \inf\{l \geq m : V_l = X_l\}$ is the optimal stopping time in $\mathfrak{S}_m$ and

$$V_m = \boldsymbol{E}_m X_{\tilde{\tau}_{1,m}}. \tag{2.9}$$

If there does not exist an optimal stopping time in $\mathfrak{S}_m$, then the sequence $\{V_m\}$ can be used to construct an ε-optimal ($\varepsilon > 0$) stopping time:

$$\tau_{1,m}(\varepsilon) = \inf\{l \geq m : X_l \geq V_l - \varepsilon\}.$$

By condition (2.2), it follows that $\boldsymbol{P}(\tau_{1,m}(\varepsilon) < \infty)$ and $\boldsymbol{E}_m X_{\tau_{1,m}(\varepsilon)} \geq V_m - \varepsilon$.

It can be proved that the following lemma gives possible variations of multiple-stopping rules in $\mathfrak{S}_{\boldsymbol{m}_i}$.

Lemma 2.1. *Let $\tau', \tau'' \in \mathfrak{S}_{\boldsymbol{m}_i}$ and $A \in \mathcal{F}_{\boldsymbol{m}_i}$. Then*

$$\bar{\tau}(\omega) = \tau' \mathbb{I}(A) + \tau'' \mathbb{I}(\bar{A}) \in \mathfrak{S}_{\boldsymbol{m}_i},$$

where $\mathbb{I}(\cdot)$ is the indicator function.

The following theorem gives the existence conditions and the structure of an optimal multiple-stopping rule in $\mathfrak{S}_m$.

Theorem 2.2 (Nikolaev (1980)). *Let condition* (2.2) *be fulfilled. We put*

$$\tau_i^* = \inf\{m_i > m_{i-1} : V_{\boldsymbol{m}_i} = X_{\boldsymbol{m}_i}\}$$

for $i = 1, 2, \ldots, k$ in the set $D_{i-1} = \{\omega : \tau_1^* = m_1, \ldots, \tau_{i-1}^* = m_{i-1}\}$, where it is assumed that $\tau_i^*(\omega) = \infty$ in $\{\omega : \tau_{i-1}^*(\omega) = \infty\}$, $m_0 = m - 1$, and $D_0 = \Omega$. In that case, if the random vector $\tau^* = (\tau_1^*, \ldots, \tau_k^*)$ is finite with probability one, then $\tau^* \in \mathfrak{S}_m$ is an optimal multiple-stopping rule.

Proof: Let us show that $V_m \geq \boldsymbol{E}_m Z_\tau$ for any multiple-stopping rule $\tau \in \mathfrak{S}_m$ and $V_m = \boldsymbol{E}_m Z_\tau$ if and only if $\tau = \tau^*$. Then $V_m = \boldsymbol{E}_m Z_{\tau^*} \geq \boldsymbol{E}_m Z_\tau$ for $\tau \in \mathfrak{S}_m$ and $v \leq \boldsymbol{E} Z_{\tau^*}$.

Following Haggstrom (1967), for any multiple-stopping rule $\tau \in \mathfrak{S}_m$, we define sequences $\{t_{\boldsymbol{m}_i}\}_{m_i > m_{i-1}}$, $i = 1, 2, \ldots, k-1$, as

$$t_{\boldsymbol{m}_i} = \mathbb{I}_{\boldsymbol{m}_i} \tau_{i+1} + (m_i + 1)(1 - \mathbb{I}_{\boldsymbol{m}_i}).$$

Here by $\mathbb{I}_{\boldsymbol{m}_i}$ denotes the indicator function of the set $\{\omega : \tau_1 = m_1, \ldots, \tau_i = m_i\}$[3].

From $(\boldsymbol{m}_{k-1}, t_{\boldsymbol{m}_{k-1}}) \in \mathfrak{S}_{\boldsymbol{m}_{k-1}, m_{k-1}+1}$ and τ_k^* coinciding with the stopping time $\tau_{\boldsymbol{m}_{k-1}, m_{k-1}+1}$ (2.6) on the set $\{\omega : \tau_1^* = m_1, \ldots, \tau_{k-1}^* = m_{k-1}\}$, we have

$$\mathbb{I}_{\boldsymbol{m}_{k-1}} V_{\boldsymbol{m}_{k-1}, m_{k-1}+1} \geq \mathbb{I}_{\boldsymbol{m}_{k-1}} \boldsymbol{E}_{\boldsymbol{m}_{k-1}, m_{k-1}+1} Z_{\boldsymbol{m}_{k-1}, t_{\boldsymbol{m}_{k-1}}}.$$

[3]For $\tau = \tau^*$ inequalities become equalities throughout this proof.

Combining this result with (2.4), we get

$$\mathbb{I}_{m_{k-1}} X_{m_{k-1}} \geq \mathbb{I}_{m_{k-1}} \boldsymbol{E}_{m_{k-1}} Z_{m_{k-1},t_{m_{k-1}}}.$$

We have

$$X_{m_{i-1},t_{m_{i-1}}} = B_{m_{i-1},t_{m_{i-1}}} - (B_{m_{i-1},t_{m_{i-1}}} - X_{m_{i-1},t_{m_{i-1}}}),$$

where the random variables B_{m_i} are defined in (2.7). Then, it can be shown that

$$\mathbb{I}_{m_{k-2}} \boldsymbol{E}_{m_{k-2},m_{k-2}+1} X_{m_{k-2},t_{m_{k-2}}} \geq \mathbb{I}_{m_{k-2}} \boldsymbol{E}_{m_{k-2},m_{k-2}+1} Z_{m_{k-2},\tilde{t}_{m_{k-2}}},$$

where $\tilde{t}_{m_i} = \mathbb{I}_{m_i}(\tau_{i+1}, \ldots, \tau_k) + (m_i + 1, \ldots, m_i + k - i)(1 - \mathbb{I}_{m_i}), 1 \leq i < k - 1$. Combining this with

$$\mathbb{I}_{m_{k-2}} V_{m_{k-2},m_{k-2}+1} \geq \mathbb{I}_{m_{k-2}} \boldsymbol{E}_{m_{k-2},m_{k-2}+1} X_{m_{k-2},t_{m_{k-2}}},$$

where inequality is reduced to equality under $\boldsymbol{\tau}_{k-1} = \boldsymbol{\tau}_{k-1}^* = (\tau_1^*, \ldots, \tau_{k-1}^*)$, we have

$$\begin{aligned}
\mathbb{I}_{m_{k-2}} X_{m_{k-2}} &= \mathbb{I}_{m_{k-2}} \boldsymbol{E}_{m_{k-2}} V_{m_{k-2},m_{k-2}+1} \\
&\geq \boldsymbol{E}_{m_{k-2}} \mathbb{I}_{m_{k-2}} \mathbf{E}_{m_{k-2},m_{k-2}+1} X_{m_{k-2},t_{m_{k-2}}} \\
&\geq \boldsymbol{E}_{m_{k-2}} \mathbb{I}_{m_{k-2}} Z_{m_{k-2},\tilde{t}_{m_{k-2}}} = \mathbb{I}_{m_{k-2}} \boldsymbol{E}_{m_{k-2}} Z_{m_{k-2},\tilde{t}_{m_{k-2}}}.
\end{aligned}$$

Similarly, we obtain $\mathbb{I}_{m_i} X_{m_i} \geq \mathbb{I}_{m_i} \boldsymbol{E}_{m_i} Z_{m_i,\tilde{t}_{m_i}}$, $i = k-3, \ldots, 2, 1$. Thus,

$$\boldsymbol{E}_m X_{\tau_1} = \boldsymbol{E}_m B_{\tau_1} - \boldsymbol{E}_m \sum_{m_1=m}^{\infty} \mathbb{I}_{m_1}(B_{m_1} - X_{m_1}) \geq \boldsymbol{E}_m Z_{\boldsymbol{\tau}}.$$

On the other hand, since $V_m \geq \boldsymbol{E}_m X_{\tau_1}$ with equality under $\tau_1 = \tau_1^*$ (see (2.3) and (2.9)), then $V_m \geq \boldsymbol{E}_m Z_{\boldsymbol{\tau}}$ and $V_m = \mathbf{E}_m Z_{\boldsymbol{\tau}^*}$. This completes the proof. □

Let us now define a random vector $\boldsymbol{\tau}_m(\varepsilon) = (\tau_{1,m}(\varepsilon), \ldots, \tau_{i-1,m}(\varepsilon))$ as

$$\tau_{i,m}(\varepsilon) = \inf\left\{m_i > m_{i-1} \,:\, X_{m_i} \geq V_{m_i} - \frac{\varepsilon}{k}\right\} \tag{2.10}$$

in the set $\{\omega : \tau_{1,m}(\varepsilon) = m_1, \ldots, \tau_{i-1,m}(\varepsilon) = m_{i-1}\}$, where $\tau_{j+1,m}(\varepsilon) = \infty$ on $\{\omega : \tau_{j,m}(\varepsilon) = \infty\}$, $m_0 = m - 1$. As mentioned above, $\boldsymbol{\tau}_m(\varepsilon)$ with components defined by (2.10) is a multiple-stopping rule in $\mathfrak{S}_m$ and

$$\boldsymbol{E}_m Z_{\boldsymbol{\tau}_m(\varepsilon)} \geq V_m - \varepsilon. \tag{2.11}$$

The following theorem gives a characterization of the m-value v_m using the sequence $\{V_m\}$.

Theorem 2.3. *If the condition* (2.2) *holds, then* $v_m = \boldsymbol{E} V_m$.

Proof: It follows from Theorem 2.2 that $V_m \geq \boldsymbol{E}_m Z_{\boldsymbol{\tau}}$ for any multiple-stopping rule $\boldsymbol{\tau} \in \mathfrak{S}_m$. Therefore, $\boldsymbol{E} V_m \geq v_m$. The opposite inequality follows from (2.11) since $\varepsilon > 0$ it is arbitrary. □

2.2.1 Necessary and sufficient conditions for optimality

In this section, we will present a theorem that gives necessary and sufficient conditions for optimality of multiple-stopping rules under a condition stronger than (2.2):

$$(\bar{A}^+) : \boldsymbol{E}\left(\sup_{m_1} \bar{A}_{m_1}\left(\sup_{\boldsymbol{m}_k} Z^+_{\boldsymbol{m}_k}\right)\right) < \infty.$$

Theorem 2.4 (Nikolaev (1981)). *Let condition $(\bar{A}^+)$ is fulfilled. A multiple-stopping rule $\tilde{\tau} = (\tilde{\tau}_1, \dots, \tilde{\tau}_k)$ is optimal if and only if*

(a_i) $X_{\boldsymbol{m}_i} \geq \boldsymbol{E}_{\boldsymbol{m}_i} X_{\boldsymbol{m}_{i-1},\tau_i}$ *on the set* $\{\omega : \tilde{\tau}_1 = m_1, \dots, \tilde{\tau}_i = m_i, \tau_i > m_i\}$,
(b_i) $X_{\boldsymbol{m}_i} \leq \boldsymbol{E}_{\boldsymbol{m}_i} X_{\boldsymbol{m}_{i-1},\tilde{\tau}_i}$ *on the set* $\{\omega : \tilde{\tau}_1 = m_1, \dots, \tilde{\tau}_{i-1} = m_{i-1}, \tilde{\tau}_i > m_i\}$,

for $i = 1, 2, \dots, k$.

Proof: First, we will prove the necessary condition. For any i-multiple-stopping rule $\boldsymbol{\tau}_i = (\tau_1, \dots, \tau_i)$, let us define $X_{\boldsymbol{\tau}_i}$ in the following way:

$$X_{\boldsymbol{\tau}_i} = \begin{cases} X_{\boldsymbol{m}_i}, & \text{if } \tau_1 = m_1, \dots, \tau_i = m_i, 1 \leq m_1 < \cdots < m_i, \\ -\infty, & \text{if at least one inequality holds } \tau_j(\omega) \geq \tau_{j+1}(\omega) \text{ or } \tau_i(\omega) = \infty. \end{cases}$$

Similarly, $B_{\boldsymbol{\tau}_i}$ is defined. From condition $(\bar{A}^+)$ and (2.7), we obtain

$$\boldsymbol{E}X_{\boldsymbol{\tau}_i} \leq \boldsymbol{E}\left(\sup_{\boldsymbol{m}_i} X_{\boldsymbol{m}_i}\right) < \infty,$$

$$\boldsymbol{E}B_{\boldsymbol{\tau}_i} \leq \boldsymbol{E}\left(\sup_{\boldsymbol{m}_i} B_{\boldsymbol{m}_i}\right) < \infty.$$

In addition, we see that $\boldsymbol{P}(X_{\boldsymbol{\tau}_i} \leq B_{\boldsymbol{\tau}_i}) = 1, i = 1, 2, \dots, k$.

Assume the converse, that is, at least one of the conditions (a_i), (b_i), $i = 1, 2, \dots, k$, does not hold. For fixed $m_1, \dots, m_j$, $1 \leq m_1 < \cdots < m_j$, $j \in \{1, 2, \dots, k\}$, let

$$D_{1j} = \{\omega : \tilde{\tau}_1 = m_1, \dots, \tilde{\tau}_j = m_j, \tau_j > m_j\} \cap \{\omega : X_{\boldsymbol{m}_j} < \boldsymbol{E}_{\boldsymbol{m}_j} X_{\boldsymbol{m}_{j-1},\tau_j}\}$$

and $\boldsymbol{P}(D_{1j}) > 0$. It follows from Lemma 2.1 that the random vector

$$(\tilde{\boldsymbol{\tau}}_{j-1}, s'_j) = \begin{cases} (\tilde{\boldsymbol{\tau}}_{j-1}, \tau_j), & \omega \in D_{1j}, \\ \tilde{\boldsymbol{\tau}}_j, & \omega \notin D_{1j} \end{cases}$$

is a j-multiple-stopping rule. Then

$$\begin{aligned} \boldsymbol{E}X_{\tilde{\boldsymbol{\tau}}_{j-1},s'_j} &= \boldsymbol{E}\left[\mathbb{I}(D_{1j})\boldsymbol{E}_{\boldsymbol{m}_j} X_{\boldsymbol{m}_{j-1},\tau_j} + \mathbb{I}(\bar{D}_{1j}) X_{\tilde{\boldsymbol{\tau}}_j}\right] \\ &> \boldsymbol{E}\left[\mathbb{I}(D_{1j}) X_{\boldsymbol{m}_j} + \mathbb{I}(\bar{D}_{1j}) X_{\tilde{\boldsymbol{\tau}}_j}\right] = \boldsymbol{E}X_{\tilde{\boldsymbol{\tau}}_j}. \end{aligned} \tag{2.12}$$

Suppose that for fixed $m_1, \dots, m_n$, $1 \leq m_1 < \cdots < m_n$, $n \in \{1, 2, \dots, k\}$,

$$\begin{aligned} D_{2n} = &\{\omega : \tilde{\tau}_1 = m_1, \dots, \tilde{\tau}_{n-1} = m_{n-1}, \tilde{\tau}_n > m_n\} \\ &\cap \{\omega : X_{\boldsymbol{m}_n} > \boldsymbol{E}_{\boldsymbol{m}_n} X_{\boldsymbol{m}_{n-1},\tilde{\tau}_n}\} \end{aligned}$$

and $\boldsymbol{P}(D_{2n}) > 0$. Put

$$(\tilde{\boldsymbol{\tau}}_{n-1}, s''_n) = \begin{cases} \boldsymbol{m}_n, & \omega \in D_{2n}, \\ \tilde{\boldsymbol{\tau}}_n, & \omega \notin D_{2n}. \end{cases}$$

Then

$$\begin{aligned} \boldsymbol{E}X_{\tilde{\boldsymbol{\tau}}_{n-1},s''_n} &= \boldsymbol{E}\left[\mathbb{I}(D_{2n})X_{\boldsymbol{m}_n} + \mathbb{I}(\bar{D}_{2n})X_{\tilde{\boldsymbol{\tau}}_n}\right] \\ &> \boldsymbol{E}\left[\mathbb{I}(D_{2n})\boldsymbol{E}_{\boldsymbol{m}_n}X_{\boldsymbol{m}_{n-1},\tau_n} + \mathbb{I}(\bar{D}_{2n})X_{\tilde{\boldsymbol{\tau}}_n}\right] = \boldsymbol{E}X_{\tilde{\boldsymbol{\tau}}_n}. \end{aligned} \tag{2.13}$$

Let us show that inequality (2.12) (as well as (2.13)) leads to a contradiction. Define

$$\begin{aligned} t_i &= \mathbb{I}_{\boldsymbol{m}_i}\tilde{\tau}_{i+1} + (m_i+1)(1-\mathbb{I}_{\boldsymbol{m}_i}), \\ t_{i,r} &= \mathbb{I}_{\boldsymbol{m}_i}(\tilde{\tau}_{i+1},\ldots,\tilde{\tau}_r) + (m_i+1,\ldots,m_i+r-i)(1-\mathbb{I}_{\boldsymbol{m}_i}), \quad r > i+1, \end{aligned}$$

where $\mathbb{I}_{\boldsymbol{m}_i} = \mathbb{I}\{\omega : \tilde{\tau}_1 = m_1,\ldots,\tilde{\tau}_i = m_i\}$. We can see that $(\boldsymbol{m}_i, t_i) \in \mathfrak{S}_{\boldsymbol{m}_i,m_i+1}$. From (2.4) and (2.5), we have

$$\mathbb{I}_{\boldsymbol{m}_i}X_{\boldsymbol{m}_i} \geq \mathbb{I}_{\boldsymbol{m}_i}\boldsymbol{E}_{\boldsymbol{m}_i}X_{\boldsymbol{m}_i,t_i}.$$

It follows from the monotone convergence theorem for the conditional expectation that

$$\mathbb{I}_{\boldsymbol{m}_{i-1}}\boldsymbol{E}_{\boldsymbol{m}_{i-1}}X_{\boldsymbol{m}_{i-1},t_{i-1}} \geq \mathbb{I}_{\boldsymbol{m}_{i-1}}\boldsymbol{E}_{\boldsymbol{m}_{i-1}}X_{\boldsymbol{m}_{i-1},t_{i-1,i+1}}.$$

Hence, similarly, we get

$$\mathbb{I}_{\boldsymbol{m}_{i-2}}\boldsymbol{E}_{\boldsymbol{m}_{i-2}}X_{\boldsymbol{m}_{i-2},t_{i-2,i}} \geq \mathbb{I}_{\boldsymbol{m}_{i-2}}\boldsymbol{E}_{\boldsymbol{m}_{i-2}}X_{\boldsymbol{m}_{i-2},t_{i-2,i+1}}.$$

Continuing this process, we obtain

$$\mathbb{I}_{m_1}\boldsymbol{E}_{m_1}X_{m_1,t_{1,i}} \geq \mathbb{I}_{m_1}\boldsymbol{E}_{m_1}X_{m_1,t_{1,i+1}}.$$

Therefore,

$$\boldsymbol{E}_1X_{\tilde{\boldsymbol{\tau}}_i} = \boldsymbol{E}_1B_{\tilde{\boldsymbol{\tau}}_i} - \boldsymbol{E}_1\sum_{m_1=1}^{\infty} I_{m_1}(B_{m_1,t_{1,i}} - X_{m_1,t_{1,i}}) \geq \boldsymbol{E}_1X_{\tilde{\boldsymbol{\tau}}_{i+1}}. \tag{2.14}$$

Then

$$V_1 \geq \boldsymbol{E}_1X_{\tilde{\tau}_1} \geq \boldsymbol{E}_1X_{\tilde{\tau}_1,\tilde{\tau}_2} \geq \cdots \geq \boldsymbol{E}_1X_{\tilde{\boldsymbol{\tau}}_{k-1}} \geq \boldsymbol{E}_1Z_{\tilde{\boldsymbol{\tau}}_k}.$$

We have

$$V_1 = \boldsymbol{E}_1Z_{\tilde{\boldsymbol{\tau}}}. \tag{2.15}$$

Thus,

$$\boldsymbol{E}_1X_{\tilde{\boldsymbol{\tau}}_{i-1}} = \boldsymbol{E}_1X_{\tilde{\boldsymbol{\tau}}_i}, \quad i = 2,\ldots,k. \tag{2.16}$$

Combining this result with (2.12), we get

$$\boldsymbol{E}X_{\tilde{\boldsymbol{\tau}}_{j-1},s'_j} > \boldsymbol{E}X_{\tilde{\boldsymbol{\tau}}_{j-1}}, \quad j > 1,$$

which contradicts (2.14). Finally, it follows from inequality (2.12) for $j = 1$ that $\boldsymbol{E}X_{s'_1} > \boldsymbol{E}X_{\tilde{\tau}_1}$. However, using (2.15) and (2.16), we have $\boldsymbol{E}V_1 = \boldsymbol{E}X_{\tilde{\tau}_1}$, which contradicts (2.8). This means that $\boldsymbol{P}(D_{1i}) = 0$ and $\boldsymbol{P}(D_{2i}) = 0$, for each $i = 1, 2, \ldots, k$. This proves the necessary condition of the theorem.

Now we will prove the sufficient condition. From conditions (a_k), (b_k), and the monotone convergence theorem, it can be shown that

$$\mathbb{I}_{\boldsymbol{m}_{k-1}} \boldsymbol{E}_{\boldsymbol{m}_{k-1}, m_{k-1}+1} Z_{\boldsymbol{m}_{k-1}, \tilde{\tau}_k} \geq \mathbb{I}_{\boldsymbol{m}_{k-1}} \boldsymbol{E}_{\boldsymbol{m}_{k-1}, m_{k-1}+1} Z_{\boldsymbol{m}_{k-1}, \tau_k}.$$

for any multiple stopping rule $(\boldsymbol{m}_{k-1}, \tau_k) \in \mathfrak{S}_{\boldsymbol{m}_{k-1}, m_{k-1}+1}$. It follows from the definition of $V_{\boldsymbol{m}_{k-1}, m_{k-1}+1}$ that

$$\mathbb{I}_{\boldsymbol{m}_{k-1}} \boldsymbol{E}_{\boldsymbol{m}_{k-1}, m_{k-1}+1} Z_{\boldsymbol{m}_{k-1}, \tilde{\tau}_k} = \mathbb{I}_{\boldsymbol{m}_{k-1}} V_{\boldsymbol{m}_{k-1}, m_{k-1}+1}.$$

Therefore,

$$\mathbb{I}_{\boldsymbol{m}_{k-1}} X_{\boldsymbol{m}_{k-1}} = \mathbb{I}_{\boldsymbol{m}_{k-1}} \boldsymbol{E}_{\boldsymbol{m}_{k-1}} Z_{\boldsymbol{m}_{k-1}, \tilde{\tau}_k}. \tag{2.17}$$

Similarly, using conditions (a_{k-1}) and (b_{k-1}), we have

$$\mathbb{I}_{\boldsymbol{m}_{k-2}} \boldsymbol{E}_{\boldsymbol{m}_{k-2}} X_{\boldsymbol{m}_{k-2}, t_{k-2}} = \mathbb{I}_{\boldsymbol{m}_{k-2}} X_{\boldsymbol{m}_{k-2}}.$$

Using (2.17), it can be shown that

$$\mathbb{I}_{\boldsymbol{m}_{k-2}} X_{\boldsymbol{m}_{k-2}} = \mathbb{I}_{\boldsymbol{m}_{k-2}} \boldsymbol{E}_{\boldsymbol{m}_{k-2}} X_{\boldsymbol{m}_{k-2}, t_{k-2}} = \mathbb{I}_{\boldsymbol{m}_{k-2}} \boldsymbol{E}_{\boldsymbol{m}_{k-2}} Z_{\boldsymbol{m}_{k-2}, t_{k-2,k}}.$$

In the same way,

$$\mathbb{I}_{\boldsymbol{m}_i} X_{\boldsymbol{m}_i} = \mathbb{I}_{\boldsymbol{m}_i} \boldsymbol{E}_{\boldsymbol{m}_i} Z_{\boldsymbol{m}_i, t_{i,k}}, \quad i = k-3, \ldots, 1.$$

Hence,

$$\begin{aligned} \boldsymbol{E}_1 X_{\tilde{\tau}_1} &= \boldsymbol{E}_1 B_{\tilde{\tau}_1} - \boldsymbol{E}_1 \sum_{m_1=1}^{\infty} \mathbb{I}_{m_1} (B_{m_1} - X_{m_1}) \\ &= \boldsymbol{E}_1 B_{\tilde{\tau}_1} - \boldsymbol{E}_1 \sum_{m_1=1}^{\infty} \mathbb{I}_{m_1} (B_{m_1} - \boldsymbol{E}_{m_1} Z_{m_1, t_{1,k}}) = \boldsymbol{E}_1 Z_{\tilde{\boldsymbol{\tau}}_k}. \end{aligned} \tag{2.18}$$

It follows from conditions (a_1) and (b_1) that $\boldsymbol{E}_1 X_{\tilde{\tau}_1} \geq \boldsymbol{E}_1 X_{\tau_1}$ for $\tau_1 \in \mathfrak{S}_1$. Then $V_1 \leq \boldsymbol{E}_1 X_{\tilde{\tau}_1}$. Combining this with (2.18), we have $v = \boldsymbol{E}V_1 = \boldsymbol{E}Z_{\tilde{\tau}}$, indicating that $\tilde{\tau} = (\tilde{\tau}_1, \ldots, \tilde{\tau}_k)$ is an optimal multiple stopping rule. This proves the sufficient condition of the theorem. □

2.2.2 Multiple-stopping for Markov processes

There are many multiple optimal stopping problems for which the gain $Z_{\boldsymbol{m}_k}$ can be represented as a function of $\xi_{m_1}, \ldots, \xi_{m_k}$, $Z_{\boldsymbol{m}_k} = g(\xi_{m_1}, \ldots, \xi_{m_k})$, where $\{\xi_n\}$ is a Markov sequence. In this section, we will describe multiple optimal stopping rules for Markov processes.

Let us formulate the problem in the Markov case. We are given

(a) a probability space $(\Omega, \mathcal{F}, \boldsymbol{P})$;

(b) a Markov chain $\{\xi_m, \mathcal{F}_m\}_{m\geq 1}$ in a space $(\mathbb{E}, \mathcal{B})$, where $\{\mathcal{F}_m\}_{m\geq 1}$ is a sequence of σ-subalgebras of σ-algebra $\mathcal{F}$;

(c) a nondecreasing sequence of σ-subalgebras $\{\mathcal{F}_{m_1,\dots,m_i}\}_{m_i > m_{i-1}}$ of σ-algebra $\mathcal{F}$ such that
$$\mathcal{F}_{m_1,\dots,m_{i-1}} \subseteq \mathcal{F}_{m_1,\dots,m_i} \subseteq \mathcal{F}_{m_i},$$
for all $i = 1, 2, \dots, k$, $0 \equiv m_0 < m_1 < \cdots < m_{i-1}$;

(d) a random process
$$\{Z_{m_1,\dots,m_{k-1},m_k}, \mathcal{F}_{m_1,\dots,m_{k-1},m_k}\}_{m_k > m_{k-1}}$$
for any fixed integers $m_1, \dots, m_{k-1}$, $1 \leq m_1 < m_2 < \cdots < m_{k-1}$, where $Z_{\boldsymbol{m}_k} = g(\xi_{m_1}, \dots, \xi_{m_k})$, and $g(\cdot)$ is a real $\mathcal{B}^k$-measurable function defined on $\mathbb{E}^k$.

Following Definition 2.2, the m-value of the game in the Markov case will become
$$v_m = \sup_{\boldsymbol{\tau}\in\mathfrak{S}_m} \boldsymbol{E} Z_{\boldsymbol{\tau}} = \sup_{\boldsymbol{\tau}\in\mathfrak{S}_m} \boldsymbol{E} g(\xi_{\tau_1}, \dots, \xi_{\tau_k}).$$

Let $\hat{\mathfrak{S}}_m$ be a class of multiple-stopping rules $\boldsymbol{\tau}$ such that $\boldsymbol{\tau} \in \mathfrak{S}_m$ and there exist $B_i \in \mathcal{B}$ such that
$$\begin{aligned} \tau_1 &= \inf\{m_1 \geq m : \xi_{m_1} \in B_1\}, \\ \tau_i &= \inf\{m_i \geq m_{i-1} : \xi_{m_i} \in B_i\}, \qquad i = 2, \dots, k, \end{aligned}$$
where $B_i = B_i(\tau_1, \dots, \tau_{i-1}, \xi_{\tau_1}, \dots, \xi_{\tau_{i-1}})$. The class $\hat{\mathfrak{S}}_m$ can be interpreted as a class of memoryless multiple-stopping rules when the states of the Markov chain are only remembered at times $\tau_1, \dots, \tau_k$.

Intuitively, we can see that in the Markov case, the decision to stop should be based on the present state and on the states at previous stopping times. This result is formulated in the following theorem. The proof of the theorem can be found in Nikolaev (1998).

Theorem 2.5 (Nikolaev (1998)). *Assume that condition* (2.2) *is fulfilled. Then*
$$v_m = \sup_{\boldsymbol{\tau}\in\hat{\mathfrak{S}}_m} \boldsymbol{E} g(\xi_{\tau_1}, \dots, \xi_{\tau_k}).$$

2.3 A Finite Horizon Case

Assume now that we observe a finite sequence of random variables $\xi_1, \xi_2, \dots, \xi_N$. Let
$$\{Z_{\boldsymbol{m}_k}, 1 \leq m_1 \leq N_1, m_1 < m_2 \leq N_2(m_1), \dots, \\ m_{k-1} < m_k \leq N_k(m_1, \dots, m_{k-1})\}$$

be a family of random variables, where N_1, $N_i(\cdot)$, $i = 2, \ldots, k$, are natural numbers, which represent the latest possible times we are allowed to stop. For example, $N_1 = N - k + 1, N_2 = N - k + 2, \ldots, N_k = N$. As in the general theory of optimal stopping (see Chow et al. (1971)), we define the sequence $V_{\boldsymbol{m}_i}$ by backward induction from the recurrent equations:

$$V_{\boldsymbol{m}_{i-1}, N_i(m_1,\ldots,m_{i-1})} = X_{\boldsymbol{m}_{i-1}, N_i(m_1,\ldots,m_{i-1})}, \tag{2.19}$$

$$V_{\boldsymbol{m}_i} = \max\{X_{\boldsymbol{m}_i}, \boldsymbol{E}_{\boldsymbol{m}_i} V_{\boldsymbol{m}_{i-1}, m_i+1}\}, \tag{2.20}$$

$$X_{\boldsymbol{m}_k} = Z_{\boldsymbol{m}_k},$$

for $1 \le m_1 \le N_1$, $m_{i-1} < m_i \le N_i(m_1, \ldots, m_{i-1})$, $i = 2, \ldots, k$. After having observed the first observations m_i, $X_{\boldsymbol{m}_i}$ can be interpreted as the conditionally expected gain if we stop at time m_i and use optimal stopping times for the later stages; $V_{\boldsymbol{m}_i}$ represents the maximum gain that is possible to receive at time m_i. If $m_i = N_i(m_1, \ldots, m_{i-1})$, then the gain is $V_{\boldsymbol{m}_{i-1}, N_i(m_1,\ldots,m_{i-1})} = X_{\boldsymbol{m}_{i-1}, N_i(m_1,\ldots,m_{i-1})}$. If $m_i < N_i(m_1, \ldots, m_{i-1})$, we can either stop or continue. If we stop, the gain is $X_{\boldsymbol{m}_i}$, and if we continue, the gain is $\boldsymbol{E}_{\boldsymbol{m}_i} V_{\boldsymbol{m}_{i-1}, m_i+1}$.

Using Theorem 2.2, we define the optimal multiple-stopping rule τ^*. From Theorem 2.3, and formulas (2.19) and (2.20), we obtain the value of the game $v_{\boldsymbol{m}}$.

Krasnosielska-Kobos (2015) considered a method allowing to transform a multiple-stopping problem with a random horizon to a multiple-stopping problem with discounting and a fixed horizon. In particular, the author considered a multiple-stopping problem with a Poisson process and a random horizon and a problem with a discount factor and a random number of offers.

2.4 An Infinite Horizon Case

As we can see from the previous section, in a finite horizon case, the value of the game $v_{\boldsymbol{m}}$ can be obtained using the recurrent equations (2.19) and (2.20). Here we focus on a natural extension: how $v_{\boldsymbol{m}}$ can be obtained in the case of an infinite horizon.

Following Nikolaev (1978, 1998), we will consider a problem with the finite horizon N and then explore the case when N goes to infinity. Suppose we are given a family of random variables

$$\{Z_{\boldsymbol{m}_k}, 1 \le m_1 < m_2 < \ldots < m_k \le N\}.$$

For $1 \le m \le N-k+1$, we will use the following notations in this section:

$$\begin{aligned}
\mathfrak{S}_m^N &= \{\tau \in \mathfrak{S}_m : \boldsymbol{P}(m \le \tau_1 < \tau_2 < \ldots < \tau_k \le N) = 1\}, \\
v_m^N &= \sup_{\tau \in \mathfrak{S}_m^N} \boldsymbol{E} Z_\tau, \\
\mathfrak{S}_{m_1}^N &= \{\tau_1 \in \mathfrak{S}_{m_1} : \boldsymbol{P}(\tau_1 \le N-k+1) = 1\}, \\
\mathfrak{S}_{\boldsymbol{m}_i}^N &= \big\{\boldsymbol{\tau}_i \in \mathfrak{S}_{\boldsymbol{m}_i} : \boldsymbol{P}(\tau_i \le N-k+i \mid \tau_1 = m_1, \ldots, \tau_{i-1} = m_{i-1}) = 1\big\}, \\
& \quad i = 2, \ldots, k, \\
V_{\boldsymbol{m}_i}^N &= \boldsymbol{E} \sup_{\boldsymbol{\tau}_i \in \mathfrak{S}_{\boldsymbol{m}_i}^N} \boldsymbol{E}_{\boldsymbol{m}_i} X_{\boldsymbol{\tau}_i}^N, \\
X_{\boldsymbol{m}_{i-1}}^N &= \boldsymbol{E}_{\boldsymbol{m}_{i-1}} V_{\boldsymbol{m}_{i-1}, m_{i-1}+1}^N, \quad i = k, k-1, \ldots, 1,
\end{aligned}$$

where $X_{\boldsymbol{\tau}_k}^N \equiv Z_{\boldsymbol{\tau}_k}$, $X_{m_0} \equiv 0$.

The sequence $\{V_{\boldsymbol{m}_i}^N\}$ can be obtained from (2.19) and (2.20):

$$\begin{aligned}
V_{\boldsymbol{m}_{i-1}, N-k+1}^N &= X_{\boldsymbol{m}_{i-1}, N-k+1}^N, \\
V_{\boldsymbol{m}_i}^N &= \max\left\{X_{\boldsymbol{m}_i}^N, \boldsymbol{E}_{\boldsymbol{m}_i} V_{\boldsymbol{m}_{i-1}, m_i+1}^N\right\},
\end{aligned} \tag{2.21}$$

with $m_{i-1} < m_i \le N-k+i-1$, $i = k, k-1, \ldots, 1$. It is clear that

$$\mathfrak{S}_m^N \subset \mathfrak{S}_m^{N+1} \subset \ldots \subset \mathfrak{S}_m.$$

Therefore,

$$v_m^N \le v_m^{N+1} \le \ldots \le v_m,$$

and, similarly,

$$V_m^N \le V_m^{N+1} \le \ldots \le V_m.$$

This means that there exist limits

$$v_m^* = \lim_{N \to \infty} v_m^N \text{ and } V_m^* = \lim_{N \to \infty} V_m^N,$$

with $v_m^* \le v_m$ and $V_m^* \le V_m$. From Theorem 2.3, we can find v_m^N using the recurrent equations (2.21). Then the question is under what conditions v_m^* coincides with v_m^N. The following assertions answer this question.

Let $\{\xi_n, \widetilde{\mathcal{F}}_n, n \ge 1\}$ be a sequence of random variables defined on the probability space $(\Omega, \mathcal{F}, \boldsymbol{P})$, $\widetilde{\mathcal{F}}_n \subseteq \widetilde{\mathcal{F}}_{n+1} \subseteq \mathcal{F}$. Denote

$$f_n = \boldsymbol{E} \sup_{t \in \mathfrak{C}_n} \boldsymbol{E}(\xi_t | \widetilde{\mathcal{F}}_n),$$

where $\mathfrak{C}_n$ is a class of stopping times t such that $\boldsymbol{P}(t \ge n) = 1$ and $\boldsymbol{E}\left(\sup_n \xi_n^+\right) < \infty$. It is known that (see, for example, Chow et al. (1971) and Haggstrom (1966))

$$f_n = \max\{\xi_n, \boldsymbol{E}(f_{n+1} | \widetilde{\mathcal{F}}_n)\}.$$

Lemma 2.6. *Assume* $\boldsymbol{E}\left(\sup_n \xi_n^-\right) < \infty$ *and let* $\{g_n, \widetilde{\mathcal{F}}_n, n \geq 1\}$ *be a sequence of random variables such that*

(a) $g_n \leq f_n$,

(b) $g_n = \max\{\xi_n, \boldsymbol{E}(g_{n+1}|\widetilde{\mathcal{F}}_n)\}$.

Then $g_n = f_n$, $n = 1, 2, \ldots$.

Proof: Let $A \in \widetilde{\mathcal{F}}_n$. Then, for $t \in \mathfrak{C}_n$,

$$\begin{aligned}
\int\limits_{A\cap\{t\geq n\}} g_n\, d\boldsymbol{P} &\geq \int\limits_{A\cap\{t=n\}} g_n\, d\boldsymbol{P} + \int\limits_{A\cap\{t>n\}} g_{n+1}\, d\boldsymbol{P} \\
&= \int\limits_{A\cap\{t=n\}} g_n\, d\boldsymbol{P} + \int\limits_{A\cap\{t=n+1\}} g_{n+1}\, d\boldsymbol{P} + \int\limits_{A\cap\{t>n+1\}} g_{n+1}\, d\boldsymbol{P} \\
&\geq \cdots \geq \sum_{j=n}^{r} \int\limits_{A\cap\{t=j\}} \xi_j\, d\boldsymbol{P} + \int\limits_{A\cap\{t>r\}} g_r\, d\boldsymbol{P}.
\end{aligned}$$

Therefore,

$$\sum_{j=n}^{r} \int\limits_{A\cap\{t=j\}} \xi_j\, d\boldsymbol{P} \leq \int\limits_{A\cap\{t\geq n\}} g_n\, d\boldsymbol{P} + \int\limits_{A\cap\{t>r\}} \sup_n \xi_n^-\, d\boldsymbol{P}.$$

Taking the limit $r \to \infty$, we obtain

$$\int\limits_{A\cap\{t\geq n\}} \xi_t\, d\boldsymbol{P} \leq \int\limits_{A\cap\{t\geq n\}} g_n\, d\boldsymbol{P}.$$

This means that $g_n \geq \boldsymbol{E}(\xi_t \mid \widetilde{\mathcal{F}}_n)$. Hence $g_n \geq \boldsymbol{E}\sup_{t\in\mathfrak{C}_n} \boldsymbol{E}(\xi_t \mid \widetilde{\mathcal{F}}_n)$ and $g_n = f_n$, $n = 1, 2, \ldots$. This concludes the proof of the lemma. □

In addition to condition (2.2), let us introduce another condition

$$(A^-) : \boldsymbol{E}\left(\sup_{\boldsymbol{m}_1} A_{\boldsymbol{m}_1}\left(\sup_{\boldsymbol{m}_k} Z_{\boldsymbol{m}_k}^-\right)\right) < \infty, \tag{2.22}$$

which is needed for the following theorem.

Theorem 2.7 (Nikolaev (1978)). *Let conditions* (2.2) *and* (2.22) *be fulfilled. Then, for* $i = 1, 2, \ldots, k$,

$$V_{\boldsymbol{m}_i} = \lim_{N\to\infty} V_{\boldsymbol{m}_i}^N, \quad X_{\boldsymbol{m}_i} = \lim_{N\to\infty} X_{\boldsymbol{m}_i}^N.$$

Proof: The proof is by backward induction on i. From (2.21),

$$V^N_{\boldsymbol{m}_k} = \max\left\{X^N_{\boldsymbol{m}_k}, \boldsymbol{E}_{\boldsymbol{m}_k} V^N_{\boldsymbol{m}_{k-1}, m_k+1}\right\},$$

with $X^N_{\boldsymbol{m}_k} = Z_{\boldsymbol{m}_k}$. Taking into account the definition of $V^N_{\boldsymbol{m}_k}$ and the fact that $\mathfrak{S}^N_{\boldsymbol{m}_k} \subset \mathfrak{S}^{N+1}_{\boldsymbol{m}_k}$, we get $V^N_{\boldsymbol{m}_k} \leq V^{N+1}_{\boldsymbol{m}_k}$. Then there exists a limit

$$V^*_{\boldsymbol{m}_k} = \lim_{N\to\infty} V^N_{\boldsymbol{m}_k},$$

where $V^*_{\boldsymbol{m}_k} \leq V_{\boldsymbol{m}_k}$. Also, it follows from condition (2.2) that $V_{\boldsymbol{m}_k} < \infty$. Next, using condition (2.22) and the monotone convergence theorem, we have

$$\lim_{N\to\infty} \boldsymbol{E}_{\boldsymbol{m}_k} V^N_{\boldsymbol{m}_{k-1}, m_k+1} = \boldsymbol{E}_{\boldsymbol{m}_k} V^*_{\boldsymbol{m}_{k-1}, m_k+1}.$$

Hence,

$$V^*_{\boldsymbol{m}_k} = \max\left\{Z_{\boldsymbol{m}_k}, \mathbf{E}_{\boldsymbol{m}_k} V^*_{\boldsymbol{m}_{k-1}, m_k+1}\right\},$$

and, from Lemma 2.6,

$$V^*_{\boldsymbol{m}_k} = V_{\boldsymbol{m}_k}.$$

It follows easily that

$$\lim_{N\to\infty} X^N_{\boldsymbol{m}_{k-1}} = \lim_{N\to\infty} \boldsymbol{E}_{\boldsymbol{m}_{k-1}} V^N_{\boldsymbol{m}_{k-1}, m_{k-1}+1} = X_{\boldsymbol{m}_{k-1}}.$$

Suppose inductively that, for $2 \leq i < k$,

$$\lim_{N\to\infty} V^N_{\boldsymbol{m}_i} = V_{\boldsymbol{m}_i}.$$

It follows that

$$\lim_{N\to\infty} X^N_{\boldsymbol{m}_{i-1}} = X_{\boldsymbol{m}_{i-1}}.$$

From (2.21), it can be shown in the same way as above that

$$V^*_{\boldsymbol{m}_{i-1}} = \max\left\{X_{\boldsymbol{m}_{i-1}}, \boldsymbol{E}_{\boldsymbol{m}_{i-1}} V^*_{\boldsymbol{m}_{i-2}, m_{i-1}+1}\right\}.$$

Using Jensen's inequality, we have

$$X^-_{\boldsymbol{m}_{i-1}} \leq \boldsymbol{E}_{\boldsymbol{m}_{i-1}} V^-_{\boldsymbol{m}_{i-1}, m_{i-1}+1}.$$

From condition (2.22), we can see that

$$\boldsymbol{E}\left(\sup_{m_{i-1}} X^-_{\boldsymbol{m}_{i-1}}\right) \leq \boldsymbol{E}\left(\sup_{m_{i-1}} A_{\boldsymbol{m}_{i-1}}\left(\sup_{\boldsymbol{m}_k} Z^-_{\boldsymbol{m}_k}\right)\right) < \infty.$$

Finally, taking into account Lemma 2.6, we obtain $V^*_{\boldsymbol{m}_{i-1}} = V_{\boldsymbol{m}_{i-1}}$. This completes the proof of the theorem. □

We can see from Theorem 2.7 that if conditions (2.2) and (2.22) hold, then the m-value of the game

$$v_m = \lim_{N\to\infty} \mathbf{E} V^N_m.$$

2.4.1 The triple limit theorem

Chow et al. (1971) considered the triple limit theorem that could be extended to a multiple stopping. This theorem gives us a different way of constructing sequences $V_{\boldsymbol{m}_i}$, $X_{\boldsymbol{m}_i}$, and the m-value v_m.

Let $-\infty < a \leq 0 \leq b < \infty$,

$$Z_{\boldsymbol{m}_k}(a) = \max\{Z_{\boldsymbol{m}_k}, a\},$$
$$Z_{\boldsymbol{m}_k}(b) = \min\{Z_{\boldsymbol{m}_k}, b\},$$
$$Z_{\boldsymbol{m}_k}(a,b) = \begin{cases} b, & Z_{\boldsymbol{m}_k} > b, \\ Z_{\boldsymbol{m}_k}, & a \leq Z_{\boldsymbol{m}_k} \leq b, \\ a, & Z_{\boldsymbol{m}_k} < a. \end{cases}$$

From (2.19) and (2.20), we can define the corresponding sequences $X_{\boldsymbol{m}_i}(a)$, $V_{\boldsymbol{m}_i}(a)$, $X_{\boldsymbol{m}_i}(b)$, $V_{\boldsymbol{m}_i}(b)$, and $X_{\boldsymbol{m}_i}(a,b)$, $V_{\boldsymbol{m}_i}(a,b)$, where $X_{\boldsymbol{m}_k}(a) = Z_{\boldsymbol{m}_k}(a)$, $X_{\boldsymbol{m}_k}(b) = Z_{\boldsymbol{m}_k}(b)$, and $X_{\boldsymbol{m}_k}(a,b) = Z_{\boldsymbol{m}_k}(a,b)$.

The following lemma will play an important role in this section.

Lemma 2.8 (Siegmund (1967)). *Assume* $\boldsymbol{E}(\sup_n \xi_n^+) < \infty$ *and a stochastic sequence* $\left\{g_n, \widetilde{\mathcal{F}}_n, n \geq 1\right\}$ *satisfies the following conditions:*

(a) $f_n \leq g_n \leq \boldsymbol{E}\left(\eta \mid \widetilde{\mathcal{F}}_n\right)$ *for some integrable random variable* η,

(b) $g_n = \max\left\{\xi_n, \boldsymbol{E}\left(g_{n+1} \mid \widetilde{\mathcal{F}}_n\right)\right\}$,

(c) $\limsup_n g_n = \limsup_n \xi_n$.

Then $g_n = f_n$ *for* $n = 1, 2, \ldots$.

Theorem 2.9 (Nikolaev (1978)). *Assume condition* (2.2) *holds. Then for* $i = 1, 2, \ldots, k$

$$\lim_{a \to -\infty} V_{\boldsymbol{m}_i}(a) = V_{\boldsymbol{m}_i}, \qquad \lim_{a \to -\infty} X_{\boldsymbol{m}_i}(a) = X_{\boldsymbol{m}_i}.$$

Proof: The proof is by backward induction on i. Note that the sequences $V_{\boldsymbol{m}_i}(a)$ and $X_{\boldsymbol{m}_i}(a)$ are non-increasing over all a, and, therefore, there exist the following limits

$$\overline{V}^*_{\boldsymbol{m}_i} = \lim_{a \to -\infty} V_{\boldsymbol{m}_i}(a), \qquad \overline{X}^*_{\boldsymbol{m}_i} = \lim_{a \to -\infty} X_{\boldsymbol{m}_i}(a).$$

It is readily seen that

$$V_{\boldsymbol{m}_i} \leq \overline{V}^*_{\boldsymbol{m}_i} \leq \boldsymbol{E}_{\boldsymbol{m}_i}\left(\sup_{m_{i+1}} A_{\boldsymbol{m}_{i+1}}\left(\sup_{\boldsymbol{m}_k} Z^+_{\boldsymbol{m}_k}\right)\right). \tag{2.23}$$

Using (2.5), we get

$$V_{\boldsymbol{m}_k}(a) = \max\left\{Z_{\boldsymbol{m}_k}(a), \boldsymbol{E}_{\boldsymbol{m}_k} V_{\boldsymbol{m}_{k-1}, m_k+1}(a)\right\}.$$

We see that

$$\boldsymbol{E}V^{+}_{\boldsymbol{m}_k}(a) \leq \boldsymbol{E}\left(\sup_{\boldsymbol{m}_k} Z^{+}_{\boldsymbol{m}_k}\right) < \infty.$$

By the monotone convergence theorem, it follows that

$$\overline{V}^{*}_{\boldsymbol{m}_k} = \max\left\{Z_{\boldsymbol{m}_k}, \boldsymbol{E}_{\boldsymbol{m}_k}\overline{V}^{*}_{\boldsymbol{m}_{k-1},m_k+1}\right\}.$$

Using Theorem 3.5 of Snell (1952), we get

$$\limsup_{m_k\to\infty} V_{\boldsymbol{m}_k}(a) = \limsup_{m_k\to\infty} Z_{\boldsymbol{m}_k}(a)$$

for all $a \leq 0$ and $m_{k-1} > \cdots > m_1 \geq 1$. Hence,

$$\limsup_{m_k\to\infty} \overline{V}^{*}_{\boldsymbol{m}_k} \leq \limsup_{m_k\to\infty} Z_{\boldsymbol{m}_k}(a) = \max\left\{\limsup_{m_k\to\infty} Z_{\boldsymbol{m}_k}, a\right\}.$$

Also, as $a \to -\infty$,

$$\max\left\{\limsup_{m_k\to\infty} Z_{\boldsymbol{m}_k}, a\right\} \to \limsup_{m_k\to\infty} Z_{\boldsymbol{m}_k}.$$

It follows that

$$\limsup_{m_k\to\infty} \overline{V}^{*}_{\boldsymbol{m}_k} = \limsup_{m_k\to\infty} Z_{\boldsymbol{m}_k}. \tag{2.24}$$

By Lemma 2.8, we get $\overline{V}^{*}_{\boldsymbol{m}_k} = V_{\boldsymbol{m}_k}$ and

$$\begin{aligned}\overline{X}^{*}_{\boldsymbol{m}_{k-1}} &= \lim_{a\to-\infty} \boldsymbol{E}_{\boldsymbol{m}_{k-1}} V_{\boldsymbol{m}_{k-1},m_{k-1}+1}(a)\\ &= \boldsymbol{E}_{\boldsymbol{m}_{k-1}} V_{\boldsymbol{m}_{k-1},m_{k-1}+1} = X_{\boldsymbol{m}_{k-1}}.\end{aligned}$$

Assume that the following statement holds for some i, $2 \leq i < k$,

$$\lim_{a\to-\infty} V_{\boldsymbol{m}_i}(a) = V_{\boldsymbol{m}_i}. \tag{2.25}$$

We need to show that

$$\lim_{a\to-\infty} V_{\boldsymbol{m}_{i-1}}(a) = V_{\boldsymbol{m}_{i-1}}.$$

From (2.25), we have

$$\lim_{a\to-\infty} X_{\boldsymbol{m}_{i-1}}(a) = X_{\boldsymbol{m}_{i-1}},$$

and, as above,

$$\overline{V}^{*}_{\boldsymbol{m}_{i-1}} = \max\left\{X_{\boldsymbol{m}_{i-1}}, \boldsymbol{E}_{\boldsymbol{m}_{i-1}}\overline{V}^{*}_{\boldsymbol{m}_{i-1},m_{i-1}+1}\right\}.$$

By the inequality $X_{\boldsymbol{m}_{i-1}} \leq V_{\boldsymbol{m}_{i-1}}$, it follows that $X^{+}_{\boldsymbol{m}_{i-1}} \leq V^{+}_{\boldsymbol{m}_{i-1}}$. Combining this with (2.23) and condition (2.2), we obtain

$$\boldsymbol{E}\left(\sup_{m_{i-1}} X^{+}_{\boldsymbol{m}_{i-1}}\right) \leq \boldsymbol{E}\left(\sup_{m_{i-1}} A_{\boldsymbol{m}_{i-1}}\left(\sup_{\boldsymbol{m}_k} Z^{+}_{\boldsymbol{m}_k}\right)\right) < \infty.$$

Finally, similar to (2.24), it can be proved that

$$\limsup_{m_{i-1}\to\infty} \overline{V}^*_{m_{i-1}} = \limsup_{m_{i-1}\to\infty} X_{m_{i-1}}.$$

By Lemma 2.8, it follows that $\overline{V}^*_{m_{i-1}} = V_{m_{i-1}}$. This completes the proof of the theorem. □

The following result is a corollary of the theorem.

Remark 5. *Assume condition* (2.2) *holds. Then*

(1) $V_{m_i} = \lim_{a\to-\infty}\lim_{N\to\infty} V^N_{m_i}(a)$, $X_{m_i} = \lim_{a\to-\infty}\lim_{N\to\infty} X^N_{m_i}(a)$;

(2) $v_m = \lim_{a\to-\infty}\lim_{N\to\infty} EV^N_m(a) = \lim_{a\to-\infty} EV_m(a)$.

The following theorem is a counterpart of Theorem 2.9 with a similar proof.

Theorem 2.10 (Nikolaev (1978)). *Assume condition* (2.22) *holds. Then*

$$\lim_{b\to\infty} V_{m_i}(b) = V_{m_i}, \qquad \lim_{b\to\infty} X_{m_i}(b) = X_{m_i}.$$

Proof: Both $V_{m_i}(b)$ and $X_{m_i}(b)$ are nondecreasing sequences over all b. Then there exist the limits

$$V_{m_i} \le V^{**}_{m_i} = \lim_{b\to\infty} V_{m_i}(b), \qquad X_{m_i} \le X^{**}_{m_i} = \lim_{b\to\infty} X_{m_i}(b).$$

The proof of $V^{**}_{m_i} = V_{m_i}$ and $X^{**}_{m_i} = X_{m_i}$ is by backward induction on i. From (2.5), we have

$$V_{m_k}(b) = \max\left\{Z_{m_k}(b), E_{m_k} V_{m_{k-1},m_k+1}(b)\right\}.$$

By condition (2.22), it follows that

$$EV^-_{m_k}(b) \le E\left(\sup_{m_k} Z^-_{m_k}\right) < \infty.$$

Using the monotone convergence theorem, we get

$$V^{**}_{m_k} = \max\left\{Z_{m_k}, E_{m_k} V^{**}_{m_{k-1},m_k+1}\right\}.$$

By Lemma 2.6, we see that $V^{**}_{m_k} = V_{m_k}$. Therefore,

$$\lim_{b\to\infty} X_{m_{k-1}}(b) = \lim_{b\to\infty} E_{m_{k-1}} V_{m_{k-1},m_{k-1}+1}(b) = X_{m_{k-1}}.$$

Assume that the following statement holds for some i, $2 \le i < k$,

$$\lim_{b\to\infty} V_{m_i}(b) = V_{m_i}. \tag{2.26}$$

Let us show that

$$\lim_{b\to\infty} V_{\boldsymbol{m}_{i-1}}(b) = V_{\boldsymbol{m}_{i-1}}.$$

From (2.26), we get

$$\lim_{b\to\infty} X_{\boldsymbol{m}_{i-1}}(b) = X_{\boldsymbol{m}_{i-1}}.$$

It follows that

$$V^{**}_{\boldsymbol{m}_{i-1}} = \max\left\{X_{\boldsymbol{m}_{i-1}}, \boldsymbol{E}_{\boldsymbol{m}_{i-1}} V^{**}_{\boldsymbol{m}_{i-2},m_{i-1}+1}\right\}.$$

From Jensen's inequality, $X^{-}_{\boldsymbol{m}_{i-1}} \leq \boldsymbol{E}_{\boldsymbol{m}_{i-1}} V^{-}_{\boldsymbol{m}_{i-1},m_{i-1}+1}$. Combining this with condition (2.22), we have

$$\boldsymbol{E}\left(\sup_{m_{i-1}} X^{-}_{\boldsymbol{m}_{i-1}}\right) \leq \boldsymbol{E}\left(\sup_{m_{i-1}} A_{\boldsymbol{m}_{i-1}}\left(\sup_{\boldsymbol{m}_k} Z^{-}_{\boldsymbol{m}_k}\right)\right) < \infty.$$

Finally, using Lemma 2.6, we get

$$V^{**}_{\boldsymbol{m}_{i-1}} = \lim_{b\to\infty} V_{\boldsymbol{m}_{i-1}}(b) = V_{\boldsymbol{m}_{i-1}}.$$

This concludes the proof of the theorem. □

The result below directly follows from the theorem.

Remark 6. *Assume condition* (2.22) *holds.*

(1) Then

$$V_{\boldsymbol{m}_i} = \lim_{b\to\infty}\lim_{N\to\infty} V^{N}_{\boldsymbol{m}_i}(b) = \lim_{N\to\infty}\lim_{b\to\infty} V^{N}_{\boldsymbol{m}_i}(b),$$
$$X_{\boldsymbol{m}_i} = \lim_{b\to\infty}\lim_{N\to\infty} X^{N}_{\boldsymbol{m}_i}(b) = \lim_{N\to\infty}\lim_{b\to\infty} X^{N}_{\boldsymbol{m}_i}(b).$$

(2) If condition (2.2) *also holds, then*

$$v_{m} = \lim_{b\to\infty}\lim_{N\to\infty} \boldsymbol{E} V^{N}_{m}(b) = \lim_{N\to\infty}\lim_{b\to\infty} \boldsymbol{E} V^{N}_{m}(b).$$

Let condition (2.2) be fulfilled for $Z_{\boldsymbol{m}_k}$. For $N \geq 2$, put

$$\overline{V}^{N}_{\boldsymbol{m}_{k-1},N} = \boldsymbol{E}_{\boldsymbol{m}_{k-1},N}\left(\sup_{r\geq N} Z_{\boldsymbol{m}_{k-1},r}\right),$$
$$\overline{V}^{N}_{\boldsymbol{m}_k} = \max\left\{Z_{\boldsymbol{m}_k}, \boldsymbol{E}_{\boldsymbol{m}_k}\overline{V}^{N}_{\boldsymbol{m}_{k-1},m_k+1}\right\}, \quad m_{k-1} < m_k < N. \tag{2.27}$$

It can be proved by induction that $\overline{V}^{m_k}_{\boldsymbol{m}_k} \geq \overline{V}^{m_k+1}_{\boldsymbol{m}_k} \geq \ldots$. Let us show that $\overline{V}^{m_k}_{\boldsymbol{m}_k} \geq \overline{V}^{m_k+1}_{\boldsymbol{m}_k}$. We have

$$\overline{V}^{m_k+1}_{\boldsymbol{m}_k} = \max\left\{Z_{\boldsymbol{m}_k}, \boldsymbol{E}_{\boldsymbol{m}_k}\left(\sup_{r\geq m_k+1} Z_{\boldsymbol{m}_{k-1},r}\right)\right\}$$
$$\leq \boldsymbol{E}_{\boldsymbol{m}_k}\left(\max\left\{Z_{\boldsymbol{m}_k}, \sup_{r\geq m_k+1} Z_{\boldsymbol{m}_{k-1},r}\right\}\right) = \overline{V}^{m_k}_{\boldsymbol{m}_k}.$$

Then there exists a limit $\overline{V}_{\boldsymbol{m}_k} = \lim_{N\to\infty} \overline{V}^N_{\boldsymbol{m}_k}$. Moreover,

$$\overline{V}^N_{\boldsymbol{m}_k} \leq \overline{V}^{m_k}_{\boldsymbol{m}_k} \leq \boldsymbol{E}_{\boldsymbol{m}_k}\left(\sup_{m_k} Z^+_{\boldsymbol{m}_k}\right). \tag{2.28}$$

By the monotone convergence theorem, it follows that

$$\lim_{N\to\infty} \boldsymbol{E}_{\boldsymbol{m}_{k-1}} \overline{V}^N_{\boldsymbol{m}_{k-1},m_{k-1}+1} = \boldsymbol{E}_{\boldsymbol{m}_{k-1}} \overline{V}_{\boldsymbol{m}_{k-1},m_{k-1}+1} \equiv \overline{X}_{\boldsymbol{m}_{k-1}}.$$

For $m_{k-2} < m_{k-1} < N-1$, define

$$\begin{aligned}\overline{V}^{N-1}_{\boldsymbol{m}_{k-2},N-1} &= \boldsymbol{E}_{\boldsymbol{m}_{k-2},N-1}\left(\sup_{r\geq N-1} \overline{X}_{\boldsymbol{m}_{k-2},r}\right),\\ \overline{V}^{N-1}_{\boldsymbol{m}_{k-1}} &= \max\left\{\overline{X}_{\boldsymbol{m}_{k-1}}, \boldsymbol{E}_{\boldsymbol{m}_{k-1}} \overline{V}^{N-1}_{\boldsymbol{m}_{k-2},m_{k-1}+1}\right\}.\end{aligned} \tag{2.29}$$

Continuing this process, we finally get, for $1 \leq m_1 < N-k+1$,

$$\begin{aligned}\overline{V}^{N-k+1}_{N-k+1} &= \boldsymbol{E}_{N-k+1}\left(\sup_{r\geq N-k+1} \overline{X}_r\right),\\ \overline{V}^{N-k+1}_{m_1} &= \max\left\{\overline{X}_{m_1}, \boldsymbol{E}_{m_1} \overline{V}^{N-k+1}_{m_1+1}\right\}.\end{aligned}$$

Theorem 2.11 (Nikolaev (1978)). *Let conditions* (2.2) *and* (2.22) *be fulfilled. Then* $\overline{V}_{\boldsymbol{m}_i} = V_{\boldsymbol{m}_i}$ *and* $\overline{X}_{\boldsymbol{m}_i} = X_{\boldsymbol{m}_i}$.

Proof: For the sake of brevity, let us consider $k = 2$ stops. The general case can be shown in the similar way.

For any $\varepsilon > 0$, consider random variables $(\bar{\tau}_{1\varepsilon}, \bar{\tau}_{2\varepsilon})$ such that

$$\begin{aligned}\bar{\tau}_{1\varepsilon} &= \inf\left\{m_1 \geq m : \overline{V}_{m_1} \leq \overline{X}_{m_1} + \frac{\varepsilon}{2}\right\},\\ \bar{\tau}_{2\varepsilon} &= \inf\left\{k > m_1 : k \geq n, \overline{V}_{m_1,k} \leq Z_{m,k} + \frac{\varepsilon}{2}\right\}\end{aligned}$$

on the set $\{\omega : \bar{\tau}_{1\varepsilon} = m_1\}$, with $\bar{\tau}_{2\varepsilon} = \infty$ on $\{\omega : \bar{\tau}_{1\varepsilon} = \infty\}$. It can be shown that (see, for example, Shiryaev (1978)) $\boldsymbol{P}(\bar{\tau}_{1\varepsilon} < \infty) = 1$ and $\boldsymbol{P}(\bar{\tau}_{2\varepsilon} < \infty \mid \bar{\tau}_{1\varepsilon} = m) = 1$, $m \geq 1$. From (2.28), we have

$$\boldsymbol{E}\left(\overline{V}^N_{mn}\right)^+ \leq \boldsymbol{E}\left(\sup_n Z^+_{mn}\right) < \infty.$$

and

$$\begin{aligned}\overline{V}^N_m \leq \overline{V}^{m+1}_m &= \boldsymbol{E}_m\left(\sup_{k\geq m} \overline{X}_k\right) \leq \boldsymbol{E}_m\left(\sup_{k\geq m} \boldsymbol{E}_k \overline{V}^{k+1}_{k,k+1}\right) \leq\\ &\leq \boldsymbol{E}_m\left(\sup_{k\geq m} \boldsymbol{E}_k\left(\sup_n Z^+_{kn}\right)\right).\end{aligned}$$

Taking the limit as $N \to \infty$ in (2.27) and (2.29), we obtain

$$\begin{aligned} \overline{V}_{mn} &= \max\left\{Z_{mn}, \boldsymbol{E}_{mn}\overline{V}_{m,n+1}\right\}, \quad n > m, \\ \overline{V}_m &= \max\left\{\overline{X}_m, \boldsymbol{E}_m\overline{V}_{m+1}\right\}, \quad m \geq 1. \end{aligned} \tag{2.30}$$

Assume $B \in \mathcal{F}_{mn}$. Using (2.30), we get

$$\begin{aligned} \int\limits_{B\{\bar{\tau}_{1\varepsilon}=m,\bar{\tau}_{2\varepsilon}\geq n\}} \overline{V}_{mn}\, d\boldsymbol{P} &= \int\limits_{B\{\bar{\tau}_{1\varepsilon}=m,\bar{\tau}_{2\varepsilon}=n\}} \overline{V}_{mn}\, d\boldsymbol{P} \\ &+ \int\limits_{B\{\bar{\tau}_{1\varepsilon}=m,\bar{\tau}_{2\varepsilon}>n\}} \overline{V}_{mn}\, d\boldsymbol{P} = \ldots \\ &= \int\limits_{B\{\bar{\tau}_{1\varepsilon}=m,n\leq\bar{\tau}_{2\varepsilon}<k\}} \overline{V}_{m,\bar{\tau}_{2\varepsilon}}\, d\boldsymbol{P} \\ &+ \int\limits_{B\{\bar{\tau}_{1\varepsilon}=m,\bar{\tau}_{2\varepsilon}\geq k\}} \overline{V}_{mk}\, d\boldsymbol{P}. \end{aligned}$$

We see that

$$\overline{V}_{mk} \leq \overline{V}^k_{mk} \leq \boldsymbol{E}_{mk}\left(\sup_n Z^+_{mn}\right)$$

and $\boldsymbol{P}(\bar{\tau}_{2\varepsilon} = \infty \mid \bar{\tau}_{1\varepsilon} = m) = 0$. Taking the limit as $k \to \infty$, we have

$$\overline{V}_{mn} \leq \boldsymbol{E}_{mn}\overline{V}_{m,\bar{\tau}_{2\varepsilon}} \leq \boldsymbol{E}_{mn}Z_{m,\bar{\tau}_{2\varepsilon}} + \frac{\varepsilon}{2} \leq V_{mn} + \frac{\varepsilon}{2}.$$

Since ε, $\varepsilon > 0$, is arbitrary, it implies that $\overline{V}_{mn} \leq V_{mn}$. Therefore, by Lemma 2.6, it follows that $\overline{V}_{mn} = V_{mn}$ and, therefore, $\overline{X}_m = X_m$. Similarly,

$$\overline{V}_m \leq \overline{V}^{m+1}_m \leq \boldsymbol{E}_m\left(\sup_k \overline{X}^+_k\right) \leq \boldsymbol{E}_m\left(\sup_k \boldsymbol{E}_k\left(\sup_n Z^+_{mn}\right)\right),$$

and $\boldsymbol{P}(\bar{\tau}_{1\varepsilon} < \infty) = 1$. Then $\overline{V}_m \leq V_m$, and using Lemma 2.6, we get $\overline{V}_m = V_m$. This concludes the proof of the theorem. □

The following result follows directly from Theorem 2.11.

Remark 7. *Let conditions* (2.2) *and* (2.22) *be fulfilled. Then*

$$v_m = \lim_{N\to\infty} \boldsymbol{E}\overline{V}^N_m.$$

The following triple limit theorem summarizes all results considered above.

Theorem 2.12 (Nikolaev (1978)). *Let condition* (2.2) *be fulfilled. Then*

$$v_m = \begin{cases} \lim_{a\to-\infty} \lim_{b\to\infty} \lim_{N\to\infty} EV_m^N(a,b), \\ \lim_{a\to-\infty} \lim_{N\to\infty} \lim_{b\to\infty} EV_m^N(a,b), \\ \lim_{a\to-\infty} \lim_{b\to\infty} \lim_{N\to\infty} E\overline{V}_m^N(a,b), \\ \lim_{a\to-\infty} \lim_{N\to\infty} \lim_{b\to\infty} E\overline{V}_m^N(a,b). \end{cases}$$

2.5 Odds Theorem in Multiple-Stopping Problems

Bruss and Paindaveine (2000) discussed the problem of selecting the last successes k. It is a multiple-stopping version of the secretary problem. The interesting issue in the paper is the application of an extension of the odds algorithm. Tamaki (2010) proved a multiplicative odds theorem which deals with the problem of stopping at any of the last ℓ successes. A tight lower bound of the probability of win is obtained by Matsui and Ano (2014).

Next, Matsui and Ano (2017) discussed a problem of selecting k from the last ℓ successes and obtained a tight lower bound of the probability of victory. When $\ell = k = 1$, the problem is equivalent to Bruss' odds problem. If $\ell = k \geq 1$, the problem is equivalent to that in Bruss and Paindaveine (2000). A problem discussed by Tamaki (2010) is obtained by setting $\ell \geq k = 1$.

3

Multilateral Multiple-Stopping: Game Theory Approach

Motto:
Harmonising Objectives, Balancing Uncertainty: Navigating Multilateral Decision-Making with Stochastic Process Observation.

A MULTILATERAL decision model related to observation of stochastic processes is a method for making decisions in situations where multiple agents or decision-makers are involved, and the outcome of the decision is uncertain due to the presence of random variables or stochastic processes. The goal of the model is to determine the optimal decision for each agent based on their objectives and constraints, as well as the group's collective goal. This is often done by using mathematical optimization techniques, such as linear programming or dynamic programming, to find the optimal solution that balances the conflicting objectives and constraints of the agents. The model can be applied to a wide range of decision-making problems, such as resource allocation, risk management, and control of complex systems.

An example of a multilateral decision problem that can be modelled using a Markov Decision Process (MDP) is a resource allocation problem in which multiple agents need to decide how to allocate a limited resource (e.g. money, time, energy) among themselves.

In this example, let us say that there are 3 agents, each with their own objectives and constraints, and the resource they need to allocate is money. Agent 1 wants to invest in a new project, Agent 2 wants to save for a rainy day, and Agent 3 wants to buy a new car. Each agent has a different level of priority for their objectives and constraints. The MDP could be defined as follows:

- The states of the MDP are the different possible allocations of money among the agents.
- The actions of each agent are the amounts of money that they can choose to allocate to themselves.
- The transition probabilities depend on the actions of all agents and the objective and constraints of each agent.
- The reward for each agent is based on their individual objectives and constraints, and it depends on his and adversarial decisions.

DOI: 10.1201/9781003407102-3

For example, Agent 1's reward would be based on the expected return of the new project, Agent 2's reward would be based on the amount of money saved and Agent 3's reward would be based on the value of the new car.

By solving MDP, we can find the optimal allocation of money that maximizes the collective goal of the group, while also taking into account the individual objectives and constraints of each agent. Special cases appear when the sets of DM's strategies are stopping times.

An example of a multilateral decision problem where strategies are stopping times is a problem where multiple agents need to decide when to stop a process. This can be relevant in many different fields, such as manufacturing, finance, and energy.

For example, consider a group of wind turbine operators who need to decide when to shut down their turbines during a storm. Each operator has different costs associated with shutting down their turbines, and they also want to minimize the amount of energy that is lost due to shutting down the turbines too early or too late.

In this case, the multilateral decision problem can be modelled as follows:

- The states of the problem are the different wind speeds and the time until the storm is expected to pass
- The actions of each agent are the stopping times, i.e when the agent chooses to shut down the turbine
- The transition probabilities depend on the wind speed and the time remaining before the storm
- The reward for each agent is the expected cost of shutting down the turbine at a certain time and the lost energy.

By solving this multilateral decision problem, the optimal stopping times for each agent can be determined, which would minimize the total expected cost and lost energy.

As mentioned in Section 1.5, a multilateral stopping problem can be referred to as a stopping game, when multiple agents or players make the decision to stop, and the outcome of each player's decision depends on the decisions of the other players. This creates a strategic interaction among the players, as each player's decision affects the payoffs of the other players.

In a stopping game, each player's strategy is a stopping time, i.e., a time at which they choose to stop a process. Players may have different objectives or rewards associated with their stopping times and must take into account the actions of other players when choosing their stopping times.

Examples of stopping games include the wind turbine operators example, which is mentioned before, where the operators are trying to minimize the expected costs and lost energy, but their effect of decisions depend on the actions of the other operators. Another example can be a stock market where investors are trying to decide when to buy or sell a stock, but their decisions depend on the decisions of other investors.

In both examples, the stopping times chosen by each player affect the payoffs of the other players, creating a strategic interaction that makes the problem a stopping game.

3.1 Multilateral Stopping Decisions Modelling

Several classical stopping game models have been widely studied in the literature. Some of the most notable examples include:

1. *The Secretary Problem* is a classic problem in which a single decision-maker must choose the best candidate from a random sequence of applicants. The decision-maker can only observe the candidates one at a time and must decide when to stop interviewing and hire the best candidate seen so far.
2. *The St. Petersburg Paradox* is a game in which the player must decide when to stop rolling a fair coin. The player wins the amount of money equal to 2^n dollars where n is the number of rolls required to obtain the first heads. The game has an infinite expected value, but most people are unwilling to pay a lot to play.
3. *The Wald's sequential analysis* is a statistical decision theory problem that deals with the optimal stopping rule for a sequence of trials in the presence of uncertainty, intending to minimize the expected loss.
4. *The Multi-armed bandit problem* is a class of problems in which a player must decide which arm of a machine to pull to maximize their expected payout. The player must balance the exploration of different arms with the exploitation of the arm that has the highest payout.
5. A standard (B, S)-*securities market* (v. Section 1.6.10) consists of a nonrandom (riskless) component B_n, which is described as a savings account (or the price of a bond) at time n with an interest r, and of a random (risky) component S_n, which can be described as the price of a stock at time n. The basic types of contracts of these types are the European option, the contract with fixed and given exercise moment. American options are a type of financial option that can be exercised at any time before the expiration date. The creation of American options as a financial instrument is more of an evolutionary process rather than being attributed to a single individual or institution. The problem of fair pricing of American options in the (B, S)-securities market leads to the optimal stopping of certain stochastic processes. Therefore, the American option problem involves determining the optimal exercise strategy for an American option, taking into account the current price of the underlying asset and the volatility of the asset's price. In fact, it is a multilateral stopping problem, as the decision-makers are the investors in the burs. However, from an option issuer point of view, the problem is to determine the adequate price. So, he should formulate the optimal stopping problem to predict the seller's behavior. This optimization problem helps the issuer to determine the rational price.
6. An *Israeli option* was introduce by Brenner and Galai (1989). The mathematical model and the pricing of such an option was proposed by Kifer (2000). His model treats the pricing problem as the game between the issuer (seller) and the buyers. The seller of an option can sell the contract at any time n. In this case, the buyer's gain is the sum $(K - S_n)^+ + \delta_n$ in the put option and $(S_n - K)^+ + \delta_n$ in the

call option case, where $\delta_n \geq 0$ is a certain penalty paid by the seller and K is some specific price to sell (put option) or buy (call option) the stock at any time n. Pricing of these options leads to a game version of the optimal stopping problem introduced in the discrete-time case by Dynkin (1969). Kifer (2000) considers basic problems concerning the extension of option pricing theory to game options. It is named Israeli options to put them in line with other names of securities such as the American and European ones.

These are just a few examples of classical stopping game models, and many other variations and extensions of these models have been studied in the literature.

3.2 Classic Models of Multi-Person Stopping Problems

It is worth starting the meta-analysis of these models with a discussion of the basic features that we encounter in this issue. The basis is the observed process, which is to bring a measurable effect to the interested parties, let us call them managers, as a result of choosing the moment to stop the observation and realize (consume) the effects of the selected state of the process. One of the natural methods of solving such a problem is to scalarize the problem and agree on a joint decision that maximizes the criterion. The individual effect is then a secondary matter which we will solve by a proper division of the obtained global result. Then, determining the right moment to end an experiment, observation, or action is then reduced to the task of optimal stopping. This approach requires communication between the participants in the decision-making process and the establishment of rules of conduct after the observation of the process.

3.2.1 Cooperative vs. non-cooperative games

It is much more difficult to model a case where the participants in the decision-making process cannot communicate with each other or communication is possible to a limited extent, with the use of intermediaries or under the direction of people outside the beneficiaries of the decision-making process. Therefore, the basic division of such models boils down to tasks, let us say games, non-cooperative, and cooperative. An example of a cooperative approach is the joint determination of the stopping point and the secondary distribution of the effect. The variants of modelling cooperation in stop games are presented in Section 3.3.

In the Bayesian perspective of decision theory, committee decision-making plays a pivotal role, illustrating the essence of game theory. Ferguson (2005) delves into the task of the committee of selecting a candidate for a position, mirroring the well-known house search problem introduced by MacQueen and Miller (1960), Derman and Sacks (1960), and Chow and Robbins (1961). Each committee member holds a unique perspective on a candidate's merit, which may align or conflict with others. Candidates emerge sequentially and undergo committee voting, with no option for reconsideration once rejected, shaping this as a multistage game involving numerous players. Expert teams take similar considerations when dealing with important managerial tasks that are perceived differently in terms of significance and consequences. Therefore, different methods of rationalization are needed for individual decisions compared to collective ones.

Initially explored by Sakaguchi (1973) in the unanimous consent scenario, the problem evolved to accommodate Poisson arrivals and multiple selections (v. Sakaguchi (1978, 1978)). Kurano et al. (1980) further extended these concepts to a majority rule among many players, while Yasuda et al. (1982) broadened the scope to encompass arbitrary voting rules. Section 3.3.4 examined the problem in the context of payoffs derived from a homogeneous Markov chain following the results of a paper by Szajowski and Yasuda (1996).

3.2.2 Non-cooperative two vs. multi-persons games

For models of non-cooperative games, in which the moments of action selection are the strategies, it is important to determine whether all participants are treated as equal partners (opponents) or not and there are significant differences between the players. Let us note that if the basis of the information on the process is its observation, then equal treatment of participants in the decision-making process means that everyone has the same knowledge and their decisions are autonomous, which in the colloquial language is defined as players having independent choices of stopping times. Then, if the players have full observation of the process, then the strategies of all are the stopping times with respect to the filtration generated by the process. For a complete picture of the situation, it is still necessary to decide on the effects of the strategies chosen by the players on the results obtained by them. This is a technical issue, the detailed discussion of which requires reference to the subject of modelling. If we agree that the effect is modelled by a real function, then the predicted effect is usually measured by a functional of the observed process. The construction of this function depends on *a priori* knowledge and the importance of observations for the assessment of the predicted value of the decision.

The given aspects lead to detailed models in which the adopted assumptions are the result of specifying the appropriate model for the real situation considered. In the following classifications, we try to get a division into mutually exclusive classes. We can classify the problem as games, which is

(CA1) antagonistic (non-cooperative);
(CA2) cooperative.

Due to the detail of the model of the observed process:

(CB1) full information
(CB2) incomplete information (the level of which may vary)

Due to the knowledge of individual decision-makers:

=>with players having the same access to information
=>with players of different levels see the process that decides the outcome

=>players' choice of strategy is independent and simultaneous;

=>In individual stages, players make decisions at significantly different times, and the stage effect is visible after everyone has made a decision. This causes an additional asymmetry for the players when making decisions at various stages, where those who make decisions later know the decisions that the players made earlier.

=>Given the above, a decision problem should be distinguished in which the players define their way of proceeding by referring only to the states of the observed process, and in which, apart from the state of the process, the behavior of other participants in the decision-making process is also taken into account.

It should be noted that, depending on the decision model analyzed, with a large number of agents, their decisions are the result of the strategies adopted. Strategy is a rule, a function that, based on the state of the game, indicates the decision to be made. For games where stopping times are the zones, the stopping time is the decision. Depending on the type of analysis and its purpose, the stopping moment is a random variable or its implementation. For the model of the stopping game analyzed, it is also possible to create auxiliary dynamic models in which the strategic moment of stopping is transformed into a sequence of decision variables, and the stopping game analyzed into a stochastic (sequential) game with a specific set of strategies.

A player's strategy is any of the options that they choose in a setting where the outcome depends not only on their own actions but also on the actions of others. Therefore, it should mainly be concerned with the action of a player in a game that affects the behavior or actions of other players. A player's strategy will determine the action the player will take at any stage of the game.

The strategy concept is a complete algorithm to play the game, telling a player what to do for every possible situation throughout the game. It is helpful to think of a **strategy** as a list of directions, and a **move** as a single turn on the list of directions itself. This strategy is based on the reward or outcome of each action. The goal of each agent is to consider their payoff based on a competitors' action.

On the other hand, the analysis of the multilateral decision problem leads to the concept of a strategy profile (sometimes called a strategy combination) – a set of strategies for all players, which fully specifies all actions in a game. A strategy profile must include one single strategy for each player.

It follows from the above that a decision problem involving many people, where each of the participants expresses their decisions by accepting the state of the observed process (which gives a recipe for how it will behave in a given state of the game), can be analyzed as a sequential game using the specificity of the formulation of the decision problem and the allowed strategy classes. This interpretation shows that, during the game, players can make decisions simultaneously or in some order. Simultaneous games are games in which players make their moves at the same time, or where the players who move later are unaware of the actions taken by those who moved earlier, effectively making the moves simultaneous.

Sequential games (or dynamic games) are games in which later players have some knowledge about earlier actions. This need not be perfect information about every action of the earlier players; it might be very little knowledge. For instance, a player

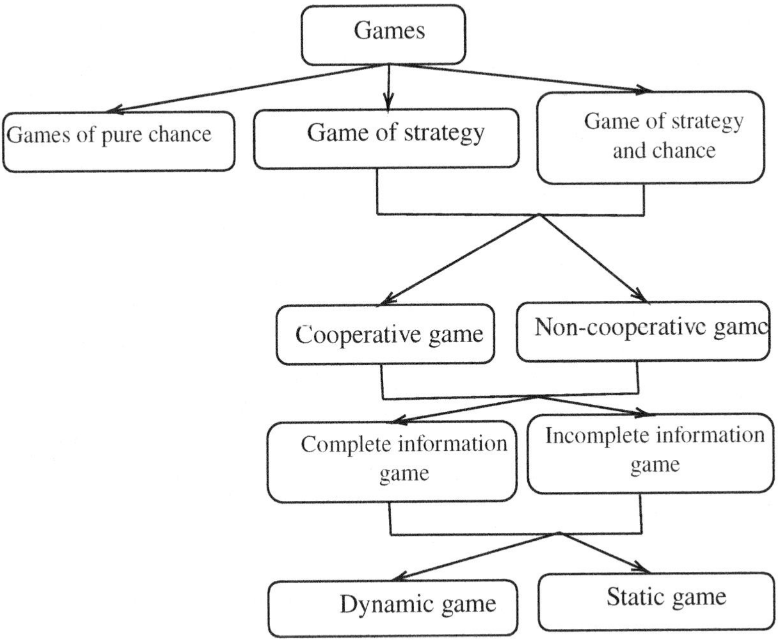

FIGURE 3.1
General classification of games[1].

may know that an earlier player did not perform one particular action, while they do not know which of the other available actions the first player actually performed.

The difference between simultaneous and sequential games is captured in the different representations discussed above. Often, the normal form is used to represent simultaneous games, while the extensive form is used to represent sequential ones. The transformation of extensive to normal form is one way, meaning that multiple extensive form games correspond to the same normal form. Consequently, notions of equilibrium for simultaneous games are insufficient for reasoning about sequential games (see subgame perfection).

This causes games with stopping processes to require extra attention and analysis. In the following, we will illustrate these detailed models that cause stopping games to require additional assumptions before general concepts of solutions (rationality) can be applied to them.

[1]The mathematical model of conflict depends on numerous conditions regarding the information available to the players, their sets of strategies, and their mutual relations. The division presented in Figure 3.1 is not simplified, but the authors often adapt the presentation of their models to such a presentation of the conditions of the analyzed conflicts or cooperation of decision-makers (v. e.g. Mehta and Kwak (2010), Farooqui and Niazi (2016), `https://www.economicsdiscussion.net/game-theory/5-types-of-games-in-game-theory-with-diagram/3827`, `https://collegedunia.com/exams/game-theory-classification-applications-and-theories-articleid-5051`).

3.2.3 Stopping game vs. repeated game

Stopping games and repeated games are two types of strategic interactions between players in game theory ...

A stopping game is a one-shot game where each player must decide whether to stop or continue playing at a certain point in time. The game ends when one player decides to stop or when a predetermined stopping time is reached. Examples of stopping games include auctions, investment decisions, and bargaining situations. In a stopping game, players must take into account the value of future payoffs when deciding whether to stop or continue playing.

On the other hand, a repeated game is a game in which the same players play the same game repeatedly over time. In a repeated game, players can use information from previous rounds to influence their decisions in subsequent rounds. The rewards in each round may depend on the results of previous rounds, and players can use strategies to build reputations or punish or reward their opponents. Examples of repeated games include pricing competition among firms, labor negotiations, and international trade.

The key differences between stopping games and repeated games are:

Time horizon: Stopping games have a finite time horizon, while repeated games have an indefinite time horizon.

Information: Stopping games are usually played under incomplete information, where players may not know the payoffs or the strategies of their opponents. In contrast, repeated games involve complete or imperfect information, where players can learn from previous interactions and use that information to inform their decisions in subsequent rounds.

Strategy: In stopping games, players must choose their strategy based on the value of future payoffs. In repeated games, players can use strategies that involve building reputations or punishing/rewarding opponents.

Outcome: In a stopping game, the outcome is determined by a single decision made by each player. In repeated games, the outcome is determined by a series of decisions made over time, and the final outcome may depend on the players' strategies in each round.

Overall, **stopping games** and **repeated games** are two distinct types of **strategic interactions**, with different characteristics and dynamics.

Understanding the differences between them can help in designing effective strategies in different contexts.

Sakaguchi (1998)
A non-zero-sum repeated game – criminal vs. police

3.3 Multilateral Stopping Decisions

As we mentioned in Section 1.5, the theory of stopping games started with the paper by Dynkin (1969). He proposed a zero-sum game based on slight modification of an optimal stopping problem for discrete time stochastic processes $\{\xi_n\}_{n=0}^{N}$, $N \in \overline{\mathbb{N}}$. The game analyzed by Dynkin (1969) is a stochastic game, and these are a natural extension of the models of controlled processes. Stochastic games, like controlled processes, strategically cover situations in which the environment changes over time in response to the actions of a larger number of decision-makers (players). The choices made by the players have two effects. First, along with the current state, the actions of the players determine the instant payout that each player receives. Second, the current state and actions of players have an impact on the future behavior of the process on which future payouts will depend. Therefore, each player must observe the current payouts and take into account the possible evolution of the situation. This is analogous to single-player decision problems, but the presence of additional players with their own goals complicates the analysis of the situation. Stochastic games were introduced in Shapley (1953) (v. Jaśkiewicz and Nowak (2017)).

Define $\mathcal{F}_n = \sigma(\xi_0, \xi_1, \ldots, \xi_n)$. If each of the two players chooses their strategy, λ and μ (both are Markov times), respectively, the payoff is given by $R(\lambda, \mu) = \xi_{\lambda\wedge\mu}$. The aim of the first player is to maximize the expected value of $R(\lambda, \mu)$, while the other player wants to minimize it. Dynkin (1969) assumes a restriction on the moves in this game. That is, the strategies of the players are such that Player 1 can stop at odd time moments n and Player 2 can choose even moments. Under this assumption, Dynkin proved the existence of the game value and the equilibrium strategies with respect to the expected payoff.

Kifer (1971) obtained another existence condition in a similar formulation. In the absence of exchange of information between the decision-makers, it is usually not possible for both decision-makers to maximize their payouts. Knowing that the utility functions of players depend on the actions of others, one can only analyze potential behaviors using strategic analytical skills. If we have two participants, then the wish to maximize utility regardless of the opponent's choices leads to the conclusion that they both choose a strategy that goes hand in hand with one that realizes the saddle point of the expected utility. Consequently, as a solution to the double optimal stopping problem with payouts adding up to a constant, we adopt a pair of equilibrium strategies. Then the payouts are the expected value (expected utility) of the pair in equilibrium.

Neveu (1975) modified Dynkin's game by changing the payoff function as follows. There are two preassigned stochastic sequences $\{\xi_n\}_{n=0}^{\infty}$, $\{\eta_n\}_{n=0}^{\infty}$, measurable with respect to some increasing sequence of σ-fields $\mathcal{F}_n$. We say that the sequences ξ_n and η_n are adapted to filtration $\mathcal{F}_n$. The players' strategies $\mathfrak{S}_i$, $i = 1, 2$, are stopping times with respect to $\{\mathcal{F}_n\}_{n=0}^{\infty}$. The payoff is equal to

$$R(\lambda, \mu) = \begin{cases} \xi_\lambda & \text{on } \{\lambda \leq \mu\}, \\ \eta_\mu & \text{on } \{\lambda > \mu\}, \end{cases}$$

with the condition

$$\xi_n \leq \eta_n \text{ for each } n. \tag{3.1}$$

Under some regularity condition, Neveu proved the existence of the value of the game and the strategies which form the ε-equilibrium.

Let us formulate the normal form of the zero-sum stopping game. To this end, define the sets of strategies $\mathfrak{S}_1$ and $\mathfrak{S}_2$, which are sets of stopping times with respect to the filtration.

Definition 3.1. *The triplet $(\mathfrak{S}_1, \mathfrak{S}_2, R(\cdot, \cdot))$ is the normal form of the zero-sum stopping game.*

The restriction given by (3.1) was suppressed in some cases by Yasuda (1985). Let ξ_n, η_n and ζ_n be $\mathcal{F}_n$–adapted stochastic sequences. He considered a zero sum stopping game with the following payout

$$R(\lambda, \mu) = \xi_\lambda \mathbb{I}_{\{\lambda<\mu\}} + \zeta_\lambda \mathbb{I}_{\{\lambda=\mu\}} + \eta_\mu \mathbb{I}_{\{\lambda>\mu\}},$$

where $\mathbb{I}$ is an indicator function. To solve the game, the set of strategies has been extended to a class of randomized strategies.

Definition 3.2. *We are given $(\mathfrak{S}_1, \mathfrak{S}_2, R(\cdot, \cdot))$. The stopping game has value v if*

$$v = \sup_{\lambda \in \mathfrak{S}_1} \inf_{\mu \in \mathfrak{S}_2} \boldsymbol{E}R(\lambda, \mu) = \inf_{\mu \in \mathfrak{S}_2} \sup_{\lambda \in \mathfrak{S}_1} \boldsymbol{E}R(\lambda, \mu). \tag{3.2}$$

The game has a solution if there exists a pair $(\lambda^\star, \mu^\star) \in \mathfrak{S}_1 \times \mathfrak{S}_2$ such that $v = \boldsymbol{E}R(\lambda^\star, \mu^\star)$.

A version of Dynkin's game for Markov chains was considered by Frid (1969). A more general version of the stopping game for discrete-time Markov processes (DTMP) was solved by Elbakidze (1976). The following assumptions provide the basis for several models of stopping games for Markov processes.

Assumption 3.1 (DTMP). *We are given*

(i) a probability space $(\Omega, \mathcal{F}, \boldsymbol{P})$;
(ii) a measurable space $(\mathbb{E}, \mathcal{B})$;
(iii) a filtration $\{\mathcal{F}_m\}_{m\geq 1}$, *a nondecreasing sequence of* σ*-subalgebras of* σ*-algebra* $\mathcal{F}$;
(iv) a sequence $\xi_m : \Omega \to \mathbb{E}$, $m \geq 0$ *of* $\{\mathcal{F}_m\}_{m\geq 0}$ *adapted random variables;*
(v) a measure $\boldsymbol{P}_x(\cdot)$ *such that for every* $A \in \mathcal{B}$ *and* $x \in \mathbb{E}$, *we have* $\boldsymbol{P}_x(A) = \boldsymbol{P}(\xi_1 \in A|\xi_0 = x)$.

The triplet $(\xi_n, \mathcal{F}_n, \boldsymbol{P}_x)$, $n \in \mathbb{N} \cup \{0\}$, $x \in \mathbb{E}$ *forms a homogeneous Markov chain with state space* $(\mathbb{E}, \mathcal{B})$.

Let g, G, e and C be certain functions of measurable real value $\mathcal{B}$. There are two players. The process can be stopped at any instant $n \geq 0$. If the process is stopped by the first, the second or by the two players simultaneously, then the payoffs of the player are $g(\xi_n)$, $G(\xi_n)$ and $e(\xi_n)$, respectively. For an unlimited duration of the game, the payoff of the first player is equal to $\limsup_{n\to\infty} C(\xi_n)$. The strategies of the first and second players are given by Markov moments relative to $\{\mathcal{F}_n\}_{n=0}^{\infty}$.

Assumption 3.2. *Let* $\mathcal{L}$ *denote a class of* $\mathcal{B}$*-measurable functions* f *such that* $\boldsymbol{E}_x\{\sup_n |f(\xi_n)|\} < \infty$. *It is assumed that*

$$g(x) \leq e(x) \leq G(x),\ g(x) \leq C(x) \leq G(x),\ x \in \mathbb{E} \text{ and } g, G \in \mathcal{L}.$$

Under these assumptions, the value of the game and ε-equilibrium strategies are constructed.

Ohtsubo (1986) showed the construction of lower and upper value for the two person zero-sum game. The condition for closedness of the game, i.e. when the value of the game exists, was given. Independently Morimoto (1986) gave condition for closing the game under similar assumption (v. also paper by Bensoussan and Friedman (1977) for related results). Stopping times which are ϵ-optimal for the both players in this zero-sum game are found under conditions which guarantee existence of the game value. The discussion of the two person non-zero sum game of Neveu (1975) lead to various observation and extension by Ohtsubo (1986). Further investigation of the two-person non-zero-sum stopping games gave, among others, weakening the conditions under which the value of the game exists (v.Ohtsubo (1987, 1986)). To present the idea of the research on such games, let $\{\xi_n^i\}_{n=0}^{\infty}$, $\{\eta_n^i\}_{n=0}^{\infty}$ and $\{\zeta_n^i\}_{n=0}^{\infty}$, $i = 1, 2$, be six sequences of random variables of real value defined in a fixed probability space and adapted to $\{\mathcal{F}_n\}_{n=0}^{\infty}$. It is assumed that

(i) $\min\{\xi_n^i, \eta_n^i\} \leq \zeta_n^i \leq \max\{\xi_n^i, \eta_n^i\}$ for each $i = 1, 2$;
(ii) $\boldsymbol{E}\left(\sup_n |\xi_n^i|\right) < \infty$ and $\boldsymbol{E}\left(\sup_n |\eta_n^i|\right) < \infty$ for each $i = 1, 2$.

The sets of players' strategies, $\mathfrak{S}_1$, $\mathfrak{S}_2$, are sets of stopping times with respect to $\{\mathcal{F}_n\}_{n=0}^{\infty}$. If the first and the second players choose stopping times $\tau_1 \in \mathfrak{S}_1$ and

$\tau_2 \in \mathfrak{S}_2$, respectively, then the i-th player, $i \in \{1,2\}$, gets the reward

$$\begin{aligned} g_i(\tau_1,\tau_2) &= \xi^i_{\tau_i}\mathbb{I}_{(\tau_i<\tau_j)} + \eta^i_{\tau_j}\mathbb{I}_{(\tau_j<\tau_i)} \\ &\quad \zeta^i_{\tau_i}\mathbb{I}_{(\tau_i=\tau_j<\infty)} + \limsup_n \zeta^i_n\mathbb{I}_{(\tau_i=\tau_j<\infty)}, \quad i,j=1,2, j\neq i. \end{aligned}$$

Under the above assumption, the Nash equilibrium for the game is constructed. Ohtsubo (1987) gave the solution for a version of the game for Markov processes (see Assumption 3.1). Ferenstein (1993) solved a version of the nonzero-sum Dynkin's game with a different special payoff structure.

Definition 3.3. *The game* $(\mathfrak{S}_1,\mathfrak{S}_2,g_1(\cdot,\cdot),g_2(\cdot,\cdot))$ *has a solution if there exists a pair* $(\tau_1^\star,\tau_2^\star)\in\mathfrak{S}_1\times\mathfrak{S}_2$ *such that*

$$\boldsymbol{E}g_1(\tau_1^\star,\tau_2^\star) \geq \boldsymbol{E}g_1(\tau_1,\tau_2^\star) \text{ for every } \tau_1\in\mathfrak{S}_1, \tag{3.3}$$

$$\boldsymbol{E}g_2(\tau_1^\star,\tau_2^\star) \geq \boldsymbol{E}g_2(\tau_1^\star,\tau_2) \text{ for every } \tau_2\in\mathfrak{S}_2. \tag{3.4}$$

The pair $(v_1^\star,v_2^\star) = (\boldsymbol{E}g_1(\tau_1^\star,\tau_2^\star),\boldsymbol{E}g_2(\tau_1^\star,\tau_2^\star))$ *is called the game's value (or* the value of the game*),* $(\tau_1^\star,\tau_2^\star)$ *is a pair of equilibrium strategies (*the solution of the game*).*

Continuous-time versions of such a game problem were studied by Bensoussan and Friedman (1974,1977), Krylov (1971), Bismut (1977), Stettner (1982), and Lepeltier and Maingueneau (1984), among many others.

While all discrete-time stopping games are within the formula of this monograph, we will focus our attention on those models that seemingly are far from the model of the game formulated in Neveu's (1975) monograph. Supplementing the strategy set with sequences of stopping moments, and taking into account the specific features of the model in the payout function, often allows the stochastic game to be reduced to a stopping game. While all discrete-time stopping games are within the formula of this monograph, we will focus our attention on those models that seemingly are far from the model of the game formulated in the Neveux monograph. Supplementing the strategy set with sequences of stopping moments and taking into account the specific features of the model in the payout function often allows the stochastic game to be reduced to a game with process stoppage. It is also worth remembering that limiting the set of strategies to stopping moments opens the research apparatus resulting from the fact that the stopping moments create a complete lattice (v. Vieille (2002)).

There has been extensive research into games with moment-to-stop strategies. In these games, a question often appears in a situation when two (or more) observers are hunting for the states of the same process. The conflict in which more than one player wants to accept the current state, requires delicate research, not only about when to choose but also how to reconcile the interests of the parties in such a way that the rules established before the game are consistent with the accepted methods of dividing limited goods. The solution we will discuss in more detail here is based on the different powers of the players. This, in turn, can lead to additional asymmetries in players. For example, one of the players reveals his intentions earlier, and the other has the possibility of additional conclusions about the process based on the opponents' behavior. Another interesting case arises when players can only accept one state that meets the expectations of those who choose it (for example, the selected state is shared among those who choose it). We focus our attention on a version of the stopping game called the random priority stopping game. The zero-sum version of the problem is considered in Section 3.3.1 and the nonzero-sum case is presented in Section 3.3.2. The single-stop play model agreed between the parties is discussed in Section 3.3.4.

On learning . . .

Adaptive and non adaptive players . . .

In Bei and Zhang (2022)'s work, they use the secretary problem (cf. Ferguson (1989)) to illustrate possible changes to the assumptions in this problem to create a model that better reflects real decision problems when the problem is dynamic and there are at least two decision-makers. The secretary problem is treated in the literature as a canonical configuration for studying algorithms online. The secretary problem has also been generalized in different directions (see Freeman (1983), Ajtai et al. (2001).

At Bei and Zhang's (2022) work, a group of managers is looking for the best candidate from the pool of candidates for their needs. This posing of the problem transforms the online decision-making problem into a multi-person problem, for which models we are looking for among those known in game theory. As with the single decision-maker problem, players have a stream of candidates arriving one by one. After the arrival of the next job applicant, all employers determine relative ranks, and then each of them decides to hire. The objective of each employer's choice is to maximize the likelihood of hiring the best candidate. Such a two-employer model has appeared in works by Dynkin (1969), Szajowski (1993, 1995), Fushimi (1981), although the goal is still the best candidate, players try to include the second employer through behaviors that guarantee balance. These jobs differ in their assumptions about pay structure and conflict resolution when both employers make offers to the same candidate. An overview of these models can be found in Sakaguchi (1995, 1995), Nowak and Szajowski (1999) and recently in paper by Jaśkiewicz and Nowak (2017). The problem with more employers is considered e.g. by Immorlica et al. (2006), where there are many indistinguishable employers and the prospect of submitting bids (when the earliest bid is submitted) decreases as the number of employers increases. Karlin and Lei (2015) considered such a game with ordered employers and showed how to determine perfect Nash

> equilibrium strategies in a subgame. Ezra et al. (2020) considered a variant of the arrangement that allows deferred elections at employers. Note that all of these papers assume that decisions are made when the next candidate arrives. The candidate's arrival triggers the process of his/her assessment by employers, which can be carried out simultaneously by all employers, or according to a specific scheme (e.g. in a fixed order or in a random order). It is also assumed that employers are identical in their preferences as to the qualifications of candidates. This formulation implies scenarios where candidates possess certain intrinsic values that all employers can observe during an interview, or receive such information relevant to making decisions about selecting a candidate from some standardized test.
>
> Bei and Zhang (2022)
> *Competing Employers on Random Edge Arrivals*

---

1. In typical multilateral selection problem, which can be considered as the auction, if the intrinsic value is objective and accepted by all players, the accessibility of candidates for **DM** participants is one of the elements that are usually determined before the selection starts and the rules are known.
2. The appearance of a candidate to work at the moment begins the procedure that will cause there will be no interest in the candidate from the players and everyone will remain in the game, or players will want to accept the candidate.
3. In the latter case, one of the players will receive new work and will cease to participate in the next selection process. The method of choosing a beneficiary should be determined and public, but not necessarily deterministic.

The conditions that have been described mean that the mathematical model for the task in this form is not a non-zero Dynkin game. You can create a supporting model that will allow you to define a solution to the basic problem in which all elements of the Dynkin game model with a non-zero sum will be defined. However, if at individual stages, when another candidate appears, players will not have identical information, then creating such an auxiliary model will be much more difficult.

Adaptive At any time t, each player i in the game knows how many players are still in the hiring process. Her strategy can be based on this and all the historical information. For player i, an adaptive threshold strategy can be represented by a sequence of thresholds r_{ij}, and the strategy is that player i plays the r_{ij} -threshold strategy as long as the number of remaining players (including herself) is j.

Nonadaptive is the case where we assume that players are unaware whether and when other players have made their hiring decisions and have to each stick to a single threshold strategy throughout the game.

3.3.1 Bilateral zero-sum stopping games

Let us consider a homogeneous Markov process according to Assumption 3.1. The decision-makers, henceforth called Player 1 and Player 2, observe the process sequentially. They want to accept the most profitable state of the process from their point of view.

We adopt the zero-sum game model for the problem (see Definition 3.1). Due to this approach, the preferences of each player are described by a gain function $f : \mathbb{E} \times \mathbb{E} \to \Re$, which depends on the state chosen by both players. Natural sets of strategies have stopping times concerning $\{\mathcal{F}_n\}_{n=0}^N$ but only if the players can obtain the state that they want. Since there is only one random sequence $\{\xi_n\}_{n=0}^N$ in a trial, therefore, at each moment n only one player can obtain the realization x_n of ξ_n. The problem of assigning an object (state of the process) to the players when both want to accept the same one at the same moment is solved by adopting a random mechanism; for example, a lottery chooses the player who benefits. The player chosen by the lottery obtains the realization x_n, and thus the player deprived of the acceptance of x_n at $n < N$ can select any subsequent realization. The realization can only be accepted when it appears without being allowed to recall. We can think about the decision process as an investigation of objects with characteristics described by the Markov process. Both players together can accept at most two objects.

The decision model described above is a generalization of the problems considered by Szajowski (1993) and Radzik and Szajowski (1990). Related questions, when Player 1 has the permanent priority, have been considered by many authors in the zero-sum game or in the settings of the non-zero-sum game.[2] Many papers on the subject were inspired by the BCP (or secretary), and prediction of results in horse competitions; see the papers by Enns and Ferenstein (1985), Fushimi (1981), Majumdar (1986), Sakaguchi (1989, 1991), Ravindran and Szajowski (1992), and Szajowski (1992), where non-zero sum versions of games have been investigated. Sakaguchi (1991) considered the two-person nonzero-sum game related to full information Best Choice Problem with random priority. A review of these problems one can find in Ravindran and Szajowski (1992). For the original secretary problem and its extension, the reader is referred to Gilbert and Mosteller (1966), Freeman (1983), Rose (1982) and Ferguson (1989). We recall Best Choice Problem in Section 3.3.2.

A formal model of random priority is derived, and then the lottery is taken into account in the sets of the strategies of the players. The very interesting question concerns the influence of the level of priority on the value of the problem or the probability of obtaining the required state of the process (or, in other words, the required object). A small part of the problem is shown by an example related to the secretary problem. The simplest problem with asymmetric aims of the players is considered. The first player's aim is to choose the best applicant (**BA**) and the second player wants to accept the best or the second best (**BOS**) but a better one than the opponent. The numerical solution provides that the game is fair when Player 1 has

[2]Further examples will be provided at the end of the chapter. For example, papers published by Ano (1990), Enns and Ferenstein (1987), Ferenstein (1992), Sakaguchi (1991) and Chen et al. (1997).

priority $p \cong 0.7579$ (in the limiting case when $N \to \infty$). More examples and further considerations can be found in Szajowski (1995).

3.3.2 A sequential game with an asymmetric information structure

An incomplete information random priority stopping game.

The model considered in Section 3.3.1 shows that the class of all stopping strategies based on the basic filtering, with which the game-defining process complies, does not match the limit of the admissible strategies. This is because of the natural constraint that only one process is observed, which states are indivisible, and each player tends to accept the same state of the process. By choosing a leader at random at any time, the follower has different options and therefore different strategies. This section considers the construction of adequate sets of strategies. Due to the random choice of the leader, the players' knowledge is modified and supplemented at each stage, hence the solution defined in 3.2 or 3.3 will be called Bayesian equilibrium.

In this section, we consider a construction of Bayesian Nash equilibria for a random priority finite horizon two-person nonzero-sum game involving stopping Markov processes. Let $(\xi_n, \mathcal{F}_n, \boldsymbol{P}_x)$, $n = 0, 1, \ldots, N$, $x \in \mathbb{E}$, be a homogeneous Markov process as in Assumption 3.1. At each moment $n = 1, 2, \ldots, N$, the decision-makers are able to observe the Markov chain. Each player has his utility function $g_i : \mathbb{E} \to \Re$, $i = 1, 2$ (the payoff function for the single-player game)[3], and at any moment n each player decides separately to accept or reject x_n, a realization of ξ_n. We assume that g_i are measurable and bounded. In case if both players have selected the same moment n to accept x_n, a random assignment mechanism is applied; see, for example, Szajowski (1994), Radzik and Szajowski (1990). The players are of various types which are characterized by their appeal of the solicited state. It is assumed that if only i-th player would like to accept the state then their proposal will be accepted with probability β_i, and with probability $1 - \beta_i$ the state will be assigned to the opponent. If a player has not chosen any realization of the Markov process, he or she gets $g_i^* = \inf_{x \in \mathbb{E}} g_i(x)$. The aim of each player is to choose a realization that maximizes their expected utility. The problem with permanent priority for Player 1 (i.e. $\beta_n = 1$, $n = 1, 2, \ldots$) has been solved by Ferenstein (1992).

A very interesting illustration of the problem of stopping games with incomplete information is the best choice problem for two people considered by Fushimi (1981) in a generalized version proposed by Szajowski (1994) when one player has a fixed type and the second player's type is unspecified. It can be described as follows.

Two companies (Player 1 and Player 2) interview a sequence of applicants one by one (as in the best choice problem) every morning independently of the other company, and the results of the interviews are communicated to the applicant in the

[3]In the multilateral stopping problem, the player's payoff depends on the strategy of all parties. In order to avoid any confusion in this regard, we distinguish between the value of the retained state (utility) and the player's payout that also depends on the choice of other players. In other words, utility is relative, and it changes with the performance of the other players. Therefore, in the multilateral sequential decision problem, players have payoff functions that depend on the state chosen by all parties and also on the stopping moments taken by all of them.

afternoon. If only the first company decides to accept the applicant, he or she (the applicant) agrees to accept the position but he or she randomly choose the agent (with probability β the Player 1 and with Probability $1 - \beta$ Player 2). The described model is an illustration of the candidates' behavior when they know that two managers are competing to fill the empty position. The candidate may have a higher preference for the player who proposed a job in the selection process (this corresponds to $\beta > 0.5$ for the player wishing to hire). Different preferences of the players by the candidates may be constant due to the fact that the department represented by the manager is more attractive. The amount of information about the type of the first player makes this problem rather different from the classical problems. We will discuss this problem in detail later. The other company, which is not accepted by the applicant, is informed of this fact and continues the interview process. If, on the other hand, both companies decide to accept the applicant, he or she selects one of them at random (the chance for Player 1 is α and $1 - \alpha$ for the second) and the other company can continue interviewing and employ another applicant. The offer of the second player is accepted at once. In Fushimi (1981), the parameters $\beta = 1$ and $\alpha = 0.5$, the threshold strategies for the players were admitted. It was shown that equilibrium strategies for players in the model are different. One of the players should behave more hastily than in the original secretary problem, and they should start asking 0.2865 for the limiting version of the problem. There are two Nash equilibria in the considered set of strategies for this game with values $(0.2865, 0.2963)$ and $(0.2963, 0.2865)$, respectively.

The construction of the random priority stopping game model, with $\beta_n = 1$ for every n, was given by Szajowski (1995). This construction is presented here for the sake of completeness. Let $\mathfrak{S}^N$ be the aggregation of Markov times with respect to $\{\mathcal{F}_n\}_{n=0}^N$. We assume that $\boldsymbol{P}_x(\tau \leq N) < 1$ for some $\tau \in \mathfrak{S}^N$ (i.e. there is a positive probability that the Markov chain will not be stopped). The elements of $\mathfrak{S}^N$ are possible strategies for the players with the restriction that Player 1 and Player 2 cannot stop at the same moment. If both players declare their willingness to accept the same object, the random device $\{\varsigma_i\}_{i=1}^N$ decides who is endowed.

Let us formalize the problem. Denote $\mathfrak{S}_k^N = \{\tau \in \mathfrak{S}^N : \tau \geq k\}$. Let $\mathfrak{S}_{1k}^N$ and $\mathfrak{S}_{2k}^N$ be copies of $\mathfrak{S}_k^N$ ($\mathfrak{S}^N = \mathfrak{S}_0^N$). The sets of strategies for Player 1 and Player 2 are:

$$\tilde{\Lambda}^N = \{(\lambda, \{\sigma_n^1\}) : \lambda \in {\mathfrak{S}_1}^N, \sigma_n^1 \in {\mathfrak{S}_1}_{n+1}^N \text{ for every } n\}$$

and

$$\tilde{M}^N = \{(\mu, \{\sigma_n^2\}) : \mu \in {\mathfrak{S}_2}^N, \sigma_n^2 \in {\mathfrak{S}_2}_{n+1}^N \text{ for every } n\},$$

respectively. Denote $\tilde{\mathcal{F}}_n = \sigma(\mathcal{F}_n, \varsigma_1, \varsigma_2, \ldots, \varsigma_n)$ and let $\tilde{\mathfrak{S}}^N$ be the set of stopping times with respect to $\{\tilde{\mathcal{F}}_n\}_{n=0}^N$. Define

$$\tau_1 = \lambda \mathbb{I}_{\{\lambda<\mu\}} + (\lambda \mathbb{I}_{\{\varsigma_\lambda \leq \alpha_\lambda\}} + \sigma_\mu^1 \mathbb{I}_{\{\varsigma_\lambda > \alpha_\lambda\}})\mathbb{I}_{\{\lambda=\mu\}} + \sigma_\mu^1 \mathbb{I}_{\{\lambda>\mu\}}$$

and

$$\tau_2 = \mu \mathbb{I}_{\{\lambda>\mu\}} + (\mu \mathbb{I}_{\{\varsigma_\mu > \alpha_\mu\}} + \sigma_\lambda^2 \mathbb{I}_{\{\varsigma_\mu \leq \alpha_\mu\}})\mathbb{I}_{\{\lambda=\mu\}} + \sigma_\lambda^2 \mathbb{I}_{\{\lambda\leq\mu\}}.$$

Remark 8. *Let us propose modelling of the random priority similarly to the randomized stopping times (see Definition 3.4). Define the random sequence* $\{\alpha_n\}_{n=0}^N$ *which is observable by both players measurable with respect to the filtration* $\{\tilde{\mathcal{F}}_n\}_{n=0}^N$ *(such that for every* n*, we have* $\mathcal{F}_n \subset \tilde{\mathcal{F}}_n$*). When the lottery is used to decide on assignment, the sample* ς_n *from the standard uniform distribution is taken. If* $\varsigma_n \leq \alpha_n$*, then Player 1 is benefited.*

One can also think about priority as the sequence of binary random variables, $\{\tilde{\mathcal{F}}_n\}_{n=0}^N$ *adapted with the distribution given by the sequence of numbers* $\{\alpha_n\}_{n=0}^N$.

Filtration $\{\tilde{\mathcal{F}}_n\}_{n=0}^N$ is common knowledge also in the incomplete information stopping game considered here. Similarly to the random priority model in Remark 8, the behavior of the applicant after the proposal from Player 1 is coded by the random sequence $\{\beta_n\}_{n=0}^N$. This gives a chance that the proposal of Player 1 will be accepted. The sequence $\{\beta_n\}_{n=0}^N$ is observable by Player 1, whose information is the filtration $\{\mathcal{K}_n\}_{n=0}^N$, where $\mathcal{K}_n = \sigma\{\tilde{\mathcal{F}}_n, \beta_1, \ldots, \beta_n\}$. To encode this information, a sequence of i.i.d. random variables with $U[0,1]$ distribution, $\{\eta_n\}_{n=0}^N$, is used. The distribution of parameters β_n is common knowledge. Denote $\hat{\beta}_n = \boldsymbol{E}(\beta_n|\tilde{\mathcal{F}}_n)$. The ultimate stopping time for the players are

$$\sigma_1 = \lambda \mathbb{I}_{\{\lambda<\mu,\eta_\lambda\leq\beta_\lambda\}} + (\lambda \mathbb{I}_{\{\varsigma_\lambda\leq\alpha_\lambda\}} + \sigma_\mu^1 \mathbb{I}_{\{\varsigma_\lambda>\alpha_\lambda\}})\mathbb{I}_{\{\lambda=\mu\}} + \sigma_\mu^1 \mathbb{I}_{\{\lambda>\mu,\eta_\lambda<\beta_\lambda\}}$$

and

$$\sigma_2 = \mu \mathbb{I}_{\{\lambda>\mu,\eta_\lambda<\beta_\lambda\}} + (\mu \mathbb{I}_{\{\varsigma_\mu>\alpha_\mu\}} + \sigma_\lambda^2 \mathbb{I}_{\{\varsigma_\mu\leq\alpha_\mu\}})\mathbb{I}_{\{\lambda=\mu\}} + \sigma_\lambda^2 \mathbb{I}_{\{\lambda\leq\mu,\eta_\lambda\geq\beta_\lambda\}}.$$

Similarly, as mentioned in the remark above, the appealing of Player 1 can be seen as binary random variables $\{\beta_n\}_{n=0}^N$.

Lemma 3.1. *(Szajowski (1993)) Random variables* σ_1 *and* σ_2 *are Markov times with respect to* $\{\mathcal{K}_n\}_{n=0}^N$ *and* $\sigma_1 \neq \sigma_2$.

Remark 9. *The second player's ultimate stopping time is not Markovian with respect to the player's information structure, the filtration* $\{\tilde{\mathcal{F}}_n\}_{n=0}^N$.

The payoff functions and strategies

In the problem of optimal stopping, the basic class of strategies $\mathcal{T}^N$ are Markov times with respect to the σ-fields $\{\mathcal{F}_n\}_{n=1}^N$. We assume that $\boldsymbol{P}(\tau \leq N) < 1$ for some $\tau \in \mathcal{T}^N$. The class of these strategies, described in the previous paragraph, is not sufficient in the nonzero-sum stopping game. To extend the class of strategies, we consider a class of randomized stopping times. It is assumed that the probability space is rich enough to admit the following constructions.

Definition 3.4 (see Yasuda (1985)). *A strategy for each player is a random sequence* $p = (p_n) \in \mathcal{P}^N$ *or* $q = (q_n) \in \mathcal{Q}^N$ *such that, for each* n,

(i) p_n, q_n *are adapted to* $\mathcal{F}_n$;
(ii) $0 \leq p_n, q_n \leq 1$ *(*$\boldsymbol{P}$*-a.s.).*

If the random variables p_n, q_n *are equal to* 0 *or* 1*, we call it a* pure strategy.

Let $A_1, A_2, \ldots, A_N$ and $B_1, B_2, \ldots, B_N$ be random i.i.d. variables having the uniform distribution on $[0,1]$ and independent of the Markov process $(\xi_n, \mathcal{F}_n, \boldsymbol{P}_x)_{n=0}^N$. Let $\mathcal{H}_n$ be the field σ generated by $\mathcal{F}_n$, $\{A_1, A_2, ..., A_n\}$ and $\{B_1, B_2, ..., B_n\}$

The randomized Markov times $\lambda(p)$ for strategy $p = (p_n) \in \mathcal{P}^N$ and $\mu(q)$ for strategy $q = (q_n) \in \mathcal{Q}^N$ are defined by $\lambda(p) = \inf\{N \geq n \geq 1 : A_n \leq p_n\}$ and $\mu(q) = \inf\{N \geq n \geq 1 : B_n \leq q_n\}$, respectively.

We denote by $\mathfrak{S}_1^N(\mathcal{P})$ and $\mathfrak{S}_2^N(\mathcal{Q})$ the sets of all randomized strategies of Player 1 and Player 2. Clearly, if each p_n is either zero or one, then the strategy is pure and $\lambda(p)$ is, in fact, a $\{\mathcal{F}_n\}$-Markov time. In particular, a $\{\mathcal{F}_n\}$ -Markov time λ corresponds to the strategy $p = (p_n)$ with $p_n = \mathbb{I}_{\{\lambda=n\}}$, where $\mathbb{I}_A$ is an indicator function of the set A.

Denote $\mathfrak{S}_k^N = \{\tau \in \mathfrak{S}^N : \tau \geq k\}$. One can define the set of strategies $\mathfrak{L}^N = \{(p, \{\sigma_n^1\}) : p \in \mathcal{P}^N, \{\sigma_n^1\} \in \mathfrak{S}_{n+1}^N$ for every $n\}$ and let $\mathfrak{M}^N = \{(q, \{\sigma_n^2\}) : q \in \mathcal{Q}^N, \{\sigma_n^2\} \in \mathfrak{S}_{n+1}^N$ for every $n\}$ for Player 1 and Player 2, respectively.

Let $\{\boldsymbol{\alpha}_n\}_{n=0}^N$ be the priority of the players, independent of $\bigvee_{n=1}^N \mathcal{H}_n$, and let $\{\boldsymbol{\beta}_n\}_{n=0}^N$ be the appealing of Player 1. Denote $\tilde{\mathcal{H}}_n = \sigma\{\mathcal{H}_n, \boldsymbol{\alpha}_n, \boldsymbol{\beta}_n\}$ and let $\mathfrak{L}^N$ be the set of Markov times with respect to $(\tilde{\mathcal{H}}_n)_{n=0}^N$ and $\mathfrak{M}^N$ be the set of Markov times with respect to $(\tilde{\mathcal{H}}_n)_{n=0}^N$. For every pair (s,t) such that $s \in \mathfrak{L}^N$, $t \in \mathfrak{M}^N$, we define

$$\tau_1(s,t) = \lambda(p)\boldsymbol{\beta}_\lambda \mathbb{I}_{\{\lambda(p)<\mu(q)\}} + (\lambda(p)\boldsymbol{\alpha}_\lambda + (1-\boldsymbol{\alpha}_\lambda)\sigma^1_{\mu(q)}) \times \mathbb{I}_{\{\lambda(p)=\mu(q)\}} + \sigma^1_{\mu(q)}\mathbb{I}_{\{\lambda(p)>\mu(q)\}}$$

and

$$\tau_2(s,t) = \mu(q)\mathbb{I}_{\{\lambda(p)>\mu(q)\}} + ((1-\boldsymbol{\alpha}_\mu)\mu(q) + \boldsymbol{\alpha}_\mu\sigma^2_{\lambda(p)}) \times \mathbb{I}_{\{\lambda(p)=\mu(p)\}} + \boldsymbol{\beta}_\lambda\sigma^2_{\lambda(p)}\mathbb{I}_{\{\lambda(p)<\mu(q)\}}.$$

We can check that the random variables $\tau_1(s,t), \tau_2(s,t) \in \tilde{\mathcal{T}}^N$ for every $s \in \mathfrak{L}$ and $t \in \mathfrak{M}$.

Definition 3.5. *The Markov times $\tau_1(s,t)$ and $\tau_2(s,t)$ are the selection times of Player 1 and Player 2 when they use strategies $s \in \mathfrak{L}$ and $t \in \mathfrak{M}$, respectively, and the priority lottery is $\bar{\alpha}$ and the appealing of Player 1 is $\bar{\beta}$.*

For each $(s,t) \in \mathfrak{L}^N \times \mathfrak{M}^N$ and a given $\bar{\alpha}$ (the numbers that characterize the system of priorities) and $\bar{\beta}$ (the characterization of the appealing of Player 1), the payoff function (the gain function) for the player i-th is defined as $f_i(s,t) = g_i(\xi_{\tau_i(s,t)})$. Let $\tilde{R}_1(x,s,t,\bar{\beta}) = \boldsymbol{E}_x f_1(s,t) = \boldsymbol{E}_x g_1(\xi_{\tau_1(s,t)})$ be the expected gain of the first player and $\tilde{R}_2^{B_2}(x,s,t) = \boldsymbol{E}_x\boldsymbol{E}_{\bar{\beta}} f_1(s,t) = \boldsymbol{E}_x\boldsymbol{E}_{\bar{\beta}} g_2(\xi_{\tau_2(s,t)})$ be the second player gain, where $\bar{\beta} = (\boldsymbol{\beta}_1, \ldots, \boldsymbol{\beta}_N)$, if the players use (s,t). We have defined the game in the normal form $(\tilde{\mathfrak{S}}_1^N, \tilde{\mathfrak{S}}_2^N, \tilde{R}_1, \tilde{R}_2)$. This random priority incomplete information game will be denoted by $\mathcal{G}_{rp}^{B_2}$ (to emphasize that the second player is the Bayesian one).

Definition 3.6. *A pair (s^*,t^*) of strategies such that $s^* \in \mathfrak{L}^N$ and $t^* \in \mathfrak{M}^N$ is called a Bayesian Nash equilibrium in $\mathcal{G}_{rp}^{B_2}$ if, for all $x \in \mathbb{E}$ and all $\bar{\beta} \in [0,1]^N$,*

$$\begin{aligned} v_1(x,\bar{\beta}) &= \tilde{R}_1(x,s^*,t^*,\bar{\beta}) \geq \tilde{R}_1(x,s,t^*,\bar{\beta}) \text{ for every } s \in \mathfrak{L}^N, \\ v_2^{B_2}(x) &= \tilde{R}_2^{B_2}(x,s^*,t^*) \geq \tilde{R}_2^{B_2}(x,s^*,t) \text{ for every } t \in \mathfrak{M}^N. \end{aligned}$$

Denote

$$h_i(n,\xi_n) = \operatorname{ess\,sup} \tau \in \mathcal{T}_n^N \boldsymbol{E}_{\xi_n} g_i(\xi_\tau)$$

and σ^{*i} a stopping time such that $h_i(0,x) = \boldsymbol{E}_x g_i(\xi_{\sigma^{*i}})$ for every $x \in \mathbb{E}$, $i = 1,2$. Let $\Gamma_n^i = \{x \in \mathbb{E} : h_i(n,x) = g_i(x)\}$. We have $\sigma^{*i} = \inf\{n : \xi_n \in \Gamma_n^i\}$ (see Shiryaev (1978)). Denote $\sigma^{*i}_k = \inf\{n > k : \xi_n \in \Gamma_n^i\}$. Taking into account the above definition of $\mathcal{G}_{rp}^{B_2}$, one can conclude that the Nash equilibrium payoffs of this game are the same as in the auxiliary game $\mathcal{G}_{wp}^{B_2}$ with the sets of strategies of the players $\mathcal{P}^N$, $\mathcal{Q}^N$ and payoff functions (see Yasuda (1985)):

$$\begin{aligned} \varphi_1(p,q,\bar{\beta}) = {} & g_1(\xi_{\lambda(p)})\beta_{\lambda}(p)\mathbb{I}_{\{\lambda(p)<\mu(q)\}} \\ & + \tilde{h}_1(\mu(q),\xi_{\mu(q)})\mathbb{I}_{\{\lambda(p)>\mu(q)\}} \\ & + \left[g_1(\xi_{\lambda(p)})\alpha_{\lambda(p)} + \tilde{h}_1(\lambda(p),\xi_{\lambda(p)})(1-\alpha_{\lambda(p)})\right] \\ & \times \mathbb{I}_{\{\lambda(p)=\mu(q)\}}, \end{aligned} \tag{3.5}$$

$$\begin{aligned} \varphi_2^{B_2}(p,q) = {} & \boldsymbol{E}_{\bar{\beta}}\big[g_2(\xi_{\mu(q)})\mathbb{I}_{\{\mu(q)<\lambda(p)\}} \\ & + \tilde{h}_2(\lambda(p),\xi_{\lambda(p)})\beta_{\lambda}(p)\mathbb{I}_{\{\mu(q)>\lambda(p)\}} \\ & + \left[g_2(\xi_{\lambda(p)})(1-\alpha_{\lambda(p)}) + \tilde{h}_2(\lambda(p),\xi_{\lambda(p)})\alpha_{\lambda(p)}\right] \\ & \times \mathbb{I}_{\{\lambda(p)=\mu(q)\}}\big], \end{aligned} \tag{3.6}$$

for each $p \in \mathcal{P}$, $q \in \mathcal{Q}$, where

$$\tilde{h}_i(n,\xi_n) = \operatorname{ess\,sup} \tau \in \mathcal{T}_{n+1}^N \boldsymbol{E}_{\xi_n} g_i(\xi_\tau) = \boldsymbol{E}_{\xi_n} h_i(n+1,\xi_{n+1}).$$

Denote $R_1(x,p,q,\bar{\beta}) = \boldsymbol{E}_x\varphi_1(p,q,\bar{\beta})$ and $R_2^{B_2}(x,p,q) = \boldsymbol{E}_x\boldsymbol{E}_{\bar{\beta}}\varphi_2^{B_2}(p,q)$ for every $x \in \mathbb{E}$, and $\bar{\boldsymbol{\beta}} : \Omega \to [0,1]^N$ with given distribution B (known to the both players).

Let

$$\mathcal{P}_n^N = \{p = (p_n) \in \mathcal{P} : p_1 = \ldots = p_{n-1} = 0, p_N = 1\}$$

and

$$\mathcal{Q}_n^N = \{q = (q_n) \in \mathcal{Q} : q_1 = \ldots = q_{n-1} = 0, q_N = 1\}.$$

We will use the following convention: if $p \in \mathcal{P}^N$ then (p_n,p) is p with the n-th coordinate replaced by p_n $((p_n,p) = (p_{-n},p_n))$.

Definition 3.7. *Let $\bar{\beta} : \Omega \to [0,1]^N$ with a given distribution B be given. A pair $(p^{\bar{\beta}^*}, q^{B_2{}^*}) \in \mathcal{P}_n^N \times \mathcal{Q}_n^N$ is called a Bayesian equilibrium point of $\mathcal{G}_{wp}^{B_2}$ at n,*

$$v_1(n, \xi_n, \bar{\beta}) = \boldsymbol{E}_{\xi_n} \varphi_1(p^{\bar{\beta}^*}, q^{B_2{}^*}, \bar{\beta}) \geq \boldsymbol{E}_{\xi_n} \varphi_1(p, q^{B_2{}^*}, \bar{\beta})$$

for every $p \in \mathcal{P}_n^N$ ($\boldsymbol{P}_x$-a.s.),

$$v_2^{B_2}(n, \xi_n) = \boldsymbol{E}_{\xi_n} \boldsymbol{E}_{\bar{\beta}} \varphi_2^{B_2}(p^{\bar{\beta}^*}, q^{B_2{}^*}) \geq \boldsymbol{E}_{\xi_n} \boldsymbol{E}_{\bar{\beta}} \varphi_2^{B_2}(p^{\bar{\beta}^*}, q)$$

for every $q \in \mathcal{Q}_n^N$ ($\boldsymbol{P}_x$-a.s.).

A Nash equilibrium point at $n = 0$ is a solution of $\mathcal{G}_{wp}^{B_2}$.

Theorem 3.2. *(Szajowski (1994)) There exists a Bayesian Nash equilibrium $(p^{\bar{\beta}^*}, q^{B_2{}^*})$ in the game $\mathcal{G}_{wp}^{B_2}$. The Nash equilibrium payoffs can be calculated recursively.*

The solution of the game $\mathcal{G}_{rp}^{B_2}$ can be constructed based on the solution (p^*, q^*) of the corresponding game $\mathcal{G}_{wp}^{B_2}$.

Theorem 3.3 (Szajowski (1994)). *The game $\mathcal{G}_{rp}^{B_2}$ with the given distribution of the appealing for Player 1, has a solution. The pair (s^*, t^*), where $s^* = (p^{\bar{\beta}^*}, \{\sigma^{*1}_n\}) \in \mathfrak{L}^N$ and $t^* = (q^{B_2{}^*}, \{\sigma^{*2}_n\}) \in \mathfrak{M}^N$, is a Bayesian Nash equilibrium point with Bayesian equilibrium payoffs $v_1(0, x, \bar{\beta})$, where $\bar{\beta}$ is the appealing characteristics of Player 1, and $v_2^{B_2}(0, x)$.*

In fact, the players play optimally $\mathcal{G}_{rp}^{B_2}$ using a Nash equilibrium strategy from $\mathcal{G}_{wp}^{B_2}$. If the strategy of both players indicates that they are stopping at moment n and neither player has stopped earlier, then the lottery chooses one of them. If Player 1 solicited the applicant, only he or she knows the chance that his proposal will be accepted and his equilibrium strategy depends on this value (the probability to be accepted). The player, who has not been selected, will accept any future realization according to the adequate optimal strategy in the optimization problem.

3.3.3 The Stackelberg Equilibrium of the Two-Person Stopping Game

Description of the model.

The very well-known secretary problem also has many modifications. Ferguson (1989) has made a review of the concepts of the best-choice problem (**BCP**) going back to the age of Kepler and the paper by Cayley (1875). Presman and Sonin (1972) considered so called no-information problem in which the appearing objects come from the rank distribution, that is, the objects are observable, the decision-maker can rank them, and all permutations of the appearing objects are equally possible.

Gilbert and Mosteller (1966) presented various models, also such that the exact value of the object is observable and the distribution of the object is known (it is assumed to be a uniform distribution in the interval $[0, 1]$). Both ideas can be described as the optimal stopping of a Markov chain. In both, there is only one decision-maker and there is no competition concept.

The game theory approach to the secretary problem was introduced by Dynkin (1969). The problem of choosing the best object is later used to show the role of information that decision-makers have. As in the work Szajowski and Skarupski (2019), we are dealing with a bilateral decision problem related to the observation of the Markov process by decision-makers. The information provided to the players is based on the aggregation of the observation data. Acceptable strategies are moment of hold related to the available information. The payouts are the result of the selected state at the time that the decision-maker stopped observing. As in the cited work, the decision-makers are not identical (symmetric). They differ in access to information and have different rights to access the observed state, similar to the model of von Stackelberg (2009). The considered model is an extension of Dynkin's (1969) game but is also closely related to the models in the works by Nakai (1997), Ferenstein (1992), Szajowski (1995), Porosiński and Szajowski (2000). Further, examples can be found in Mazalov's (1987, 2014) book. Another way to agree on acceptable behavior in stopping games has been proposed Kurano et al. (1980), Enns and Ferenstein (1985), Majumdar (1988), Vinnichenko and Mazalov (1989) (also v. Bruss et al. (1998)).

Business motivation.

Consider a two companies A and B. Both of them are interested in buying a bundle of certain products on a commodity exchange. Company A is a large corporation and knows the actual value of the product on the market. In addition, it knows the previous values of the objects and can compare them. The problem of company B is that it does not have information about the actual value of the good. However, the owner of company B can compare the actual position of the good in the market with the previous observations. Both players want to choose the very best object overall without the possibility of recall. The number of objects is fixed and finite. A very good example can be described from the reliability position. Consider two buyers of the same item. Both want to buy the most reliable object. Buyer A has the ability to know the values of the reliability function derived by experts and quality controllers. The player B has no such contact and intelligence, so he must rely on his basic knowledge and the knowledge of the previous observation, i.e. he can judge whether the object is better or worse than the previous one. We can say that the buyers of the objects are two types: the first is a business, and the decision problem is preceded by unilateral consideration. The form of the optimal strategy in the decision problem is the inspiration for the mean-value formulation. Threshold strategies are crucial tools for optimal stopping problems. The simplest case related to the observation of a sequence of random variables can be found, e.g., in Szajowski (1984) or Porosiński (1990). The bilateral extension of these models can be found in Neumann et al. (1996). Two players, **I** and **II**, observe sequentially a known finite number (or a number having a geometric distribution) of independent and identically distributed random variables. They must choose the largest. Variables cannot be perfectly observed. When a random variable is sampled, the sampler is informed only whether it is greater or less than some level that he has specified. Each player can choose at most one observation. After the sampling, the players decide on acceptance or rejection of the observation. If both accept the same observation, Player **I** has priority. The class of adequate strategies and a gain function are constructed. In the finite case, the game has a solution in pure strategies. In the case of a geometric distribution, Player **I** has a pure equilibrium strategy, and Player **II** has either a pure equilibrium strategy or a mixture of two pure strategies. The game is symmetric, as the players are watching the same string to the same extent. Increasing opposing interests is possible by completely different preferences of the players. Evaluation of the same object by two decision-makers can mean that players observe the different coordinates of the vector and formulate their expectations for their realization. When players' aim is to achieve a minimum level of the observed rate, then the problem can be reduced to a game in which strategies are setting of just levels. Discussion of such issues can be found in the work by Sakaguchi (1973). However, in those tasks, though, the information players are incomplete, lacking clear asymmetry players. The payoffs of the players are function of the thresholds, and the perfect comparison of the observed variable with these defined levels is guaranteed. Asymmetric tools in measure of the observed r.v. are presented by Sakaguchi and Szajowski (2000). However, for private random variables. These players with asymmetric tools applied to there same sequence are the subject of consideration in the paper.

3.3.4 Multilateral stopping problem with players' cooperation

Following the results of Szajowski and Yasuda (1996) the multilateral stopping of a Markov chain problem can be described in the terms of the notation used in the non-cooperative game theory; see Nash (1951), Dresher (1981), Moulin (1982), Owen (2013). To this end, the process and utilities of its states should be specified.

Definition 3.8 (ISS-An individual stopping set). *Let a homogeneous Markov chain* $(\boldsymbol{X}_n, \mathcal{F}_n, \boldsymbol{P}_x)$, $n = 0, 1, 2, \ldots, N$, *with the state space* $(\mathbb{E}, \mathcal{B})$ *be given.*

- *The players can observe the Markov chain sequentially. The horizon can be finite or infinite:* $N \in \mathbb{N} \cup \{\infty\}$.
- *Each player has their utility function* $f_i : \mathbb{E} \to \Re$, $i = 1, 2, \ldots, p$, *such that* $\boldsymbol{E}_x|f_i(\boldsymbol{X}_1)| < \infty$ *and the cost function* $c_i : \mathbb{E} \to \Re$, $i = 1, 2, \ldots, p$.
- *If the process is not stopped at the moment* n, *then each player, based on* $\mathcal{F}_n$, *can independently declare their willingness to stop the observation of the process.*

Definition 3.9 (see Yasuda et al. (1982)). *ISS of Player i is the sequence of random variables $\{\sigma_n^i\}_{n=1}^N$, where $\sigma_n^i : \Omega \to \{0,1\}$ such that σ_n^i is $\mathcal{F}_n$-measurable.*

The strategy is interpreted as follows: If $\sigma_n^i = 1$, Player i indicates a desire to stop the process and accept the completion of X_n.

Definition 3.10 (SS–An individual stopping set (the aggregate function)). *Denote*

$$\sigma^i = (\sigma_1^i, \sigma_2^i, \ldots, \sigma_N^i)$$

and let $\mathscr{S}^i$ be the set of ISSs of Player i, $i = 1, 2, \ldots, p$. Define $\mathscr{S} = \mathscr{S}^1 \times \mathscr{S}^2 \times \ldots \times \mathscr{S}^p$. The element $\sigma = (\sigma^1, \sigma^2, \ldots, \sigma^p)^T \in \mathscr{S}$ will be called the stopping strategy (SS).

The stopping strategy $\sigma \in \mathscr{S}$ acts as a random matrix. Its rows represent ISSs, while the columns capture players' decisions over time. The strategy's application, determining the actual stopping of the observation process and players' payoff realization, leverages a p-variate logical function. In this stopping game model, let $\delta : \{0,1\}^p \to \{0,1\}$ be the aggregation function. The stopping strategy is essentially a list of declarations made by individual players. The aggregate function δ then translates these declarations into an effective stopping time.

Definition 3.11 (aSS–an aggregated SS). *A stopping time $\tau_\delta(\sigma)$ generated by the SS $\sigma \in \mathscr{S}$ and the aggregate function δ is defined by*

$$\tau_\delta(\sigma) = \inf\{1 \le n \le N : \delta(\sigma_n^1, \sigma_n^2, \ldots, \sigma_n^p) = 1\},$$

$\inf(\emptyset) = \infty$. *Since δ is fixed during the analysis we skip index δ and write $\tau(\sigma) = \tau_\delta(\sigma)$.*

Definition 3.12 (Process and utilities of its states). *Let us declare the denotations*

- $\{\omega \in \Omega : \tau_\delta(\sigma) = n\} = \bigcap_{k=1}^{n-1}\{\omega \in \Omega : \delta(\sigma_k^1, \sigma_k^2, \ldots, \sigma_k^p) = 0\} \cap \{\omega \in \Omega : \delta(\sigma_n^1, \sigma_n^2, \ldots, \sigma_n^p) = 1\} \in \mathcal{F}_n$;
- $\tau_\delta(\sigma)$ *is a stopping time with respect to* $\{\mathcal{F}_n\}_{n=1}^N$;
- *for any stopping time $\tau_\delta(\sigma)$ and* i $\in \{1, 2, \ldots, p\}$, *the payoff of Player i is defined as follows (see Shiryaev (1978)):*

$$f_i(X_{\tau_\delta(\sigma)}) = f_i(X_n)\mathbb{I}_{\{\tau_\delta(\sigma)=n\}} + \limsup_{n\to\infty} f_i(X_n)\mathbb{I}_{\{\tau_\delta(\sigma)=\infty\}}.$$

Definition 3.13 (An equilibrium strategy). *Let the aggregate rule δ be fixed (see Szajowski and Yasuda (1996)). The strategy $^*\sigma = ({}^*\sigma^1, {}^*\sigma^2, \ldots, {}^*\sigma^p)^T \in \mathscr{S}$ is an equilibrium strategy with respect to δ if for each* i $\in \{1, 2, \ldots, p\}$ *and any $\sigma^i \in \mathscr{S}^i$, we have*

$$\boldsymbol{E}_x[f_i(\boldsymbol{X}_{\tau_\delta({}^*\sigma)}) + \sum_{k=1}^{\tau_\delta({}^*\sigma)} c_i(\boldsymbol{X}_{k-1})] \leq \boldsymbol{E}_x[f_i(\boldsymbol{X}_{\tau_\delta({}^*\sigma(i))}) + \sum_{k=1}^{\tau_\delta({}^*\sigma(i))} c_i(\boldsymbol{X}_{k-1})].$$

Equilibria in voting stopping game.

The set of stopping strategies, denoted by $\mathscr{S}$, along with the vector of utility functions $\boldsymbol{f} = (f_1, f_2, \ldots, f_p)$ and the monotone rule δ, collectively define the non-cooperative game $\mathcal{G} = (\mathscr{S}, \boldsymbol{f}, \delta)$. The construction of the equilibrium strategy, when the players observe the Markov sequence, was detailed for the first time by Szajowski and Yasuda (1996). It will be denoted $^*\sigma \in \mathscr{S}$ in $\mathcal{G}$.

Definition 3.14. *An individual stopping set of Player i gives the sequence of stopping events $D_n^i = \{\omega : \sigma_n^i = 1\}$. Each aggregate rule δ defines the corresponding set value function $\Delta : \mathcal{F} \to \mathcal{F}$ such that*

$$\delta(\sigma_n^1, \sigma_n^2, \ldots, \sigma_n^p) = \delta\{\mathbb{I}_{D_n^1}, \mathbb{I}_{D_n^2}, \ldots, \mathbb{I}_{D_n^p}\} = \mathbb{I}_{\Delta(D_n^1, D_n^2, \ldots, D_n^p)}.$$

Corollary 3.7. *The important class of ISS and the stopping events are defined by subsets $C^i \in \mathcal{B}$ of the state space $\mathbb{E}$. These subsets are called the stopping sets for Player i at moment n if $D_n^i = \{\omega : X_n \in C^i\}$ is the stopping event.*

By properties of the logical function, we have

$$\delta(\boldsymbol{x}) = x^i \cdot \delta(\boldsymbol{x}_{-i}, \overset{i}{\breve{1}}) + \overline{x}^i \cdot \delta(\boldsymbol{x}_{-i}, \overset{i}{\breve{0}}). \tag{3.8}$$

where

$$(a, \boldsymbol{x}_{-i}) = (\boldsymbol{x}_{-i}, a) = (x_1, \ldots, x_{i-1}, a, x_{i+1}, \ldots, x_n). \tag{3.9}$$

This implies that for $D^i \in \mathcal{F}$ set

$$\Delta(D^1, \ldots, D^p) = \{D^i \cap \Delta(D^1, \ldots, \overset{i}{\breve{\Omega}}, \ldots, D^p)\} \cup \{\overline{D}^i \cap \Delta(D^1, \ldots, \overset{i}{\breve{\emptyset}}, \ldots, D^p)\}.$$

The form of stopping sets.

Let f_i and g_i be real-valued integrable defined on $\mathbb{E}$. For fixed D_n^j, $j = 1, 2, \ldots, p$, $j \neq i$, and $C^i \in \mathcal{B}$ define

$$\psi(C^i) = \boldsymbol{E}_x\left[f_i(X_1)\mathbb{I}_{{}^iD_1(D_1^i)} + g_i(X_1)\mathbb{I}_{\overline{{}^iD_1(D_1^i)}}\right],$$

where ${}^iD_1(A) = \Delta(D_1^1, \ldots, D_1^{i-1}, A, D_1^{i+1}, \ldots, D_1^p)$ and $D_1^i = \{\omega : X_n \in C^i\}$.

Lemma 3.4. *For integrable function f_i, g_i, $h_i(x) = f_i(x) - g_i(x)$, and sets $C^j \in \mathcal{B}$, $j = 1, 2, \dots, p$, $j \neq i$, we have*

$$\psi({}^*C^i) = \sup_{C^i \in \mathcal{B}} \psi(C^i) = \boldsymbol{E}_x h_i^+(X_1)\mathbb{I}_{{}^{i*}D_1(\Omega)} - \boldsymbol{E}_x h_i^-(X_1)\mathbb{I}_{{}^{i*}D_1(\Omega)} + \boldsymbol{E}_x g_i(X_1).$$

where the set ${}^*C^i = \{x \in \mathbb{E} : h_i(x) \geq 0\} \in \mathcal{B}$.

Using Lemma 3.4, we derive the recursive formulae that define the equilibrium point and the equilibrium value for the finite horizon game.

The finite horizon game.

In the finite horizon game, the construction of equilibria is based on the backward induction. Let the horizon N be finite, and let us denote the equilibrium strategy ${}^*\sigma$.

- If the equilibrium strategy ${}^*\sigma$ exists, then we denote $v_{i,N}(x) = \boldsymbol{E}_x f_i(X_{t({}^*\sigma)})$ the equilibrium payoff of i-th player when $X_0 = x$.
- Let $\mathscr{S}_n^i = \{\{\sigma_k^i\}, k = n, \dots, N\}$ be the set of ISS for moments $n \leq k \leq N$ and $\mathscr{S}_n = \mathscr{S}_n^1 \times \mathscr{S}_n^2 \times \dots \times \mathscr{S}_n^p$.
- SS for moments not earlier than n is ${}^n\sigma = ({}^n\sigma^1, {}^n\sigma^2, \dots, {}^n\sigma^p) \in \mathscr{S}_n$, where ${}^n\sigma^i = (\sigma_n^i, \sigma_{n+1}^i, \dots, \sigma_N^i)$.
- $t_n = t_n(\sigma) = t({}^n\sigma) = \inf\{n \leq k \leq N : \delta(\sigma_k^1, \sigma_k^2, \dots, \sigma_k^p) = 1\}$, a stopping time not earlier than n.

Definition 3.15 (Equilibrium in $\mathscr{S}_n$). *The stopping strategy*[a]

$${}^{n*}\sigma = ({}^{n*}\sigma^1, {}^{n*}\sigma^2, \dots, {}^{n*}\sigma^p)$$

is an equilibrium in $\mathscr{S}_n$ if $\boldsymbol{P}_x$-a.s.

$$\boldsymbol{E}_x[f_i(X_{t_n({}^*\sigma)}) + \sum_{k=1}^{t_n({}^*\sigma)} c_i(\boldsymbol{X}_{k-1})] \leq \boldsymbol{E}_x[f_i(X_{t_n(({}^*\sigma_{-i},\sigma))}) + \sum_{k=1}^{t_n(({}^*\sigma_{-i},\sigma))} c_i(\boldsymbol{X}_{k-1})]$$

for every $i \in \{1, 2, \dots, p\}$.

[a] The vector of individual strategies $({}^{n*}\sigma_{-i}, {}^n\sigma)$ in some circumstances will be denoted ${}^{n*}\sigma(i) = ({}^{n*}\sigma^1, \dots, {}^{n*}\sigma^{i-1}, {}^n\sigma^i, {}^{n*}\sigma^{i+1}, \dots, {}^{n*}\sigma^p)$

In Theorem 3.5, we show the construction of a solution compliant with Definition 3.13 for the case of a finite and infinite horizon.

Denote

$$\begin{aligned} v_{i,N-n+1}(X_{n-1}) &= \boldsymbol{E}_x[f_i(X_{t_n({}^*\sigma)}) + \sum_{k=n}^{t_n({}^*\sigma)} c_i(\boldsymbol{X}_{k-1}) | \mathcal{F}_{n-1}] \\ &= \boldsymbol{E}_{X_{n-1}}[f_i(X_{t_n({}^*\sigma)}) + \sum_{k=n}^{t_n({}^*\sigma)} c_i(\boldsymbol{X}_{k-1})]. \end{aligned}$$

At moment $n = N$ the players have to declare to stop and $v_{i,0}(x) = f_i(x)$. Let us assume that the process is not stopped up to moment n, the players are using the equilibrium strategies ${}^*\sigma_k^i$, $i = 1, 2, \ldots, p$, at moments $k = n+1, \ldots, N$. Choose player i and assume that other players are using the equilibrium strategies ${}^*\sigma_n^j$, $j \neq i$, and player i is using strategy σ_n^i defined by a stopping set C^i.

The expected payoff $\varphi_{N-n}(X_{n-1}, C^i)$ of player i in the game starting at moment n, when the state of the Markov chain at moment $n-1$ is X_{n-1}, is equal to

$$\varphi_{N-n}(X_{n-1}, C^i) = \boldsymbol{E}_{X_{n-1}}\left[f_i(X_n)\mathbb{I}_{{}^{i*}D_n(D_n^i)} + v_{i,N-n}(X_n)\mathbb{I}_{\overline{{}^{i*}D_n(D_n^i)}}\right],$$

where ${}^{i*}D_n(A) = \Delta({}^*D_n^1, \ldots, {}^*D_n^{i-1}, A, {}^*D_n^{i+1}, \ldots, {}^*D_n^p)$.

By Lemma 3.4 the conditional expected gain $\varphi_{N-n}(X_{N-n}, C^i)$ attains the maximum on the stopping set ${}^*C_n^i = \{x \in \mathbb{E} : f_i(x) - v_{i,N-n}(x) \le 0\}$ and

$$\begin{aligned}(v_{i,N-n+1} - c_i)(X_{n-1}) \;=\;& \boldsymbol{E}_x[(f_i - v_{i,N-n})^+(X_n)\mathbb{I}_{{}^{i*}D_n(\Omega)}|\mathcal{F}_{n-1}] \\ &- \boldsymbol{E}_x[(f_i - v_{i,N-n})^-(X_n)\mathbb{I}_{{}^{i*}D_n(\emptyset)}|\mathcal{F}_{n-1}] \\ &+ \boldsymbol{E}_x[v_{i,N-n}(X_n)|\mathcal{F}_{n-1}]\end{aligned}$$

$\boldsymbol{P}_x$−a.e.. It allows us to formulate the following construction of the equilibrium strategy and the equilibrium value for the game $\mathcal{G}$.

Theorem 3.5. *[Solution of the finite horizon stopping game based on voting.] In the game $\mathcal{G}$ with finite horizon N, we have the following solution.*

(i) *The equilibrium value $v_i(x)$, $i = 1, 2, \ldots, p$, of the game $\mathcal{G}$ can be calculated recursively as follows:*

1. $v_{i,0}(x) = f_i(x)$;

2. *For $n = 1, 2, \ldots, N$, we have $\boldsymbol{P}_x$−a.e.*

$$\begin{aligned}(v_{i,n} - c_i)(X_{N-n}) =\;& \boldsymbol{E}_x[(f_i - v_{i,n-1})(X_{N-n+1}))^+\mathbb{I}_{{}^{i*}D_{N-n+1}(\Omega)}|\mathcal{F}_{N-n}] \\ &- \boldsymbol{E}_x[(f_i - v_{i,n-1})(X_{N-n+1}))^-\mathbb{I}_{{}^{i*}D_{N-n+1}(\emptyset)}|\mathcal{F}_{N-n}] \\ &+ \boldsymbol{E}_x[v_{i,n-1}(X_{N-n+1})|\mathcal{F}_{N-n}],\end{aligned}$$

for $i = 1, 2, \ldots, p$.

(ii) *The equilibrium strategy ${}^*\sigma \in \mathscr{S}$ is defined by the SS of the players ${}^*\sigma_n^i$, where ${}^*\sigma_n^i = 1$ if $X_n \in {}^*C_n^i$, and ${}^*C_n^i = \{x \in \mathbb{E} : f_i(x) - v_{i,N-n}(x) \le 0\}$, $n = 0, 1, \ldots, N$.*

We have $v_i(x) = v_{i,N}(x)$, and $\boldsymbol{E}_x f_i(X_{t({}^*\sigma)}) = v_{i,N}(x)$, $i = 1, 2, \ldots, p$.

Proof: Part (i) is a consequence of the Lemma 3.4. We have $v_{i,1}(x) = \boldsymbol{E}_x f_i(X_1)$ and $v_{i,1}(X_{N-1}) = \boldsymbol{E}_{X_{N-1}} f_i(X_{t({}^{N*}\sigma)})$ $\boldsymbol{P}_x$−a.e.. Assume that

$$v_{i,N-n-1}(X_{n+1}) = \boldsymbol{E}_{X_{n+1}} f_i(X_{t({}^{(n+2)*}\sigma)})\boldsymbol{P}_x\text{-a.e.} \tag{3.1}$$

and define ${}^*D_{n+1} = \Pi({}^*D^1_{n+1}, \ldots, {}^*D^i_{n+1}, \ldots, {}^*D^p_{n+1})$. We have

$$\begin{aligned}
\boldsymbol{E}_{X_n} f_i(X_{t((n+1)*\sigma)}) &= \boldsymbol{E}_{X_n}[f_i(X_{t((n+1)*\sigma)})\mathbb{I}_{{}^*D_{n+1}} + f_i(X_{t((n+1)*\sigma)})\mathbb{I}_{\overline{{}^*D_{n+1}}}] \\
&\overset{(3.1)}{=} \boldsymbol{E}_{X_n} f_i(X_{n+1})\mathbb{I}_{{}^*D_{n+1}} + \boldsymbol{E}_{X_n} v_{i,N-n+1}(X_{n+1})\mathbb{I}_{\overline{{}^*D_{n+1}}} \\
&\overset{(L.\,3.4)}{=} \boldsymbol{E}_{X_n}(f_i(X_{n+1}) - v_{i,N-n+1}(X_{n+1}))^+\mathbb{I}_{{}^{i*}D_{n+1}(\Omega)} \\
&- \boldsymbol{E}_{X_n}(f_i(X_{n+1}) - v_{i,N-n+1}(X_{n+1}))^-\mathbb{I}_{{}^{i*}D_{n+1}(\emptyset)} \\
&+ \boldsymbol{E}_{X_n} v_{i,N-n+1}(X_{n+1}).
\end{aligned}$$

We show that ${}^*\sigma$ is an equilibrium point in the game $\mathcal{G}$. To this end, let us assume that $p-1$ players are using the strategies ${}^*\sigma^i$ and one of the players, say the player 1, the strategy $\sigma^1_{\{n\}} = (\sigma^1_1, \sigma^1_2, \ldots, \sigma^1_n, {}^*\sigma^1_{n+1}, \ldots, {}^*\sigma^1_N)$, $(\sigma^1_{\{0\}} = {}^*\sigma^1)$. We have

$$\boldsymbol{E}_x f_1(X_{t(\sigma\{n\})}) \leq \boldsymbol{E}_x f_1(X_{t(\sigma\{n-1\})}),$$

where $\sigma\{n\} = (\sigma^1_{\{n\}}, {}^*\sigma^2, \ldots, {}^*\sigma^p)^T$. It means that the player 1 is using some fixed strategy at moments $1, 2, \ldots, n$ and the strategy defined by (ii) in further moments. We get

$$\begin{aligned}
\boldsymbol{E}_x f_1(X_{t(\sigma\{n\})}) &= \boldsymbol{E}_x f_1(X_{t(\sigma\{n\})})\mathbb{I}_{\{t(\sigma\{n\})<n\}} \\
&\quad + \boldsymbol{E}_x f_1(X_{t(\sigma\{n\})})\mathbb{I}_{\{t(\sigma\{n\})\geq n\}} \\
&= \boldsymbol{E}_x f_1(X_{t(\sigma\{n-1\})})\mathbb{I}_{\{t(\sigma\{n-1\})<n\}} \\
&\quad + \boldsymbol{E}_x f_1(X_{t(\sigma\{n\})})\mathbb{I}_{\{t(\sigma\{n\})\geq n\}}
\end{aligned}$$

and

$$\begin{aligned}
\boldsymbol{E}_x f_1(X_{t(\sigma\{n\})})\mathbb{I}_{\{t(\sigma\{n\})\geq n\}} &= \boldsymbol{E}_x f_1(X_{t_n(*\sigma)})\mathbb{I}_{\{t(\sigma\{n\})\geq n\}} \\
&\leq \boldsymbol{E}_x \boldsymbol{E}_x[f_1(X_n)\mathbb{I}_{{}^*D_n}|\mathfrak{F}_{n-1}]\mathbb{I}_{\{t(\sigma\{n\})\geq n\}} \\
&\quad + \boldsymbol{E}_x \mathbf{E}_x[v_{i,N-n}(X_n)\mathbb{I}_{\overline{{}^*D_n}}|\mathfrak{F}_{n-1}]\mathbb{I}_{\{t(\sigma\{n\})\geq n\}} \\
&= \boldsymbol{E}_x v_{i,N-n+1}(X_{n-1})\mathbb{I}_{\{t((n-1)*\sigma)\geq n\}}.
\end{aligned}$$

Hence

$$\begin{aligned}
\boldsymbol{E}_x f_1(X_{t(\sigma\{n\})}) &\leq \boldsymbol{E}_x f_1(X_{t(\sigma\{n-1\})})\mathbb{I}_{\{t(\sigma\{n-1\})<n\}} \\
&\quad + \boldsymbol{E}_x v_{i,N-n+1}(X_{n-1})\mathbb{I}_{\{t((n-1)*\sigma)\geq n\}} \\
&= \boldsymbol{E}_x f_1(X_{t(\sigma\{n-1\})})\mathbb{I}_{\{t(\sigma\{n-1\})<n\}} \\
&\quad + \boldsymbol{E}_x f_1(X_{t(\sigma\{n-1\})})\mathbb{I}_{\{t(\sigma\{n-1\})\geq n\}} \\
&= \boldsymbol{E}_x f_1(X_{t(\sigma\{n-1\})}).
\end{aligned}$$

This ends the proof of the theorem. □

Remark 10. *For the case of i.i.d. Ferguson (2005) has gotten interesting results.*

3.3.5 The infinite horizon game

Let us assume that a solution $(w_1(x), w_2(x), \ldots, w_p(x))$ of the equations exists:

$$\begin{aligned} w_i(x) - c_i(x) &= \boldsymbol{E}_x(f_i(X_1) - w_i(X_1))^+ \mathbb{I}_{{}^{i*}D_1(\emptyset)} \\ &\quad - \boldsymbol{E}_x(f_i(X_1) - w_i(X_1))^- \mathbb{I}_{{}^{i*}D_1(\Omega)} + \boldsymbol{E}_x w_i(X_1), \end{aligned}$$

$i = 1, 2, \ldots, p$. Consider the stopping game with the following payoff function for $i = 1, 2, \ldots, p$.

$$\phi_{i,N}(x) = \begin{cases} f_i(x) & \text{if } n < N, \\ v_i(x) & \text{if } n \geq N. \end{cases}$$

Lemma 3.6. *Let ${}^*\sigma \in \mathscr{S}_f^*$ be an equilibrium strategy in the infinite horizon game $\mathcal{G}$. For every N, we have*

$$\boldsymbol{E}_x \phi_{i,N}(X_{t^*}) = v_i(x).$$

Proof: For $N = 1$, we have $\boldsymbol{E}_x \phi_{i,1}(X_{t^*}) = v_1(x)$ by Lemma 3.4. Let us assume for the induction that for $k = 1, 2, \ldots, N$

$$\boldsymbol{E}_x \phi_{i,k}(X_{t^*}) = v_i(x). \tag{3.2}$$

We have

$$\begin{aligned} \boldsymbol{E}_x \phi_{i,N+1}(X_{t^*}) &= \boldsymbol{E}_x \phi_{i,N+1}(X_{t^*}) \mathbb{I}_{\{t^* > 1\}} + \boldsymbol{E}_x \phi_{i,N+1}(X_{t^*}) \mathbb{I}_{\{t^* = 1\}} \\ &= \boldsymbol{E}_x [\phi_{i,N+1}(X_{t^*}) \mathbb{I}_{{}^*D_1} + \phi_{i,N+1}(X_{t^*}) \mathbb{I}_{\overline{{}^*D_1}}] \\ &= \boldsymbol{E}_x [f_i(X_1) \mathbb{I}_{{}^*D_1} + \boldsymbol{E}_{X_1} \phi_{i,N+1}(X_{t^*}) \mathbb{I}_{\overline{{}^*D_1}}] \\ &\overset{(3.2)}{=} \boldsymbol{E}_x [f_i(X_1) \mathbb{I}_{{}^*D_1} + v_i(X_1) \mathbb{I}_{\overline{{}^*D_1}}] \\ &\overset{(\text{L. }3.4)}{=} \boldsymbol{E}_x (f_i(X_1) - v_i(X_1))^+ \mathbb{I}_{{}^{i*}D_1(\Omega)} \\ &\quad - \boldsymbol{E}_x (f_i(X_1) - v_i(X_1))^- \mathbb{I}_{{}^{i*}D_1(\emptyset)} + \boldsymbol{E}_x v_i(X_1) \\ &\overset{(3.1)}{=} v_i(x), \end{aligned}$$

where ${}^*D_1 = \Pi({}^*D_1^1, \ldots, {}^*D_1^i, \ldots, {}^*D_1^p)$. It ends the proof of Lemma 3.6. □

Let us assume that for $i = 1, 2, \ldots, p$ and every $x \in \mathbb{E}$, we have

$$\begin{aligned} \boldsymbol{E}_x[\sup\nolimits_{n\in\mathbb{N}} f_i^+(X_n)] &< \infty, \\ \boldsymbol{E}_x[\sup\nolimits_{n\in\mathbb{N}} c_i^+(X_n)] &< \infty. \end{aligned} \tag{3.3}$$

Theorem 3.7. *Let $(X_n, \mathcal{F}_n, \boldsymbol{P}_x)_{n=0}^{\infty}$ be a homogeneous Markov chain and let the payoff functions of the players fulfil (3.3). If $t^* = t({}^*\sigma)$, ${}^*\sigma \in \mathscr{S}_f^*$, then $\boldsymbol{E}_x f_i(X_{t^*}) = v_i(x)$.*

Proof: By Lemma 3.6 and by (3.3), we have

$$\begin{aligned} |\boldsymbol{E}_x f_i(X_{t^*}) &- \boldsymbol{E}_x \phi_{i,N}(X_{t^*})| = \left|\boldsymbol{E}_x (f_i(X_{t^*}) - \mathbf{v}_i(X_{t^*})) \mathbb{I}_{\{t^* > N\}}\right| \\ &\leq \boldsymbol{P}_x\{t^* > N\} \left|\boldsymbol{E}_x[(f_i(X_{t^*})|\mathbb{I}_{\{t^* > N\}}] - \boldsymbol{E}_x[v_i(X_{t^*})|\mathbb{I}_{\{t^* > N\}}]\right| \\ &\leq \boldsymbol{P}_x\{t^* > N\}\{\boldsymbol{E}_x[\sup_{n\geq N}(f_i^+(X_{t^*})] + v_i(x)\}. \end{aligned}$$

To complete the proof let $N \to \infty$. □

Theorem 3.8. *Let the stopping strategy $^*\sigma \in \mathscr{S}_f^*$ be defined by the stopping sets $^*C_n^i = \{x \in \mathbb{E} : f_i(x) \leq v_i(x)\}$, $i = 1, 2, \ldots, p$, then $^*\sigma$ is the equilibrium strategy in the infinite stopping game $\mathcal{G}$.*

Proof: We show that if $p - 1$ players are using the strategies $^*\sigma^j$ and one player, say the i-th player, uses the strategy such as $\sigma^i_{\{n\}} = (\sigma^i_1, \sigma^i_2, \ldots, \sigma^i_n, {}^*\sigma^i_{n+1}, {}^*\sigma^i_{n+2}, \ldots)$, $n \geq 1$, $\sigma^i_{\{0\}} = {}^*\sigma^i$, then

$$\boldsymbol{E}_x f_1(X_{t(\sigma\{n\})}) \leq \boldsymbol{E}_x f_1(X_{t(\sigma\{n-1\})}),$$

where $\sigma_{\{n\}} = ({}^*\sigma^1, \ldots, {}^*\sigma^{i-1}, \sigma^i_{\{n\}}, {}^*\sigma^{i+1} \ldots, {}^*\sigma^p)^T$. Denote $t(\sigma_{\{n\}}) = t(n)$. Since from moment $n - 1$ the strategies $t(n)$ and $t(n-1)$ coincide, then

$$\boldsymbol{E}_x f_i(X_{t(n)}) = \boldsymbol{E}_x f_i(X_{t(n-1)})\mathbb{I}_{\{t(n-1)<n\}} + \boldsymbol{E}_x f_i(X_{t(n)})\mathbb{I}_{\{t(n)\geq n\}}.$$

Moreover

$$\begin{aligned}
& \boldsymbol{E}_x f_i(X_{t(n)})\mathbb{I}_{\{t(n)\geq n\}} \\
= \ & \boldsymbol{E}_x \mathbf{E}_x [f_i(X_{t(n)})|\mathfrak{F}_{n-1}]\mathbb{I}_{\{t(n)\geq n\}} \\
= \ & \boldsymbol{E}_x \mathbf{E}_{X_{n-1}} f_i(X_{t(n)})\mathbb{I}_{\{t(n)\geq n\}} \\
= \ & \boldsymbol{E}_x \{\boldsymbol{E}_{X_{n-1}}[f_i(X_n)\mathbb{I}_{D_n} + f_i(X_{t(n)})\mathbb{I}_{\overline{D_n}}]\}\mathbb{I}_{\{t(n)\geq n\}} \\
\leq \ & \boldsymbol{E}_x \{\boldsymbol{E}_{X_{n-1}}[f_i(X_n)\mathbb{I}_{{}^*D_n} + v_i(X_{t(n)})\mathbb{I}_{\overline{{}^*D_n}}]\}\mathbb{I}_{\{t(\sigma\{n\})\geq n\}} \\
= \ & \boldsymbol{E}_x v_i(X_{n-1})\mathbb{I}_{\{t(n-1)\geq n\}}.
\end{aligned}$$

Hence

$$\begin{aligned}
\boldsymbol{E}_x f_i(X_{t(n)}) & \leq \boldsymbol{E}_x [f_i(X_{t(n-1)})\mathbb{I}_{\{t(n-1)<n\}} + v_i(X_{n-1})\mathbb{I}_{\{t(n)\geq n\}}] \\
& = \boldsymbol{E}_x f_i(X_{t(n-1)}).
\end{aligned}$$

We have

$$\boldsymbol{E}_x f_i(X_{t(n)}) \leq \boldsymbol{E}_x f_i(X_{t(n-1)}) \leq \boldsymbol{E}_x f_i(X_{t(0)}) = \boldsymbol{E}_x f_i(X_{t^*}).$$

We show that if player $j = 1, 2, \ldots, p$, $j \neq i$, use the strategies $^*\sigma^j$ and the player i any strategy σ^i then $\mathbf{E}_x f_i(X_{t(\widetilde{\sigma})}) \leq \boldsymbol{E}_x f_i(X_{t({}^*\sigma)})$, where

$$\widetilde{\sigma} = ({}^*\sigma^1, \ldots, {}^*\sigma^{i-1}, \sigma^i, {}^*\sigma^{i+1} \ldots, {}^*\sigma^p)^T.$$

Let us consider the difference

$$\begin{aligned}
& \boldsymbol{E}_x f_i(X_{t(\widetilde{\sigma})}) - \boldsymbol{E}_x f_i(X_{t(n)}) \\
= \ & \boldsymbol{E}_x [f_i(X_{t(\widetilde{\sigma})}) - f_i(X_{t(n)})]\mathbb{I}_{\{t(n)>n\}} \\
= \ & \boldsymbol{P}_x\{t(\widetilde{\sigma}) > n\}\left\{\boldsymbol{E}_x[(f_i(X_{t(\widetilde{\sigma})})|\mathbb{I}_{\{t(\widetilde{\sigma})>n\}}] - \boldsymbol{E}_x[v_i(X_{t(n)}))|\mathbb{I}_{\{t(n)>n\}}]\right\} \\
\leq \ & \boldsymbol{P}_x\{t(\widetilde{\sigma}) > n\}\{\boldsymbol{E}_x[\sup_{k\geq n}(f_i^+(X_k)] - v_i(x)\}
\end{aligned}$$

for every $n \in \mathbb{N}$. Since we consider SS from $\mathfrak{S}_f^*$ then

$$\boldsymbol{E}_x f_i(X_{t(\tilde{\sigma})}) - \boldsymbol{E}_x f_i(X_{t^*}) \leq 0.$$

□

3.3.6 Examples

The most important example of the monotone rule is the equal majority rule π_r. For $1 \leq r \leq p$ we define $\pi_r(x^1, x^2, \ldots, x^p) = 1$ if $\sum_{i=1}^p x^i \geq r$ and 0 otherwise. Let us consider the two-person voting game on observation of a homogeneous Markov chain with the equal majority rule. Then we have $p = 2$ and two cases $r = 1$ and $r = 2$. Assume that the Markov chain $(X_n, \mathfrak{F}_n, \boldsymbol{P}_x)$, $n = 0, 1, 2, \ldots$ with the state space $\mathbb{E} = \{0, 1\} \times \{0, 1\}$ and $\boldsymbol{P}_x = \boldsymbol{P}_{(x_1, x_2)} = \boldsymbol{P}_{x_1} \times \boldsymbol{P}_{x_2}$, where $\boldsymbol{P}_{x_i}$ is given by the transition probability matrix

$$\begin{bmatrix} p & q \\ q & p \end{bmatrix}$$

$p \geq 0, q \geq 0, p+q = 1$. The payoff functions of the players are $f_i(x) = f_i(x_1, x_2) = x_i$, $i = 1, 2$. Let us formulate the optimality equations and their solution for both cases.

[r=1] The Nash values of the game $(v_1(x), v_2(x))$ fulfill the following equation

$$\begin{aligned} v_1(x) &= \boldsymbol{E}_x(f_1(X_1) - v_1(X_1))^+ \mathbb{I}_{\{f_2(X_1) < v_2(X_1)\}} \\ &\quad + \boldsymbol{E}_x(f_1(X_1) - v_1(X_1)) \mathbb{I}_{\{f_2(X_1) \geq v_2(X_1)\}} + \boldsymbol{E}_x v_1(X_1) \end{aligned}$$

$$\begin{aligned} v_2(x) &= \boldsymbol{E}_x(f_2(X_1) - v_2(X_1))^+ \mathbb{I}_{\{f_1(X_1) < v_1(X_1)\}} \\ &\quad + \boldsymbol{E}_x(f_2(X_1) - v_2(X_1)) \mathbb{I}_{\{f_1(X_1) \geq v_1(X_1)\}} + \boldsymbol{E}_x v_2(X_1). \end{aligned}$$

The solution of these equations are the functions

$$v_1(x_1, x_2) = v_2(x_2, x_1) = \begin{cases} \frac{q}{1-p^2} & \text{if } (x_1, x_2) = (0, 0), \\ \frac{q+2pq}{1+p} & \text{if } (x_1, x_2) = (0, 1), \\ \frac{2p}{1+p} & \text{if } (x_1, x_2) = (1, 0), \\ \frac{q^2+p^2+p}{1+p} & \text{if } (x_1, x_2) = (1, 1), \end{cases}$$

and the stopping sets are $C^1 = \{(1, 0), (1, 1)\}$ and $C^2 = \{(0, 1), (1, 1)\}$. It is easy to check that for $p \geq q$, we have $v_1(1, 0) \geq v_1(0, 1) \geq v_1(0, 0) \geq v_1(1, 1)$. For $p = q = 1/2$ the equilibrium values $v_1(x) = v_2(x) = 2/3$ for both players are equal and independent of the initial state of the Markov chain.

[r=2] In this case the equilibrium values fulfill equations

$$w_1(x) = \boldsymbol{E}_x(f_1(X_1) - w_1(X_1))^+ \mathbb{I}_{\{f_2(X_1) \geq w_2(X_1)\}} + \boldsymbol{E}_x w_1(X_1)$$
$$w_2(x) = \boldsymbol{E}_x(f_2(X_1) - w_2(X_1))^+ \mathbb{I}_{\{f_1(X_1) \geq w_1(X_1)\}} + \boldsymbol{E}_x w_2(X_1).$$

The solutions of these equations are $w_1(x) = w_2(x) = 1$ for every $x \in \mathbb{E}$ and the stopping sets are $C^1 = C^2 = \{(1, 1)\}$.

3.3.7 Players knowledge vs. their profit

Gensbittel et al. (2023) extended the problem of prophet inequality in a competitive setting (v. Rinott and Samuel-Cahn (1992), Kennedy and Kertz (1997), Hill and Kertz (1992)) for references to the paper formulated the problem). At every period, a new realization of a random variable with a known distribution arrives, which is publicly observed. Then, two players simultaneously decide whether to pick an available value or to pass and wait until the next period (ties are broken uniformly at random). As soon as a player gets a value, he leaves the market, and his payoff is the value of this realization. In the first variant, namely the "no recall" case, agents can only bid at each period for the current value. In a second variant, the "full recall" case, agents can also bid for any of the previous realizations that have not already been selected. For each variant, we study the subgame-perfect Nash equilibrium payoffs of the corresponding game. More specifically, we give a full characterization in the full-recall case and show in particular that the expected payoffs of the players at any equilibrium are always equal, whereas in the no-recall case the set of equilibrium payoffs typically has full dimension. Regarding the welfare at equilibrium, surprisingly the best equilibrium payoff a player can have may be strictly higher in the no-recall case. However, the sum of equilibrium payoffs is weakly larger when the players have full recall. Finally, we show that in the case of 2 arrivals and arbitrary distributions, the prices of Anarchy and Stability in the no-recall case are at most 4/3, and this bound is tight.

Shmaya and Solan (2004) considered two player discrete time nonzero-sum-game with randomize stopping times (v. Yasuda (1985), Rosenberg et al. (2001), Neumann et al. (2002)) as the players' strategies, i.e. two players are using randomized stopping strategies $x = (x_n)$ and $y = (y_n)$, respectively. The expected payoff of the i-th player (under the strategies (x, y)) is equal to $\gamma_i(x, y) = \boldsymbol{E}_{x,y}\left\{R^i_{Q,\theta}1_{\{\theta<\infty\}}\right\}$, where $R^i_{Q,n}$ are the coordinates of the process R_n, $n \geq 0$ that appears in the definition of the game, θ is the first stage in which at least one player stops, and Q is the set of players that stop at θ. The authors proved that if $\sup_{n\geq 0}\|R_n\| \in L^1$, then there is for every $\varepsilon > 0$ an ε-Nash equilibrium, i.e. there is a pair $(x^*(\varepsilon), y^*(\varepsilon))$ such that $\gamma_1((x^*(\varepsilon), y^*(\varepsilon)) \geq \gamma_1(x, y^*(\varepsilon)) - \varepsilon$ and $\gamma_2((x^*(\varepsilon), y^*(\varepsilon)) \geq \gamma_2(x^*(\varepsilon), y) - \varepsilon$, for every (x, y). The proof is based on a stochastic version of Ramsey's (1929) theorem, which allows a reduction of the problem to a simple game with finite state space (cf. Shmaya et al. (2003)).

Part III

Applications of Multiple-Stopping Models

I should like to see this problem (v. Problem 1.1) solved before I die.

Herbert Robbins, 26 June 1990[a]

[a]International Conference on Search and Selection in Real Time, Amherst, MA, 21-27 June 1990

A MULTIPLE stopping problem for stochastic processes is a type of decision-making problem where an agent must decide when to stop a stochastic process based on observations of the process. The agent can stop the process at multiple points in time, and the goal is to choose the stopping times that optimize some measure of performance. These problems can arise in various fields, such as finance, engineering, and operations research. For example, in finance, a multiple-stopping problem might involve deciding when to sell a stock based on its price movements, while in engineering, it might involve deciding when to shut down a manufacturing process based on the quality of the products it produces. The solutions to these problems are often based on dynamic programming or optimal stopping theory.

An example of a multiple-stopping problem in finance is a portfolio management problem, where an investor must decide when to sell a stock to maximize their returns. The investor has a portfolio of a single stock and wants to sell it at some point in time. The price of the stock follows a stochastic process and the investor can observe the stock price at different points in time. The investor's goal is to choose the time to sell the stock to maximize the expected return. The investor can stop the process of holding the stock at any time and sell it. The investor has the option to wait or sell the stock at any time. The performance measure is the expected profit.

The solutions of these problems are based on the concept of optimal stopping theory, where the value of the portfolio is modeled as a function of time and stock price, and the investor must find the optimal time to sell the stock that maximizes the expected value of the portfolio. The value of the portfolio, at any point in time, depends on the current stock price, the expected future stock price, and the risk involved with keeping the stock for longer.

The multiple-stopping problem in engineering refers to the scenario where there are multiple possible points at which a process or system can be stopped and the goal is to determine the optimal stopping point that maximizes some performance metric. One example of this in the field of mechanical engineering is the optimal stopping of a drilling operation. In drilling, the goal is to reach a certain depth as efficiently as possible while minimizing wear and tear on the drilling equipment. There may be multiple-stopping points at different depths where the drilling operation could be halted, and the decision of which point to stop at will depend on factors such as the rate of drilling progress and the condition of the equipment. An optimal stopping algorithm could be used to determine the best point at which to stop the drilling operation in order to maximize overall efficiency and equipment lifespan.

Another example of a multiple-stopping problem in management is the decision on when to launch a new product. A company may have multiple options for when to release a new product, such as during a specific season, after a competitor's product has been released, or after certain product features have been developed. The decision of when to launch the product will depend on factors such as market conditions, the readiness of the product, and the company's resources. An optimal stopping algorithm could be used to determine the best time to launch the product based on factors such as the expected demand for the product, the potential for market saturation, and the impact on the company's resources. For example, launching too early could result in low demand and insufficient resources, while launching too late could result in lost market share. An optimal stopping algorithm would help identify the point at which the potential benefits of launching the product outweigh the potential costs in order to maximize the overall success of the product launch.

In operations research, the multiple-stopping problem is often encountered in the context of dynamic programming and Markov decision processes. An example of this is the inventory management problem. A company may have multiple options for when to reorder inventory, such as when inventory levels reach a certain threshold or when the lead time for a new order is shorter. The decision of when to reorder inventory will depend on factors such as inventory cost, stockout cost, and holding costs.

An optimal stopping algorithm could be used to determine the best time to reorder inventory based on these factors (v. Lovejoy (1991) for review of Markov Decision Processes models with partial observation, which can be used to solve multiple optimal stopping problems, when the description of the process is not complete and the observer can reconstruct the model from past observations). For example, reordering too early could result in high holding costs, while reordering too late could result in stockouts and lost sales. An optimal stopping algorithm would help identify the point at which the potential benefits of reordering inventory outweigh the potential costs, in order to maximize the overall efficiency of the inventory management system.

Statistical analysis can be used in the context of multiple-stopping of stochastic processes by providing estimates of the underlying probability distributions and parameters that govern the process. These estimates can then be used to inform the decision-making process and help identify the optimal stopping point. However, a change-point analysis is a statistical method used to identify significant changes in the behavior of a time series or other data set. It is also called the *disorder problem*. Multiple-stopping methods can be applied to the stochastic disorder problem to determine the number and location of change points in a data set. There are several different stopping methods that can be used to determine the number and location of change points in a data set. These include:

1. Fixed sample size: A fixed number of data points are used to determine the change points.
2. Information Criteria: A statistical measure such as Akaike's Information Criterion (AIC) or Bayesian Information Criterion (BIC) is used to determine the number and location of change points.

3. Cross-validation: The data are split into a training set and a validation set, and the change points are identified using the training set. The identified change points are then evaluated using the validation set.
4. False discovery rate (FDR): A threshold for **FDR** is set and change points are identified based on this threshold.
5. Bootstrap: A bootstrap procedure is used to estimate the uncertainty of the change points.

Each method has its advantages and disadvantages, and the choice of which method to use will depend on the specific characteristics of the data set and the goals of the analysis. It is also important to note that depending on the problem and data you may be able to use other more specific methods such as Bayesian change point analysis, segmentation methods, likelihood ratio tests, etc.

Intruder detection and the disorder problem, also known as the change-point problem, are two distinct but related concepts. Intruder detection refers to the process of identifying the presence of an unauthorized person or object in a secure area. This can be done using various methods such as surveillance cameras, motion sensors, or other forms of monitoring. The goal of intruder detection is to identify and respond to potential security breaches in real time. Most of the technical devices are the source of the signals of stochastic character.

The disorder problem, on the other hand, is a statistical method used to identify significant changes in the behavior of a time series or other data set. It is used to detect changes in a system or process, such as changes in the mean, variance, or distribution of a data set. The goal of the disorder problem is to identify when a change has occurred and to estimate the parameters of the new system or process. The solution to the disorder problem can be the basic tools for the analysis of signals from the surveillance system.

In such a way these concepts are related in certain situations, such as in the case of detecting an intrusion in a system is identified by change point detection in signal processes coming from the surveillance devices. For example, an intruder detection system could be used to detect changes in the behavior of a monitored process, such as an unusual increase in temperature or pressure, which could indicate an intrusion or malfunction. The models of Partial Observation Markov Decision Processes have been proposed to create the operational supporting of the automatic defender of the network system. The recent articles in these direction by Hammar and Stadler (2022) has proposed significant efforts to automate security frameworks and the process of obtaining effective security policies. Examples of this research include: automated creation of threat models (v. Johnson et al. (2018)), computation of defender policies using dynamic programming and control theory proposed by Rasouli et al. (2014), Miehling et al. (2019), computation of exploits and corresponding defenses through evolutionary methods (cf. Bronfman-Nadas et al. (2018)), identification of infrastructure vulnerabilities through attack simulations and threat intelligence as it is shown by Wagner et al. (2016a) and Wagner et al. (2016b), computation of defender policies through game-theoretic methods (v. Alpcan and Başar (2011), Sarıtaş et al. (2019), and use of machine learning techniques to estimate model parameters and policies as by Hammar and Stadler (2021, 2020). The idea should strengthen the classical approach in the organization's security strategy, which has traditionally been defined, implemented, and updated by domain experts as it is shown by Fuchsberger (2005).

These problems can be solved by solving the Hamilton-Jacobi-Bellman (HJB) equation which is a partial differential equation (PDE) that describes the evolution of the value function over time. The solution to this equation gives the optimal stopping rule for the decision-makers i.e. investors, engineers. The Hamilton-Jacobi-Bellman (HJB) equation is typically used to analyze problems in continuous time, where the process being observed or controlled evolves continuously over time. The equation describes the evolution of the value function, which is a measure of the performance of the system, over time. However, the HJB equation can also be adapted to discrete-time models by replacing the derivatives with finite differences. In this case, instead of a continuous-time variable, the HJB equation is defined on a discrete-time grid. It is called a "Discrete Time HJB" equation (v. Seierstad (2009), Bäuerle and Rieder (2011)).

For a multiple-stopping problem in a discrete-time model, the HJB equation can be solved using dynamic programming techniques, such as backward induction. The value function is calculated at each point in time, by considering all possible stopping actions and the corresponding expected future rewards.

In summary, the HJB equation can be used to analyze multiple-stopping problems in both continuous- and discrete-time models; however, the equation and the solution methods will be different in each case. The equation is a result of the dynamic programming theory pioneered in the 1950s by Richard Bellman and co-workers (v. Bellman (1954, 1957), Bellman and Dreyfus (1962). The connection to the Hamilton–Jacobi equation from classical physics was first drawn by Rudolf Kalman (1963). In discrete-time problems, the corresponding difference equation is usually referred to as the Bellman equation.

4

Sequential Methods of Statistics

Motto:

Statistical tasks are key source of stochastic optimization problems. Through Sequential Steps to Deeper Understanding: Uncovering Connections in Statistics with Sequential Methods

In STATISTICAL investigation the size of sample is a crucial element of successful inference on reasonable cost. The quality of the investigation of the stochastic phenomena based on sample depends on the data. This observation was the birth of the sequential methods (v. Section 1.6.1). The sequential method of statistics involves collecting data, analyzing it as it becomes available, and making decisions on drawing conclusions based on intermediate results or continue collecting observation of the phenomena. This approach is particularly useful in situations where resources are limited or when time is of the essence, such as clinical trials (see Sect. 1.6.5), quality control in manufacturing or online data analysis. One common application of the sequential method is sequential hypothesis testing, where hypotheses are tested as data become available, allowing for early termination of experiments if significant results are observed.

The advantages of the sequential method include increased flexibility, the ability to adapt to changing conditions, and potentially faster decision-making compared to traditional fixed-sample methods. The size of the sample refers to the number of observations or data points collected for analysis in a statistical investigation. Larger sample sizes generally provide more reliable estimates of population parameters and reduce the variability of estimates. Increasing the sample size can improve the precision of statistical estimates and increase the power of hypothesis tests, making it easier to detect true effects or differences. However, larger sample sizes also typically require more resources (time, money, manpower) for data collection and analysis.

In a natural way, the idea of the sequential method interacts with the idea of the sample size. The sequential method can be applied regardless of the sample size, but the impact of the sample size on the effectiveness of sequential methods varies. In situations where sample sizes are relatively small, the sequential method can be particularly advantageous. Allows adaptive sampling strategies, where additional data can be collected if initial results are inconclusive or ambiguous. With larger sample sizes, the benefits of the sequential method may be less pronounced, as the increased precision from a large sample can make interim analysis less necessary.

It is not difficult to observe that there are tasks where decisions need to be made at different stages of data collection, which means dealing with different sample sizes. In practice, sequential methods are used in situations where data collection occurs step by step and decisions are made based on interim results, with the possibility of further

DOI: 10.1201/9781003407102-4

adjusting the course of the study. This approach is applicable in various fields such as clinical trials, quality control, financial risk management, or production process optimization. In clinical trials, researchers can use sequential methods to monitor the effectiveness of a new drug, making decisions about continuing the study, stopping it, or changing the protocol based on interim results. In quality control in manufacturing, sequential methods can be applied to monitor production processes, making decisions about adjusting production parameters depending on interim results.

In each of these cases, decisions need to be made at various stages of the data collection process, and sequential methods allow for flexible adjustment of the study based on the results obtained. However, even with large samples, the sequential method can still be valuable for certain applications, such as monitoring processes for quality control or detecting rare events in real-time data streams.

In conclusion, the sequential method of statistics offers flexibility and efficiency in data analysis, particularly in situations with limited resources or time constraints. The size of the sample influences the precision and power of statistical estimates, and larger samples generally provide more reliable results. The interaction between the sequential method and the sample size depends on the specific context of the statistical investigation and the objectives of the analysis.

Some sequential inference models naturally include the problem of selecting stopping points as a proper element of the procedure's construction. In the next part of this chapter, selected examples will be provided. In Section 4.1, we will discuss the problem of observation selection for statistical inference purposes, followed by presenting the task of limiting the number of Bernoulli trials in cases of inferring the probability of success in Section 4.2. In Section 4.3, we will discuss the task of change-point detection, followed by the generalized problem of detecting multiple changes in real time. Sections 4.3 and 4.4 contain detailed formulations of the problem and its transformation into the task of choosing optimal stopping points, which serve as change-point estimators.

4.1 Sequential Sampling

In statistics, population sampling is an essential part of statistical analysis. A sampling method should be fast, reliable, and of low cost. One of the possible approaches is sequential sampling methods, which can offer efficiency and flexibility in various data collection scenarios. Here, we will show how sampling methods can be considered as multiple-stopping problems.

Consider a finite population of units, $1, 2, \dots, N$. A sample can be described by the vector $\boldsymbol{s} = (s_1, \dots, s_N)$, where $s_j = 1$ if the unit j is in the sample and $s_j = 0$ otherwise. Let p_j, $j = 1, 2, \dots, N$, be inclusion probabilities, where p_j is the probability that a unit j is included in $\boldsymbol{s}$, that is, $p_j = \boldsymbol{P}(s_j = 1)$. For a sample of size k, we have $\sum_{j=1}^{N} p_j = k$. From the multiple stopping point of view, every inclusion of a unit can be treated as a stop.

Example 4.1 (Bernoulli sampling). *Assume that all units of the population have an equal probability of being included in the sample, that is, $p_j = p$ for $j = 1, 2, \dots, N$. If a sampling process is organized in such a way that for each unit j of the population, an independent Bernoulli trial is performed which determines whether the unit will be included in the sample or not. This sampling process is called Bernoulli sampling. The sample size k follows a binomial distribution. In this case, the multiple-stopping rule $\boldsymbol{\tau} = (\tau_1, \dots, \tau_k)$ can be written as follows*

$$\begin{aligned} \tau_1 &= \min\{m_1 : 1 \leq m_1 \leq N, s_{m_1} = 1\}, \\ \tau_i &= \min\{m_i : \tau_{i-1} < m_i \leq N, s_{m_i} = 1\}, \quad i = 2, \dots, k. \end{aligned} \tag{4.1}$$

Bernoulli sampling can also be considered for infinite populations. This means that the sample size k is fixed and τ_k follows a Negative Binomial (Pascal) distribution, where

$$\begin{aligned} \tau_1 &= \min\{m_1 : 1 \leq m_1, s_{m_1} = 1\}, \\ \tau_i &= \min\{m_i : \tau_{i-1} < m_i, s_{m_i} = 1\}, \quad i = 2, \dots, k. \end{aligned}$$

Example 4.2 (Poisson sampling). *If a sampling process is organized in such a way that for each unit j of the population, an independent Bernoulli trial with different probability p_j is performed which determines whether the unit will be included in the sample or not. This sampling process is called Poisson sampling. The corresponding multiple-stopping rule $\boldsymbol{\tau} = (\tau_1, \dots, \tau_k)$ would be the same as in (4.1).*

Using this approach, it is possible to express other well-known sampling methods, such as stratified and systematic sampling. The application of multiple-stopping rules gives the advantage of having the possibility to optimize a predefined gain function. The gain function may include a cost of observation and some other characteristics like recruiting a well spread sample (see, for example, Hof et al. (2014)).

Let $\xi_1, \xi_2, \dots$ be a sequence of independent random variables with a known distribution, which are observed sequentially. We would like to choose k of the variables to make a sample so that a statistical distance between the empirical distribution function based on the sample and the known distribution is minimal. For example, if we would like to minimize the Kullback-Leibler (or cross-entropy) divergence, the problem will be equivalent to maximizing the corresponding log-likelihood. After k, $k \geq 2$, stops at times $m_1, m_2, \dots, m_k$, $1 \leq m_1 < m_2 < \dots < m_k$, we get a gain

$$Z_{m_k} = \ln f(\xi_{m_1}) + \ln f(\xi_{m_2}) + \dots + \ln f(\xi_{m_k}),$$

where $f(\cdot)$ is a common pdf (or pf) of random variable ξ. If we would like to incorporate a cost per observation of one unit c, then the gain function may be written in the following way

$$Z_{m_k} = \ln f(\xi_{m_1}) + \ln f(\xi_{m_2}) + \dots + \ln f(\xi_{m_k}) - c m_k.$$

In this case, we would be able to construct a sequential procedure that would allow us to sample from any stream of data in an optimal way.

Example 4.3 (Sampling from a normal population). *Suppose that we would like to sequentially sample from a finite normal population $N(\mu, \sigma^2)$ with the aim of maximizing the log-likelihood as described above. In this case,*

$$\begin{aligned} Z_{\boldsymbol{m}_k} &= \ln f(\xi_{m_1}) + \ln f(\xi_{m_2}) + \cdots + \ln f(\xi_{m_k}) \\ &= \sum_{i=1}^{k} \ln \left(\frac{1}{\sqrt{2\pi}\sigma} \exp\left\{ -\frac{(\xi_{m_i} - \mu)^2}{2\sigma^2} \right\} \right) \\ &= -k \ln\left(\sqrt{2\pi}\sigma\right) - \frac{1}{2} \sum_{i=1}^{k} \left(\frac{\xi_{m_i} - \mu}{\sigma} \right)^2, \end{aligned}$$

which is equivalent to maximizing

$$Z_{\boldsymbol{m}_k} = -\chi^2_{1,m_1} - \cdots - \chi^2_{1,m_k},$$

where χ^2_1 follows a χ^2_1-distribution. Then

$$\begin{aligned} v^{n,1} &= \frac{2}{\sqrt{\pi}} \gamma\left(\frac{3}{2}, v^{n-1,1}\right) - \frac{v^{n-1,1}}{\sqrt{\pi}} \gamma\left(\frac{1}{2}, v^{n-1,1}\right) + v^{n-1,1}, \\ v^{n,k-i+1} &= \frac{2}{\sqrt{\pi}} \gamma\left(\frac{3}{2}, v^{n-1,k-i+1} - v^{n-1,k-i}\right) \\ &\quad - \frac{v^{n-1,k-i+1} - v^{n-1,k-i}}{\sqrt{\pi}} \gamma\left(\frac{1}{2}, v^{n-1,k-i+1} - v^{n-1,k-i}\right) + v^{n-1,k-i+1}, \end{aligned}$$

where γ is the lower incomplete gamma function, $v^{k-i,k-i+1} = \infty$, $k - i + 1 \le n \le N$, $i = k, \ldots, 1$. The optimal multiple-stopping rule $\boldsymbol{\tau}^* = (\tau_1^*, \ldots, \tau_k^*)$ is

$$\begin{aligned} \tau_1^* &= \min\{m_1 \,:\, 1 \le m_1 \le N - k + 1, \chi^2_{1,m_1} \le v^{N-m_1,k} - v^{N-m_1,k-1}\}, \\ \tau_i^* &= \min\{m_i \,:\, m_{i-1} < m_i \le N - k + i, \chi^2_{1,m_i} \le v^{N-m_i,k-i+1} - v^{N-m_i,k-i}\}, \\ &\quad i = 2, \ldots, k-1, \\ \tau_k^* &= \min\{m_k \,:\, m_{k-1} < m_k \le N, \chi^2_{1,m_k} \le v^{N-m_k,1}\}. \end{aligned}$$

Table 4.1 presents the values $v^{n,j}$ with $n \le 10$ steps and $j \le 7$ stops left for the chi-squared distribution with 1 degree of freedom.

4.2 Success Runs in Bernoulli Trials

In this section, we will consider an extension of the problem studied by Starr (1972) to a case when multiple stops are allowed (see Nikolaev (1998)). Let $\xi_1, \ldots, \xi_N$ be independent and identically distributed Bernoulli random variables, $\boldsymbol{P}(\xi_i = 1) = p = 1 - \boldsymbol{P}(\xi_i = 0) = 1 - q$, $0 < p < 1$. Let R^i_j denote the length of an interval of

TABLE 4.1
The values for χ_1^2.

n	$v^{n,1}$	$v^{n,2}$	$v^{n,3}$	$v^{n,4}$	$v^{n,5}$	$v^{n,6}$	$v^{n,7}$
1	1.0000						
2	0.5849	2.0000					
3	0.4032	1.2972	3.0000				
4	0.3009	0.9476	2.0747	4.0000			
5	0.2357	0.7354	1.5764	2.8924	5.0000		
6	0.1910	0.5930	1.2582	2.2623	3.7377	6.0000	
7	0.1585	0.4911	1.0365	1.8446	2.9897	4.6031	7.0000
8	0.1341	0.4151	0.8734	1.5453	2.4793	3.7486	5.4841
9	0.1152	0.3565	0.7488	1.3202	2.1054	3.1519	4.5324
10	0.1002	0.3100	0.6507	1.1449	1.8191	2.7070	3.8553

successful trials, that is, the number of successive ones counting backward from i-th observation,

$$R_j^i = \sum_{n=j}^{i} \mathbb{I}(\xi_n = \cdots = \xi_i = 1).$$

Define the total gain after k, $k \geq 2$, stops in the following way

$$Z_{\boldsymbol{m}_k} = R_1^{m_1} + R_{m_1+1}^{m_2} + \cdots + R_{m_{k-1}+1}^{m_k} - cm_k, \quad 1 \leq m_1 < m_2 < \cdots < m_k,$$

where c is a cost paid for each observation, $c < 1$. If $c \geq 1$, then $Z_{\boldsymbol{m}_k} \leq 0$ and the optimal strategy is trivial: stop at the beginning of the sequence. Put

$$v^N = \sup_{\tau \in \mathfrak{S}_1^N} \boldsymbol{E} Z_\tau.$$

Let us find an optimal stopping rule $\boldsymbol{\tau}^* = (\tau_1^*, \ldots, \tau_k^*) \in \mathfrak{S}_1^N$ and the value of the problem v^N. Let us rewrite $Z_{\boldsymbol{m}_k}$, the total gain after k stops, as

$$Z_{\boldsymbol{m}_k} = \sum_{i=1}^{k} g_{m_{i-1}+1}^{m_i},$$

where $g_{m_{i-1}+1}^{m_i} = R_{m_{i-1}+1}^{m_i} - c(m_i - m_{i-1})$ is the gain after i-th series, $m_0 = 0$.

Let $\mathcal{F}_{m_i}$ denote σ-algebra generated by random variables $\xi_1, \ldots, \xi_{m_i}$. Using equations (2.19) and (2.20), we can write

$$V_{\boldsymbol{m}_k}^N = \max\left\{\sum_{i=1}^{k-1} g_{m_{i-1}+1}^{m_i} + g_{m_{k-1}+1}^{m_k}, \boldsymbol{E}_{\boldsymbol{m}_k} V_{\boldsymbol{m}_{k-1}, m_k+1}^N\right\}. \tag{4.2}$$

From the independence of the random variables, we have

$$V_{\boldsymbol{m}_k}^N = \sum_{i=1}^{k-1} g_{m_{i-1}+1}^{m_i} + u_1^N(m_k, R_{m_{k-1}+1}^{m_k}),$$

with, for $m_{k-1} < m_k \leq N$,

$$u_1^N(m_k, R) = \max\left\{g_{m_{k-1}+1}^{m_k}, u_1^N(m_k+1, R+1)p + u_1^N(m_k+1, 0)q\right\}, \quad (4.3)$$

where $u_1^N(N+1, R+1) = -\infty$.

Next, from (2.19), (2.20), and the independence of $\xi_1, \ldots, \xi_N$, we obtain

$$\begin{aligned} X_{\boldsymbol{m}_{k-1}}^N &= \sum_{i=1}^{k-1} g_{m_{i-1}+1}^{m_i} + \boldsymbol{E}u_1^N\left(m_{k-1}+1, R_{m_{k-1}+1}^{m_{k-1}+1}\right) \\ &= \sum_{i=1}^{k-1} g_{m_{i-1}+1}^{m_i} + u_1^N(m_{k-1}+1), \end{aligned}$$

where, for $m_{k-2} < m_{k-1} \le N-1$,

$$\begin{aligned} u_1^N(m_{k-1}+1) &= \boldsymbol{E}u_1^N\left(m_{k-1}+1, R_{m_{k-1}+1}^{m_{k-1}+1}\right) \\ &= u_1^N(m_{k-1}+1, 1)p + u_1^N(m_{k-1}+1, 0)q. \end{aligned} \quad (4.4)$$

Similarly,

$$\begin{aligned} V_{\boldsymbol{m}_i}^N &= \sum_{j=1}^{i-1} g_{m_{j-1}+1}^{m_j} + u_{k-i+1}^N\left(m_i, R_{m_{i-1}+1}^{m_i}\right), \\ X_{\boldsymbol{m}_{i-1}}^N &= \sum_{j=1}^{i-1} g_{m_{j-1}+1}^{m_j} + u_{k-i+1}^N(m_{i-1}+1), \end{aligned}$$

where

$$\begin{aligned} u_{k-i+1}^N(m_i, R) =& \max\{g_{m_{i-1}+1}^{m_i} + u_{k-i}^N(m_i+1), \\ & u_{k-i+1}^N(m_i+1, R+1)p + u_{k-i+1}^N(m_i+1, 0)q\}, \quad (4.5) \\ u_{k-i}^N(m_i+1) =& \boldsymbol{E}u_{k-i}^N(m_i+1, R_{m_i+1}^{m_i+1}) \\ =& u_{k-i}^N(m_i+1, 1)p + u_{k-i}^N(m_i+1, 0)q, \quad (4.6) \end{aligned}$$

$u_{k-i+1}^N(N-k+i+1, R+1) = -\infty$, $m_{i-1} < m_i \le N-k+i$, $m_{i-2} < m_{i-1} \le N-k+i-1$, $i = k-1, \ldots, 2$,

Finally, if $i = 1$, then

$$V_{m_1}^N = u_k^N(m_1, R_1^{m_1}),$$

where

$$\begin{aligned} u_k^N(m_1, R) =& \max\{g_1^{m_1} + u_{k-1}^N(m_1+1), \\ & u_k^N(m_1+1, R+1)p + u_k^N(m_1+1, 0)q\}, \end{aligned} \quad (4.7)$$

$u_k^N(N-k+2, R+1) = -\infty$, $1 \le m_1 \le N-k+1$. From Theorem 2.3, we obtain the value of the problem

$$v^N = \boldsymbol{E}V_1^N = u_k^N(1,1)p + u_k^N(1,0)q.$$

This means that the sequences $\{V_{m_i}^N\}$ and $\{X_{m_i}^N\}$ are described by the auxiliary sequences $\{u_l^N(n)\}$, which can be calculated from (4.3), (4.4) and (4.6). After having seen n observations, $u_l^N(n)$ can be interpreted as the optimal expected gain if l stops are left.

We will now describe the multiple optimal stopping rule. Let

$$\Gamma_1 = \{(t,R) : R \le t \le N-k+1, t \ge 1, R \ge R_1^*\},$$

where

$$R_1^* = V^N(t+1, R+1)p + V^N(t+1, 0)q - u_{k-1}^N(t+1) + ct.$$

Similarly, for $i = 2, \dots, k$,

$$\Gamma_i = \{(t,R) : R \le t-j+1 \le N-k-j+i+1, i \le j \le t \le N-k+i, R \ge R_i^*\},$$

where

$$R_i^* = u_{k-i+1}^N(t+1, R+1)p + u_{k-i+1}^N(t+1,0)q + c(t-j+1) - u_{k-i}^N(t+1), \quad u_0^N = 0.$$

Then, from Theorem 2.2, (4.2), (4.3), (4.5) and (4.7), we obtain the multiple optimal stopping rule $\boldsymbol{\tau}^* = (\tau_1^*, \dots, \tau_k^*)$:

$$\tau_i^* = \min\{m_i > m_{i-1} : (m_i, R_{m_{i-1}+1}^{m_i}) \in \Gamma_i\}, \quad m_0 = 0, \quad i = 1, 2, \dots, k, \tag{4.8}$$

on the set $F_{i-1} = \{\omega : \tau_1^* = m_1, \dots, \tau_{i-1}^* = m_{i-1}\}$, $F_0 = \Omega$.

Finally, we will give a formula that allows us to calculate the limiting expected gain.

Theorem 4.1. *Let* $R = \min\{R' > 0 : p^{R'+1} \le c\}$. *Then*

$$\lim_{N\to\infty} v^N = kR - \frac{kc(1-p^R)}{qp^R}.$$

The theorem can be proved using the following results. First, it was shown by Starr (1972) that

$$\lim_{N\to\infty} u_1^N = R - \frac{c(1-p^R)}{qp^R}.$$

Second, we can see that, for $2 \le m \le k$,

$$u_m^N = \sup_{1\le t\le N-m+1} \boldsymbol{E}(R_1^t - ct + u_{m-1}^{N-t}) \le \sup_{1\le t\le N} \boldsymbol{E}(R_1^t - ct) + u_{m-1}^{N-t} \le u_1^N + u_{m-1}^N.$$

Finally, it is clear that, for large enough N,

$$u_m^N \ge u_1^{\lfloor \frac{N}{2} \rfloor} + u_{m-1}^{\lfloor \frac{N}{2} \rfloor},$$

where $u_l^n \equiv u_l^n(1)$ is the optimal expected gain for l series of successful runs in an n–step problem, $\lfloor \frac{N}{2} \rfloor$ is the floor function (or integer part) of $\frac{N}{2}$.

4.2.1 The double choice problem in Bernoulli trials

Let us consider in more detail the problem of selecting two success runs in Bernoulli trials.

Note that $u_1^N(t, R_1^t)$ coincides with $x_N(t, R_1^t)$ (see p. 1885 of Starr (1972)), conditional expected gain if the optimal policy for an N-step problem in state (t, R_1^t) is used, $1 \le t \le N$. If $p > c$,

$$x_N = \sup_{0\le t\le N} \boldsymbol{E}(R_1^t - ct) = \sup_{1\le t\le N} \boldsymbol{E}(R_1^t - ct) = x_N(1,1)p + x_N(1,0)q, \quad (4.9)$$

where $R_1^0 = 0$. Then, from (4.4), we have

$$x_N = u_1^N(1) \equiv u_1^N, \quad p > c.$$

In what follows, we can assume that $p > c$. In the converse case, if $p \le c$, then $x_N = 0$. Therefore,

$$v^N \le \sup_{0\le t<N} \boldsymbol{E}(R_1^t - ct + u_1^N(t+1)) \le \sup_{0\le t\le N} \boldsymbol{E}(R_1^t - ct) + u_1^N = 2x_N = 0.$$

Let τ_n^* be the optimal strategy for an n–step problem with one stop. It was shown in Starr (1972) that

$$\tau_n^* = \min\left\{t \ge 1 : R_1^t \ge \frac{p-c}{q} + u_1^{n-t-1}\right\}, \quad (4.10)$$

where $u_1^{-1} = -\frac{p-c}{q}$, $u_1^0 = 0$, and $\tau_n^* = 0$ only if $p \le c$. Let $R^{(n)}$ denote the length of the optimal success run in an n-step problem. Then

$$R^{(n)} \ge \frac{p-c}{q} + u_1^{n-m-1} \text{ on } \{\tau_n^* = m\},\ m \ge 1, \quad (4.11)$$

with

$$R^{(n)} = \min\left\{R : R \text{ is integer},\ R \ge \frac{p-c}{q} + u_1^{n-m-1}\right\} \text{ if } \tau_n^* = m. \quad (4.12)$$

The following lemmas will give several properties of $R^{(n)}$ and sequences u_1 and u_2, which will be needed for constructing stopping sets Γ_1 and Γ_2.

Lemma 4.2. *For* $n = 1, 2, \ldots, N-1$, $0 \le R^{(n+1)} - R^{(n)} \le 1$.

Proof: First, we will show that $u_1^{n+1} - u_1^n \le 1$. Indeed, taking into account that $R_1^t - 1 \le R_1^{t-1}$, $R_1^0 \equiv 0$, we get

$$u_1^n \le \sup_{1\le t\le n} \boldsymbol{E}(R_1^{t-1} - c(t-1) + 1) = \sup_{1\le t\le n-1} \boldsymbol{E}(R_1^t - ct) + 1 = u_1^{n-1} + 1.$$

Using (4.11), (4.12), and the facts that $u_1^m \ge u_1^{m-1}$ and $\boldsymbol{P}(\tau_n^* \le \tau_{n+1}^*) = 1$, we obtain

$$R^{(n+1)} - R^{(n)} < u_1^{n-m} - u_1^{n-l-1} + 1 \le u_1^{n-l} - u_1^{n-l-1} + 1 \le 2$$

on the set $\{\omega : \tau_n^* = l, \tau_{n+1}^* = m\}$, $1 \le l \le m$. Then $R^{(n+1)} - R^{(n)} \le 1$, which proves the right inequality.

Now we will look at the left inequality. Denote $A = \{\omega : R^{(n+1)} < R^{(n)}\}$, $\boldsymbol{P}(A) > 0$. It is clear that $\tau_{n+1}^*(\omega) < \tau_n^*(\omega)$ on A, but $\boldsymbol{P}(\tau_n^* \le \tau_{n+1}^*) = 1$. This means that $\boldsymbol{P}(A) = 0$, and therefore $R^{(n+1)} - R^{(n)} \ge 0$. This completes the proof. □

Lemma 4.3. *For* $n = 1, 2, \ldots, N$, $u_1^n - u_1^{n-1} \le u_1^1$, *where* $u_1^0 = 0$.

Proof: The proof is by induction on n. For $n = 1$, the inequality is obvious. Suppose that the inequality holds for n, $n \le N - 1$, that is, $u_1^n - u_1^{n-1} \le u_1^1$.

Let us show that $u_1^{n+1} - u_1^n \le u_1^1$. From Lemma 4.2, $0 \le R^{(n+1)} - R^{(n)} \le 1$. Put $R^{(n)} = R$. Using (4.3), (4.4) and (4.8), we have

$$\begin{aligned} u_1^n =& u_1^n(1,1)p + u_1^n(1,0)q, \\ u_1^n(1,1) =& u_1^n(2,2)p + u_1^n(2,0)q, \\ &\vdots \\ u_1^n(R-1,R-1) =& u_1^n(R,R)p + u_1^n(R,0)q, \\ u_1^n(R,R) =& R - cR. \end{aligned}$$

Taking into account that $u_1^n(R,0) = u_1^{n-R} - cR$ (see p. 1887 of Starr (1972)), we obtain

$$u_1^n = (R - cR)p^R + u_1^n(R,0)p^{R-1}q + \ldots + u_1^n(2,0)pq + u_1^{n-1}q - cq.$$

Put now $R^{(n+1)} = R$,

$$u_1^{n+1} = (R - cR)p^R + u_1^{n+1}(R,0)p^{R-1}q + \ldots + u_1^n q - cq.$$

Since $n + 1 - R \le n$,

$$\begin{aligned} u_1^{n+1} - u_1^n &= [u_1^{n+1}(R,0) - u_1^n(R,0)]p^{R-1}q + \ldots + (u_1^n - u_1^{n-1})q \\ &= (u_1^{n+1-R} - u_1^{n-R})p^{R-1}q + \ldots + (u_1^n - u_1^{n-1})q \\ &\le u_1^1 p^{R-1}q + \ldots + u_1^1 q = u_1^1(1 - p^R) < u_1^1. \end{aligned}$$

Next, put $R^{(n+1)} = R + 1$. Then

$$\begin{aligned} u_1^{n+1} &= [R + 1 - c(R+1)]p^{R+1} + u_1^{n+1}(R+1,0)p^R q \\ &\quad + u_1^{n+1}(R,0)p^{R-1}q + \ldots + u_1^n q - cq, \end{aligned}$$

and

$$\begin{aligned} u_1^{n+1} - u_1^n &= [R + 1 - c(R+1)]p^{R+1} - (R - cR)p^R \\ &\quad + u_1^{n+1}(R+1,0)p^R q + (u_1^{n+1-R} - u_1^{n-R})p^{R-1}q + \ldots + (u_1^n - u_1^{n-1})q \\ &\le p^R\left[(R + 1 - c(R+1))p + u_1^{n-R}q - c(R+1)q - (R - cR)\right] \\ &\quad + u_1^1(1 - p^R). \end{aligned} \tag{4.13}$$

By the inductive assumption, $u_1^{n-R} - u_1^{n-R-1} \leq u_1$. From (4.11), we have

$$R - cR \geq (R+1)p - c(R+1) + u_1^{n-R-1}q.$$

Hence,

$$(R + 1 - c(R+1))p - (R - cR) + u_1^{n-R}q - c(R+1)q \leq u_1^1.$$

Combining this result with (4.13), we get $u_1^{n+1} - u_1^n \leq u_1^1$. This concludes the proof. □

The following lemma is closely related to lemmas (A), (B), and (C) in Starr (1972), where similar results are shown for a problem with one stop.

Lemma 4.4. *For the state (m, R_1^m) of an N-step problem, the following results hold:*

(1) $u_2^N(m, 0) = v^{N-m} - cm$,

(2) $u_2^N(m, R_1^m) \leq u_2^{N+1}(m, R_1^m)$,

(3) $u_2^N(m, 0) \geq u_2^N(m+1, 0)$, $m < N - 1$.

Proof: Each part of theorem is proved separately.

(1) This equality reflects the fact that if in an N-step problem our m-th observation is 0, then our expected gain is the same as the expected gain of an $(N-m)$-step problem minus the cost paid for m observations.
More formally, we have

$$\begin{aligned} u_2^N(m,0) &= \sup_{m \leq t < N} \boldsymbol{E}_m(R_1^t - ct + u_1^N(t+1)) \\ &= \sup_{0 \leq t < N-m} \boldsymbol{E}(R_1^t - ct + u_1^{N-m}(t+1)) - cm. \end{aligned}$$

From Lemma 4.3,

$$\begin{aligned} v^{N-m} &= \sup_{1 \leq t < N-m} \mathbf{E}(R_1^t - ct + u_1^{N-m}(t+1)) \\ &= u_2^{N-m}(1,1)p + u_2^{N-m}(1,0)q \geq u_1^1 + u_1^{N-m-1} \geq u_1^{N-m}. \end{aligned}$$

Hence,

$$u_2^N(m,0) = \sup_{1 \leq t < N-m} \mathbf{E}(R_1^t - ct + u_1^{N-m}(t+1)) - cm = v^{N-m} - cm,$$

which proves the first part of the lemma.

(2) The inequality shows that the expected gain is a nondecreasing function of the number of observations (or steps) in the problem. This result does not depend on the state of the problem, and easily follows from the definition of u_2.

(3) It can easily be checked that this inequality follows from parts (1) and (2) of this lemma.

□

The following theorem shows how to construct stopping sets Γ_1 and Γ_2 in the double choice problem.

Theorem 4.5. *(1) State $(t, R) \in \Gamma_2$ if and only if*

$$R \geq \frac{p-c}{q} + u_1^{N-t-1}, \quad u_1^{-1} = -\frac{p-c}{q}, u_1^0 = 0.$$

(2) If $c \geq 2 - 1/p$, then state $(t, R) \in \Gamma_1$ if and only if

$$R \geq \frac{p-c}{q} + \frac{u_1^{N-t-1}p - u_1^{N-t}}{q} + v^{N-t-1}, \tag{4.14}$$

where $u_1^0 = v_1^0 = 0$.

Proof: We shall only prove the second part of the theorem. The first statement can be proved in a similar way. Suppose that $(t, R) \in \Gamma_1$. This means that, for $0 \leq R \leq t$,

$$\begin{aligned} R - ct + u_1^{N-t} &\geq u_2^N(t+1, R+1)p + u_2^N(t+1, 0)q \\ &\geq (R + 1 - c(t+1) + u_1^{N-t-1})p + u_2^N(t+1, 0)q. \end{aligned}$$

Using Lemma 4.4(1), we have

$$R - ct + u_1^{N-t} \geq (R + 1 - c(t+1) + u_1^{N-t-1})p + [v^{N-t-1} - c(t+1)]q, \tag{4.15}$$

which is equivalent to

$$R(1-p) \geq p + ct - c(t+1)(p+q) + u_1^{N-t-1}p - u_1^{N-t} + v^{N-t-1}.$$

This proves (4.14).

Now, we need to show that $(t, R) \in \Gamma_1$ under condition (4.14). Note that the inequalities (4.14) and (4.15) are equivalent. We will prove this part of the theorem by induction on t.

It is clear that $(N-1, R) \in \Gamma_1$. Suppose (4.15) holds for $t = N-2$. This means that, for $0 \leq R \leq N-2$,

$$\begin{aligned} R - c(N-2) + u_1^2 &\geq (R + 1 - c(N-1) + u_1^1)p + u_2^N(N-1, 0)q \\ &= (R + 1 - c(N-1) + p - c)p + u_2^N(N-1, 0)q \\ &= (R + 1 + p - cN)p + u_2^N(N-1, 0)q. \end{aligned}$$

From (4.6),

$$u_2^N(N-1, R+1) = R + 1 + p - cN.$$

Therefore,

$$R - c(N-2) + u_1^2 \geq u_2^N(N-1, R+1)p + u_2^N(N-1, 0)q,$$

indicating that $(N-2, R) \in \Gamma_1$.

Suppose that, for $0 < t \leq N - 2$, if (4.15) holds, then $(t, R) \in \Gamma_1$. Assume that inequality (4.15) holds for $t - 1$. Then, for $0 \leq R \leq t - 1$,

$$R - c(t-1) + u_1^{N-t+1} \geq (R + 1 - ct + u_1^{N-t})p + u_2^N(t, 0)q. \qquad (4.16)$$

Hence, using Lemma 4.4(3), we obtain

$$\begin{aligned} R + 1 - ct + u_1^{N-t} - 1 + c + u_1^{N-t-1} - u_1^{N-t} + p(1-c) - (u_1^{N-t} - u_1^{N-t-1})p \\ \geq (R + 2 - c(t+1) + u_1^{N-t-1})p + u_2^N(t, 0)q \\ \geq (R + 2 - c(t+1) + u_1^{N-t-1})p + u_2^N(t+1, 0)q. \end{aligned}$$

Taking into account Lemma 4.3 and the condition $c \geq 2 - 1/p$, we have

$$-1 + p + cq + u_1^{N-t-1} - u_1^{N-t} - p(u_1^{N-t} - u_1^{N-t-1}) \leq 0.$$

This implies that

$$R + 1 - ct + u_1^{N-t} \geq (R + 2 - c(t+1) + u_1^{N-t-1})p + u_2^N(t+1, 0)q.$$

Therefore, (4.15) holds for state $(t, R+1)$. Since $(t, R+1) \in \Gamma_1$, then

$$u_2^N(t, R+1) = R + 1 - ct + u_1^{N-t}.$$

Substituting this expression in (4.16), we obtain

$$R - c(t-1) + u_1^{N-t+1} \geq u_2^N(t, R+1)p + u_2^N(t, 0)q,$$

which means that $(t-1, R) \in \Gamma_1$. This completes the proof. □

Corollary 4.17. *If* $c \geq 2 - 1/p$ *and* $(t, R) \in \Gamma_1$, *then* $(t+1, R+1) \in \Gamma_1$, $t + 1 < N$.

Proof: If $(t, R) \in \Gamma_1$, then

$$\begin{aligned} R - ct + u_1^{N-t} &\geq (R + 1 - c(t+1) + u_1^{N-t-1})p + u_2^N(t+1, 0)q \\ &\geq (R + 1 - c(t+1) + u_1^{N-t-1})p + u_2^N(t+2, 0)q. \end{aligned}$$

Therefore,

$$\begin{aligned} R + 1 - c(t+1) + u_1^{N-t-1} + u_1^{N-t} - u_1^{N-t-1} \\ - (u_1^{N-t-1} - u_1^{N-t-2})p + p - cp - 1 + c \\ \geq (R + 2 - c(t+2) + u_1^{N-t-2})p + u_2^N(t+2, 0)q. \end{aligned}$$

From Lemma 4.3, for $c \geq 2 - 1/p$,

$$R + 1 - c(t+1) + u_1^{N-t-1} \geq (R + 2 - c(t+2) + u_1^{N-t-2})p + u_2^N(t+2, 0)q.$$

Hence, $(t+1, R+1) \in \Gamma_1$. □

Remark 11. *Similarly, it can be shown that if* $(t, R) \in \Gamma_2$, *then* $(t+1, R+1) \in \Gamma_2$.

Figures 4.1a and 4.1b show the expected gain v^N and the limiting expected gains for $p = q = 0.5$ and two choices of cost $c = 0.25$ and $c = 0.125$. We can see that in both cases the sequences v_N rapidly converge to their limits.

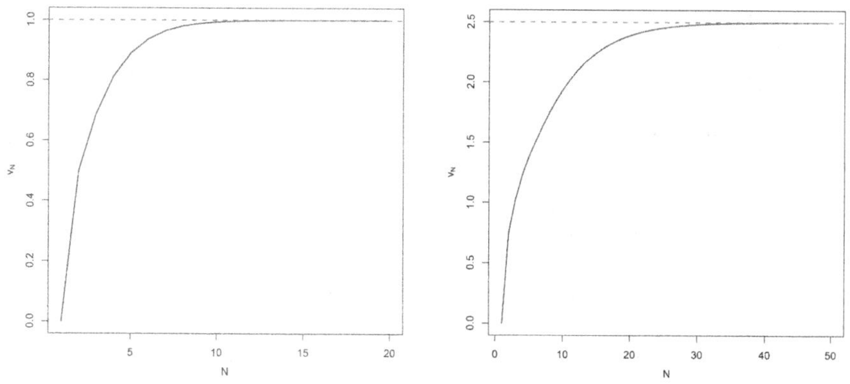

(a) For $p = 0.5, c = 0.25$; $\lim_{N\to\infty} v^N = 1$. **(b)** For $p = 0.5, c = 0.125$; $\lim_{N\to\infty} v^N = 2.5$.

FIGURE 4.1
The expected gain v^N (blue line) and the asymptotic value (in red).

4.3 Change Point Detection

Motto:
Unveiling Hidden Shifts: Navigating stochastic models to detect significant changes

Stochastic models of dynamic phenomena allow one to forecast their course in the future. Such use is possible when the modeled process has a stable source of randomness. This formulation is imprecise, but in practice often not much more can be said about the nature of the phenomenon. The aforementioned stability means in practice the possibility of fitting a relatively simple mathematical model to the observations. If the observations are correlated with another phenomenon, then a wider class of useful models can be obtained. It is enough to relate the model under study to the state of the process that influences it. However, suppose that we are unable to observe the phenomenon that governs the observed process. In that case, we can still conclude which model is appropriate for the observation as long as these changes are not too frequent. Let us follow the consideration of Sections 1.6.1 and the formulation of Problem 1.7. Suppose that the observations follow the adopted model up to some random moment θ, and then follow a different model. Moment θ is an additional parameter that allows the model to be more consistent with the data. The problem, however, is that without observing the related phenomenon, we do not know θ. Therefore, we can look for procedures that will allow θ to be assessed with the greatest possible precision and reliability.

Let us register a stochastic sequence affected by one disorder. Monitoring of the sequence is made in the circumstances when there is not complete information about distributions before and after the change is available. The initial problem of disorder detection transforms into an optimal stopping of the observed sequence. The formula for optimal decision functions is derived.

The section is focused on sequential detection using the Bayesian approach[1]. A.N. Kolmogorov formulated the disorder problem in this framework at the end of the 1950s of the XX century (v. Shiryaev (2006)) and solved by Shiryaev (1961a). The next turning point is a paper by Peskir and Shiryaev (2002), where the authors provide a complete solution to the basic problem. Since then, many publications have provided new solutions and generalizations in sequential detection. Some of them are articles by Karatzas (2003) and Bayraktar et al. (2005). For the discrete-time case, there is a detailed analysis in the papers by Bojdecki (1979), Bojdecki and Hosza (1984), Moustakides (1998), Yakir (1994), Yoshida (1983) and Szajowski (1996). The section focuses attention on models under assumption of uncertainty about distribution before or after the change. The example of such models can be found in research by Dube and Mazumdar (2001) with application to the detection of traffic anomalies in networks or in paper by Sarnowski and Szajowski (2008). The solution of a single disorder model with an unspecified distribution of the observed sequence is presented. First, the details of the investigated model are specified. Next, the transformation of the optimization task related to the statistical problem into the optimal stopping problem for the specific stochastic process is considered. A construction of the optimal estimator of the disorder moment is given in Section 4.3.4. The technical parts of the investigations are moved to a separate section. It is worth mentioning that the construction of an optimal procedure is difficult and there are some problematic proofs in the literature (v. Mei (2006)).

4.3.1 Description of the model

Basic notations.

For further considerations it will be convenient to introduce the following notation which will make our formulas more compact and clear

$$\begin{aligned}
\underline{x}_{k,n} &= (x_k, x_{k+1}, \dots, x_{n-1}, x_n),\ k \le n, \\
\mathfrak{L}_m^{i,j}(\underline{x}_{k,n}) &= \prod_{r=k+1}^{m-1} f_{x_{r-1}}^{0,i}(x_r) \prod_{r=m}^{n} f_{x_{r-1}}^{1,j}(x_r), \\
\mathfrak{R}_{m,n}^{i,j}(\underline{x}_{m-1,n}) &= \prod_{r=s}^{n} \frac{f_{x_{r-1}}^{0,i}(x_r)}{f_{x_{r-1}}^{1,j}(x_r)}, \\
\underline{A}_{k,n} &= A_k \times A_{k+1} \times \ldots \times A_n,
\end{aligned}$$

[1]The terminology of this issue was developed in the mid-twenty century. We distinguish between the issues of studying the homogeneity of data (compliance with the established model) or data analysis in the real-time of the flow, i.e. sequential detection of the point of change in the character of the data. The area of these issues in the literature is called the disorder problem. For sequential tasks, we are talking about the change point detection problem. The very nature of the problem requires that the phenomenon be dynamic. Generalizations to random fields, due to the very nature of such observations, require a proper definition of the dynamics of increasing the data set to talk about the disorder problem.

where: $\prod_{r=m_1}^{m_2} u_r = 1$ for $m_1 > m_2$ and $u_r \in \Re$, $A_i \in \mathcal{B}$, $k \leq i \leq n$. The function $\mathfrak{L}^{\cdot,\cdot}_{\cdot}(\cdot)$ is the likelihood function and $\mathfrak{R}^{\cdot,\cdot}_{\cdot,\ldots}(\cdot)$ is the likelihood ratio function.

It will be convenient to write $\underline{\beta} = (\beta_1, \beta_2)$ and denote by $\overline{\alpha} = (\alpha_{11}, \ldots, \alpha_{1l_1}, \ldots, \alpha_{l_01}, \ldots, \alpha_{l_0l_1})$ any $\overline{l_0} \times l_1$ matrix

$$\hat{\alpha} = (\alpha_{ij})_{\substack{i=1,\ldots,l_1 \\ j=1,\ldots,l_2}} = \begin{bmatrix} \alpha_{11} & \alpha_{12} & \cdots & \alpha_{1l_1} \\ \alpha_{21} & \alpha_{22} & \cdots & \alpha_{2l_1} \\ \vdots & \vdots & \ddots & \vdots \\ \alpha_{l_01} & \alpha_{l_02} & \cdots & \alpha_{l_0l_1} \end{bmatrix}$$

In a consequence the vectors $\overline{\pi}$, $\overline{b}$, $\overline{p}$ are representation of $\hat{\pi} = (\pi_{ij})_{\substack{i=1,\ldots,l_1 \\ j=1,\ldots,l_2}}$, $\hat{b} = (b_{ij})_{\substack{i=1,\ldots,l_1 \\ j=1,\ldots,l_2}}$, $\hat{p} = (p_{ij})_{\substack{i=1,\ldots,l_1 \\ j=1,\ldots,l_2}}$.

We need also a short notation for the vector (the matrix) of densities $f_x^{0,i}(y)$ with dimension l_1 ($l_0 \times l_1$) organized as follows:

$$\overline{f}_x^{0,i}(y) = \underbrace{f_x^{0,i}(y), \ldots, f_x^{0,i}(y)}_{l_1 \text{ times}}$$

and

$$\widehat{f_x^0}(y) = \underbrace{(\overline{f}_x^{0,1}(y), \overline{f}_x^{0,2}(y), \ldots, \overline{f}_x^{0,l_0}(y))}_{l_0 \text{ vectors}}^T,$$

where $x, y \in \mathbb{E}$ stands behind:

$$\overline{f}_x^0(y) = (\underbrace{f_x^{0,1}(y), \ldots, f_x^{0,1}(y)}_{l_1 \text{ times}}, \ldots, \underbrace{f_x^{0,l_0}(y), \ldots, f_x^{0,l_0}(y)}_{l_1 \text{ times}}).$$

The componentwise product of vectors and matrices is denoted "$\circ$". For vectors $\overline{\alpha}$ and $\overline{\beta}$, we have

$$\overline{\alpha} \circ \overline{\beta} = (\alpha_{11}\beta_{11}, \ldots, \alpha_{1l_1}\beta_{1l_1}, \ldots, \alpha_{l_01}\beta_{l_01}, \ldots, \alpha_{l_0l_1}\beta_{l_0l_1}).$$

Change point problem

Let $(X_n)_{n\in\mathbb{N}}$ be sequence of observable random variables defined on $(\Omega, \mathcal{F}, \boldsymbol{P})$ with value in $(\mathbb{E}, \mathcal{B})$, $\mathbb{E} \subset \Re$. Sequence (X_n) generates filtration $\mathcal{F}_n = \sigma(X_0, X_1, ..., X_n)$. On the same space there are also defined variables θ, β_1 and β_2. θ takes values in $\{1, 2, 3, \ldots\}$. Variables β_1, β_2 are valued in $I_k = \{1, 2, \ldots, l_k\}$, where $l_k \in \mathbb{N}$, $k = 0, 1$. Let us assume the following parametrization:

$$\boldsymbol{P}(\beta_1 = i, \beta_2 = j) = b_{ij} \tag{4.17}$$

$$\boldsymbol{P}(\theta = n|\beta_1 = i, \beta_2 = j) = \begin{cases} \pi_{ij}, & \text{if } n = 1, \\ (1 - \pi_{ij})p_{ij}^{n-2}q_{ij}, & \text{if } n > 1, \end{cases} \tag{4.18}$$

where $i \in I_0, j \in I_1, \sum_{i\in I_0, j\in I_1} b_{ij} = 1, b_{ij} \geq 0, \pi_{ij} \in [0,1], p_{ij} = 1-q_{ij} \in (0,1)$. We have

$$\sum_{k=1}^{\infty}\sum_{i\in I_0}\sum_{j\in I_1} \boldsymbol{P}(\theta = k, \beta_1 = i, \beta_2 = j) = 1$$

It means that there are two models of dynamics of the states given by the conditional densities. This law of motion changes at the random moment θ. However, the conditional distribution before and after change are not predetermined but they are elements of some finite sets of distributions. The distribution of models between and after the disorder is described by probabilities $b_{ij} = \boldsymbol{P}(\beta_1 = i, \beta_2 = j)$. For completeness it will be assumed that the state of β_1 is stable before θ and the same at the moment 0. The conditional distribution of θ given $\underline{\beta} = (i,j)$ has a form (4.18) and the marginal distribution of the θ is

$$\begin{aligned} \boldsymbol{P}(\theta = k) &= \sum_{i,j} \boldsymbol{P}(\theta = k, \beta_1 = i, \beta_2 = j) \\ &= \begin{cases} \sum_{i,j} \pi_{ij}\cdot b_{ij} & \text{if } k = 1, \\ \sum_{i,j}(1-\pi_{ij})p_{ij}^{k-2}q_{ij}b_{ij} & \text{if } k > 1. \end{cases} \end{aligned} \tag{4.19}$$

The assumption about the model allows to say that the observed sequence has a form

$$X_n = X_n^{0,i}\cdot \mathbb{I}_{\{\theta>n,\ \beta_1=i\}} + X_{n-\theta+1}^{1,j}\cdot \mathbb{I}_{\{\theta\leq n,\ \beta_2=j\ ,X_0^{1,j}=X_{\theta-1}^{0,i}\}}, \tag{4.20}$$

where $(X_n^{r,i}, \mathcal{G}_n^{r,i}, \boldsymbol{P}_x^{r,i})$, $r = 0,1$, are Markov processes and σ-fields: $\mathcal{G}_n^{r,i} = \sigma(X_0^{r,i}, X_1^{r,i}, \ldots, X_n^{r,i})$, with $i \in I_0$, $j \in I_1$, $r = 0,1$ and $n \in \{0,1,2,\ldots\}$. Variables θ, β_1 and β_2 are not measurable w.r.t $\mathcal{F}_n$.

On the space $(\mathbb{E}, \mathcal{B})$ there are σ-additive measures $\mu(\cdot)$ and measures $\mu_x^{\bullet,\bullet}$ absolutely continuous with respect to μ. It is assumed that the measures $\boldsymbol{P}_x^{k,i}(\cdot)$, $i = 1,2,\ldots,l_k$, $k = 0,1$, have following representation:

$$\begin{aligned} \boldsymbol{P}_x^{k,i}(\{\omega : X_1^{k,i} \in B\}) &= \boldsymbol{P}(X_1^{k,i} \in B | X_0^{k,i} = x) \\ &= \int_B f_x^{k,i}(y)\mu(dy) = \int_B \mu_x^{k,i}(dy) = \mu_x^{k,i}(B). \end{aligned}$$

for any $B \in \mathcal{B}$. The conditional densities $f_x^{k,1}(\cdot), \ldots, f_x^{k,l_k}(\cdot)$ are different and the support of all measures $\mu_x^{\cdot,\cdot}$ are there same for given $x \in \mathbb{E}$. Moreover, to avoid analytical problem it is assumed that the every probability from the first group is absolutely continuous with respect any measure of the second group.

It is the model of the following random phenomenon. At the beginning we register process $\{X_n^{0,i}, n \in \mathbb{N}\}$, where $i \in I_0$ is unknown. At random moment θ initial process is switched on $\{X_n^{1,j}, n \in \mathbb{N}\}$ where $j \in I_1$ is unknown. It can be interpreted as disorder of $\{X_n, n \in \mathbb{N}\}$ causing change in distribution of $\{X_n\}_{n\in\mathbb{N}}$. We monitor the process and we wish to detect the change as close θ as possible. However our knowledge about densities before and after the change moment θ is limited generally to the information about sets of possible conditional densities only: $\{f_x^{0,i}(y), i \in I_0\}$

and $\{f_x^{1,j}(y), j \in I_1\}$ respectively. We also know probabilities of distribution pairs b_{ij} and parameters (π_{ij}, p_{ij}).

For $i \in I_0, j \in I_1$ let us introduce functions $\Psi^{i,j}$, $\widetilde{\Psi}^{i,j}$, $\Lambda^{i,j}$, $\widetilde{\Lambda}^{i,j}$ defined on the product $\mathbb{N} \times (\times_{i=1}^n \mathbb{E}) \times [0,1]$ with values in $\Re$:

$$\Psi^{i,j}(l, \underline{x}_{0,l+1}, \alpha) = (1-\alpha)\left[q_{ij}\sum_{k=0}^{l} p_{ij}^{l-k} L_{k+1}^{i,j}(\underline{x}_{0,l+1}) + p_{ij}^{l+1} L_0^{i,j}(\underline{x}_{0,l+1})\right] + \alpha L_{l+1}^{i,j}(\underline{x}_{0,l+1}) \tag{4.21}$$

$$\widetilde{\Psi}^{i,j}(l, \underline{x}_{0,l+1}, \alpha) = (1-\alpha)\left[q_{ij}\sum_{k=1}^{l} p_{ij}^{l-k} L_k^{i,j}(\underline{x}_{0,l+1}) + p_{ij}^{l} L_0^{i,j}(\underline{x}_{0,l+1})\right] + \alpha L_{l+1}^{i,j}(\underline{x}_{0,l+1}), \tag{4.22}$$

$$\Lambda^{i,j}(l, \underline{x}_{0,l+1}, \alpha) = \Psi^{i,j}(l, \underline{x}_{0,l+1}, \alpha) - (1-\alpha)p_{ij}^{l+1} L_0^{i,j}(\underline{x}_{0,l+1}),$$
$$\widetilde{\Lambda}^{i,j}(l, \underline{x}_{0,l+1}, \alpha) = \widetilde{\Psi}^{i,j}(l, \underline{x}_{0,l+1}, \alpha) - (1-\alpha)p_{ij}^{l} L_0^{i,j}(\underline{x}_{0,l+1}).$$

Next let us define on $\mathbb{N} \times (\times_{i=1}^n \mathbb{E}) \times (\times_{i=1}^{l_1 l_2}[0,1]) \times (\times_{i=1}^{l_1 l_2}[0,1])$ function $S, \widetilde{S}$:

$$S(k, \underline{x}_{0,k+1}, \overline{\gamma}, \overline{\delta}) = \sum_{i,j} \gamma_{ij}\, \Psi^{i,j}(k, \underline{x}_{0,k+1}, \delta_{ij}), \tag{4.23}$$

$$\widetilde{S}(k, \underline{x}_{0,k+1}, \overline{\gamma}, \overline{\delta}) = \sum_{i,j} \gamma_{ij}\, \widetilde{\Psi}^{i,j}(k, \underline{x}_{0,k+1}, \delta_{ij}). \tag{4.24}$$

For any $D_n = \{\omega : X_i \in B_i, i = 1, 2, \ldots, n\}$, where $B_i \in \mathcal{B}$ and any $x \in \mathbb{E}$ define:

$$\begin{aligned} \boldsymbol{P}_x(D_n) = \boldsymbol{P}(D_n | X_0 = x) &= \int_{\times_{i=1}^n B_i} \widetilde{S}(n-1, \underline{x}_{0,n}, \overline{b}, \overline{\pi})\mu(d\underline{x}_{1,n}) \\ &= \int_{\times_{i=1}^n B_i} \mu_{x_0}(d\underline{x}_{1,n}) = \mu_{x_0}(\times_{i=1}^n B_i). \end{aligned}$$

For the process (4.20) the set of estimators for the disorder moment θ is $\mathfrak{S}^X$ – the set of stopping times with respect to $\{\mathcal{F}_n\}_{n\in\mathbb{N}\cup\{0\}}$. The construction of the optimal estimator is to find a stopping time $\boldsymbol{\tau}^* \in \mathfrak{S}^X$ such that for any $x \in \mathbb{E}$

$$\boldsymbol{P}_x(|\theta - \boldsymbol{\tau}^*| \le d) = \sup_{\tau \in \mathfrak{S}^X} \boldsymbol{P}_x(|\theta - \tau| \le d), \tag{4.25}$$

where $d \in \{0, 1, 2, \ldots\}$ is fixed level of detection precision.

Example 4.4 (Markov chain with one transition probability switch.). *Let us consider for example the Markov sequence $\{\xi_k\}_{k=0}^M$ having the probability transition matrices P for $k = 0, 1, \ldots, \theta - 1$ and Q for $k = \theta, \ldots, M-1$. Define*

$$P = \begin{bmatrix} \frac{16}{34} & \frac{2}{34} & \frac{3}{34} & \frac{13}{34} \\ \frac{5}{34} & \frac{11}{34} & \frac{10}{34} & \frac{8}{34} \\ \frac{9}{34} & \frac{7}{34} & \frac{6}{34} & \frac{12}{34} \\ \frac{4}{34} & \frac{14}{34} & \frac{15}{34} & \frac{1}{34} \end{bmatrix} \quad Q = \begin{bmatrix} \frac{1}{19} & \frac{2}{19} & \frac{3}{19} & \frac{13}{19} \\ 0 & \frac{1}{19} & \frac{10}{19} & \frac{8}{19} \\ 0 & 0 & \frac{6}{18} & \frac{12}{18} \\ 0 & 0 & \frac{15}{16} & \frac{1}{16} \end{bmatrix} \tag{4.26}$$

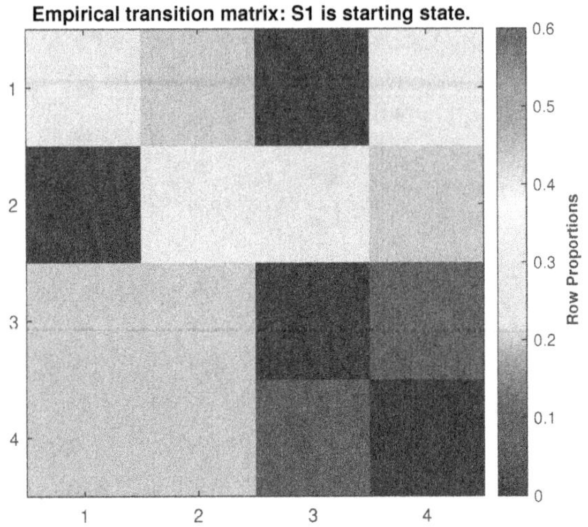

(**a**) Empirical transition matrix for $k < \theta$.

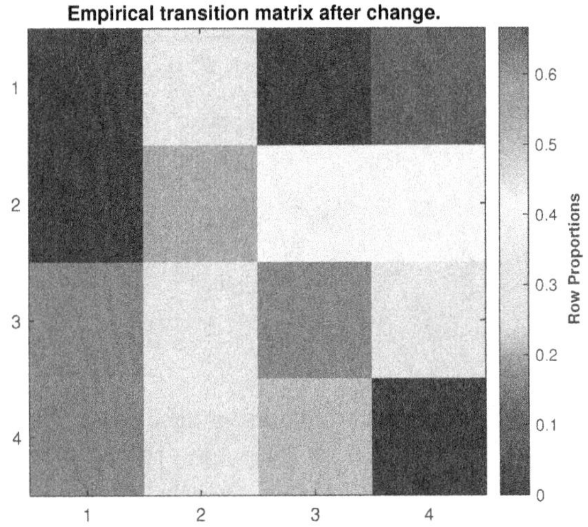

(**b**) Empirical transition matrix for $k \geq \theta$.

FIGURE 4.2
The empirical transition matrices before/after the switching moment.

The switching moment θ is random or control variable. For $M = 20$ and $\theta = 10$ the empirical transition fraction matrices for the random walk before and after the switching moment θ (the transition matrices P and Q, respectively) are shown on Figure 4.2.

Example 4.5 (Markov chain with two switching of the transition probabilities.). *Let $M = 60$ and there are two switching moments $\theta_1 = 20$ and $\theta_2 = 40$. Let us simulate the process with the transition matrix P in the first segment, Q in the second, and return to P in the third segment (v. Figure 4.3).*

4.3.2 Existence of solution

In this section, we are going to show that there exists a solution of the problem (4.25). Let us define:

$$\begin{aligned} Z_n &= \boldsymbol{P}(|\theta - n| \leq d \mid \mathcal{F}_n),\ n = 1, 2, \ldots, \\ V_n &= \operatorname*{ess\,sup}_{\{\tau \in \mathfrak{S}^X,\ \tau \geq n\}} \boldsymbol{P}(|\theta - n| \leq d \mid \mathcal{F}_n),\ n = 0, 1, 2, \ldots \\ \tau_0 &= \inf\{n : Z_n = V_n\}. \end{aligned} \tag{4.27}$$

Notice that, if $Z_\infty = 0$, then $Z_\tau = \boldsymbol{P}(|\theta - \tau| \leq d \mid \mathcal{F}_\tau)$ for $\tau \in \mathfrak{S}^X$. Because $\mathcal{F}_n \subseteq \mathcal{F}_\tau$ (when $n \leq \tau$), we obtain

$$\begin{aligned} V_n &= \operatorname*{ess\,sup}_{\tau \geq n} \boldsymbol{P}(|\theta - \tau| \leq d \mid \mathcal{F}_n) \\ &= \operatorname*{ess\,sup}_{\tau \geq n} \boldsymbol{E}(\boldsymbol{E}(\mathbb{I}_{\{|\theta-\tau|\leq d\}} | \mathcal{F}_\tau) \mid \mathcal{F}_n) \\ &= \operatorname*{ess\,sup}_{\tau \geq n} \boldsymbol{E}(Z_\tau \mid \mathcal{F}_n) \end{aligned}$$

The following lemma states that solution exists.

Lemma 4.6. *Stopping time τ_0 given by (4.27) is a solution of the problem (4.25).*

Proof: Applying Theorem 1 from Bojdecki (1979) it is enough to show that $\lim_{n\to\infty} Z_n = 0$. For all n, k, where $n \geq k$ we have:

$$Z_n = \boldsymbol{E}(\mathbb{I}_{\{|\theta-n|\leq d\}} \mid \mathcal{F}_n) \leq \boldsymbol{E}(\sup_{j\geq k} \mathbb{I}_{\{|\theta-j|\leq d\}} \mid \mathcal{F}_n)$$

Basing on Levy's theorem we get

$$\limsup_{n\to\infty} Z_n \leq \boldsymbol{E}(\sup_{j\geq k} \mathbb{I}_{\{|\theta-j|\leq d\}} \mid \mathcal{F}_\infty)$$

where $\mathcal{F}_\infty = \sigma\left(\bigcup_{n=1}^\infty \mathcal{F}_n\right)$. We have: $\limsup_{j\geq k,\ k\to\infty} \mathbb{I}_{\{|\theta-j|\leq d\}} = 0$ *a.s.* Basing on dominated convergence theorem we get

$$\lim_{k\to\infty} \boldsymbol{E}(\sup_{j\geq k} \mathbb{I}_{\{|\theta-j|\leq d\}} \mid \mathcal{F}_\infty) = 0 \text{ a.s.}$$

what ends the proof. □

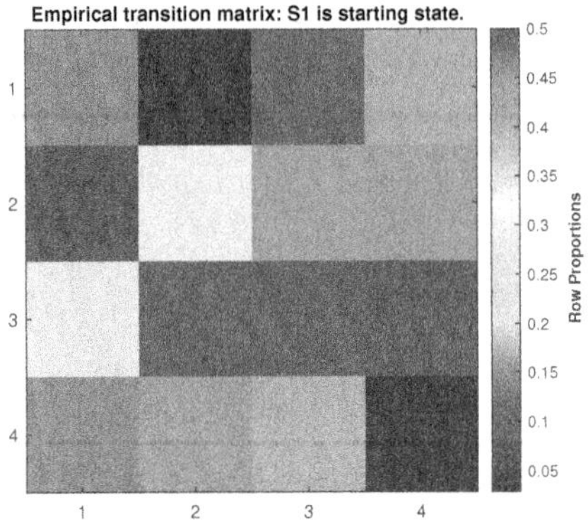

(a) Empirical transition matrix for $k < \theta_1$.

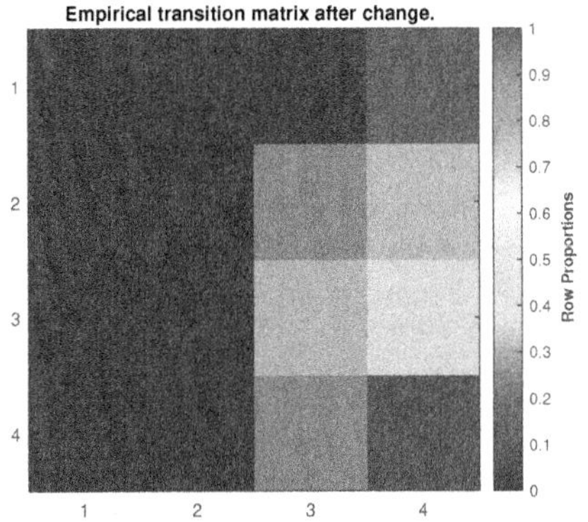

(b) Empirical transition matrix for $\theta_1 \leq k < \theta_2$.

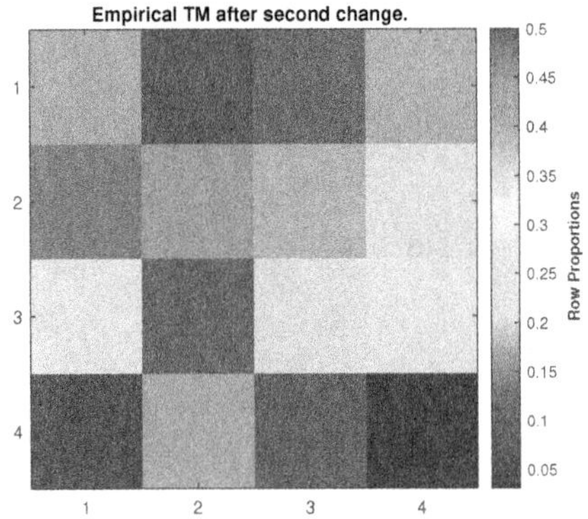

(c) Empirical transition matrix for $k \geq \theta_2$.

FIGURE 4.3
The empirical transition matrices for two switching moment.

It turns out that we need at least d observations to detect disorder in optimal way:

Lemma 4.7. *Let τ be stopping rule in the problem (4.25). Then rule $\tilde{\tau} = \max(\tau, d+1)$ is at least as good as τ (in the sense of (4.25)).*

Proof: For $\tau \geq d+1$ the rules the same. Let us consider case when $\tau < d+1$. Then $\tilde{\tau} = d+1$ and:

$$\begin{aligned} \boldsymbol{P}(|\theta - \tau| \leq d) &= \boldsymbol{P}(\tau - d \leq \theta \leq \tau + d) = \boldsymbol{P}(1 \leq \theta \leq \tau + d) \\ &\leq \boldsymbol{P}(1 \leq \theta \leq 2d + 1) = \boldsymbol{P}(\tilde{\tau} - d \leq \theta \leq \tilde{\tau} + d) \\ &= \boldsymbol{P}(|\theta - \tilde{\tau}| \leq d). \end{aligned}$$

□

4.3.3 Estimator of the disorder moment

Function and processes

Let us fix parameters $\overline{\pi}$, $\overline{b}$ and set initial state of X_n: $\boldsymbol{P}(X_0 = x) = 1$. We denote $\varphi = (\overline{\pi}, \overline{b}, x)$ and we will write $\boldsymbol{P}^{\varphi}(\bullet)$ to emphasis that the probability of the events defined by the process are dependent on these *a priori* set parameters. Let us define the following crucial posterior processes for $n \in \mathbb{N}$, $i \in I_0, j \in I_1$, where $\tilde{\mathcal{F}}_n^{i,j} = \sigma(\mathcal{F}_n, \mathbb{I}_{\{\underline{\beta}=(i,j)\}})$:

$$\begin{aligned} \Pi_{k,n}^{i,j} &= \boldsymbol{P}^{\varphi}(\theta \leq n + k | \underline{\beta} = (i,j), \mathcal{F}_n) \\ &= \boldsymbol{P}^{\varphi}(\theta \leq n + k | \tilde{\mathcal{F}}_n^{ij}) \\ \Pi_n^{i,j} &= \Pi_{0,n}^{i,j} \\ B_n^{i,j} &= \boldsymbol{P}^{\varphi}(\underline{\beta} = (i,j) | \mathcal{F}_n). \end{aligned} \tag{4.28}$$

Process $\Pi_n^{i,j}$ is designed for updating information about disorder distribution. $B_n^{i,j}$ in turn refreshes information about distributions of variables $\underline{\beta}$. Notice that $\Pi_n^{i,j}$, $B_n^{i,j}$ starts from following states: $\Pi_0^{i,j} = 0$, $B_0^{i,j} = b_{ij}$ Dynamics of $\Pi_n^{i,j}$ and $B_n^{i,j}$ are characterized by formulas (4.58), (4.60). The above notations hold also for (4.28):

$$\overline{\Pi}_n = \left(\Pi_n^{1,1}, \ldots, \Pi_n^{1,l_1}, \ldots, \Pi_n^{l_0,1}, \ldots, \Pi_n^{l_0 l_1}\right), \tag{4.29}$$

$$\overline{B}_n = \left(B_n^{1,1}, \ldots, B_n^{1,l_1}, \ldots, B_n^{l_0,1}, \ldots, B_n^{l_0 l_1}\right), \tag{4.30}$$

$$\overline{\Xi}_n = \left(\overline{\Pi}_n, \overline{B}_n\right). \tag{4.31}$$

At the end of section let us define auxiliary functions $\Pi^{\cdot,\cdot}(\cdot,\cdot,\cdot)$, $\Gamma^{\cdot,\cdot}(\cdot,\cdot,\cdot,\cdot)$. For $x, y \in \mathbb{E}$, $\alpha, \gamma_{ij}, \delta_{ij} \in [0,1]$, $i \in I_0$, $j \in I_1$ put:

$$\Pi^{i,j}(k, \underline{x}_{0,n}, \alpha) = \frac{\Lambda^{i,j}(k, \underline{x}_{0,n}, \alpha)}{\Psi^{i,j}(k, \underline{x}_{0,n}, \alpha)} \tag{4.32}$$

$$\Gamma^{i,j}(k, \underline{x}_{0,n}, \overline{\gamma}, \overline{\delta}) = \frac{\gamma_{ij} \Psi^{i,j}(k, \underline{x}_{0,n}, \delta_{ij})}{S(k, \underline{x}_{0,n}, \overline{\gamma}, \overline{\delta})}. \tag{4.33}$$

Let $D_n = \{\omega : \underline{X}_{1,n} \in \underline{B}_{1,n}\}$, $X_0 = x$ and $B_i \in \mathcal{B}$, where $\underline{B}_{k,n} = \times_{i=k}^{n} B_i$. We have

$$\int_{D_n^{ij}} \mathbb{I}_{\{\theta>n\}} d\boldsymbol{P}_x^{\varphi} = \boldsymbol{P}_x^{\varphi}(\theta > n, \beta = (i,j), D_n) \tag{4.34}$$

$$= \int_{D_n} \mathbb{I}_{\{\underline{\beta}=(i,j)\}} \mathbb{I}_{\{\theta>n\}} d\boldsymbol{P}_x^{\varphi}$$

$$\int_{\underline{B}_{0,n}} \frac{(1-\pi_{ij})p_{ij}^{n-1} L_0^{ij}(\underline{x}_{0,n})}{S_n^{i,j}(\underline{x}_{0,n})} \frac{b_{ij} S_n^{i,j}(\underline{x}_{0,n})}{S_n(\underline{x}_{0,n})} S_n(\underline{x}_{0,n}) \mu(d\underline{x}_{1,n}) \tag{4.35}$$

$$= \int_{D_n} (1 - \Pi_n^{i,j}) B_n^{i,j} d\boldsymbol{P}_x^{\varphi},$$

where

$$\begin{aligned} S_n^{i,j}(\underline{x}_{0,n}) &= \pi_{ij} L_n^{i,j}(\underline{x}_{0,n}) + (1-\pi_{ij}) p_{ij}^{n-1} L_0^{i,j}(\underline{x}_{0,n}) \\ &\quad + (1-\pi_{ij}) \sum_{s=2}^{n} p_{ij}^{s-2} q_{ij} L_{n-s+1}^{i,j}(\underline{x}_{0,n}) \\ &= \Psi^{i,j}(n-1, \underline{x}_{0,n}, \pi_{ij}) \end{aligned} \tag{4.36}$$

and $S_n(\underline{x}_{0,n}) = \sum_{i,j} b_{ij} S_n^{i,j}(\underline{x}_{0,n}) = S(n-1, \underline{x}_{0,n}, \bar{b}, \bar{\pi})$.

4.3.4 Solution

According to Shiryaev's (1978) methodology, we are going to find a solution reducing initial problem (4.25) to the case of stopping Random Markov Function with special payoff function. Using (4.28) we are able to cast initial problem (4.25) to the case of stopping Random Markov Function with special payoff.

Lemma 4.8. *For* $n \geq d+1$

$$\boldsymbol{P}^{\varphi}(|\theta - n| \leq d) = \begin{cases} \boldsymbol{E}^{\varphi}\left[h(\underline{X}_{n-1-d,n}, \overline{\Pi}_n, \overline{B}_n)\right], & \textit{if } n > d+1, \\ \boldsymbol{E}^{\varphi}\left[\widetilde{h}(\overline{\Pi}_{d+1}, \overline{B}_{d+1})\right], & \textit{if } n = d+1. \end{cases} \tag{4.37}$$

where

$$h(\underline{x}_{1,d+2}, \overline{\gamma}, \overline{\delta}) = \sum_{i,j} \left(1 - p_{ij}^d + q_{ij} \sum_{k=1}^{d+1} \frac{L_k^{i,j}(\underline{x}_{1,d+2})}{p_{ij}^k \, L_0^{i,j}(\underline{x}_{1,d+2})}\right)(1-\gamma_{ij})\delta_{ij}, \tag{4.38}$$

$$\widetilde{h}(\overline{\gamma}, \overline{\delta}) = \sum_{i,j} \left(1 - p_{ij}^d (1-\gamma_{ij})\right) \delta_{ij}, \tag{4.39}$$

$x_1, ..., x_{d+2} \in \mathbb{E}$, $\gamma_{ij}, \delta_{ij} \in [0,1]$, $i \in I_0$, $j \in I_1$.

Proof: Let us rewrite initial criterion as expectation:

$$\boldsymbol{P}^{\varphi}(|\theta - n| \leq d) \quad = \quad \boldsymbol{E}^{\varphi}\left[\boldsymbol{P}^{\varphi}(|\theta - n| \leq d \mid \mathcal{F}_n)\right]. \tag{4.40}$$

Let us analyze conditional probability under expectation in equation (4.40) using total probability formula

$$\begin{aligned}\boldsymbol{P}^{\varphi}\left(|\theta - n| \leq d \mid \mathcal{F}_n\right) &= \boldsymbol{P}^{\varphi}(\theta \leq n + d \mid \mathcal{F}_n) \qquad (4.41)\\ &-\boldsymbol{P}^{\varphi}(\theta \leq n - d - 1 \mid \mathcal{F}_n)\\ &= \sum_{i,j} \Pi^{ij}_{n+d} B^{i,j}_n - \sum_{i,j} \Pi^{ij}_{n-d-1} B^{i,j}_n, \qquad (4.42)\end{aligned}$$

because

$$\begin{aligned}\boldsymbol{P}^{\varphi}(\theta \leq n + d|\mathcal{F}_n) &= \boldsymbol{E}^{\varphi}(\mathbb{I}_{\theta \leq n+d}|\mathcal{F}_n)\\ &= \sum_{i,j} \boldsymbol{E}^{\varphi}(\mathbb{I}_{\{\theta \leq n+d\}}\mathbb{I}_{\{\underline{\beta}=(i,j)\}}|\mathcal{F}_n)\\ &= \sum_{i,j} \boldsymbol{E}^{\varphi}(\boldsymbol{E}^{\varphi}(\mathbb{I}_{\{\theta \leq n+d\}}\mathbb{I}_{\{\underline{\beta}=(i,j)\}}|\tilde{\mathcal{F}}^{i,j}_n)|\mathcal{F}_n)\\ &= \sum_{i,j} \boldsymbol{E}^{\varphi}(\mathbb{I}_{\{\underline{\beta}=(i,j)\}}\boldsymbol{E}^{\varphi}(\mathbb{I}_{\{\theta \leq n+d\}}|\tilde{\mathcal{F}}^{i,j}_n)|\mathcal{F}_n)\\ &= \sum_{i,j} \boldsymbol{E}^{\varphi}(\mathbb{I}_{\{\theta \leq n+d\}}|\tilde{\mathcal{F}}^{i,j}_n)\boldsymbol{E}^{\varphi}(\mathbb{I}_{\{\underline{\beta}=(i,j)\}}|\mathcal{F}_n).\end{aligned}$$

The last equality is a consequence of the very special form of the extended σ-field $\tilde{\mathcal{F}}^{i,j}_n$. The random variable measurable with respect of $\tilde{\mathcal{F}}^{i,j}_n$ is also $\mathcal{F}_n$ measurable. Putting $n = d + 1$ in lemma 4.11 we get: $\boldsymbol{P}^{\varphi}(\theta \leq n - d - 1 \mid \mathcal{F}_n, \underline{\beta} = (i,j)) = 0$, for $i \in I_0, j \in I_1$. Hence

$$\boldsymbol{P}^{\varphi}(|\theta - n| \leq d \mid \mathcal{F}_n) = \sum_{i,j} \boldsymbol{P}^{\varphi}(\theta \leq n + d \mid \tilde{\mathcal{F}}_n)\boldsymbol{P}^{\varphi}(\underline{\beta} = (i,j) \mid \mathcal{F}_n).$$

Lemma 4.10 implies that:

$$\boldsymbol{P}^{\varphi}(|\theta - n| \leq d) = \boldsymbol{E}^{\varphi}\left[\tilde{h}(\overline{\Pi}_{d+1}, \overline{B}_{d+1})\right].$$

Now let $n > d + 1$. Basing on lemma 4.10 probability $\boldsymbol{P}^{\varphi}(\theta \leq n + d \mid \tilde{\mathcal{F}}_n)$ is given by (4.52). From lemma 4.11 we know that $\boldsymbol{P}^{\varphi}(\theta \leq n - d - 1 \mid \tilde{\mathcal{F}}_n)$ is expressed by equation (4.55). Formula (4.55) reveals connection between payoff function (4.37) and posterior process at instants n and $n - d - 1$, i.e $\Pi^{i,j}_n$, $\Pi^{i,j}_{n-d-1}$ for $i \in I_0, j \in I_1$. Dependence on $\Pi^{i,j}_{n-d-1}$ can be rule out by expressing $\Pi^{i,j}_{n-d-1}$ in terms of $\Pi^{i,j}_n$. By Lemma 4.12 and (4.57), using denotation

$$\begin{aligned}G^{i,j}(d, \underline{x}_{0,d+1}, \alpha) = (1 - \alpha)\big(&q_{ij}\sum_{k=0}^{d} p_{ij}^{d-k} L^{i,j}_{k+1}(\underline{x}_{0,d+1}) - L^{i,j}_{d+1}(\underline{x}_{0,d+1})\big)\\ &-\alpha p_{ij}^{d+1} L^{i,j}_0(\underline{x}_{0,d+1}) \qquad (4.43)\end{aligned}$$

we get

$$\Pi^{i,j}_{n-d-1} = \left[p_{ij}^{d+1} L_0^{i,j}(\underline{X}_{n-d-1,n}) - \Pi^{i,j}_n \left(\sum_{k=0}^{d} p_{ij}^{d-k} L_{k+1}^{i,j}(\underline{X}_{n-d-1,n})\right)\right. \tag{4.44}$$
$$\left. + q_{ij} \sum_{k=0}^{d} p_{ij}^{d-k} L_{k+1}^{i,j}(\underline{X}_{n-d-1,n})\right] (G^{i,j}(d, \underline{X}_{n-d-1,n}, \Pi^{i,j}_n))^{-1}$$

The result above and formula (4.55) lead us to:

$$\boldsymbol{P}^{\varphi}(\theta \leq n-d-1 \mid \mathcal{F}_n, \underline{\beta} = (i,j)) \tag{4.45}$$
$$= \frac{p_{ij}^{d+1} L_0^{i,j}(\underline{X}_{n-d-1,n})\Pi^{i,j}_n - q_{ij}\sum_{k=0}^{d} p_{ij}^{d-k} L_{k+1}^{i,j}(\underline{X}_{n-d-1,n})(1-\Pi^{i,j}_n)}{p_{ij}^{d+1} L_0^{i,j}(\underline{X}_{n-d-1,n})}.$$

Applying equations (4.52) and (4.45) in formula (4.41) we get the thesis. □

Notice that for $n \geq d+1$ function h depends on process $\eta_n = (\underline{X}_{n-d-1,n}, \overline{\Pi}_n, \overline{B}_n)$ It turns out that $\{\eta_n\}$ is Markov Random Function (see lemma 4.16 in appendix 4.3.5). We do not care about $\{\eta_n\}$ for $n < d+1$. It is a consequence of discussion in Lemma 4.7 which leads to the conclusion that under the considered payoff function (criterion) it is not optimal to stop before instant $d+1$. The decision-maker can start his decision based on at least $d+1$ observations $X_0, \ldots, X_{d+1}$.

Lemmas 4.8 and 4.16 imply that initial problem can be reduced to the optimal stopping of Markov Random Function $(\eta_n, \mathcal{F}_n, \boldsymbol{P}^{\varphi}_{\underline{y}})_{n=1}^{\infty}$, where $\underline{y} = (\underline{x}_{n-d-1,n}, \overline{\gamma}, \overline{\delta}) \in \overrightarrow{\Xi} = \mathbb{E}^{d+2} \times [0,1]^{l_1 l_2} \times [0,1]^{l_1 l_2}$ with payoff described by (4.38). However, the new problem is no longer homogeneous one as it is emphasize by the definition of $\underline{y}$. It is a consequence of the fact that the process $\{\eta_n\}$ for $n < d+1$ has formally different structure than for $n \geq d+1$. In result the payoffs for instances $n = 0, \ldots, d+1$ are different. Lemma 2 gives a justification to work on the homogeneous part of the process in construction the optimal estimator of the disorder moment.

To solve the maximization problem (4.37), for any Borel function $u : \overrightarrow{\Xi} \longrightarrow \Re$ let us define operators:

$$\begin{aligned}
\mathbf{T}u(\underline{x}_{1,d+2}, \overline{\gamma}, \overline{\delta}) &= \boldsymbol{E}^{\varphi}_{(\underline{x}_{1,d+2}, \overline{\gamma}, \overline{\delta})}\left[u(\underline{X}_{n-d,n+1}, \overline{\Pi}_{n+1}, \overline{B}_{n+1})\right], \\
\mathbf{Q}u(\underline{x}_{1,d+2}, \overline{\gamma}, \overline{\delta}) &= \max\{u(\underline{x}_{1,d+2}, \overline{\gamma}, \overline{\delta}), \mathbf{T}u(\underline{x}_{1,d+2}, \overline{\gamma}, \overline{\delta})\}.
\end{aligned}$$

Operators $\mathbf{T}$ and $\mathbf{Q}$ act on function h and they determine the shape of optimal stopping rule $\boldsymbol{\tau}^{\star}$. Recursive formulas are given by lemma 4.17, which is presented in appendix 4.3.5. Lemma 4.17 characterizes structure of sequence of functions $s_k(\underline{x}_{1,d+2}, \overline{\gamma}, \overline{\delta})$, where $\underline{x} \in \mathbb{E}^{d+2}$, $\overline{\gamma}, \overline{\delta} \in [0,1]^{l_1 l_2}$, which is used in the theorem stated below.

Theorem 4.9. *The solution of problem (4.25) is the following stopping rule (see denotation* (4.29), (4.30) *and* (4.31)*):*

$$\boldsymbol{\tau}^* = \begin{cases} \inf\left\{ n \geq d+2 : (\underline{X}_{n-1-d,n}, \Xi_n) \in D^\star \right\}, \text{ if } \widetilde{h}(\Xi_{d+1}) < s^*(\underline{X}_{1,d+2}, \Xi_{d+2}), \\ d+1, \qquad \text{ if } \widetilde{h}(\Xi_{d+1}) \geq s^*(\underline{X}_{1,d+2}, \Xi_{d+2}), \end{cases} \tag{4.46}$$

where the stopping area $D^\star$*:*

$$D^\star = \left\{ (\underline{x}_{1,d+2}, \overline{\gamma}, \overline{\delta}) \in \vec{\Xi} : h(\underline{x}_{1,d+2}, \overline{\gamma}, \overline{\delta}) \geq s^*(\underline{x}_{1,d+2}, \overline{\gamma}, \overline{\delta}) \right\},$$

and $s^*(\underline{x}_{1,d+2}, \overline{\gamma}, \overline{\delta}) = \lim_{k\longrightarrow\infty} s_k(\underline{x}_{1,d+2}, \overline{\gamma}, \overline{\delta})$.

Proof: First let us consider subproblem of finding the optimal rule $\widetilde{\tau}^\star \in \mathfrak{F}^X_{d+2}$:

$$\boldsymbol{E}^\varphi\left[h(\underline{X}_{\widetilde{\tau}^\star-d-1,\widetilde{\tau}^\star}, \overline{\Pi}_{\widetilde{\tau}^\star}, \overline{B}_{\widetilde{\tau}^\star})\right] = \sup_{\tau\in\mathfrak{F}^X_{d+2}} \boldsymbol{E}^\varphi\left[h(\underline{X}_{\tau-d-1,\tau}, \overline{\Pi}_\tau, \overline{B}_\tau)\right]. \tag{4.47}$$

Then, basing on lemmas 4.6, 4.7 and according to optimal stopping theory (see Shiryaev (1978)) it is known that τ_0 defined by (4.27) can be expressed as

$$\tau_0 = \inf\{n \geq d+2 : h(\underline{X}_{n-1-d,n}, \overline{\Pi}_n, \overline{B}_n) \geq h^*(\underline{X}_{n-1-d,n}, \overline{\Pi}_n, \overline{B}_n)\},$$

where $h^*(\underline{x}_{1,d+2}, \overline{\gamma}, \overline{\delta}) = \lim_{k\longrightarrow\infty} \mathbf{Q}^k h(\underline{x}_{1,d+2}, \overline{\gamma}, \overline{\delta})$. The limit exists according to Lebesgue's theorem and structure of functions h and s_k. Lemma 4.17 implies that:

$$\begin{aligned} \tau_0 &= \inf\left\{n \geq d+2 : h(\underline{X}_{0,d+1}, \overline{\Pi}_n, \overline{B}_n)\right. \\ &\qquad \left.\geq \max\left\{h(\underline{X}_{n-1-d,n}, \overline{\Pi}_n, \overline{B}_n), s^*(\underline{X}_{n-1-d,n}, \overline{\Pi}_n, \overline{B}_n)\right\}\right\} \\ &= \inf\left\{n \geq d+2 : h(\underline{X}_{n-1-d,n}, \overline{\Pi}_n, \overline{B}_n) \geq s^*(\underline{X}_{n-1-d,n}, \overline{\Pi}_n, \overline{B}_n)\right\}. \end{aligned}$$

According to the optimality principle, rule $\widetilde{\tau}^*$ solves the maximization problem of (4.37) if only at $n = d+1$ the payoff $\widetilde{h}$ will be smaller than expected payoff in successive periods (for $n > d+1$). Thus, another word:

$$\boldsymbol{\tau}^\star = \widetilde{\tau}^*, \quad \text{if } \widetilde{h}(\underline{X}_{0,d+1}, \overline{\Pi}_{d+1}, \overline{B}_{d+1}) < s^*(\underline{X}_{1,d+2}, \overline{\Pi}_{d+2}, \overline{B}_{d+2}).$$

In opposite case $\boldsymbol{\tau}^\star = d+1$. This ends the proof of formula (4.46). □

4.3.5 Lemmata

The Appendix is devoted to the technical details mentioned in the text of the main part of the paper. We present formulae and lemmata which help to understand the details of the derivation of the optimal indicator of the disorder moment, which is a solution of the problem (4.25). However, a part of these considerations has no explicit application.

Let us introduce the n-dimensional distribution of the first n random variables when disorder appears before or after moment n, i.e. when $\theta \leq n$ and $\theta > n$, respectively. These densities with respect to some σ-finite measure have a form:

$$f_x^{\theta\leq n}(x_{1,n}) = \sum_{i,j}(\pi_{ij}\prod_{s=1}^{n} f_{x_{s-1}}^{1,i}(x_s) + \sum_{k=1}^{n}\bar{\pi}_{ij}p_{ij}^{k-1}q_{ij}\mathfrak{L}_k^{i,j}(\underline{x}_{0,n})\beta_{ij}$$

$$f_x^{\theta>n}(x_{1,n}) = \sum_{ij}\bar{\pi}_{ij}p_{ij}^{n}\prod_{s=1}^{n} f_{x_{s-1}}^{0,i}(x_s)\beta_{ij}.$$

Let us define the sequence of functions $S_n : \times_{i=1}^{n}\mathbb{E} \to \Re$ as follows: $S_0(x_0) = 1$ and for $n \geq 1$

$$\begin{aligned} S_n(x_{1,n}) &= f_x^{\theta\leq n}(x_{1,n}) + f_x^{\theta>n}(x_{1,n}) \\ &= \sum_{i,j}\left(\pi_{ij}\mathfrak{L}_0^{ij}(\underline{x}_{0,n}) + \bar{\pi}_{ij}(q_{ij}\sum_{k=1}^{n}p_{ij}^{k-1}\mathfrak{L}_k^{ij}(\underline{x}_{0,n}) + p_{ij}^{n}\mathfrak{L}_{n+1}^{ij}(\underline{x}_{0,n}))\right)\beta_{ij}. \end{aligned}$$

Remark 12. *For $n \geq l \geq 0$, $k > 0$, $i \in I_0$, $j \in I_1$, on the set $\{\omega : \underline{X}_{0,l} \in \underline{A}_{0,l}, A_0 = \{x\}, A_i \in \mathcal{F}_i, i \leq l\}$ the following equations hold:*

$$\begin{aligned} &\boldsymbol{P}^{\varphi}(\theta = n+k \mid \underline{X}_{0,l} \in \underline{A}_{0,l}, \underline{\beta} = (i,j), \theta > n) \\ &\qquad = \begin{cases} p_{ij}^{k-1}q_{ij}, & \text{if } n,k > 0, \\ \pi_{ij}, & \text{if } n = 0, k = 1, \\ (1-\pi_{ij})p_{ij}^{k-2}q_{ij}, & \text{if } n = 0, k > 1, \end{cases} \end{aligned} \tag{4.48}$$

Remark 13. *Let us observe that*

1. *as a simple consequence of the formula (4.48) we get*

$$\begin{aligned} &\boldsymbol{P}^{\varphi}(\theta > n+k \mid \underline{X}_{0,l} \in \underline{A}_{0,l}, \underline{\beta} = (i,j), \theta > n) \\ &\qquad = \begin{cases} p_{ij}^{k}, & \text{if } n,k > 0, \\ (1-\pi_{ij})p_{ij}^{k-1}, & \text{if } n = 0, k > 0. \end{cases} \end{aligned} \tag{4.49}$$

2. *and, for $k = 1$ by (4.49), we have:*

$$\begin{aligned} &\boldsymbol{P}^{\varphi}(\theta \neq n+1 \mid \underline{X}_{0,l} \in \underline{A}_{0,l}, \underline{\beta} = (i,j), \theta > n) \\ &\qquad = \boldsymbol{P}^{\varphi}(\theta > n+1 \mid \underline{X}_{0,l} \in \underline{A}_{0,l}, \underline{\beta} = (i,j), \theta > n) \\ &\qquad = \begin{cases} p_{ij}, & \text{if } n > 0, \\ (1-\pi_{ij}), & \text{if } n = 0. \end{cases} \end{aligned} \tag{4.50}$$

Remark 14. *Let us observe that conditional probabilities of the events* $\{\omega : \theta(\omega) > n\}$ *and* $\{\omega : \underline{\beta}(\omega) = (i,j)\}$ *for* $n \geq 1$, $i \in \mathbb{I}_0$, $j \in I_1$ *have form*

$$\boldsymbol{P}_x^{\varphi}(\theta > n \mid \tilde{\mathcal{F}}_n) = \bar{\pi}_{ij} p_{ij}^n \prod_{s=1}^{n} f_{x_{s-1}}^{0,i}(x_s)(S_n(x_{0,n}))^{-1},$$

$$B_n^{i,j} = b_{ij} S_n^{i,j}(\underline{x}_{0,n})(S_n(\underline{x}_{0,n}))^{-1}, \tag{4.51}$$

and $B_0^{i,j} = b_{ij}$.

Proof: For any $B_i \in \mathcal{B}$, $1 \leq i \leq n+1$ and $X_0 = x$ denote $D_n^{ij} = \{\omega : \underline{X}_{1,n}(\omega) \in \underline{B}_{1,n},\ \underline{\beta} = (i,j)\}$. We have

$$\int_{D_n^{ij}} \mathbb{I}_{\{\omega:\theta(\omega)>n\}} d\boldsymbol{P}_x^{\varphi} = \int_{D_n^{ij}} \boldsymbol{E}_x^{\varphi}(\mathbb{I}_{\{\omega:\theta>n\}}|\tilde{\mathcal{F}}_n) d\boldsymbol{P}_x^{\varphi} = \boldsymbol{P}_x^{\varphi}(\theta > n, D_n^{ij})$$

$$= \int_{\underline{B}_{1,n}} \bar{\pi}_{ij} p_{ij}^n \mathfrak{L}_{n+1}^{ij}(\underline{x}_{0,n})(S_n(\underline{x}_{0,n}))^{-1} \mu_x(d\underline{x}_{1,n}).$$

For verification (4.51) let us check first that

$$B_0^{i,j} = \boldsymbol{P}_x^{\varphi}(\underline{\beta} = (i,j) \mid \mathcal{F}_0) = \boldsymbol{P}^{\varphi}(\underline{\beta} = (i,j)) = b_{ij}.$$

On the set $D_n = \{\omega : \underline{X}_{1,n} \in \underline{B}_{1,n}\}$ when $X_0 = x$, we have

$$\int_{D_n} \mathbb{I}_{\{\underline{\beta}=(i,j)\}} d\boldsymbol{P}_x^{\varphi} = \int_{D_n} \boldsymbol{E}_x^{\varphi}(\mathbb{I}_{\{\underline{\beta}=(i,j)\}}|\mathcal{F}_n) d\boldsymbol{P}^{\varphi} = \boldsymbol{P}_x^{\varphi}(\underline{\beta} = (i,j), D_n)$$

$$= \int_{\underline{B}_{1,n}} \frac{b_{ij} S_n^{i,j}(\underline{x}_{0,n})}{S_n(\underline{x}_{0,n})} \mu_x(d\underline{x}_{1,n}) = \int_{D_n} \frac{b_{ij} S_n^{i,j}(\underline{X}_{0,n})}{S_n(\underline{X}_{0,n})} d\boldsymbol{P}_x^{\varphi}.$$

□

Equations (4.48)-(4.50) will be used in proofs of another lemmas collected in appendix.

Lemma 4.10. *For* $n > 0$, $k \geq 0$, $i \in I_0$, $j \in I_1$ *the following equation is satisfied:*

$$\boldsymbol{P}^{\varphi}(\theta \leq n+k \mid \tilde{\mathcal{F}}_n) = 1 - p_{ij}^k(1 - \Pi_n^{i,j}). \tag{4.52}$$

Proof: Let us suppose that $B_i \in \mathcal{B}$, $1 \leq i \leq n+1$ and $X_0 = x$. Denote $D_n = \{\omega : \underline{X}_{1,n}(\omega) \in \underline{B}_{1,n}\}$ and $D_n^{ij} = \{\omega : \underline{X}_{1,n}(\omega) \in \underline{B}_{1,n},\ \underline{\beta} = (i,j)\}$. For $A_i = \{\omega : X_i \in B_i\} \in \mathcal{F}_i$, $1 \leq i \leq n+1$, taking into account that $\underline{\beta}$ and θ are independent, we have

$$\begin{aligned}\boldsymbol{P}_x^\varphi(\theta > n+k, D_n^{ij}) &= \int\limits_{D_n^{ij}} \mathbb{I}_{\{\omega:\theta>n+k\}} d\boldsymbol{P}_x^\varphi = \int\limits_{D_n^{ij}} \boldsymbol{E}_x^\varphi(\mathbb{I}_{\{\omega:\theta>n+k\}}|\tilde{\mathcal{F}}_n) d\boldsymbol{P}_x^\varphi \\ &= \int\limits_{\{\omega:\underline{X}_{0,n}\in\underline{B}_{0,n}\}} \boldsymbol{E}_x^\varphi(\mathbb{I}_{\{\omega:\beta=(i,j)\}}\boldsymbol{E}_x^\varphi(\mathbb{I}_{\{\omega:\theta>n+k\}}|\tilde{\mathcal{F}}_n)|\mathcal{F}_n) d\boldsymbol{P}_x^\varphi \\ &= \int\limits_{\{\omega:\underline{X}_{0,n}\in\underline{B}_{0,n}\}} \boldsymbol{P}_x^\varphi(\mathbb{I}_{\{\omega:\beta=(i,j)\}}|\mathcal{F}_n)\boldsymbol{P}_x^\varphi(\mathbb{I}_{\{\omega:\theta>n+k\}}|\tilde{\mathcal{F}}_n) d\boldsymbol{P}_x^\varphi.\end{aligned} \tag{4.53}$$

By direct computation on the state space we get

$$\begin{aligned}\boldsymbol{P}_x^\varphi(\theta > n+k, D_n^{ij}) &= \int\limits_{\underline{B}_{1,n}} \sum_{s=n+k+1}^{\infty} (1-\pi_{ij}) q_{ij} b_{ij} p_{ij}^{s-2} \mathfrak{L}_0^{ij}(\underline{x}_{0,n})\mu(d\underline{x}_{1,n}) \\ &= p_{ij}^k \int\limits_{\underline{B}_{1,n}} \frac{(1-\pi_{ij}) b_{ij} p_{ij}^{n-1} \mathfrak{L}_0^{ij}(\underline{x}_{0,n})}{S_n^{i,j}(\underline{x}_{0,n})} \frac{S_n^{i,j}(\underline{x}_{0,n})}{S_n(\underline{x}_{0,n})} \mu_x(d\underline{x}_{1,n}) \\ &= p_{ij}^k \int\limits_{\{\omega:\underline{X}_{1,n}\in\underline{B}_{1,n}\}} \mathbb{I}_{\{\omega:\theta>n\}}\mathbb{I}_{\{\omega:\beta=(i,j)\}} d\boldsymbol{P}_x^\varphi \\ &= p_{ij}^k \int\limits_{\{\omega:\underline{X}_{1,n}\in\underline{B}_{1,n}\}} (1-\Pi_n^{i,j}) B_n^{i,j} d\boldsymbol{P}_x^\varphi\end{aligned} \tag{4.54}$$

Henceforth we have

$$\boldsymbol{E}^\varphi(\mathbb{I}_{\{\omega:\beta=(i,j)\}}\boldsymbol{E}^\varphi(\mathbb{I}_{\{\omega:\theta>n+k\}}|\tilde{\mathcal{F}}_n)|\mathcal{F}_n) = p_{ij}^k \boldsymbol{P}^\varphi(\beta=(i,j)|\mathcal{F}_n)\boldsymbol{P}^\varphi(\theta>n|\tilde{\mathcal{F}}_n).$$

Comparison of (4.53) and (4.54) implies 4.52 and this ends the proof of lemma.

□

Lemma 4.11. *For $n > k \geq 0$, $i \in I_0$, $j \in I_1$ it is true that*

$$\boldsymbol{P}^\varphi(\theta \leq n-k-1 \mid \tilde{\mathcal{F}}_n) = 1 - (1-\Pi_n^{i,j})\left(1 + q_{ij}\sum_{s=1}^{k+1} \frac{\mathfrak{L}_s^{i,j}(\underline{X}_{n-s+1,n})}{p_{ij}^s \mathfrak{L}_0^{i,j}(\underline{X}_{n-s+1,n})}\right). \tag{4.55}$$

Proof: If $n = k+1$ then

$$\boldsymbol{P}^\varphi(\theta \leq n-k-1 \mid \tilde{\mathcal{F}}_n) = \boldsymbol{P}^\varphi(\theta \leq 0 \mid \tilde{\mathcal{F}}_{k+1}) = 0.$$

Because of the fact that $\theta > 0$ a.s.:

$$\Pi_{n-k-1}^{i,j} = \Pi_0^{i,j} = \boldsymbol{P}^\varphi(\theta \leq 0 \mid \tilde{\mathcal{F}}_0) = 0. \tag{4.56}$$

Hence formula (4.55) holds. The case where $n > k + 1$, we have

$$\boldsymbol{P}^{\varphi}(\theta > n - k - 1 \mid \tilde{\mathcal{F}}_n) = \boldsymbol{P}^{\varphi}(\theta > n \mid \tilde{\mathcal{F}}_n) + \sum_{s=1}^{k+1} \boldsymbol{P}^{\varphi}(\theta = n - s \mid \tilde{\mathcal{F}}_n).$$

On the set $D_n = \{\omega : \underline{\beta} = (i, j), \underline{X}_{0,n} \in \underline{B}_{0,n}, B_0 = \{x\}\}$, we have

$$\boldsymbol{P}^{\varphi}(\theta = n - s, D_n) = \int_{D_n} \mathbb{I}_{\{\theta = n-s\}} d\boldsymbol{P}^{\varphi} = \int_{D_n} \boldsymbol{P}^{\varphi}(\theta = n - s|\tilde{\mathcal{F}}_n) d\boldsymbol{P}^{\varphi}$$

$$= \int_{\times_{r=1}^{n} B_r} (1 - \pi_{ij}) p_{ij}^{n-s-2} q_{ij} b_{ij} \mathfrak{L}_{s+1}^{i,j}(\underline{x}_{n-s,n}) d\mu(\underline{x}_{0,n})$$

$$= \int_{\times_{r=1}^{n} B_r} \frac{q_{ij} \mathfrak{L}_{s+1}^{i,j}(\underline{x}_{n-s,n})}{p_{ij}^{s+1} \mathfrak{L}_0^{i,j}(\underline{x}_{n-s,n})} \frac{(1 - \pi_{ij}) p_{ij}^{n-1} b_{ij} \mathfrak{L}_0^{i,j}(\underline{x}_{0,n})}{S_n^{i,j}(\underline{x}_{0,n})} \frac{S_n^{i,j}(\underline{x}_{0,n})}{S_n(\underline{x}_{0,n})} S_n(\underline{x}_{0,n}) d\mu(\underline{x}_{0,n})$$

$$= p_{ij}^{k} \int_{\{\omega : \underline{X}_{0,n} \in \underline{B}_{0,n}\}} \frac{q_{ij} \mathfrak{L}_{s+1}^{i,j}(\underline{X}_{n-s,n})}{p_{ij}^{s+1} \mathfrak{L}_0^{i,j}(\underline{X}_{n-s,n})} (1 - \Pi_n^{ij}) B_n^{i,j} d\boldsymbol{P}^{\varphi}.$$

Therefore

$$\boldsymbol{P}^{\varphi}(\theta = n - s|\tilde{\mathcal{F}}_n) = \frac{q_{ij} \mathfrak{L}_{s+1}^{i,j}(\underline{X}_{n-s,n})}{p_{ij}^{s+1} \mathfrak{L}_0^{i,j}(\underline{X}_{n-s,n})} (1 - \Pi_n^{i,j})$$

and

$$\boldsymbol{P}^{\varphi}(\theta > n - k - 1|\tilde{\mathcal{F}}_n) = \left(1 + q_{ij} \sum_{s=0}^{k} \frac{\mathfrak{L}_{s+1}^{i,j}(\underline{X}_{n-s,n})}{p_{ij}^{s+1} \mathfrak{L}_0^{i,j}(\underline{X}_{n-s,n})}\right) (1 - \Pi_n^{i,j}).$$

□

Lemma 4.12. *For $n > l \geq 0$, $i \in I_0$, $j \in I_1$ following equation holds:*

$$\Pi_n^{i,j} = \begin{cases} \Pi^{i,j}(l, \underline{X}_{n-l-1,n}, \Pi_{n-l-1}^{i,j}), & \text{if } n > l + 1, \\ \frac{\widetilde{A}(l, \underline{X}_{n-l-1,n}, \pi_{ij})}{\widetilde{\Psi}^{i,j}(l, \underline{X}_{0,l+1}, \pi_{ij})}, & \text{if } n = l + 1. \end{cases} \tag{4.57}$$

Remark 15. *In particular, taking $l = 0$, we get equation characterizing "one-step" dynamics of the process $\Pi_n^{i,j}$:*

$$\Pi_n^{i,j} = \begin{cases} \frac{f_{X_{n-1}}^{1,j}(X_n)(q_{ij} + p_{ij}\Pi_{n-1}^{i,j})}{f_{X_{n-1}}^{1,j}(X_n)(q_{ij} + p_{ij}\Pi_{n-1}^{i,j}) + f_{X_{n-1}}^{0,i}(X_n) p_{ij}(1 - \Pi_{n-1}^{i,j})}, & \text{if } n > 1, \\ \frac{f_{X_0}^{1,j}(X_1)\pi_{ij}}{f_{X_0}^{1,j}(X_1)\pi_{ij} + f_{X_0}^{0,i}(X_1)(1 - \pi_{ij})}, & \text{if } n = 1, \end{cases} \tag{4.58}$$

with initial condition $\Pi_0^{i,j} = 0$.

Remark 16. *For $l > 0$ recursive structure defined in equation (4.57) requires vector of initial states $\Pi_0^{i,j}, \Pi_1^{i,j}, \dots, \Pi_l^{i,j}$. State $\Pi_0^{i,j}$ is given above. To obtain remaining states $\Pi_1^{i,j}, \dots, \Pi_l^{i,j}$ it is enough to apply formula (4.58).*

Proof: (of Lemma 4.12) Condition $\Pi_0^{ij} = 0$ has been shown in lemma 4.11 (equation (4.56)). We have following recursive relation:

$$
\begin{aligned}
&S_{n-s-1}^{i,j}(\underline{x}_{0,n-s-1})\,\Psi^{i,j}(\underline{x}_{n-s-1,n}, \Pi^{i,j}(n-s, \underline{x}_{n-s-1,n}, \pi_{ij}))\\
&= S_{n-s-1}^{i,j}(\underline{x}_{0,n-s-1})\Pi_{n-s-1}^{i,j}\mathfrak{L}_{l+1}(\underline{x}_{n-s-1,n}) + S_{n-s-1}(\underline{x}_{0,n-s-1})(1-\Pi_{n-s-1}^{i,j})\\
&\times\left[q_{ij}\sum_{k=0}^{s} p_{ij}^{s-k}\mathfrak{L}_{k+1}(\underline{x}_{n-s-1,n}) + p_{ij}^{s+1}\mathfrak{L}_0(\underline{x}_{n-s-1,n})\right]\\
&= \left(\pi_{ij}\mathfrak{L}_{n-s}^{i,j}(\underline{x}_{0,n-s-1}) + (1-\pi_{ij})q_{ij}\sum_{k=1}^{n-s-1} p_{ij}^{k-1}\mathfrak{L}_{n-s-k}(\underline{x}_{0,n-s-1})\right)\\
&\times\mathfrak{L}_{s+1}(\underline{x}_{n-s-1,n}) + (1-\pi_{ij})p_{ij}^{n-s-1}\mathfrak{L}_0(\underline{x}_{0,n-s-1})\\
&\times\left(\sum_{k=0}^{l} p_{ij}^{s-k}q_{ij}\mathfrak{L}_{k+1}(\underline{x}_{n-s-1,n}) + p_{ij}^{s+1}\mathfrak{L}_0(\underline{x}_{n-s-1,n})\right)\\
&= \pi_{ij}\mathfrak{L}_n^{i,j}(\underline{x}_{0,n}) + (1-\pi_{ij})\left[\sum_{k=1}^{n} p_{ij}^{k-1}q_{ij}\mathfrak{L}_{n-k+1}(\underline{X}_{0,n}) + p_{ij}^{n}\mathfrak{L}_0(\underline{X}_{0,n})\right] = S_n^{i,j}(\underline{x}_{0,n}).
\end{aligned}
$$

Now, on the set $D_n = \{\omega : \underline{X}_{0,n} \in \underline{B}_{0,n}\}$, $X_0 = x$ and $B_i \in \mathcal{B}$, we have by (4.34):

$$
\begin{aligned}
\boldsymbol{P}(\theta > n, \beta = (i,j), D_n) &= \int_{\{\beta=(i,j),D_n\}} \mathbb{I}_{\{\theta>n\}} d\boldsymbol{P}^{\varphi} = \int_{\{\beta=(i,j),D_n\}} \boldsymbol{P}^{\varphi}(\theta > n|\tilde{\mathcal{F}}_n) d\boldsymbol{P}^{\varphi}\\
&= \int_{\underline{B}_{0,n}} \frac{p_{ij}^{s-1}\mathfrak{L}_0^{ij}(\underline{x}_{n-s-1,n})}{\Psi^{i,j}(n-s, \underline{x}_{n-s-1,n}, \Pi^{i,j}(n-s, \underline{x}_{n-s-1,n}, \pi_{ij}))}\\
&\times\frac{(1-\pi_{ij})p_{ij}^{n-s-2}\mathfrak{L}_0^{i,j}(\underline{x}_{0,n-s-1})}{S_{n-s-1}^{i,j}(\underline{x}_{0,n-s-1})}\frac{b_{ij}S_{n-s-1}^{i,j}(\underline{x}_{0,n-s-1})}{S_n(\underline{x}_{0,n})}S_n(\underline{x}_{0,n})\mu(d\underline{x}_{1,n})\\
&= \int_{D_n} \frac{p_{ij}^{s-1}\mathfrak{L}_0^{ij}(\underline{X}_{n-s-1,n})}{\Psi_{n-s-1}^{i,j}(\underline{X}_{n-s-1,n}, \Pi_{n-s-1}^{i,j})}(1-\Pi_{n-s-1}^{i,j})B_n^{i,j} d\boldsymbol{P}^{\varphi}.
\end{aligned}
$$

This follows

$$
\boldsymbol{P}^{\varphi}(\theta > n|\tilde{\mathcal{F}}_n) = \frac{p_{ij}^{s-1}\mathfrak{L}_0^{ij}(\underline{X}_{n-s-1,n})}{\Psi_{n-s-1}^{i,j}(\underline{X}_{n-s-1,n}, \Pi_{n-s-1}^{i,j})}(1-\Pi_{n-s-1}^{i,j}).
$$

In the case where $n = s + 1$ the proof is similar. □

Lemma 4.13. *For $i \in I_0$, $j \in I_1$ the following equations hold:*

$$
\boldsymbol{P}^{\varphi}(\theta > n+1|\tilde{\mathcal{F}}_n) = p_{ij}(1-\Pi_n^{i,j}) \tag{4.59}
$$

Proof: On the set $D_n = \{\omega : \underline{X}_{0,n} \in \underline{B}_{0,n}\}$, $X_0 = x$ and $B_i \in \mathcal{B}$, we have by (4.34):

$$\begin{aligned}
\boldsymbol{P}(\theta > n+1, \beta = (i,j), D_n) &= \int\limits_{\{\beta=(i,j),D_n\}} \mathbb{I}_{\{\theta>n+1\}} d\boldsymbol{P}^{\varphi} \\
&= \int\limits_{\{\beta=(i,j),D_n\}} \boldsymbol{P}^{\varphi}(\theta > n+1|\tilde{\mathcal{F}}_n) d\boldsymbol{P}^{\varphi} \\
&= \int\limits_{\underline{B}_{1,n}} (1-\pi_{ij}) p_{ij}^n b_{ij} \mathfrak{L}_0^{ij}(\underline{x}_{0,n}) \mu(d\underline{x}_{1,n}) \\
&= p_{ij} \int\limits_{\{\omega:\underline{\beta}=(i,j),D_n\}} \mathbb{I}_{\{\theta>n\}} d\boldsymbol{P}^{\varphi} = p_{ij} \int\limits_{\{\omega:\underline{\beta}=(i,j),D_n\}} \boldsymbol{P}^{\varphi}(\theta > n|\tilde{\mathcal{F}}_n) d\boldsymbol{P}^{\varphi}.
\end{aligned}$$

□

Lemma 4.14. *For $n > 0$, $i \in I_0$, $j \in I_1$, we have*

$$B_n^{i,j} = \begin{cases} \Gamma^{i,j}(0, \underline{X}_{n-1,n}, \overline{B}_{n-1}, \overline{\Pi}_{n-1}), & \text{if } n > 1, \\ \dfrac{b_{n-1}^{i,j} \widetilde{\Psi}^{i,j}(0, \underline{X}_{0,1}, \pi_{ij})}{\widetilde{S}(0, \underline{X}_{0,1}, \overline{b}, \overline{\pi})}, & \text{if } n = 1. \end{cases} \tag{4.60}$$

with condition $B_0^{i,j} = b_{ij}$.

Proof: First, let us verify the initial condition:

$$B_0^{i,j} = \boldsymbol{P}^{\varphi}(\underline{\beta} = (i,j) \mid \mathcal{F}_0) = \boldsymbol{P}^{\varphi}(\underline{\beta} = (i,j)) = b_{ij}.$$

Let $n > 1$. Let us consider formula (4.60) on the set $D_n = \{\omega : \underline{X}_{0,n} \in \underline{B}_{0,n}; B_0 = \{x\}, B_i \in \mathcal{B} \text{ for } 1 \le i \le n\}$:

$$\begin{aligned}
\boldsymbol{P}^{\varphi}(\underline{\beta} = (i,j), \underline{X}_{0,n} \in \underline{B}_{0,n}) &= \int_{D_n} \mathbb{I}_{\{\underline{\beta}=(ij)\}} d\boldsymbol{P}^{\varphi} = \int_{D_n} \boldsymbol{E}^{\varphi}(\mathbb{I}_{\{\underline{\beta}=(ij)\}} | \mathcal{F}_n) d\boldsymbol{P}^{\varphi} \\
&= \int_{\underline{B}_{1,n}} \frac{b_{ij} S_n^{i,j}(\underline{x}_{0,n})}{S_n(\underline{x}_{0,n})} S_n(\underline{x}_{0,n}) \mu(d\underline{x}_{1,n}) \qquad (4.61) \\
&= \int_{D_n} \frac{b_{ij} S_n^{i,j}(\underline{X}_{0,n})}{S_n(\underline{X}_{0,n})} d\boldsymbol{P}^{\varphi}.
\end{aligned}$$

Taking into account the formulae (4.36), (4.21), (4.23) and (4.33), we have gotten (4.60) for $n > 1$. The case $n = 1$ is a consequence of (4.61) and (4.33) with (4.22) and (4.24). □

Lemma 4.15. *For $n > 0$ the following equation holds:*

$$\boldsymbol{P}^{\varphi}(X_n \in [y, y+dy) \mid \mathcal{F}_{n-1}) = \begin{cases} S(0, X_{n-1}, y, \overline{B}_{n-1}, \overline{\Pi}_{n-1}) \mu(dy), & \text{if } n > 1, \\ \widetilde{S}(0, X_0, y, \overline{b}, \overline{\pi}) \mu(dy), & \text{if } n = 1. \end{cases}$$

Proof: Assume that $n > 1$, then using similar arguments as in proof of lemma 4.8 we get

$$\begin{aligned} &\boldsymbol{P}^{\varphi}\ (\ X_n \in [y, y+dy) \mid \mathcal{F}_{n-1}) \\ &\quad = \sum_{i,j} \boldsymbol{P}^{\varphi}(X_n \in [y, y+dy) \mid \mathcal{F}_{n-1}, \underline{\beta} = (i,j)) B_{n-1}^{i,j}. \end{aligned} \tag{4.62}$$

Notice that

$$\begin{aligned} &\boldsymbol{P}^{\varphi}(X_n \in [y, y+dy) \mid \mathcal{F}_{n-1}, \underline{\beta} = (i,j)) \\ &\quad = \frac{1 - \boldsymbol{P}^{\varphi}(\theta \le n-1 \mid \mathcal{F}_{n-1}, \underline{\beta} = (i,j))}{1 - \boldsymbol{P}^{\varphi}(\theta \le n \mid \mathcal{F}_n, \underline{\beta} = (i,j))} p_{ij} f^{0,i}_{X_{n-1}}(y)\mu(dy) \\ &\quad = \frac{1 - \Pi^{i,j}_{n-1}}{1 - \Pi^{i,j}_n} p_{ij} f^{0,i}_{X_{n-1}}(y)\mu(dy). \end{aligned} \tag{4.63}$$

Formula (4.58) in lemma 4.12 reveals the connection between $\Pi^{i,j}_n$ and $\Pi^{i,j}_{n-1}$. As a consequence, combining (4.62) and (4.63) we get the thesis for $n > 1$. The case $n = 1$ is proved similarly.

□

Lemma 4.16. *Let* $\eta_n = (\underline{X}_{n-d-1,n}, \overline{\Pi}_n, \overline{B}_n)$, *where* $n \ge d+1$. *System* $(\eta_n, \mathcal{F}_n, \boldsymbol{P}^{\varphi})$ *is Markov Random Function.*

Proof: It is enough to show that η_{n+1} is a function of η_n and variable X_{n+1} as well as that conditional distribution of X_{n+1} given $\mathcal{F}_n$ depends only on η_n (see Shiryaev(1978)).

For $x_1, ..., x_{d+2}, y \in \mathbb{E}, \gamma_{ij}, \delta_{ij} \in [0,1]$, $i \in I_0, j \in I_1$ let us consider the following function

$$\begin{aligned} \varphi(\underline{x}_{1,d+2}, \overline{\gamma}, \overline{\delta}, y) = \Big(&\underline{x}_{2,d+2}, y, \Pi^{1,1}(0, x_{d+2}, y, \delta_{11}), \ldots, \Pi^{1,l_2}(0, x_{d+2}, y, \delta_{1l_2}), \ldots, \\ &\Pi^{l_1,1}(0, x_{d+2}, y, \delta_{l_1 1}), \ldots, \Pi^{l_1,l_2}(0, x_{d+2}, y, \delta_{l_1 l_2}), \\ &\Gamma^{1,1}(0, x_{d+2}, y, \overline{\gamma}, \overline{\delta}), \ldots, \Gamma^{1,l_2}(0, x_{d+2}, y, \overline{\gamma}, \overline{\delta}), \ldots, \\ &\Gamma^{l_1,1}(0, x_{d+2}, y, \overline{\gamma}, \overline{\delta}), \ldots, \Gamma^{l_1,l_2}(0, x_{d+2}, y, \overline{\gamma}, \overline{\delta})\Big). \end{aligned}$$

We will show that $\eta_{n+1} = \varphi(\eta_n, X_{n+1})$. Using formulas (4.58) and (4.60) we express $\Pi^{i,j}_{n+1}$ as a function of $\Pi^{i,j}_n$ and $B^{i,j}_{n+1}$ as function on $B^{i,j}_n$. Then:

$$\begin{aligned} \varphi(\eta_n, X_{n+1}) &= \varphi(\underline{X}_{n-d-1,n}, \overline{\Pi}_n, \overline{B}_n, X_{n+1}) \\ &= \Big(\underline{X}_{n-d,n}, X_{n+1}, \Pi^{1,1}(0, \underline{X}_{n,n+1}, \Pi^{1,1}_n), \ldots, \Pi^{1,l_2}(0, \underline{X}_{n,n+1}, \Pi^{1,l_2}_n), \ldots, \\ &\quad \Pi^{l_1,1}(0, \underline{X}_{n,n+1}, \Pi^{l_1,1}_n), \ldots, \Pi^{l_1,l_2}(0, \underline{X}_{n,n+1}, \Pi^{l_1,l_2}_n), \\ &\quad \Gamma^{1,1}(0, \underline{X}_{n,n+1}, \overline{B}_n, \overline{\Pi}_n), \ldots, \Gamma^{1,l_2}(0, \underline{X}_{n,n+1}, \overline{B}_n, \overline{\Pi}_n), \ldots, \\ &\quad \Gamma^{l_1,1}(0, \underline{X}_{n,n+1}, \overline{B}_n, \overline{\Pi}_n), \ldots, \Gamma^{l_1,l_2}(0, \underline{X}_{n,n+1}, \overline{B}_n, \overline{\Pi}_n)\Big). \\ &= (\underline{X}_{n-d,n+1}, \overline{\Pi}_{n+1}, \overline{B}_{n+1}) = \eta_{n+1}. \end{aligned}$$

Let us now consider the conditional expectation $u(X_{n+1})$ under the condition of σ field $\mathcal{F}_n$, for the Borel function $u : \mathbb{E} \longrightarrow \Re$. Applying equation (4.59) and splitting by event $\{\theta \leq n+1\}$ we get:

$$\begin{aligned}
\boldsymbol{E}^{\varphi}(u(X_{n+1}) \mid \mathcal{F}_n) &= \sum_{i,j} \boldsymbol{E}^{\varphi}(u(X_{n+1})\mathbb{I}_{\{\underline{\beta}=(i,j)\}} \mid \mathcal{F}_n) \\
&= \sum_{i,j} B_n^{i,j} \Big[\boldsymbol{E}^{\varphi}\Big(p_{ij} \int_{\mathbb{E}} u(y)(1-\Pi^{i,j}(0,y,\Pi_n^{i,j})) f_{X_n}^{0,i}(y)\mu(dy) \mid \mathcal{F}_n \Big) \\
&\quad + \boldsymbol{E}^{\varphi}\Big(\int_{\mathbb{E}} u(y)(q_{ij} + p_{ij}\Pi^{i,j}(0,y,\Pi_n^{i,j})) f_{X_n}^{1,j}(X_{n+1})\mu(dy) \mid \mathcal{F}_n \Big) \Big]
\end{aligned} \tag{4.64}$$

We see that conditional distribution of X_{n+1} given $\mathcal{F}_n$ depends only on component of η_n what ends the proof.

□

Lemma 4.17. *Let*

$$s_k(\underline{x}_{1,d+2}, \overline{\gamma}, \overline{\delta}) = \begin{cases} \mathbf{T}\ \mathbf{Q}^k h(\underline{x}_{1,d+2}, \overline{\gamma}, \overline{\delta}), & \text{if } k \geq 1, \\ \mathbf{T}\ h(\underline{x}_{1,d+2}, \overline{\gamma}), & \text{if } k = 0. \end{cases} \tag{4.65}$$

Then, for function $h(\cdot)$ given by (4.38) and $k \geq 1$, following equalities hold:

$$\begin{aligned}
\mathbf{Q}^k h(\underline{x}_{1,d+2}, \overline{\gamma}, \overline{\delta}) = \max \Bigg\{ & \sum_{i,j} \Bigg(1 - p_{ij}^d + q_{ij} \sum_{m=1}^{d+1} \frac{\mathfrak{L}_m^{i,j}(\underline{x}_{1,d+2})}{p_{ij}^m \mathfrak{L}_0^{i,j}(\underline{x}_{1,d+2})} \Bigg) \\
& \times (1-\gamma_{ij})\delta_{ij}, s_{k-1}(\underline{x}_{1,d+2}, \overline{\gamma}, \overline{\delta}) \Bigg\},
\end{aligned}$$

$$\begin{aligned}
s_k(\underline{x}_{1,d+2}, \overline{\gamma}, \overline{\delta}) = \int_{\mathbb{E}} \max \Bigg\{ & \sum_{i,j} \Bigg(1 - p_{ij}^d + q_{ij} \sum_{m=1}^{d+1} \frac{\mathfrak{L}_m^{i,j}(\underline{x}_{1,d+3})}{p_{ij}^m \mathfrak{L}_0^{i,j}(\underline{x}_{1,d+3})} \Bigg) f_{x_{d+2}}^{0,i}(x_{d+3}) p_{ij} \\
& \times (1-\gamma_{ij})\delta_{ij}, s_{k-1}(\underline{x}_{2,d+3}, \overline{\gamma}, \overline{p} \circ \overline{f} 0_{x_{d+2}}(x_{d+3}) \circ \overline{\delta}) \Bigg\} \mu(dx_{d+3}),
\end{aligned}$$

where:

$$s_0(\underline{x}_{1,d+2}, \overline{\gamma}, \overline{\delta}) = \sum_{i,j} \Bigg(1 - p_{ij}^d + q_{ij} \sum_{m=1}^{d+1} \frac{\mathfrak{L}_{m-1}^{i,j}(\underline{x}_{2,d+2})}{p_{ij}^m \mathfrak{L}_0^{i,j}(\underline{x}_{2,d+2})} \Bigg) p_{ij}(1-\gamma_{ij})\delta_{ij}.$$

Moreover for $k \geq 0$ and vector $\eta_{n+1} = (\underline{X}_{n-d,n+1}, \overline{\Pi}_{n+1}, \overline{B}_{n+1})$, function s_k has the property:

$$s_k(\underline{X}_{n-d,n+1}, \overline{\Pi}_{n+1}, \overline{B}_{n+1}) = \frac{s_k(\underline{X}_{n-d,n+1}, \overline{\Pi}_n, \overline{p} \circ \overline{f} 0_{X_n}(X_{n+1}) \circ \overline{B}_n)}{S(0, \underline{X}_{n,n+1}, \overline{B}_n, \overline{\Pi}_n)}. \tag{4.66}$$

Proof: Notice that lemmas 4.12, 4.14, formulas (4.58) i (4.60) allow us to rewrite function $h(\underline{X}_{n-d,n+1}, \overline{\Pi}_{n+1}, \overline{B}_{n+1})$ in the following way:

$$h(\underline{X}_{n-d,n+1},\overline{\Pi}_{n+1},\overline{B}_{n+1}) = \sum_{i,j}\left(1 - p_{ij}^d + q_{ij}\sum_{m=1}^{d+1}\frac{\mathfrak{L}_m^{i,j}(\underline{X}_{n-d,n+1})}{p_{ij}^m\mathfrak{L}_0^{i,j}(\underline{X}_{n-d,n+1})}\right)$$
$$\times (1-\Pi_{n+1}^{i,j})B_{n+1}^{i,j}$$
$$= \sum_{i,j}\left(\frac{(1-p_{ij}^d)p_{ij}(1-\Pi_n^{i,j})B_n^{i,j}}{S(0,\underline{X}_{n,n+1},\overline{B}_n,\overline{\Pi}_n)}f_{X_n}^{0,i}(X_{n+1})\right.$$
$$\left.+q_{ij}\sum_{m=1}^{d+1}\frac{\mathfrak{L}_{m-1}^{i,j}(\underline{X}_{n-d,n})}{p_{ij}^m\mathfrak{L}_0^{i,j}(\underline{X}_{n-d,n})}\frac{p_{ij}(1-\Pi_n^{i,j})B_n^{i,j}}{S(0,\underline{X}_{n,n+1},\overline{B}_n,\overline{\Pi}_n)}f_{X_n}^{1,j}(X_{n+1})\right). \tag{4.67}$$

Using definition of operator $\mathbf{T}$, equation (4.67) and lemma 4.15, for $k = 0$ and $(\underline{X}_{n-1-d,n}, \overline{\Pi}_n, \overline{B}_n) = (\underline{x}_{1,d+2}, \overline{\gamma}, \overline{\delta})$ we get

$$s_0(\underline{x}_{1,d+2},\overline{\gamma},\overline{\delta}) = \boldsymbol{E}^{\varphi}(h(\underline{X}_{n-d,n},X_{n+1},\overline{\Pi}_{n+1},\overline{B}_{n+1})|\mathcal{F}_n)$$
$$= \sum_{i,j}\left(1-p_{ij}^d+q_{ij}\sum_{m=1}^{d+1}\frac{\mathfrak{L}_{m-1}^{i,j}(\underline{X}_{n-d,n})}{p_{ij}^m\mathfrak{L}_0^{i,j}(\underline{X}_{n-d,n})}\right)p_{ij}(1-\Pi_n^{i,j})B_n^{i,j}$$
$$= \sum_{i,j}\left(1-p_{ij}^d+q_{ij}\sum_{m=1}^{d+1}\frac{\mathfrak{L}_{m-1}^{i,j}(\underline{x}_{2,d+2})}{p_{ij}^m\mathfrak{L}_0^{i,j}(\underline{x}_{2,d+2})}\right)p_{ij}(1-\gamma_{ij})\delta_{ij}.$$

Hence, applying equations (4.58) and (4.60) one more time we end with

$$s_0(\underline{X}_{n-d,n+1},\overline{\Pi}_{n+1},\overline{B}_{n+1}) = \sum_{i,j}\left(1-p_{ij}^d+q_{ij}\sum_{m=1}^{d+1}\frac{\mathfrak{L}_{m-1}^{i,j}(\underline{X}_{n-d+1,n+1})}{p_{ij}^m\mathfrak{L}_0^{i,j}(\underline{X}_{n-d+1,n+1})}\right)$$
$$\times\frac{p_{ij}(1-\Pi_n^{i,j})p_{ij}f_{X_n}^{0,i}(X_{n+1})B_n^{i,j}}{S(0,\underline{X}_{n,n+1},\overline{B}_n,\overline{\Pi}_n)} = \frac{s_0(\underline{X}_{n-d,n+1},\overline{\Pi}_n,\overline{p}\circ\overline{f}0_{X_n}(X_{n+1})\circ\overline{B}_n)}{S(0,\underline{X}_{n,n+1},\overline{B}_n,\overline{\Pi}_n)}.$$

If $k = 1$, then by definition of $\mathbf{Q}$:

$$\mathbf{Q}h(\underline{x}_{1,d+2},\overline{\gamma},\overline{\delta}) = \max\left\{\sum_{i,j}\left(1-p_{ij}^d+q_{ij}\sum_{m=1}^{d+1}\frac{\mathfrak{L}_m^{i,j}(\underline{x}_{1,d+2})}{p_{ij}^m\mathfrak{L}_0^{i,j}(\underline{x}_{1,d+2})}\right)\right.$$
$$\left.\times(1-\gamma_{ij})\delta_{ij}, s_0(\underline{x}_{1,d+2},\overline{\gamma},\overline{\delta})\right\}.$$

Now, for $(\underline{X}_{n-1-d,n}, \overline{\Pi}_n, \overline{B}_n) = (\underline{x}_{1,d+2}, \overline{\gamma}, \overline{\delta})$, taking into account link between $\Pi_{n-1}^{i,j}$ and $\Pi_n^{i,j}$ as well as between $B_{n-1}^{i,j}$ and $B_n^{i,j}$ given by (4.58) and (4.60), we get with the support of 4.64:

$$
\begin{aligned}
s_1(\underline{x}_{1,d+2},\overline{\gamma},\overline{\delta}) &= \boldsymbol{E}^{\varphi}\Big[\max\{h(\underline{X}_{n-d,n},X_{n+1},\overline{\Pi}_{n+1},\overline{B}_{n+1}),\\
&\qquad s_0(\underline{X}_{n-d,n},X_{n+1},\overline{\Pi}_{n+1},\overline{B}_{n+1})\}\mid \mathcal{F}_n\Big]\\
&= \int_{\mathbb{E}} \max\Big\{\sum_{i,j}\Big(1-p_{ij}^d+q_{ij}\sum_{m=1}^{d+1}\frac{\mathfrak{L}_m^{i,j}(\underline{x}_{2,d+2},y)}{p_{ij}^m\mathfrak{L}_0^{i,j}(\underline{x}_{2,d+2},y)}\Big) f^{0,i}_{x_{d+2}}(y)p_{ij}\\
&\qquad \times(1-\gamma_{ij})\delta_{ij}, s_0(\underline{x}_{2,d+2},y,\overline{\gamma},\overline{p}\circ\overline{f}0_{x_{d+2}}(y)\circ\overline{\delta})\Big\}\mu(dy).
\end{aligned}
\tag{4.68}
$$

Basing on (4.68) with the help of (4.58) and (4.60) let us verify formula (4.66):

$$
\begin{aligned}
s_1(\underline{X}_{n-d,n+1},\overline{\Pi}_{n+1},\overline{B}_{n+1}) &= \int_{\mathbb{E}} \max\Big\{\sum_{i,j}\Big(1-p_{ij}^d+q_{ij}\sum_{m=1}^{d+1}\frac{\mathfrak{L}_m^{i,j}(\underline{X}_{n-d+1,n+1},y)}{p_{ij}^m\mathfrak{L}_0^{i,j}(\underline{X}_{n-d+1,n+1},y)}\Big)\\
&\times f^{0,i}_{X_{n+1}}(y)p_{ij}(1-\Pi^{i,j}_{n+1})B^{i,j}_{n+1}, s_0(\underline{X}_{n+1-d,n+1},y,\overline{\Pi}_{n+1},\overline{p}\circ\overline{f}0_{X_{n+1}}(y)\circ\overline{B}_{n+1})\Big\}\mu(dy)\\
&= \frac{s_1(\underline{X}_{n-d,n+1},\overline{\Pi}_n,\overline{p}\circ\overline{f}0_{X_n}(X_{n+1})\circ\overline{B}_n)}{S(0,\underline{X}_{n,n+1},\overline{B}_n,\overline{\Pi}_n)}.
\end{aligned}
$$

Suppose that lemma 4.17 holds for some $k>1$. We will show that equations characterizing $\mathbf{Q}^{k+1}h$ and s_{k+1} are true and that condition (4.66) for s_{k+1} is satisfied. It follows from definition of operator $\mathbf{Q}^{k+1}$ that:

$$
\begin{aligned}
\mathbf{Q}^{k+1}h(\underline{x}_{1,d+2},\overline{\gamma},\overline{\delta}) &= \max\Big\{\sum_{i,j}\Big(1-p_{ij}^d+q_{ij}\sum_{m=1}^{d+1}\frac{\mathfrak{L}_m^{i,j}(\underline{x}_{1,d+2})}{p_{ij}^m\mathfrak{L}_0^{i,j}(\underline{x}_{1,d+2})}\Big)\\
&\qquad\times(1-\gamma_{ij})\delta_{ij}, s_k(\underline{x}_{1,d+2},\overline{\gamma},\overline{\delta})\Big\}.
\end{aligned}
$$

Given $(\underline{X}_{n-1-d,n},\overline{\Pi}_n,\overline{B}_n)=(\underline{x}_{1,d+2},\overline{\gamma},\overline{\delta})$ and basing on inductive assumption, we have also:

$$
\begin{aligned}
s_{k+1}(\underline{x}_{1,d+2},\overline{\gamma},\overline{\delta}) &= \boldsymbol{E}^{\varphi}\Big[\max\Big\{h(\underline{X}_{n-d,n},X_{n+1},\overline{\Pi}_{n+1},\overline{B}_{n+1}),\\
&\qquad s_k(\underline{X}_{n-d,n},X_{n+1},\overline{\Pi}_{n+1},\overline{B}_{n+1})\Big\}\mid\mathcal{F}_n\Big]\\
&= \int_{\mathbb{E}} \max\Big\{\sum_{i,j}\Big(1-p_{ij}^d+q_{ij}\sum_{m=1}^{d+1}\frac{\mathfrak{L}_m^{i,j}(\underline{X}_{n-d,n},y)}{p_{ij}^m\mathfrak{L}_0^{i,j}(\underline{X}_{n-d,n},y)}\Big)\frac{f^{0,i}_{X_n}(y)p_{ij}(1-\Pi^{i,j}_n)B^{i,j}_n}{S(0,X_n,y,\overline{B}_n,\overline{\Pi}_n)},\\
&\qquad \frac{s_k(\underline{X}_{n-d,n},y,\overline{\Pi}_n,\overline{p}\circ\overline{f}0_{X_n}(y)\circ\overline{B}_n)}{S(0,X_n,y,\overline{B}_n,\overline{\Pi}_n)}\Big\}S(0,X_n,y,\overline{B}_n,\overline{\Pi}_n)\mu(dy)\\
&= \int_{\mathbb{E}} \max\Big\{\sum_{i,j}\Big(1-p_{ij}^d+q_{ij}\sum_{m=1}^{d+1}\frac{\mathfrak{L}_m^{i,j}(\underline{x}_{2,d+2},y)}{p_{ij}^m\mathfrak{L}_0^{i,j}(\underline{x}_{2,d+2},y)}\Big) f^{0,i}_{x_{d+2}}(y)p_{ij}\\
&\times(1-\gamma_{ij})\delta_{ij}, s_k(\underline{x}_{2,d+2},y,\overline{\gamma},\overline{p}\circ\overline{f}0_{x_{d+2}}(y)\circ\overline{\delta})\Big\}\mu(dy).
\end{aligned}
$$

Finally, we obtain:

$$
s_{k+1}(\underline{X}_{n-d,n+1},\overline{\Pi}_{n+1},\overline{B}_{n+1}) = \frac{s_{k+1}(\underline{X}_{n-d,n+1},\overline{\Pi}_n,\overline{p}\circ\overline{f}0_{X_n}(X_{n+1})\circ\overline{B}_n)}{S(0,\underline{X}_{n,n+1},\overline{B}_n,\overline{\Pi}_n)}.
$$

□

4.4 Known Procedure which can be Extended to Detect Multiple Disorders

In statistical analysis of data homogeneity, a natural problem arises: developing an algorithm for systematic identification of change points. A natural source of methods to solve this problem is the modification of algorithms that detect individual changes. The examination of homogeneity in collected data can simultaneously be treated as a hypothesis testing task. For analyzing this issue in real-time, a better approach is to formulate an estimation task. In the further part of this chapter, we will focus on tasks related to detecting disturbances in real-time.

4.4.1 CUSUM algorithms

The Cumulative Sums Method algorithm is typically used to detect small changes in the mean of a normal distribution, but it can also be applied to other types of distribution (v. p. 36 for explanation of the acronym). The CUSUM chart is based on the cumulative sum of the deviations of individual observations from a target value. The CUSUM chart is sensitive to small changes in the mean of the distribution and is less sensitive to changes in the standard deviation.

For other distributions, the CUSUM algorithm can still be applied, but it might not be as effective as for normally distributed data. In these cases, it may be necessary to make adjustments to the algorithm or use alternative methods.

In practice, CUSUM is often used with non-normal data, such as counts or proportions, by using an appropriate transformation to make the data approximately normally distributed.

4.4.2 The Shiryaev-Roberts algorithm

The *Shiryaev-Roberts*[2] algorithm is a statistical method used to detect changes in the mean of a time series. It is an extension of the classical cumulative sum (CUSUM) algorithm and is particularly useful for detecting changes in the mean of a time series with a small or unknown change point.

The algorithm works by comparing the current observation with the mean of the observations up to that point. If the current observation is significantly different from the mean, it is considered a change point.

The algorithm can be described as follows:

1. Initialize the algorithm by setting the mean and variance of the first m observations.
2. For each observation, calculate the likelihood ratio of the observation being from the current distribution compared to the distribution before the change.

[2]Roberts, S.W. –an American statistician (v. Roberts (1958, 1966)) affiliated to Bell Telephone Laboratories.

3. Compare this likelihood ratio with a threshold that will be determined by the desired false alarm rate.
4. If the likelihood ratio is greater than the threshold, the change point is detected and the new mean and variance are calculated.
5. Repeat the process for all observations.

The Shiryaev-Roberts algorithm is known to be asymptotically optimal and is more powerful than the CUSUM algorithm in certain situations, such as when the change point is small or unknown, or when the noise level is high.

It is important to note that the Shiryaev-Roberts algorithm assumes that you are detecting changes in the mean of a normal distribution, and the implementation could be different depending on the type of distribution you are dealing with.

Recently, the method has been extended to a more complicated model by Zhang and Mei (2023). In many real-world problems of real-time monitoring high-dimensional streaming data, one wants to detect an undesired event or change quickly once it occurs, but under the sampling control constraint in the sense that one might be able to only observe or use selected components' data for decision-making per time step in the resource-constrained environments. In this article, we propose to incorporate multiarmed bandit approaches into sequential change-point detection to develop an efficient bandit change-point detection algorithm based on the limiting Bayesian approach to incorporate prior knowledge of potential changes. Our proposed algorithm, termed Thompson-Sampling-Shiryaev-Roberts-Pollak (TSSRP in short), consists of two policies per time step: the adaptive sampling policy applies the Thompson sampling algorithm to balance between exploration to acquire long-term knowledge and exploitation for immediate reward gain, and the statistical decision policy fuses the local Shiryaev–Roberts–Pollak statistics to determine whether to raise a global alarm using sum shrinkage techniques. Extensive numerical simulations and case studies demonstrate the statistical and computational efficiency of our proposed TSSRP algorithm.

4.4.3 Double disorder of Markov chain

Let us continue the discussion of the disorder problem in Sections 1.6.1 and 4.3.1. A random sequence having segments which are homogeneous Markov processes is registered. Each segment has his own transition probability law and the length of the segment is unknown and random. The transition probabilities of each process are known and the joint *a priori* distribution of the disorder moments are given. Detection of the disorder is rarely precise. The decision-maker accepts some deviation in the estimation of the disorder moments. The aim is to indicate the segment of a given length between disorders with maximal probabilities.

The case with various precisions for overestimation and underestimation of the middle point is analyzed, including a situation where the disorders do not appear with positive probability and is also included. The observed sequence, when the change point is known, has Markov properties. The results explain the structure of an optimal detector under various circumstances, show new details of the solution construction, and insignificantly extend the range of application. The motivation for this investigation is modeling the selection of suspicious observations in the experiments. Such observations can be treated as outliers or disturbed. The objective is to detect such an inaccuracy immediately or in a very short time before or after its appearance with the highest probability. The problem is reformulated to provide optimal stopping of the observed sequences. A detailed analysis of the problem is presented to show the form of the optimal decision function.

The application of the results to the analysis of piecewise deterministic processes with change points appearing at the moment of jumps is shown (see Herberts and Jensen (2004), Poor and Hadjiliadis (2009), Ferenstein and Pasternak-Winiarski (2011).

Formulation of the problem.

Let $(\Omega, \mathcal{F}, \boldsymbol{P})$ be a probability space that supports a sequence of observable random variables $\{X_n\}_{n\in\mathbb{N}}$ generating filtration $\mathcal{F}_n = \sigma(X_0, X_1, ..., X_n)$. The random variables X_n take values in $(\mathbb{E}, \mathcal{B})$, where $\mathbb{E}$ is a subset of $\Re$. Space $(\Omega, \mathcal{F}, \boldsymbol{P})$ supports also unobservable random variables θ_1, θ_2 with values in $\mathbb{N}$ and the following distributions:

$$\begin{aligned} \boldsymbol{P}(\theta_1 = j) &= \mathbb{I}_{\{j=0\}}(j)\pi \\ &\quad + \mathbb{I}_{\{j>0\}}(j)\bar{\pi}p_1^{j-1}q_1, \end{aligned} \tag{4.69}$$

$$\begin{aligned} \boldsymbol{P}(\theta_2 = k \mid \theta_1 = j) &= \mathbb{I}_{\{k=j\}}(k)\rho \\ &\quad + \mathbb{I}_{\{k>j\}}(k)\bar{\rho}p_2^{k-j-1}q_2 \end{aligned} \tag{4.70}$$

where $j = 0, 1, 2, ...$, $k = j, j+1, j+2, ...$, $\bar{\pi} = 1 - \pi$, $\bar{\rho} = 1 - \rho$. Furthermore, we consider Markov processes $(X_n^i, \mathcal{G}_n^i, \boldsymbol{P}_x^i)$ on $(\Omega, \mathcal{F}, \boldsymbol{P})$, $i = 0, 1, 2$ where there σ-fields $\mathcal{G}_n^i$ are the smallest σ-fields which $(X_n^i)_{n=0}^{\infty}$, $i = 0, 1, 2$, are adapted, respectively. Let us define process $(X_n)_{n\in\mathbb{N}}$ in the following way:

$$\begin{aligned} X_n &= X_n^0 \mathbb{I}_{\{\theta_1 > n\}} + X_{n-\theta_1+1}^1 \mathbb{I}_{\{X_0^1 = x_{\theta_1-1}^0, \theta_1 \le n < \theta_2\}} \\ &\quad + X_{n-\theta_2+1}^2 \mathbb{I}_{\{X_0^2 = x_{\theta_2-\theta_1}^1, \theta_2 \le n\}}. \end{aligned}$$

We infer on θ_1 and θ_2 from the observable sequence $(X_n,\ n \in \mathbb{N})$ only. It should be emphasized that the sequence $(X_n,\ n \in \mathbb{N})$ is not Markovian under the admitted assumption, as it has been mentioned in Szajowski (1992), Yakir (1994) and Dube and Mazumdar (2001). However, the sequence satisfies the Markov property given θ_1 and θ_2 (v. Szajowski (1996) and Moustakides (1998)). Thus, for further consideration, we define filtration $\{\mathcal{F}_n\}_{n\in\mathbb{N}}$, where $\mathcal{F}_n = \sigma(X_0, X_1, ..., X_n)$, related to real observation. The variables θ_1, θ_2 do not have stopping times with respect to $\mathcal{F}_n$ and σ-fields $\mathcal{G}_n^\bullet$. Furthermore, we have knowledge about the distribution of (θ_1, θ_2) independent of any observation of the sequence $(X_n)_{n\in\mathbb{N}}$. This distribution, called *a priori distribution* of (θ_1, θ_2), is given by (4.69) and (4.70).

It is assumed that the measures $\boldsymbol{P}_x^i(\cdot)$ on $\mathcal{F}$, $i = 0, 1, 2$, have the following representation. For any $B \in \mathcal{B}$, we have

$$\boldsymbol{P}_x^i(\omega : X_1^i \in B) = \boldsymbol{P}(X_1^i \in B | X_0^i = x) = \int_B f_x^i(y)\mu(dy) = \int_B \mu_x^i(dy) = \mu_x^i(B),$$

where the functions $f_x^i(\cdot)$ are different and $f_x^i(y)/f_x^{(i+1)\mathrm{mod}3}(y) < \infty$ for $i = 0, 1, 2$ and for all $x, y \in \mathbb{E}$. We assume that the measures μ_x^i, $x \in \mathbb{E}$ are known in advance.

For any $D_n = \{\omega : X_i \in B_i,\ i = 1, \ldots, n\}$, where $B_i \in \mathcal{B}$, and any $x \in \mathbb{E}$ define

$$\begin{aligned}\boldsymbol{P}_x(D_n) = \boldsymbol{P}(D_n | X_0 = x) &= \int_{\times_{i=1}^n B_i} S_n(x, \boldsymbol{y}_n)\mu(d\boldsymbol{y}_n) = \int_{\times_{i=1}^n B_i} \mu_x(d\boldsymbol{y}_n) \\ &= \mu_x(\times_{i=1}^n B_i),\end{aligned}$$

where the sequence of functions $S_n : \times_{i=1}^n \mathbb{E} \to \Re$ is given by (4.86) in the appendix.

The presented model has the following heuristic justification: Two disorders occur in the observed sequence (X_n). They affect the distributions by changing their parameters. Disorders occur at two random times θ_1 and θ_2, $\theta_1 \leq \theta_2$. They split the sequence of observations into segments, at most three ones. The first segment is described by (X_n^0), the second one – for $\theta_1 \leq n < \theta_2$ – by (X_n^1). The third is given by (X_n^2) and is observed when $n \geq \theta_2$. When the first disorder occurs, there is a "switch" from the initial distribution to the distribution with the conditional density f_x^i with respect to the measure μ, where $i = 1$ or $i = 2$, when $\theta_1 < \theta_2$ or $\theta_1 = \theta_2$, respectively. Next, if $\theta_1 < \theta_2$, at random time θ_2, the distribution of observations becomes μ_x^2. We assume that the variables θ_1, θ_2 are not directly observable.

Let $\mathcal{S}$ denote the set of all stopping times with respect to the filtration $(\mathcal{F}_n)$, $n = 0, 1, \ldots$ and $\mathcal{T} = \{(\tau, \sigma) : \tau \leq \sigma,\ \tau, \sigma \in \mathcal{S}\}$. Two problems with three distributional segments are recalled to investigate them under the weaker assumption that there are at most three homogeneous segments.

Detection of the first change before the second one.

Our aim is to stop the observed sequence between the two disorders. This can be interpreted as a strategy for protecting against a second failure when the first has already happened. The mathematical model of this is to control the probability $\boldsymbol{P}_x(\tau < \infty, \theta_1 + d_1 \leq \tau < \theta_2 - d_2)$ by choosing the stopping time $\boldsymbol{\tau}^* \in \mathcal{S}$ for which the condition formulated in (1.19) of Section 1.8 is fulfilled.

Detection the moment of the first and second change.

Our aim is to indicate the switching moments with given precision d_1, d_2 (Problem $\mathrm{D}_{d_1 d_2}$). We want to determine a pair of stopping times $(\boldsymbol{\tau}^*, \boldsymbol{\sigma}^*) \in \mathcal{T}$ such that for every $x \in \mathbb{E}$ the condition given in (1.20) is fulfilled. The problem has been considered in Szajowski (1996) under natural simplification that there are three segments of data (i.e. there is $0 < \theta_1 < \theta_2$). In Section 4.4.3, the problem D_{00} is analyzed.

Schema of algorithm to get the estimates of disorder points.

Let us denote for $n = 0, 1, 2, \ldots$:

$$\begin{aligned} Z_n^{(d_1,d_2)} &= \boldsymbol{P}_x(\theta_1 + d_1 \le \tau < \theta_2 - d_2 \mid \mathcal{F}_n), \\ V_n^{(d_1,d_2)} &= \operatorname*{ess\,sup}_{\{\tau \in \mathfrak{S}^X,\ \tau \ge n\}} \boldsymbol{P}_x(\theta_1 + d_1 \le \tau < \theta_2 - d_2 \mid \mathcal{F}_n), \\ \tau_0 &= \inf\{n : Z_n^{(d_1,d_2)} = V_n^{(d_1,d_2)}\} \end{aligned} \tag{4.71}$$

Notice that, if $Z_\infty^{(d_1,d_2)} = 0$, then $Z_\tau^{(d_1,d_2)} = \boldsymbol{P}_x(\theta_1 + d_1 \le \tau < \theta_2 - d_2 \mid \mathcal{F}_\tau)$ for $\tau \in \mathfrak{S}^X$. Since $\mathcal{F}_n \subseteq \mathcal{F}_\tau$ (when $n \le \tau$), we have

$$\begin{aligned} V_n^{(d_1,d_2)} &= \operatorname*{ess\,sup}_{\tau \ge n} \boldsymbol{P}_x(-d_1 \le \theta - \tau \le d_2 \mid \mathcal{F}_n) \\ &= \operatorname*{ess\,sup}_{\tau \ge n} \boldsymbol{E}_x(Z_\tau^{(d_1,d_2)} \mid \mathcal{F}_n). \end{aligned}$$

The following lemma (see Bojdecki (1979), Sarnowski and Szajowski (2011)) ensures existence of the solution

Lemma 4.18. *The stopping time τ_0 defined by formula (4.71) is the solution of problem (1.19).*

The formulated problems are translated to optimal stopping problems for some Markov processes. The important part of the reformulation process is the choice of the *statistics* describing the knowledge of the decision-maker. The *a posteriori* probabilities of some events play a crucial role. Let us define the following *a posteriori* processes (cf. Yoshida (1983), Szajowski (1992)).

$$\Pi_n^i = \boldsymbol{P}_x(\theta_i \le n | \mathcal{F}_n), \tag{4.72}$$

$$\begin{aligned} \Pi_n^{12} &= \boldsymbol{P}_x(\theta_1 = \theta_2 > n | \mathcal{F}_n) \\ &= P_x(\theta_1 = \theta_2 > n | \mathcal{F}_{mn}), \end{aligned} \tag{4.73}$$

$$\Pi_{mn} = \boldsymbol{P}_x(\theta_1 = m, \theta_2 > n | \mathcal{F}_{mn}), \tag{4.74}$$

where $\mathcal{F}_{m\,n} = \mathcal{F}_n$ for $m, n = 1, 2, \ldots, m < n$, $i = 1, 2$. For a recursive representation of (4.72)–(4.74) we need the following functions:

$$\begin{aligned} \Pi^1(x, y, \alpha, \beta, \gamma) &= 1 - \frac{p_1(1-\alpha) f_x^0(y)}{\boldsymbol{H}(x, y, \alpha, \beta, \gamma)} \\ \Pi^2(x, y, \alpha, \beta, \gamma) &= \frac{(q_2\alpha + p_2\beta + q_1\gamma) f_x^2(y)}{\boldsymbol{H}(x, y, \alpha, \beta, \gamma)} \\ \Pi^{12}(x, y, \alpha, \beta, \gamma) &= \frac{p_1 \gamma f_x^0(y)}{\boldsymbol{H}(x, y, \alpha, \beta, \gamma)} \\ \Pi(x, y, \alpha, \beta, \gamma, \delta) &= \frac{p_2 \delta f_x^1(y)}{\boldsymbol{H}(x, y, \alpha, \beta, \gamma)} \end{aligned}$$

where $\boldsymbol{H}(x,y,\alpha,\beta,\gamma)=(1-\alpha)p_1f_x^0(y)+[p_2(\alpha-\beta)+q_1(1-\alpha-\gamma)]f_x^1(y)+[q_2\alpha+p_2\beta+q_1\gamma]f_x^2(y)$. In the sequel we adopt the following denotations

$$\begin{aligned}\boldsymbol{\alpha} &= (\alpha,\beta,\gamma) && (4.75)\\ \vec{\Pi}_n &= (\Pi_n^1,\Pi_n^2,\Pi_n^{12}). && (4.76)\end{aligned}$$

The basic formulae used in the transformation of the disorder problems to the stopping problems are given in the following

Lemma 4.19. *For each $x\in\mathbb{E}$ the following formulae, for $m,n=1,2,\ldots,m<n$, hold:*

$$\begin{aligned}\Pi_{n+1}^1 &= \Pi^1(X_n,X_{n+1},\Pi_n^1,\Pi_n^2,\Pi_n^{12}) && (4.77)\\ \Pi_{n+1}^2 &= \Pi^2(X_n,X_{n+1},\Pi_n^1,\Pi_n^2,\Pi_n^{12}) && (4.78)\\ \Pi_{n+1}^{12} &= \Pi^{12}(X_n,X_{n+1},\Pi_n^1,\Pi_n^2,\Pi_n^{12}) && (4.79)\\ \Pi_{m\,n+1} &= \Pi(X_n,X_{n+1},\Pi_n^1,\Pi_n^2,\Pi_n^{12},\Pi_{m\,n}) && (4.80)\end{aligned}$$

with boundary condition $\Pi_0^1=\pi$, $\Pi_0^2(x)=\bar{\pi}\rho$, $\Pi_0^{12}(x)=\bar{\pi}\rho$, and $\Pi_{m\,m}=(1-\rho)\frac{q_1f^1_{X_{m-1}}(X_m)}{p_1f^0_{X_{m-1}}(X_m)}(1-\Pi_m^1)$.

We have

$$\begin{aligned}Z_n^{(d_1,d_2)} &= \boldsymbol{P}_x(\theta_1+d_1\le\tau<\theta_2-d_2\mid\mathcal{F}_n) && (4.81)\\ &= \Pi_{n-d_1}^1-\Pi_{n+d_2}^2\\ &= u(\underline{X}_{n-d_1-1,n},\Pi_n^1,\Pi_n^2,\Pi_n^{12}).\end{aligned}$$

By Lemma 4.19 the conclusion is that the equivalent optimal stopping problem will be dependent on the segment of observations and the posterior distribution (see Ochman-Gozdek and Szajowski (2013)). The exact solution allows to get the approximate form of optimal detector.

4.4.4 A detailed description of the tailored model

Distributions of disordered samples.

Let us introduce the n-dimensional distribution for various configuration of disorders.

$$\begin{aligned}f_x^{\theta_1\le\theta_2\le n}(\boldsymbol{x}_{1,n}) &= \bar{\pi}\rho\sum_{j=1}^{n}\{p_1^{j-1}q_1\prod_{s=1}^{j-1}f^0_{x_{s-1}}(x_s)\prod_{t=j}^{n}f^2_{x_{t-1}}(x_t)\} && (4.82)\\ &+\bar{\pi}\bar{\rho}\sum_{j=1}^{n-1}\sum_{k=j+1}^{n}\{p_1^{j-1}q_1p_2^{k-j-1}q_2\prod_{s=1}^{j-1}f^0_{x_{s-1}}(x_s)\prod_{t=j}^{k-1}f^1_{x_{t-1}}(x_t)\prod_{u=k}^{n}f^2_{x_{u-1}}(x_u)\}\\ &+\bar{\pi}\rho\prod_{s=1}^{n}f^2_{x_{s-1}}(x_s)\end{aligned}$$

$$f_x^{\theta_1\le n<\theta_2}(\boldsymbol{x}_{1,n}) = \bar{\pi}\bar{\rho}\sum_{j=1}^{n}\{p_1^{j-1}q_1p_2^{n-j}\prod_{s=1}^{j-1}f^0_{x_{s-1}}(x_s)\prod_{t=j}^{n}f^1_{x_{t-1}}(x_t)\}$$
$$+ \pi\bar{\rho}\sum_{j=1}^{n}\{p_2^{j-1}q_2\prod_{s=1}^{j-1}f^1_{x_{s-1}}(x_s)\prod_{t=j}^{n}f^2_{x_{t-1}}(x_t)\} \quad (4.83)$$

$$f_x^{\theta_1=\theta_2>n}(\boldsymbol{x}_{1,n}) = \rho\bar{\pi}p_1^n\prod_{s=1}^{n}f^0_{x_{s-1}}(x_s) \quad (4.84)$$

$$f_x^{n<\theta_1<\theta_2}(\boldsymbol{x}_{1,n}) = \bar{\rho}\bar{\pi}p_1^n\prod_{s=1}^{n}f^0_{x_{s-1}}(x_s). \quad (4.85)$$

Let us define the sequence of functions $S_n : \times_{i=1}^{n}\mathbb{E} \to \Re$ as follows: $S_0(x_0) = 1$ and for $n \ge 1$

$$S_n(\boldsymbol{x}_n) = f_x^{\theta_1\le\theta_2\le n}(\boldsymbol{x}_{1,n}) + f_x^{\theta_1\le n<\theta_2}(\boldsymbol{x}_{1,n}) \quad (4.86)$$
$$+ f_x^{\theta_1=\theta_2>n}(\boldsymbol{x}_{1,n}) + f_x^{n<\theta_1<\theta_2}(\boldsymbol{x}_{1,n}).$$

Lemma 4.20. *For $n > 0$ the function $S_n(\boldsymbol{x}_{1,n})$ follows recursion*

$$S_{n+1}(\boldsymbol{x}_{1,n+1}) = \boldsymbol{H}(x_n, x_{n+1}, \vec{\Pi}_n)S_n(\boldsymbol{x}_{1,n}) \quad (4.87)$$

where

$$\boldsymbol{H}(x,y,\alpha,\beta,\gamma) = (1-\alpha)p_1f_x^0(y) + [p_2(\alpha-\beta)+q_1(1-\alpha-\gamma)]f_x^1(y) \quad (4.88)$$
$$+ [q_2\alpha + p_2\beta + q_1\gamma]f_x^2(y).$$

Proof: Let us assume $0 \le \theta_1 \le \theta_2$ and suppose that $B_i \in \mathcal{B}$, $1 \le i \le n+1$ and let us assume that $X_0 = x$ and denote $D_n = \{\omega : X_i(\omega) \in B_i, 1 \le i \le n\}$. For $A_i = \{\omega : X_i \in B_i\} \in \mathcal{F}_i$, $1 \le i \le n+1$, we have by properties of the density function $S_n(\boldsymbol{x})$ with respect to the measure $\mu(\cdot)$

$$\begin{aligned}
\int_{D_{n+1}} d\boldsymbol{P}_x &= \int_{\times_{i=1}^{n+1}B_i} S_{n+1}(\boldsymbol{x}_{n+1})\mu(d\boldsymbol{x}_{1,n+1}) \\
&= \int_{\times_{i=1}^{n}B_i}\int_{B_{n+1}} f(x_{n+1}|\boldsymbol{x}_n)\mu(dx_{n+1})S_n(\boldsymbol{x}_{0,n})\mu(d\boldsymbol{x}_{1,n}) \\
&= \int_{\times_{i=1}^{n}B_i} \boldsymbol{P}(A_{n+1}|\boldsymbol{X}_n = x_n)\mu_x(d\boldsymbol{x}_{1,n}) \\
&= \int_{D_n} \boldsymbol{P}_x(A_{n+1}|\boldsymbol{X}_{1,n})d\boldsymbol{P}_x = \int_{D_n} \boldsymbol{P}_x(A_{n+1}|\mathcal{F}_n)d\boldsymbol{P}_x \\
&= \int_{D_n} \mathbb{I}_{A_{n+1}}d\boldsymbol{P}_x
\end{aligned}$$

Now we split the conditional probability of A_{n+1} into the following parts

$$\begin{aligned}
\boldsymbol{P}_x(X_{n+1} \in A_{n+1} \mid \mathcal{F}_n)& \\
&= \boldsymbol{P}_x(n < \theta_1 < \theta_2, X_{n+1} \in A_{n+1} \mid \mathcal{F}_n) \quad (4.89)\\
&+ \boldsymbol{P}_x(\theta_1 \le n < \theta_2, X_{n+1} \in A_{n+1} \mid \mathcal{F}_n) \quad (4.90)\\
&+ \boldsymbol{P}_x(n < \theta_1 = \theta_2, X_{n+1} \in A_{n+1} \mid \mathcal{F}_n) \quad (4.91)\\
&+ \boldsymbol{P}_x(\theta_1 \le \theta_2 \le n, X_{n+1} \in A_{n+1} \mid \mathcal{F}_n) \quad (4.92)
\end{aligned}$$

In (4.89), we have:

$$\begin{aligned}
&\int_{D_n} \boldsymbol{P}_x(\theta_2 > \theta_1 > n, X_{n+1} \in A_{n+1} \mid \mathcal{F}_n) d\boldsymbol{P}_x \\
&= \int_{D_n} (\mathbb{I}_{\{\theta_1 = n+1\}} + \mathbb{I}_{\{\theta_1 > n+1\}}) \mathbb{I}_{A_{n+1}} d\boldsymbol{P}_x \\
&= \int_{\times_{i=1}^{n+1} B_i} (f_x^{n<\theta_1<\theta_2}(\boldsymbol{x}_{1,n})(p_1 f_{x_n}^0(x_{n+1}) + q_1 f_{x_n}^1(x_{n+1}))\mu(d\boldsymbol{x}_{1,n+1}) \\
&= \int_{\times_{i=1}^{n} B_i} (f_x^{n<\theta_1<\theta_2}(\boldsymbol{x}_{1,n}) \\
&\qquad \int_{B_{n+1}} (p_1 f_{x_n}^0(x_{n+1}) + q_1 f_{x_n}^1(x_{n+1}))\mu(dx_{n+1}))\mu(d\boldsymbol{x}_{1,n}) \\
&= \int_{D_n} \boldsymbol{P}_x(\theta_2 > \theta_1 > n \mid \mathcal{F}_n)[\boldsymbol{P}_{X_n}^0(A_{n+1})p_1 + q_1 \boldsymbol{P}_{X_n}^1(A_{n+1})] d\boldsymbol{P}_x.
\end{aligned}$$

In (4.90) we get by similar arguments as for (4.89)

$$\begin{aligned}
\boldsymbol{P}_x(\theta_1 \le n < \theta_2 \quad , \quad & X_{n+1} \in A_{n+1} \mid \mathcal{F}_n) \\
= \quad & \boldsymbol{P}_x(\theta_1 \le n < \theta_2, \theta_2 = n+1, X_{n+1} \in A_{n+1} \mid \mathcal{F}_n) \\
& + \boldsymbol{P}_x(\theta_1 \le n < \theta_2, \theta_2 \ne n+1, X_{n+1} \in A_{n+1} \mid \mathcal{F}_n) \\
= \quad & (\boldsymbol{P}_x(\theta_1 \le n \mid \mathcal{F}_n) - \boldsymbol{P}_x(\theta_2 \le n \mid \mathcal{F}_n)) \\
& \times [q_2 \boldsymbol{P}_{X_n}^2(A_{n+1}) + p_2 \boldsymbol{P}_{X_n}^1(A_{n+1})]
\end{aligned}$$

In (4.92) this part has the form:

$$\boldsymbol{P}_x(\theta_2 \le n, X_{n+1} \in A_{n+1} \mid \mathcal{F}_n) = \boldsymbol{P}_x(\theta_2 \le n \mid \mathcal{F}_n) \boldsymbol{P}_{X_n}^2(A_{n+1})$$

In (4.91) the conditional probability is equal to

$$\begin{aligned}
\boldsymbol{P}_x(\theta_1 = \theta_2 > n \quad , \quad & X_{n+1} \in A_{n+1} \mid \mathcal{F}_n) \\
= \quad & \boldsymbol{P}_x(\theta_1 = \theta_2 > n, \theta_2 = n+1, X_{n+1} \in A_{n+1} \mid \mathcal{F}_n) \\
& + \boldsymbol{P}_x(\theta_1 = \theta_2 > n, \theta_2 \ne n+1, X_{n+1} \in A_{n+1} \mid \mathcal{F}_n) \\
= \quad & \boldsymbol{P}_x(\theta_1 = \theta_2 > n \mid \mathcal{F}_n)[q_1 \boldsymbol{P}_{X_n}^2(A_{n+1}) + p_1 \boldsymbol{P}_{X_n}^0(A_{n+1})]
\end{aligned}$$

These formulae lead to

$$f(X_{n+1}|\boldsymbol{X}_{1,n}) = \boldsymbol{H}(X_n, X_{n+1}, \Pi_n^1, \Pi_n^2, \Pi_n^{12}).$$

which proves the lemma. □

4.4.5 Disorder Moments' Conditional Distribution

According to definition of Π_n^1, Π_n^2, Π_n^{12} we get

Lemma 4.21. *For the model described in Section 1.8 the following formulae are valid:*

1. $\boldsymbol{P}_x(\theta_2 > \theta_1 > n|\mathcal{F}_n) = 1 - \Pi_n^1 - \Pi_n^{12} = \frac{f_x^{n<\theta_1<\theta_2}(\boldsymbol{x}_{1,n})}{S_n(\boldsymbol{x}_n)}$;
2. $\boldsymbol{P}_x(\theta_2 = \theta_1 > n|\mathcal{F}_n) = \Pi_n^{12} = \frac{f_x^{\theta_1=\theta_2>n}(\boldsymbol{x}_{1,n})}{S_n(\boldsymbol{x}_n)}$;
3. $\boldsymbol{P}_x(\theta_1 \le n < \theta_2|\mathcal{F}_n) = \Pi_n^1 - \Pi_n^2$;
4. $\boldsymbol{P}_x(\theta_2 \ge \theta_1 > n|\mathcal{F}_n) = 1 - \Pi_n^1 = \frac{\bar{\pi}p_1^n \prod_{s=1}^n f_{x_{s-1}}^0(x_s)}{S_n(\boldsymbol{x}_n)}$.

Proof:

1. We have

$$\begin{aligned} \Omega &= \{\omega : n < \theta_1 < \theta_2\} \cup \{\omega : \theta_1 \le n < \theta_2\} \\ &\quad \cup \{\omega : \theta_1 \le \theta_2 \le n\} \cup \{\omega : \theta_1 = \theta_2 > n\}. \end{aligned} \tag{4.93}$$

Hence $1 = \boldsymbol{P}_x(\omega : n < \theta_1 < \theta_2|\mathcal{F}_n) + (\Pi_n^1 - \Pi_n^2) + \Pi_n^2 + \Pi_n^{12}$ and

$$\boldsymbol{P}_x(\omega : n < \theta_1 < \theta_2|\mathcal{F}_n) = 1 - \Pi_n^1 - \Pi_n^{12}.$$

Let $B_i \in \mathcal{B}$, $1 \le i \le n$, $X_0 = x$ and denote $D_n = \{\omega : X_i(\omega) \in B_i, 1 \le i \le n\}$. For $A_i = \{\omega : X_i \in B_i\} \in \mathcal{F}_i$, $1 \le i \le n$ and $D_n \in \mathcal{F}_n$, we have

$$\begin{aligned} \int_{D_n} \mathbb{I}_{\{\theta_2>\theta_1>n\}} d\boldsymbol{P}_x &= \int_{D_n} \boldsymbol{P}_x(\theta_2 > \theta_1 > n|\mathcal{F}_n) d\boldsymbol{P}_x \\ &= \int_{D_n} \boldsymbol{P}_x(\theta_2 > \theta_1 > n|\boldsymbol{X}_n) d\boldsymbol{P}_x \\ &= \boldsymbol{P}_x(\theta_2 > \theta_1 > n, D_n) \\ &= \int_{\times_{i=1}^n B_i} f_x^{n<\theta_1<\theta_2}(\boldsymbol{x}_{1,n}) \mu(d\boldsymbol{x}_{1,n}) \\ &= \int_{\times_{i=1}^n B_i} f_x^{n<\theta_1<\theta_2}(\boldsymbol{x}_{1,n}) (S_n(\boldsymbol{x}_n))^{-1} \mu_x(d\boldsymbol{x}_{1,n}) \\ &= \int_{D_n} f_x^{n<\theta_1<\theta_2}(\boldsymbol{X}_{1,n}) (S_n(\boldsymbol{X}_n))^{-1} d\boldsymbol{P}_x. \end{aligned}$$

Thus, $\boldsymbol{P}_x(\theta_2 > \theta_1 > n|\mathcal{F}_n) = \bar{\rho}\bar{\pi}p_1^n \prod_{i=1}^n f_{X_{i-1}}^0(X_i)(S_n(\boldsymbol{X}_n))^{-1}$.

2. The second formula can be obtained by a similar argument.
3. Let $\theta_1 \leq \theta_2$. Since $\{\omega : \theta_2 \leq n\} \subset \{\omega : \theta_1 \leq n\}$ it follows that $\boldsymbol{P}_x(\{\omega : \theta_1 \leq n < \theta_n\} | \mathcal{F}_n) = \boldsymbol{P}_x(\{\omega : \theta_1 \leq n\} \setminus \{\omega : \theta_2 \leq n\} | \mathcal{F}_n) = \Pi_n^1 - \Pi_n^2$.

These end the proof of the lemma. □

Remark 17. *Let $B_i \in \mathcal{B}$, $1 \leq i \leq n+1$, $X_0 = x$ and denote $D_n = \{\omega : X_i(\omega) \in B_i, 1 \leq i \leq n\}$. For $A_i = \{\omega : X_i \in B_i\} \in \mathcal{F}_i$, $1 \leq i \leq n$ and $D_n \in \mathcal{F}_n$, we have*

$$\begin{aligned} \int_{D_n} \mathbb{I}_{\{\theta_1 > n\}} d\boldsymbol{P}_x &= \int_{D_n} \boldsymbol{P}_x(\theta_1 > n | \mathcal{F}_n) d\boldsymbol{P}_x = \int_{D_n} \boldsymbol{P}_x(\theta_1 > n | \boldsymbol{X}_n) d\boldsymbol{P}_x \\ &= \boldsymbol{P}_x(\theta_1 > n, D_n) = \int_{\times_{i=1}^n B_i} p_1^n \prod_{i=1}^n f_{x_{i-1}}^0(x_i) \mu(d\boldsymbol{x}_{1,n}) \\ &= \int_{\times_{i=1}^n B_i} p_1^n \prod_{i=1}^n f_{x_{i-1}}^0(x_i) (S_n(\boldsymbol{x}_n))^{-1} \mu_x(d\boldsymbol{x}_{1,n}). \end{aligned}$$

Thus, $\boldsymbol{P}_x(\theta_1 > n | \mathcal{F}_n) = p_1^n \prod_{i=1}^n f_{X_{i-1}}^0(X_i)(S_n(\boldsymbol{X}_n))^{-1}$. Moreover

$$1 - \Pi_{n+1}^1 = p_1 f_{X_n}^0(X_{n+1})(1 - \Pi_n^1) S_n(\boldsymbol{X}_n)(S_{n+1}(\boldsymbol{X}_{n+1}))^{-1}$$

and $S_{n+1}(\boldsymbol{X}_{n+1}) = \boldsymbol{H}(X_n, X_{n+1}, \vec{\Pi}_n^1) S_n(\boldsymbol{X}_n)$. Hence

$$\Pi_{n+1}^1 = 1 - \frac{p_1 f_{X_n}^0(X_{n+1})(1 - \Pi_n^1)}{\boldsymbol{H}(X_n, X_{n+1}, \vec{\Pi}_n)}.$$

Some recursive formulae

In the derivation of the formulae in Theorem 4.19 the form of the distribution of some random vectors is taken into account.

Lemma 4.22. *For the model described in Section 1.8 the following formulae are valid:*

1. $\boldsymbol{P}_x(\theta_2 = \theta_1 > n+1 | \mathcal{F}_n) = p_1 \Pi_n^{12} = p_1 \rho (1 - \Pi_n^1)$;
2. $\boldsymbol{P}_x(\theta_2 > \theta_1 > n+1 | \mathcal{F}_n) = p_1(1 - \Pi_n^1 - \Pi_n^{12})$;
3. $\boldsymbol{P}_x(\theta_1 \leq n+1 | \mathcal{F}_n) = \boldsymbol{P}_x(\theta_1 \leq n+1 < \theta_2 | \mathcal{F}_n) + \boldsymbol{P}_x(\theta_2 \leq n+1 | \mathcal{F}_n)$;
4. $\boldsymbol{P}_x(\theta_1 \leq n+1 < \theta_2 | \mathcal{F}_n) = q_1(1 - \Pi_n^1 - \Pi_n^{12}) + p_2(\Pi_n^1 - \Pi_n^2)$;
5. $\boldsymbol{P}_x(\theta_2 \leq n+1 | \mathcal{F}_n) = q_2 \Pi_n^1 + p_2 \Pi_n^2 + q_1 \Pi_n^{12}$.
6. $\boldsymbol{P}_x(\theta_1 = m, \theta_2 > n+1 | \mathcal{F}_n) = p_2 \Pi_{m\,n}$.

Proof:

1. On the set $D = \{\omega : X_0 = x, X_1 \in A_1, X_2 \in A_2, \ldots, X_n \in A_n\} \in \mathcal{F}_n$, we have

$$\begin{aligned}
\int_D \mathbb{I}_{\{\theta_2=\theta_1>n+1\}} d\boldsymbol{P}_x &= \boldsymbol{P}_x(D)\boldsymbol{P}_x(\theta_2 = \theta_1 > n+1|D) \\
&= \rho\bar{\pi} \sum_{j=n+2}^{\infty} p_1^{j-1} q_1 \int_{\times_{i=1}^n A_i} \prod_{i=1}^{n} f^0_{x_{i-1}}(x_i)\mu(d\boldsymbol{x}_{1,n}) \\
&= p_1 \rho \bar{\pi} p_1^n \int_{\times_{i=1}^n A_i} \prod_{i=1}^{n} f^0_{x_{i-1}}(x_i)\mu(d\boldsymbol{x}_{1,n}) \\
&= p_1 \boldsymbol{P}_x(D)\boldsymbol{P}_x(\theta_2 = \theta_1 > n|D) \\
&= p_1 \int_D \mathbb{I}_{\{\theta_2=\theta_1>n\}} d\boldsymbol{P}_x.
\end{aligned}$$

By (4.73) and the definition of the conditional probability this implies $\boldsymbol{P}_x(\theta_2 = \theta_1 > n+1|\mathcal{F}_n) = p_1 \Pi_n^{12}$. Next,

$$\begin{aligned}
\int_D \mathbb{I}_{\{\theta_1>n\}} d\boldsymbol{P}_x &= \boldsymbol{P}_x(D)\boldsymbol{P}_x(\theta_1 > n|D) \\
&= \bar{\pi} \sum_{j=n+1}^{\infty} p_1^{j-1} q_1 \int_{\times_{i=1}^n A_i} \prod_{i=1}^{n} f^0_{x_{i-1}}(x_i)\mu(d\boldsymbol{x}_{1,n}) \\
&= \frac{1}{\rho}\boldsymbol{P}_x(D)\boldsymbol{P}_x(\theta_2 = \theta_1 > n|D) = \frac{1}{\rho}\int_D \mathbb{I}_{\{\theta_2=\theta_1>n\}} d\boldsymbol{P}_x.
\end{aligned}$$

These prove the part 1 of the lemma.

2. Similarly as above, we get

$$\begin{aligned}
\int_D \mathbb{I}_{\{\theta_2>\theta_1>n+1\}} d\boldsymbol{P}_x &= \boldsymbol{P}(D)\boldsymbol{P}_x(\theta_2 > \theta_1 > n+1|D) \\
&= p_1 \rho \bar{\pi} p_1^n \int_{\times_{i=1}^n A_i} \prod_{i=1}^{n} f^0_{x_{i-1}}(x_i)\mu(d\boldsymbol{x}_{1,n}) \\
&= p_1 \boldsymbol{P}(D)\boldsymbol{P}_x(\theta_2 > \theta_1 > n|D) = p_1 \int_D \mathbb{I}_{\{\theta_2>\theta_1>n\}} d\boldsymbol{P}_x
\end{aligned}$$

By point 2 of Lemma 4.21 we get the formula 2 of the lemma.

3. It is obvious by assumption $\theta_1 \leq \theta_2$.
4. On the set D, we have

$$\begin{aligned}
\int_D \mathbb{I}_{\{\theta_1\leq n+1<\theta_2\}} d\boldsymbol{P}_x &= \boldsymbol{P}(D)\boldsymbol{P}_x(\theta_1 \leq n+1 < \theta_2|D) \\
&\overset{(1.16)}{\underset{(1.17)}{=}} \sum_{j=0}^{n+1} \boldsymbol{P}(\omega : \theta_1 = j) \sum_{k=n+2}^{\infty} \bar{\rho} p_2^{k-j-1} q_2 \int_{\times_{i=1}^n A_i} \prod_{s=1}^{j-1} f^0_{x_{s-1}}(x_s) \times
\end{aligned}$$

$$\prod_{r=j}^{n} f^1_{x_{r-1}}(x_r)\mu(d\boldsymbol{x}_{1,n})$$

$$= \bar{\pi} p_1^n q_1(1-\rho)\int_{\times_{i=1}^n A_i}\prod_{s=1}^{n} f^0_{x_{s-1}}(x_s)\mu(d\boldsymbol{x}_{1,n})$$

$$+ p_2 \sum_0^n \boldsymbol{P}(\omega : \theta_1 = j) p_2^{n+1-j}\int_{\times_{i=1}^n A_i}\prod_{s=1}^{j-1} f^0_{x_{s-1}}(x_s)\times$$

$$\prod_{r=j}^{n} f^1_{x_{r-1}}(x_r)\mu(d\boldsymbol{x}_{1,n})$$

$$\overset{(L.4.21)}{=} q_1\boldsymbol{P}(D)\boldsymbol{P}_x(\theta_2 > \theta_1 > n|D) + p_2\boldsymbol{P}(D)\boldsymbol{P}_x(\theta_1 \le n < \theta_2|D)$$

$$= q_1\int_D \mathbb{I}_{\{\theta_2>\theta_1>n\}} d\boldsymbol{P}_x + p_2\int_D \mathbb{I}_{\{\theta_1\le n<\theta_2\}} d\boldsymbol{P}_x.$$

5. If we substitute n by $n+1$ in (4.93) than we obtain

$$\begin{aligned}\boldsymbol{P}_x(\theta_2 \le n+1|\mathcal{F}_n)&\\ &= 1 - \boldsymbol{P}_x(n+1 < \theta_1 = \theta_2|\mathcal{F}_n)\\ &- \boldsymbol{P}_x(n+1 < \theta_1 < \theta_2|\mathcal{F}_n) - \boldsymbol{P}_x(\theta_1 \le n+1 < \theta_2|\mathcal{F}_n)\\ &= 1 - p_1\Pi_n^{12} - p_1(1-\Pi_n^1 - \Pi_n^{12}) - q_1(1-\Pi_n^1-\Pi_n^{12})\\ &+ p_2(\Pi_n^2 - \Pi_n^1) = q_2\Pi_n^1 + p_2\Pi_n^2 + q_1\Pi_n^{12}.\end{aligned}$$

6. We have

$$\int_D \mathbb{I}_{\{\theta_1=m,\theta_2>n+1\}} d\boldsymbol{P}_x$$

$$= \boldsymbol{P}_x(D)\boldsymbol{P}_x(\theta_1 = m, \theta_2 > n+1|D)$$

$$= \bar{\pi}\bar{\rho} p_1^{m-1} q_1 \sum_{j=n+2}^{\infty} p_2^{j-m-1} q_2 \int_{\times_{i=1}^n B_i}\prod_{i=1}^{m} f^0_{x_{i-1}}(x_i)\times$$

$$\prod_{j=m+1}^{n} f^1_{x_{j-1}}(x_j)\mu(d\boldsymbol{x}_{1,n})$$

$$= p_2\bar{\pi}\bar{\rho} p_1^{m-1} q_1 p_2^{n-m}\int_{\times_{i=1}^n B_i}\prod_{i=1}^{m} f^0_{x_{i-1}}(x_i)\prod_{j=m+1}^{n} f^1_{x_{j-1}}(x_j)\mu(d\boldsymbol{x}_{1,n})$$

$$= p_2\boldsymbol{P}_x(D)\boldsymbol{P}_x(\theta_1 = m, \theta_2 > n|D) = p_2\int_D \mathbb{I}_{\{\theta_1=m,\theta_2>n\}} d\boldsymbol{P}_x.$$

By (4.74) and the definition of conditional probability this implies $\boldsymbol{P}_x(\theta_2 = m, \theta_1 > n+1|\mathcal{F}_n) = p_2\Pi_{n\ m}$. These prove the part 6 of the lemma.

□

5

Financial Applications

> Success in investing doesn't correlate with I.Q. Once you have ordinary intelligence, what you need is the temperament to control the urges that get other people into trouble in investing.
>
> Warren Buffett

MULTIPLE RATIONAL stopping points in finance refer to the concept of having several predetermined conditions or criteria that trigger a decision to stop or exit an investment or trading position. These stopping points are based on rational analysis of various factors, such as market conditions, risk tolerance, investment goals, and financial metrics. Let us emphasize some crucial issue supported by these models in practice.

In *risk management* setting multiple-stopping points helps to manage risk by providing various exit opportunities based on different risk levels. For example, an investor might have one stopping point based on a percentage loss from the initial investment, another based on a specific price level, and another based on a technical indicator signaling a trend reversal. This diversification of exit strategies helps mitigate the impact of unexpected market movements.

Supports *flexibility*. Having multiple-stopping points allows investors to adapt to changing market conditions and adjust their strategies accordingly. If the market behaves differently than expected, having alternative exit criteria enables investors to respond more effectively and protect their capital.

Profit taking needs rational stopping points to secure profits. Investors may have multiple targets to make profits at different price levels or based on different performance metrics. This approach helps investors lock in gains while still allowing for potential upside if the investment continues to perform well.

Objective decision-making is performed by establishing predetermined stopping points based on quantitative criteria. By this option, investors can make more objective decisions without being swayed by emotions or market noise. This disciplined approach helps investors stick to their investment strategy and avoid impulsive or irrational decisions.

Having multiple-stopping points encourages regular *monitoring and evaluation* investment positions. Investors can continuously assess whether the investment thesis remains valid and whether the underlying fundamentals or market dynamics have

DOI: 10.1201/9781003407102-5

changed. If any of the predetermined stopping points are activated, it prompts a reassessment of the investment and the need for potential action.

The specifications above show that multiple rational stopping points provide investors with a structured framework to manage risk, secure profits, and make objective decisions in the dynamic environment of financial markets.

5.1 Variation on Buying-Selling Problems

Motto:

Strategic Decision-Making in the Housing Market: Navigating Uncertainty with Informed Insight.

WE EXAMINE buying-selling scenarios requiring two or more stops (contracts), a subject widely explored using diverse techniques. Section 1.6.9 introduces prior research on the topic. Additionally, we provide insights into constructing a mathematical model for buying-selling within the housing market context.

In agent-based modeling of the housing market, strategies for buyers' timing can be formulated based on modeled behaviors and decision-making processes of individual agents. By simulating various scenarios and market conditions, such as changes in prices, interest rates, and demand, patterns emerge to inform buyer strategies. Data collection, particularly raw data, is crucial for this approach. Key information categories include:

Market Conditions: Optimal buying opportunities arise when prices are low or interest rates are favorable.

Personal Circumstances: Factors like financial stability influence a buyer's decision.

Market Trends: Observing trends aids in strategic buying to maximize investment or find desired properties.

Agent Interactions: Considering behaviors of other agents helps adjust strategy, e.g., timing based on observed bidding wars.

Incorporating these factors into agent-based models enables informed buyer strategies. However, mathematical models of the housing market, based on stochastic processes with statistical calibration, are abstractions. Despite attempts to include relevant variables, not all factors influencing housing availability and prices are fully captured.

These models recognize house prices as random variables, reflecting market uncertainty due to economic conditions, sentiment, and policy changes. By accounting for stochastic price movements, probabilistic analysis and scenario planning become possible. This approach allows for assessing a range of outcomes and associated probabilities, enhancing decision-making.

While valuable, these models have inherent limitations and uncertainties due to their abstraction and stochastic nature, necessitating caution in interpretation and forecasting. Accordingly, special cases of this approach supporting agent strategy will be presented in forthcoming sections.

5.1.1 Independent case

Let $\xi_1, \xi_2, \ldots, \xi_N$ be a sequence of random variables, where $\xi_n = f_n + \varepsilon_n$, $f_n = f(n)$, $n = 1, 2, \ldots, N$, is a known deterministic function (for example, a trend), $\{\varepsilon_n, n = 1, \ldots, N\}$ is a sequence of independent random variable with $\boldsymbol{E}\varepsilon_n = 0$. The random variable ξ_n can be interpreted as a value of asset at time n. We observe these random variables sequentially and have to decide when we must stop. The first stopping means a buying an asset, the second one signifies a selling an asset. Our decision to stop depends on the observations already made, but does not depend on the future which is not yet known. After two stoppings at times m_1, m_2, $1 \le m_1 < m_2 \le N$ we get again $Z_{m_1,m_2} = \xi_{m_2} - \xi_{m_1}$. If we do not buy anything until time N then we will obtain the gain $Z_{m_1,m_2} = 0$, we may assume that $m_1 = N, m_2 = N+1$. This implies that

$$\begin{aligned} &1 \le m_1 \le N_1, \quad N_1 = N, \\ &m_1 < m_2 \le N_2(m_1), \quad N_2(m_1) = \begin{cases} N & \text{if } m_1 < N, \\ N+1 & \text{if } m_1 = N. \end{cases} \end{aligned}$$

The problem consists of finding a procedure for maximizing the expected gain.

The following theorem gives an optimal double stopping rule τ^* and the value v.

Theorem 5.1 (Sofronov et al. (2006)). *Define $\{\xi_n\}$ and Z_{m_1,m_2} as above. Let $v^{L,l}$ be the value of a game with l ($l \le 2$) stoppings and L ($L \le N$) steps; then the value $v = v^{N,2}$, where*

$$\begin{aligned} v^{n,2} &= \boldsymbol{E}\big(\max\{v^{n-1,1} - \xi_{N-n+1}, v^{n-1,2}\}\big),\ 2 \le n \le N,\ v^{1,2} = 0, \\ v^{n,1} &= \boldsymbol{E}\big(\max\{\xi_{N-n+1}, v^{n-1,1}\}\big),\ 1 \le n \le N-1,\ v^{0,1} = -\infty. \end{aligned}$$

We put

$$\begin{aligned} \tau_1^* &= \min\{\min\{m_1 : 1 \le m_1 \le N-1, \xi_{m_1} \le v^{N-m_1,1} - v^{N-m_1,2}\}, N\}, \\ \tau_2^* &= \min\{\min\{m_2 : \tau_1^* < m_2 \le N, \xi_{m_2} \ge v^{N-m_2,1}\}, \{N+1 : \tau_1^* = N\}\}, \end{aligned}$$

then $\tau^* = (\tau_1^*, \tau_2^*)$ is the optimal double stopping rule.

Proof: If $m_1 = N, m_2 = N+1$, then $Z_{m_1,m_2} = 0$. It follows easily that $v^{1,2} = 0$.

Let us consider the case $1 \le m_1 \le N-1$, $m_1 < m_2 \le N$. From (2.20) and independence of $\xi_1, \ldots, \xi_N$ we obtain

$$\begin{aligned} V_{m_1,m_2} &= \max\{X_{m_1,m_2}, \mathbf{E}_{m_1,m_2} V_{m_1,m_2+1}\} \\ &= \max\{\xi_{m_2} - \xi_{m_1}, \mathbf{E}_{m_1,m_2} V_{m_1,m_2+1}\} \\ &= \max\{\xi_{m_2}, \mathbf{E}_{m_1,m_2} V_{m_1,m_2+1} + \xi_{m_1}\} - \xi_{m_1} \\ &= \max\{\xi_{m_2}, v^{N-m_2,1}\} - \xi_{m_1}, \end{aligned} \tag{5.1}$$

where

$$v^{N-m_2+1,1} = \mathbf{E}_{m_1,m_2-1}\big(\max\{\xi_{m_2}, v^{N-m_2,1}\}\big) = \boldsymbol{E}\big(\max\{\xi_{m_2}, v^{N-m_2,1}\}\big)$$

for $m_1 < m_2 < N$, $v^{N-m_2+1,1} = v^{1,1} = \boldsymbol{E}\xi_N = f_N$ for $m_2 = N$. Indeed, it follows from (2.19) that $V_{m_1,N} = X_{m_1,N}$. Hence $v^{0,1} = -\infty$.

From general recurrent equations, we have

$$\begin{aligned} X_{m_1} &= \boldsymbol{E}_{m_1} V_{m_1,m_1+1} = \boldsymbol{E}_{m_1}\left(\max\{\xi_{m_1+1}, v^{N-m_1-1,1}\} - \xi_{m_1}\right) \\ &= \boldsymbol{E}\big(\max\{\xi_{m_1+1}, v^{N-m_1-1,1}\}\big) - \xi_{m_1} = v^{N-m_1,1} - \xi_{m_1}. \end{aligned}$$

In the same way,

$$\begin{aligned} V_{m_1} &= \max\{X_{m_1}, \mathbf{E}_{m_1} V_{m_1+1}\} = \max\{v^{N-m_1,1} - \xi_{m_1}, \boldsymbol{E}_{m_1} V_{m_1+1}\} \\ &= \max\{v^{N-m_1,1} - \xi_{m_1}, v^{N-m_1,2}\}, \end{aligned} \tag{5.2}$$

where

$$\begin{aligned} v^{N-m_1+1,2} &= \boldsymbol{E}_{m_1-1}\big(\max\{v^{N-m_1,1} - \xi_{m_1}, v^{N-m_1,2}\}\big) \\ &= \boldsymbol{E}\big(\max\{v^{N-m_1,1} - \xi_{m_1}, v^{N-m_1,2}\}\big) \end{aligned}$$

for $m_1 < N-1$,$v^{N-m_1+1,2} = v^{2,2} = \boldsymbol{E}\big(\max\{f_N - \xi_{N-1}, 0\}\big)$ for $m_1 = N-1$.

Taking into account Theorem 2.3, we obtain the value $v = v^{N,2}$.

Now, using Theorem 2.2, (5.1), and (5.2), we obtain the optimal double stopping rule $\tau^* = (\tau_1^*, \tau_2^*)$:

$$\begin{aligned} \tau_1^* &= \min\big\{\min\{m_1 : 1 \le m_1 \le N-1, \xi_{m_1} \le v^{N-m_1,1} - v^{N-m_1,2}\}, N\big\}, \\ \tau_2^* &= \min\big\{\min\{m_2 : \tau_1^* < m_2 \le N, \xi_{m_2} \ge v^{N-m_2,1}\}, \{N+1 : \tau_1^* = N\}\big\}. \end{aligned}$$

This completes the proof. □

Following Sofronov et al. (2006), we discuss three examples in which we specify the distribution of the "noise" component ε. We assume that the distribution of ε is either uniform, or Laplace (double exponential), or normal and in each case we present a solution of the double optimal stopping problem.

Example 5.1 (Uniform distribution)**.** *Let $\varepsilon_1, \ldots, \varepsilon_N$ be a sequence of independent random variable having uniform distribution $U(-a, a)$, $a > 0$ is a fixed number. From Theorem 3, we have*

$$\begin{aligned} v^{n,1} &= \boldsymbol{E}\big(\max\{\xi_{N-n+1}, v^{n-1,1}\}\big) = \boldsymbol{E}\big(\max\{f_{N-n+1} + \varepsilon_{N-n+1}, v^{n-1,1}\}\big) \\ &= \boldsymbol{E}\big(\max\{\varepsilon_{N-n+1}, v^{n-1,1} - f_{N-n+1}\}\big) + f_{N-n+1} \\ &= \int_{-a}^{a} \max\{x, v^{n-1,1} - f_{N-n+1}\}(2a)^{-1}\,dx + f_{N-n+1} \\ &= \int_{-a}^{c} c(2a)^{-1}\,dx + \int_{c}^{a} x(2a)^{-1}\,dx + f_{N-n+1} \\ &= (v^{n-1,1} - f_{N-n+1} + a)^2(4a)^{-1} + f_{N-n+1}, \end{aligned}$$

where $c = v^{n-1,1} - f_{N-n+1}$, $1 \le n \le N-1$, $v^{0,1} = f_N - a$.

TABLE 5.1
The values for a uniform distribution.

n	0	1	2	3	4	5	6
$v^{n,1}$	9.6000	10.6000	10.8025	10.8918	10.9334	10.9512	
$v^{n,2}$		0.0000	0.3025	0.6050	0.8484	1.0442	1.2070

Similarly,

$$
\begin{aligned}
v^{n,2} &= \boldsymbol{E}\big(\max\{v^{n-1,1} - \xi_{N-n+1}, v^{n-1,2}\}\big) \\
&= \boldsymbol{E}\big(\max\{v^{n-1,1} - f_{N-n+1} - \varepsilon_{N-n+1}, v^{n-1,2}\}\big) \\
&= \boldsymbol{E}\big(\max\{-\varepsilon_{N-n+1}, v^{n-1,2} - v^{n-1,1} + f_{N-n+1}\}\big) + v^{n-1,1} - f_{N-n+1} \\
&= \int_{-a}^{a} \max\{-x, d\}(2a)^{-1}\,dx + v^{n-1,1} - f_{N-n+1} \\
&= \int_{-a}^{-d} -x(2a)^{-1}\,dx + \int_{-d}^{a} d(2a)^{-1}\,dx + v^{n-1,1} - f_{N-n+1} \\
&= (v^{n-1,2} - v^{n-1,1} + f_{N-n+1} + a)^2(4a)^{-1} + v^{n-1,1} - f_{N-n+1},
\end{aligned}
$$

where $d = v^{n-1,2} - v^{n-1,1} + f_{N-n+1}$, $2 \le n \le N$, $v^{1,2} = 0$.

In Table 5.1, we present some numerical results for the choice $f_n = 0.1n + 10$, $N = 6$, $a = 1$.

Thus we get the expected gain $v = v^{6,2} = 1.2070$, $\tau^* = (\tau_1^*, \tau_2^*)$*:*

$$
\begin{aligned}
\tau_1^* &= \min\big\{\min\{m_1 : 1 \le m_1 \le 5, \xi_{m_1} \le v^{6-m_1,1} - v^{6-m_1,2}\}, 6\big\}, \\
\tau_2^* &= \min\big\{\min\{m_2 : \tau_1^* < m_2 \le 6, \xi_{m_2} \ge v^{6-m_2,1}\}, \{7 : \tau_1^* = 6\}\big\}.
\end{aligned}
$$

For instance, if we observe the following sequence $\{\xi_n, n = 1, \dots, 6\}$*: 10.0384, 9.3296, 10.5656, 10.6310, 10.3470, 11.2767; then* $m_1 = 2$ *(because* $\xi_2 = 9.3296 \le v^{4,1} - v^{4,2} = 10.9334 - 0.8484 = 10.0850$*),* $m_2 = 6$ *(*$\xi_6 = 11.2767 \ge v^{0,1} = 9.6000$*). This yields the gain* $11.2767 - 9.3296 = 1.9471$.

Example 5.2 (Laplace distribution). *Suppose independent random variables* $\varepsilon_1, \dots, \varepsilon_N$ *are identically distributed with a Laplace (double exponential) distribution* $L(0, b)$. *Its probability density function is*

$$
g(x) = \frac{1}{2b} \exp\left\{-\frac{|x|}{b}\right\}, \quad x \in (-\infty, \infty), \; b > 0.
$$

As above, using Theorem 5.1, we get

$$
\begin{aligned}
v^{n,1} &= \boldsymbol{E}\big(\max\{\xi_{N-n+1}, v^{n-1,1}\}\big) = \boldsymbol{E}\big(\max\{f_{N-n+1} + \varepsilon_{N-n+1}, v^{n-1,1}\}\big) \\
&= \boldsymbol{E}\big(\max\{\varepsilon_{N-n+1}, v^{n-1,1} - f_{N-n+1}\}\big) + f_{N-n+1}
\end{aligned}
$$

TABLE 5.2
The values for a Laplace distribution.

n	0	1	2	3	4	5
$v^{n,1}$	$-\infty$	95.0000	96.3591	97.3082	98.3069	
$v^{n,2}$		0.0000	0.1839	0.4031	0.5704	0.7117

$$
\begin{aligned}
&= \int_{-\infty}^{\infty} \max\{x, v^{n-1,1} - f_{N-n+1}\} g(x)\,dx + f_{N-n+1} \\
&= \int_{-\infty}^{c} c g(x)\,dx + \int_{c}^{\infty} x g(x)\,dx + f_{N-n+1} \\
&= v^{n-1,1} + \frac{b}{2}\exp\left\{\frac{f_{N-n+1} - v^{n-1,1}}{b}\right\},
\end{aligned}
$$

where $c = v^{n-1,1} - f_{N-n+1}$, $1 \le n \le N-1$, $v^{0,1} = -\infty$.

Likewise,

$$
\begin{aligned}
v^{n,2} &= \boldsymbol{E}\big(\max\{v^{n-1,1} - \xi_{N-n+1}, v^{n-1,2}\}\big) \\
&= \boldsymbol{E}\big(\max\{v^{n-1,1} - f_{N-n+1} - \varepsilon_{N-n+1}, v^{n-1,2}\}\big) \\
&= \boldsymbol{E}\big(\max\{-\varepsilon_{N-n+1}, v^{n-1,2} - v^{n-1,1} + f_{N-n+1}\}\big) + v^{n-1,1} - f_{N-n+1} \\
&= \int_{-\infty}^{\infty} \max\{-x, d\} g(x)\,dx + v^{n-1,1} - f_{N-n+1} \\
&= \int_{-\infty}^{-d} -x g(x)\,dx + \int_{-d}^{\infty} d g(x)\,dx + v^{n-1,1} - f_{N-n+1} \\
&= v^{n-1,2} + \frac{b}{2}\exp\left\{\frac{v^{n-1,1} - v^{n-1,2} - f_{N-n+1}}{b}\right\},
\end{aligned}
$$

where $d = v^{n-1,2} - v^{n-1,1} + f_{N-n+1}$, $2 \le n \le N$, $v^{1,2} = 0$.

Table 5.2 presents some numerical results for the choice $f_n = 100 - n$, $N = 5$, $b = 1$.

It follows that we have the expected gain $v = v^{5,2} = 0.7117$, $\tau^* = (\tau_1^*, \tau_2^*)$*:*

$$
\begin{aligned}
\tau_1^* &= \min\big\{\min\{m_1 : 1 \le m_1 \le 4, \xi_{m_1} \le v^{5-m_1,1} - v^{5-m_1,2}\}, 5\big\}, \\
\tau_2^* &= \min\big\{\min\{m_2 : \tau_1^* < m_2 \le 5, \xi_{m_2} \ge v^{5-m_2,1}\}, \{6 : \tau_1^* = 5\}\big\}.
\end{aligned}
$$

In particular, if we observe the following sequence $\{\xi_n, n = 1, \ldots, 5\}$*: 99.2057, 97.1005, 96.8307, 95.6100, 95.7487; then* $m_1 = 5$, $m_2 = 6$. *We buy nothing, so the gain* $v = 0$.

Example 5.3 (Normal distribution). *Let* $\varepsilon_1, \ldots, \varepsilon_N$ *be a sequence of independent random variables having normal distribution* $N(0, \sigma^2)$ *with probability density function*

$$
h(x) = \frac{1}{\sqrt{2\pi}\sigma}\exp\left\{-\frac{x^2}{2\sigma^2}\right\}, \quad x \in (-\infty, \infty),\ \sigma > 0.
$$

TABLE 5.3
The values for a normal distribution.

n	0	1	2	3	4	5	6	7
$v^{n,1}$	$-\infty$	3.0615	5.6583	7.4079	8.1709	8.2954	8.2967	
$v^{n,2}$		0.0000	0.0015	0.0183	0.1841	0.9497	2.6573	5.2368

As before, from Theorem 5.1, we have

$$\begin{aligned} v^{n,1} &= \boldsymbol{E}\big(\max\{\xi_{N-n+1}, v^{n-1,1}\}\big) = \boldsymbol{E}\big(\max\{f_{N-n+1} + \varepsilon_{N-n+1}, v^{n-1,1}\}\big) \\ &= \boldsymbol{E}\big(\max\{\varepsilon_{N-n+1}, v^{n-1,1} - f_{N-n+1}\}\big) + f_{N-n+1} \\ &= \int_{-\infty}^{\infty} \max\{x, v^{n-1,1} - f_{N-n+1}\} h(x)\, dx + f_{N-n+1} \\ &= \int_{-\infty}^{c} c h(x)\, dx + \int_{c}^{\infty} x h(x)\, dx + f_{N-n+1} \\ &= c\Phi\left(\frac{c}{\sigma}\right) + \sigma\varphi\left(\frac{c}{\sigma}\right) + f_{N-n+1} \\ &= \sigma\psi\left(\frac{v^{n-1,1} - f_{N-n+1}}{\sigma}\right) + f_{N-n+1}, \end{aligned}$$

where $\psi(x) = \varphi(x) + x\Phi(x)$, $\varphi(x)$ is the density function of the standard normal distribution, $\Phi(x)$ is the distribution function of the standard normal distribution, $c = v^{n-1,1} - f_{N-n+1}$, $1 \le n \le N-1$, $v^{0,1} = -\infty$.

In the same way,

$$\begin{aligned} v^{n,2} &= \boldsymbol{E}\big(\max\{v^{n-1,1} - \xi_{N-n+1}, v^{n-1,2}\}\big) \\ &= \boldsymbol{E}\big(\max\{v^{n-1,1} - f_{N-n+1} - \varepsilon_{N-n+1}, v^{n-1,2}\}\big) \\ &= \boldsymbol{E}\big(\max\{-\varepsilon_{N-n+1}, v^{n-1,2} - v^{n-1,1} + f_{N-n+1}\}\big) + v^{n-1,1} - f_{N-n+1} \\ &= \int_{-\infty}^{\infty} \max\{-x, d\} h(x)\, dx + v^{n-1,1} - f_{N-n+1} \\ &= \int_{-\infty}^{-d} -x h(x)\, dx + \int_{-d}^{\infty} d h(x)\, dx + v^{n-1,1} - f_{N-n+1} \\ &= \sigma\varphi\left(\frac{d}{\sigma}\right) + d\Phi\left(\frac{d}{\sigma}\right) + v^{n-1,1} - f_{N-n+1} \\ &= \sigma\psi\left(\frac{v^{n-1,2} - v^{n-1,1} + f_{N-n+1}}{\sigma}\right) + v^{n-1,1} - f_{N-n+1}, \end{aligned}$$

where $d = v^{n-1,2} - v^{n-1,1} + f_{N-n+1}$, $2 \le n \le N$, $v^{1,2} = 0$.

In Table 5.3, we present some numerical results for the choice $f_n = 8\sin(\pi n/8)$, $N = 7$, $\sigma = 1$. It follows that the expected gain is $v = v^{7,2} = 5.2368$ and the optimal double stopping rule is $\tau^* = (\tau_1^*, \tau_2^*)$:

$$\begin{aligned} \tau_1^* &= \min\big\{\min\{m_1 : 1 \le m_1 \le 6, \xi_{m_1} \le v^{7-m_1,1} - v^{7-m_1,2}\}, 7\big\}, \\ \tau_2^* &= \min\big\{\min\{m_2 : \tau_1^* < m_2 \le 7, \xi_{m_2} \ge v^{7-m_2,1}\}, \{8 : \tau_1^* = 7\}\big\}. \end{aligned}$$

For instance, if we observe the following sequence $\{\xi_n, n = 1,\dots,7\}$*: 2.4894, 5.9438, 7.0311, 8.9202, 7.8443, 5.4808, 3.5506; then* $m_1 = 1$ *(because* $\xi_1 = 2.4894 \le v^{6,1} - v^{6,2} = 8.2967 - 2.6573 = 5.6394$*),* $m_2 = 4$ *(*$\xi_4 = 8.9202 \ge v^{3,1} = 7.4079$*). We see that the gain* $8.9202 - 2.4894 = 6.4308$.

Theorem 5.1 can easily be generalized when more than two stops are required.

Let $\xi_1, \xi_2, \dots, \xi_N$ be a sequence of random variables, where $\xi_n = f_n + \varepsilon_n$, $f_n = f(n)$ is a known deterministic function (for example, a trend) with finite values for $n = 1, 2, \dots, N$, $\{\varepsilon_n, n = 1, \dots, N\}$ is a sequence of independent random variable with $\boldsymbol{E}\varepsilon_n = 0$. The random variable ξ_n can be interpreted as a value of asset at time n. We observe these random variables sequentially and it is possible to make k, k is even, stops, taking into consideration that there is no recall allowed. The odd stops mean the buying of an asset, the next even ones signify the selling of the asset. Our decision to stop at moments $m_1, \dots, m_k$, $1 \le m_1 \le N_1$, $m_{i-1} < m_i \le N_i(m_1, \dots, m_{i-1})$, $i = 2, 3, \dots, k$,

$$N_1 = N, \quad N_i(m_1, \dots, m_{i-1}) = \begin{cases} N, & \text{if } m_{i-1} \le N-1, \\ m_{i-1}+1, & \text{if } m_{i-1} \ge N, \end{cases}$$

depends on the observations already made, but does not depend on the future which is not yet known. After k stops at times $m_1, \dots, m_k$ we get a gain

$$Z_{m_1,\dots,m_k} = \begin{cases} \sum_{i=1}^{j} (\xi_{m_{2i}} - \xi_{m_{2i-1}}), & \text{if } j \ge 1, \\ 0, & \text{if } j = 0, \end{cases} \tag{5.3}$$

where $j = \max\{i : m_{2i} \le N, i = 0, 1, \dots, k/2\}$, $m_0 = 0$. If we buy an asset at time $N-1$ or earlier, we are required to sell it no later than at time N. In particular, if $1 \le j < k/2$, we receive the gain $Z_{m_1,\dots,m_k} = \sum_{i=1}^{j} (\xi_{m_{2i}} - \xi_{m_{2i-1}})$ after making j pairs of buying-selling actions with either $1 \le m_1 < \cdots < m_{2j} \le N-1$, $m_{2j+1} = N, \dots, m_k = N + k - 2j - 1$, or $1 \le m_1 < \cdots < m_{2j} = N$, $m_{2j+1} = N+1, \dots, m_k = N + k - 2j$. If we do not buy anything before time N then we get the gain $Z_{m_1,\dots,m_k} = 0$, $m_1 = N, \dots, m_k = N + k - 1$.

The problem consists of finding a procedure for maximizing the expected gain.

The following theorem gives the multiple optimal stopping rule τ^* and the value v.

Theorem 5.2 (Sofronov 2016). *Let* $\{\xi_n\}$*,* $Z_{m_1,\dots,m_k}$ *be as in (5.3). Let* $v^{L,l}$ *be the value of a game with* l $(l \le k)$ *stops and* L $(L \le N)$ *steps. If there exist* $\boldsymbol{E}\xi_1$*,* $\boldsymbol{E}\xi_2, \dots, \boldsymbol{E}\xi_N$*, then the value* $v = v^{N,k}$*, where*

$$\begin{aligned}
v^{n,1} &= \boldsymbol{E}\big(\max\{\xi_{N-n+1}, v^{n-1,1}\}\big),\ 1 \le n \le N-k+1,\ v^{0,1} = -\infty, \\
v^{n,2i} &= \boldsymbol{E}\big(\max\{v^{n-1,2i-1} - \xi_{N-n+1}, v^{n-1,2i}\}\big),\ 2 \le n \le N-k+2i, \\
&\quad v^{1,2i} = 0,\ i = 1, 2, \dots, k/2-1, \\
v^{n,2i+1} &= \boldsymbol{E}\big(\max\{v^{n-1,2i} + \xi_{N-n+1}, v^{n-1,2i+1}\}\big),\ 1 \le n \le N-k+2i+1, \\
&\quad v^{0,2i+1} = -\infty,\ i = 1, 2, \dots, k/2-1, \\
v^{n,k} &= \boldsymbol{E}\big(\max\{v^{n-1,k-1} - \xi_{N-n+1}, v^{n-1,k}\}\big),\ 2 \le n \le N, v^{1,k} = 0.
\end{aligned}$$

We put

$$\begin{aligned}
\tau_1^* &= \min\{\min\{m_1 : 1 \le m_1 \le N-1, \xi_{m_1} \le v^{N-m_1,k-1} - v^{N-m_1,k}\}, N\},\\
\tau_{2i}^* &= \min\{\min\{m_{2i} : \tau_{2i-1}^* < m_{2i} \le N,\\
&\quad \xi_{m_{2i}} \ge v^{N-m_{2i},k-2i+1} - v^{N-m_{2i},k-2i}\}, \{N : \tau_{2i-1}^* \le N-1\},\\
&\quad \{\tau_{2i-1}^* + 1 : N \le \tau_{2i-1}^* \le N_{2i-1}(m_1,\ldots,m_{2i-2})\}\},\\
&\quad i = 1,2,\ldots,k/2-1,\\
\tau_{2i+1}^* &= \min\{\min\{m_{2i+1} : \tau_{2i}^* < m_{2i+1} \le N-1,\\
&\quad \xi_{m_{2i+1}} \le v^{N-m_{2i+1},k-2i-1} - v^{N-m_{2i+1},k-2i}\}, \{N : \tau_{2i}^* \le N-1\},\\
&\quad \{\tau_{2i}^* + 1 : N-1 \le \tau_{2i}^* \le N_{2i}(m_1,\ldots,m_{2i-1})\}\},\\
&\quad i = 1,2,\ldots,k/2-1,\\
\tau_k^* &= \min\{\min\{m_k : \tau_{k-1}^* < m_k \le N, \xi_{m_k} \ge v^{N-m_k,1}\},\\
&\quad \{N : \tau_{k-1}^* \le N-1\},\\
&\quad \{\tau_{k-1}^* + 1 : N \le \tau_{k-1}^* \le N_{k-1}(m_1,\ldots,m_{k-2})\}\},
\end{aligned}$$

where $\min\emptyset = \infty$. *Then* $\tau^* = (\tau_1^*,\ldots,\tau_k^*)$ *is the optimal multiple-stopping rule.*

Proof: From (2.20) and independence of $\xi_1,\ldots,\xi_N$ we obtain

$$\begin{aligned}
V_{\boldsymbol{m}_k} &= \max\{X_{\boldsymbol{m}_k}, \mathbf{E}_{\boldsymbol{m}_k} V_{(\boldsymbol{m})_{k-1},m_k+1}\}\\
&= \max\{-\xi_{m_1} + \xi_{m_2} - \cdots - \xi_{m_{k-1}} + \xi_{m_k}, \mathbf{E}_{\boldsymbol{m}_k} V_{(\boldsymbol{m})_{k-1},m_k+1}\}\\
&= \max\{\xi_{m_k}, v^{N-m_k,1}\} - \xi_{m_1} + \xi_{m_2} - \cdots - \xi_{m_{k-1}}, \qquad (5.4)
\end{aligned}$$

where

$$\begin{aligned}
v^{N-m_k+1,1} &= \mathbf{E}_{(\boldsymbol{m})_{k-1},m_k-1}(\max\{\xi_{m_k}, v^{N-m_k,1}\})\\
&= \boldsymbol{E}(\max\{\xi_{m_k}, v^{N-m_k,1}\})
\end{aligned}$$

for $m_{k-1} < m_k \le N$, $v^{0,1} = -\infty$. Indeed, it follows from (2.19) that $V_{(\boldsymbol{m})_{k-1},N} = X_{(\boldsymbol{m})_{k-1},N}$.

From (2.4) and (5.4), we have

$$\begin{aligned}
X_{(\boldsymbol{m})_{k-1}} &= \boldsymbol{E}_{(\boldsymbol{m})_{k-1}} V_{(\boldsymbol{m})_{k-1},m_{k-1}+1}\\
&= \boldsymbol{E}_{(\boldsymbol{m})_{k-1}}(\max\{\xi_{m_{k-1}+1}, v^{N-m_{k-1}-1,1}\} - \xi_{m_1} + \xi_{m_2} - \cdots - \xi_{m_{k-1}}\\
&= \boldsymbol{E}(\max\{\xi_{m_{k-1}+1}, v^{N-m_{k-1}-1,1}\}) - \xi_{m_1} + \xi_{m_2} - \cdots - \xi_{m_{k-1}}\\
&= v^{N-m_{k-1},1} - \xi_{m_1} + \xi_{m_2} - \cdots - \xi_{m_{k-1}}. \qquad (5.5)
\end{aligned}$$

From (2.20) and (5.5) we obtain

$$\begin{aligned}
V_{(\boldsymbol{m})_{k-1}} &= \max\{X_{(\boldsymbol{m})_{k-1}}, \mathbf{E}_{(\boldsymbol{m})_{k-1}} V_{(\boldsymbol{m})_{k-2},m_{k-1}+1}\}\\
&= \max\{v^{N-m_{k-1},1} - \xi_{m_1} + \xi_{m_2} - \cdots - \xi_{m_{k-1}},\\
&\quad \mathbf{E}_{(\boldsymbol{m})_{k-1}} V_{(\boldsymbol{m})_{k-2},m_{k-1}+1}\}\\
&= \max\{v^{N-m_{k-1},1} - \xi_{m_{k-1}}, v^{N-m_{k-1},2}\} - \xi_{m_1} + \cdots + \xi_{m_{k-2}}, \qquad (5.6)
\end{aligned}$$

where

$$\begin{aligned} v^{N-m_{k-1}+1,2} &= \mathbf{E}_{(\boldsymbol{m})_{k-2},m_{k-1}-1}\big(\max\{v^{N-m_{k-1},1} - \xi_{m_{k-1}}, v^{N-m_{k-1},2}\}\big) \\ &= \boldsymbol{E}\big(\max\{v^{N-m_{k-1},1} - \xi_{m_{k-1}}, v^{N-m_{k-1},2}\}\big) \end{aligned}$$

for $m_{k-2} < m_{k-1} \leq N-1$, $v^{1,2} = 0$.

As above, we have

$$\begin{aligned} X_{(\boldsymbol{m})_{k-2}} &= \boldsymbol{E}_{(\boldsymbol{m})_{k-2}} V_{(\boldsymbol{m})_{k-2},m_{k-2}+1} \\ &= \boldsymbol{E}_{(\boldsymbol{m})_{k-2}}\big(\max\{v^{N-m_{k-2}-1,1} - \xi_{m_{k-2}+1}, v^{N-m_{k-2}-1,2}\} \\ &\quad -\xi_{m_1} + \xi_{m_2} - \cdots + \xi_{m_{k-2}}\big) \\ &= \boldsymbol{E}\big(\max\{v^{N-m_{k-2}-1,1} - \xi_{m_{k-2}+1}, v^{N-m_{k-2}-1,2}\}\big) \\ &\quad -\xi_{m_1} + \xi_{m_2} - \cdots + \xi_{m_{k-2}} \\ &= v^{N-m_{k-2},2} - \xi_{m_1} + \xi_{m_2} - \cdots + \xi_{m_{k-2}}. \end{aligned}$$

For the even stops, we have

$$\begin{aligned} V_{(\boldsymbol{m})_{2i}} &= \max\{X_{(\boldsymbol{m})_{2i}}, \boldsymbol{E}_{(\boldsymbol{m})_{2i}} V_{(\boldsymbol{m})_{2i-1},m_{2i}+1}\} \\ &= \max\{v^{N-m_{2i},k-2i} - \xi_1 + \cdots + \xi_{m_{2i}}, \boldsymbol{E}_{(\boldsymbol{m})_{2i}} V_{(\boldsymbol{m})_{2i-1},m_{2i}+1}\} \quad (5.7) \\ &= \max\{v^{N-m_{2i},k-2i} + \xi_{m_{2i}}, v^{N-m_{2i},k-2i+1}\} - \xi_{m_1} + \cdots - \xi_{m_{2i-1}}, \end{aligned}$$

where

$$\begin{aligned} v^{N-m_{2i}+1,k-2i+1} &= \boldsymbol{E}_{(\boldsymbol{m})_{2i-1},m_{2i}-1}\big(\max\{v^{N-m_{2i},k-2i} + \xi_{m_{2i}}, \\ &\quad v^{N-m_{2i},k-2i+1}\}\big) \\ &= \boldsymbol{E}\big(\max\{v^{N-m_{2i},k-2i} + \xi_{m_{2i}}, v^{N-m_{2i},k-2i+1}\}\big) \end{aligned}$$

for $m_{2i-1} < m_{2i} \leq N$, $v^{0,k-2i+1} = -\infty$, $i = k/2-1, k/2-2, \ldots, 1$.

Similarly,

$$\begin{aligned} X_{(\boldsymbol{m})_{2i-1}} &= \boldsymbol{E}_{(\boldsymbol{m})_{2i-1}} V_{(\boldsymbol{m})_{2i-1},m_{2i-1}+1} \\ &= \boldsymbol{E}_{(\boldsymbol{m})_{2i-1}}\big(\max\{v^{N-m_{2i-1}-1,k-2i} + \xi_{m_{2i-1}+1}, v^{N-m_{2i-1}-1,k-2i+1}\} \\ &\quad -\xi_{m_1} + \cdots - \xi_{m_{2i-1}}\big) \\ &= \boldsymbol{E}\big(\max\{v^{N-m_{2i-1}-1,k-2i} + \xi_{m_{2i-1}+1}, v^{N-m_{2i-1}-1,k-2i+1}\}\big) \\ &\quad -\xi_{m_1} + \cdots - \xi_{m_{2i-1}} \\ &= v^{N-m_{2i-1},k-2i+1} - \xi_{m_1} + \cdots - \xi_{m_{2i-1}}, \end{aligned}$$

$i = k/2-1, k/2-2, \ldots, 1$.

In the same way, for the odd stops

$$\begin{aligned} V_{(\boldsymbol{m})_{2i+1}} &= \max\{X_{(\boldsymbol{m})_{2i+1}}, \boldsymbol{E}_{(\boldsymbol{m})_{2i+1}} V_{(\boldsymbol{m})_{2i},m_{2i+1}+1}\} \quad (5.8) \\ &= \max\{v^{N-m_{2i+1},k-2i-1} - \xi_1 + \cdots - \xi_{m_{2i+1}}, \boldsymbol{E}_{(\boldsymbol{m})_{2i+1}} V_{(\boldsymbol{m})_{2i},m_{2i+1}+1}\} \\ &= \max\{v^{N-m_{2i+1},k-2i-1} - \xi_{m_{2i+1}}, v^{N-m_{2i+1},k-2i}\} \\ &\quad -\xi_{m_1} + \cdots + \xi_{m_{2i}}, \end{aligned}$$

where

$$\begin{aligned} v^{N-m_{2i+1}+1,k-2i} &= \boldsymbol{E}_{(\boldsymbol{m})_{2i},m_{2i+1}-1}\big(\max\{v^{N-m_{2i+1},k-2i} - \xi_{m_{2i+1}}, \\ &\quad v^{N-m_{2i+1},k-2i}\}\big) \\ &= \boldsymbol{E}\big(\max\{v^{N-m_{2i+1},k-2i-1} - \xi_{m_{2i+1}}, v^{N-m_{2i+1},k-2i}\}\big) \end{aligned}$$

for $m_{2i} < m_{2i+1} \le N-1$, $v^{1,k-2i} = 0$, $i = k/2-2,\ldots,1$.

Similarly,

$$\begin{aligned} X_{(\boldsymbol{m})_{2i}} &= \boldsymbol{E}_{(\boldsymbol{m})_{2i}} V_{(\boldsymbol{m})_{2i},m_{2i}+1} \\ &= \boldsymbol{E}_{(\boldsymbol{m})_{2i}}\big(\max\{v^{N-m_{2i}-1,k-2i-1} - \xi_{m_{2i}+1}, v^{N-m_{2i}-1,k-2i}\} \\ &\quad -\xi_{m_1} + \cdots + \xi_{m_{2i}}\big) \\ &= \boldsymbol{E}\big(\max\{v^{N-m_{2i}-1,k-2i-1} - \xi_{m_{2i}+1}, v^{N-m_{2i}-1,k-2i}\}\big) \\ &\quad -\xi_{m_1} + \cdots + \xi_{m_{2i}} \\ &= v^{N-m_{2i},k-2i} - \xi_{m_1} + \cdots + \xi_{m_{2i}}, \end{aligned}$$

$i = k/2-2,\ldots,1$.

Finally,

$$\begin{aligned} V_{m_1} &= \max\{X_{m_1}, \boldsymbol{E}_{m_1} V_{m_1+1}\} \\ &= \max\{v^{N-m_1,k-1} - \xi_{m_1}, \boldsymbol{E}_{m_1} V_{m_1+1}\} \\ &= \max\{v^{N-m_1,k-1} - \xi_{m_1}, v^{N-m_1,k}\}, \end{aligned} \tag{5.9}$$

where

$$\begin{aligned} v^{N-m_1+1,k} &= \boldsymbol{E}_{m_1}(\max\{v^{N-m_1,k-1} - \xi_{m_1}, v^{N-m_1,k}\}) \\ &= \boldsymbol{E}(\max\{v^{N-m_1,k-1} - \xi_{m_1}, v^{N-m_1,k}\}) \end{aligned}$$

for $1 \le m_1 \le N-1$, $v^{1,k} = 0$.

Taking into account Theorem2.3, we obtain $v = v^{N,k}$.

Now, using Theorem 2.2, (5.4), (5.6), (5.7), (5.8), and (5.9) we obtain the optimal multiple-stopping rule $\tau^* = (\tau_1^*, \ldots, \tau_k^*)$:

$$\begin{aligned} \tau_1^* &= \min\{\min\{m_1 : 1 \le m_1 \le N-1, \xi_{m_1} \le v^{N-m_1,k-1} - v^{N-m_1,k}\}, N\}, \\ \tau_{2i}^* &= \min\{\min\{m_{2i} : \tau_{2i-1}^* < m_{2i} \le N, \\ &\quad \xi_{m_{2i}} \ge v^{N-m_{2i},k-2i+1} - v^{N-m_{2i},k-2i}\}, \{N : \tau_{2i-1}^* \le N-1\}, \\ &\quad \{\tau_{2i-1}^* + 1 : N \le \tau_{2i-1}^* \le N_{2i-1}(m_1,\ldots,m_{2i-2})\}\}, \\ &\quad i = 1,2,\ldots,k/2-1, \\ \tau_{2i+1}^* &= \min\{\min\{m_{2i+1} : \tau_{2i}^* < m_{2i+1} \le N-1, \\ &\quad \xi_{m_{2i+1}} \le v^{N-m_{2i+1},k-2i-1} - v^{N-m_{2i+1},k-2i}\}, \{N : \tau_{2i}^* \le N-1\}, \\ &\quad \{\tau_{2i}^* + 1 : N-1 \le \tau_{2i}^* \le N_{2i}(m_1,\ldots,m_{2i-1})\}\}, \\ &\quad i = 1,2,\ldots,k/2-1, \end{aligned}$$

TABLE 5.4
The number of times the double optimal stopping rule is reapplied for different values of horizon N, Uniform$(-1, 1)$.

N	1	2	3	4	5	6	7	8	9
10	283	555	150	12	0	0	0	0	0
50	15	156	355	308	128	33	5	0	0
100	3	74	181	326	257	116	32	10	1

TABLE 5.5
The number of times the double optimal stopping rule is reapplied for different values of horizon N, Laplace$(0, 1)$.

N	1	2	3	4	5	6	7	8
10	434	505	59	2	0	0	0	0
50	123	370	366	121	17	3	0	0
100	56	258	369	228	74	12	2	1

$$\begin{aligned}\tau_k^* = \; & \min\{\min\{m_k : \tau_{k-1}^* < m_k \le N, \xi_{m_k} \ge v^{N-m_k,1}\}, \\ & \{N : \tau_{k-1}^* \le N-1\}, \\ & \{\tau_{k-1}^* + 1 : N \le \tau_{k-1}^* \le N_{k-1}(m_1, \ldots, m_{k-2})\}\},\end{aligned}$$

This completes the proof. □

Example 5.4. *Previously, several examples were discussed in which the distribution of the noise component ε_n was specified — uniform, Laplace (double exponential), and normal — in each case a solution of the optimal double stopping problem was presented. We will discuss here three examples assuming the distribution of ε_n being either uniform, Laplace (double exponential), or normal. We compare the value from Theorem 5.2 with the gain obtained by reapplying the optimal double stopping rule (see Theorem 5.1).*

Tables 5.4, 5.5 and 5.6 display how many times the double optimal stopping rule was reapplied depending on different values of horizon N, $N = 10, 50, 100$, using 1000 simulated sequences of random variables from Uniform$(-1, 1)$, Laplace$(0, 1)$ *and* Normal$(0, 1)$ *distributions, respectively. The trend $f_n = 0$, $n = 1, 2, \ldots, N$.*

Tables 5.7, 5.8 and 5.9 show the expected gains, $v^{N,2}, \ldots, v^{N,14}$, $N = 10, 50, 100$, and the average gain (with the standard deviation in the brackets) when we allow the double optimal stopping rule to be reapplied as many times as necessary. The average gain and its standard deviation were estimated using the same

TABLE 5.6
The number of times the double optimal stopping rule is reapplied for different values of horizon N, Normal$(0, 1)$.

N	1	2	3	4	5	6	7	8
10	355	523	119	3	0	0	0	0
50	48	289	394	212	49	7	1	0
100	23	159	346	313	114	36	8	1

TABLE 5.7
The expected and the average gains for Uniform$(-1, 1)$.

N	$v^{N,2}$	$v^{N,4}$	$v^{N,6}$	$v^{N,8}$	$v^{N,10}$	$v^{N,12}$	$v^{N,14}$	Average gain
10	1.278	1.963	2.212	2.249	2.250	—	—	1.858 (1.071)
50	1.812	3.458	4.949	6.288	7.478	8.523	9.424	4.809 (1.845)
100	1.902	3.716	5.448	7.100	8.674	10.170	11.588	6.312 (2.227)

TABLE 5.8
The expected and the average gains for Laplace$(0, 1)$.

N	$v^{N,2}$	$v^{N,4}$	$v^{N,6}$	$v^{N,8}$	$v^{N,10}$	$v^{N,12}$	$v^{N,14}$	Average gain
10	2.897	3.777	4.268	4.468	4.500	—	—	3.686 (2.652)
50	5.852	8.335	10.504	12.420	14.126	15.648	17.008	9.826 (5.237)
100	7.192	10.366	13.238	15.872	18.309	20.578	22.698	14.233 (7.040)

TABLE 5.9
The expected and the average gains for Normal$(0, 1)$.

N	$v^{N,2}$	$v^{N,4}$	$v^{N,6}$	$v^{N,8}$	$v^{N,10}$	$v^{N,12}$	$v^{N,14}$	Average gain
10	2.159	3.176	3.534	3.589	3.590	—	—	3.034 (1.880)
50	3.738	6.664	9.105	11.178	12.947	14.450	15.716	7.630 (3.423)
100	4.315	7.906	11.076	13.936	16.542	18.932	21.132	10.177 (4.266)

1000 simulated sequences of random variables that were mentioned above. From these tables, it can clearly be seen that if we allow to stop several times, the expected gains based on the multiple optimal stopping rules derived in Theorem 5.2 are higher than the revenue obtained form a procedure when the optimal double stopping rule can be reapplied several times. In particular, for Uniform$(-1, 1)$, $v^{10,6}$, $v^{50,6}$ and $v^{100,8}$ (as well as the following values) are greater than the corresponding average gains. We can observe a similar feature for other two distributions, Laplace$(0, 1)$ and Normal$(0, 1)$. From the tables, it can also be seen that the greater is the variance *of the distribution of the noise component, the higher is the average gain. For comparison, the variance of* $\mathsf{Laplace}(0,1)$ *is 2, while the variances of* $\mathsf{Uniform}(-1,1)$ *and* $\mathsf{Normal}(0,1)$ *are* $\frac{1}{3}$ *and 1, respectively.*

5.1.2 Dependent case

Let $\xi_1, \xi_2, \ldots, \xi_N$ be a sequence of random variables. The random variable ξ_n can be interpreted as a value of asset at time n. We observe these random variables sequentially and have to decide when we must stop. The first stopping means a buying an asset, the second one signifies a selling an asset. Our decision to stop depends on the observations already made, but does not depend on the future which is not yet known. After two stoppings at times m_1, m_2, $1 \le m_1 < m_2 \le N$ we get a gain $Z_{m_1,m_2} = \xi_{m_2} - \xi_{m_1}$. If we do not buy anything until time N then we will obtain the gain $Z_{m_1,m_2} = 0$, we may assume that $m_1 = N$, $m_2 = N+1$. This implies that

$$\begin{aligned} &1 \le m_1 \le N_1, \quad N_1 = N, \\ &m_1 < m_2 \le N_2(m_1), \quad N_2(m_1) = \begin{cases} N & \text{if } m_1 < N, \\ N+1 & \text{if } m_1 = N. \end{cases} \end{aligned}$$

The problem consists of finding a procedure for maximizing the expected gain.

If we observe a sequence of independent random variables, then in order to find the optimal rule we actually only need to calculate unconditional expectations of the form $\boldsymbol{E}\big(\max\{c_1 + c_2 y, c_3\}\big)$, where c_1, c_2, c_3 are constants. This means that the values $v^{n,1}$ and $v^{n,2}$ can be calculated in advance, before we start observing the sequence of random variables. If the assumption of independence is not used, we need to calculate conditional expectations so that the $v^{n,1}$ and $v^{n,2}$ are functions of the previous observations. The following theorem generalizes Theorem 5.1 to a case of dependent random variables.

Theorem 5.3 (Sofronov (2018)). *Let* $\{\xi_n\}$, Z_{m_1,m_2} *be defined as above. Let* $v^{L,l}$ *be the value of a game with* l $(l \le 2)$ *stoppings and* L $(L \le N)$ *steps; then the value* $v = v^{N,2}$, *where*

$$\begin{aligned} v^{n,2} &= \boldsymbol{E}_{N-n}\big(\max\{v^{n-1,1} - \xi_{N-n+1}, v^{n-1,2}\}\big),\ 2 \le n \le N,\ v^{1,2} = 0, \\ v^{n,1} &= \boldsymbol{E}_{N-n}\big(\max\{\xi_{N-n+1}, v^{n-1,1}\}\big),\ 1 \le n \le N-1,\ v^{0,1} = -\infty. \end{aligned}$$

We define

$$\begin{aligned} \tau_1^* &= \min\big\{\min\{m_1 : 1 \le m_1 \le N-1, \xi_{m_1} \le v^{N-m_1,1} - v^{N-m_1,2}\}, N\big\}, \\ \tau_2^* &= \min\big\{\min\{m_2 : \tau_1^* < m_2 \le N, \xi_{m_2} \ge v^{N-m_2,1}\}, \{N+1 : \tau_1^* = N\}\big\}, \end{aligned}$$

then $\tau^* = (\tau_1^*, \tau_2^*)$ *is the optimal double stopping rule.*

Proof: If $m_1 = N, m_2 = N+1$, then $Z_{m_1,m_2} = 0$. It follows easily that $v^{1,2} = 0$. Let us consider the case $1 \le m_1 \le N-1, m_1 < m_2 \le N$. From (2.3) we obtain

$$\begin{aligned} V_{m_1,m_2} &= \max\{\xi_{m_2} - \xi_{m_1}, \mathbf{E}_{m_2} V_{m_1,m_2+1}\} \\ &= \max\{\xi_{m_2}, \mathbf{E}_{m_2} V_{m_1,m_2+1} + \xi_{m_1}\} - \xi_{m_1} \\ &= \max\{\xi_{m_2}, v^{N-m_2,1}\} - \xi_{m_1} \end{aligned} \tag{5.10}$$

with

$$v^{N-m_2+1,1} = \mathbf{E}_{m_2-1}\left(\max\{\xi_{m_2}, v^{N-m_2,1}\}\right), \quad m_1 < m_2 < N,$$

where $v^{N-m_2+1,1}$ may depend on $\xi_1, \dots, \xi_{m_2-1}$. If $m_2 = N$, then $v^{N-m_2+1,1} = v^{1,1} = \boldsymbol{E}_{N-1}\xi_N$ since $v^{0,1} = -\infty$.

From (2.19) and (5.10), we have

$$X_{m_1} = \boldsymbol{E}_{m_1}\left(\max\{\xi_{m_1+1}, v^{N-m_1-1,1}\} - \xi_{m_1}\right) = v^{N-m_1,1} - \xi_{m_1}. \tag{5.11}$$

From (2.20) and (5.11), we have

$$\begin{aligned} V_{m_1} &= \max\{v^{N-m_1,1} - \xi_{m_1}, \boldsymbol{E}_{m_1} V_{m_1+1}\} \\ &= \max\{v^{N-m_1,1} - \xi_{m_1}, v^{N-m_1,2}\} \end{aligned} \tag{5.12}$$

with

$$v^{N-m_1+1,2} = \boldsymbol{E}_{m_1-1}\left(\max\{v^{N-m_1,1} - \xi_{m_1}, v^{N-m_1,2}\}\right), \quad m_1 < N-1,$$

where $v^{N-m_1+1,2}$ may depend on $\xi_1, \dots, \xi_{m_1-1}$. If $m_1 = N-1$, then

$$v^{N-m_1+1,2} = v^{2,2} = \boldsymbol{E}_{N-2}\left(\max\{v^{1,1} - \xi_{N-1}, 0\}\right)$$

since $v^{1,2} = 0$.

Taking into account Theorem 2.3, we obtain the value $v = v^{N,2}$.

Now, using Theorem 2.2 together with (5.10), (5.11) and (5.12), we obtain the optimal double stopping rule $\tau^* = (\tau_1^*, \tau_2^*)$:

$$\begin{aligned} \tau_1^* &= \min\{\min\{m_1 : 1 \le m_1 \le N-1, \xi_{m_1} \le v^{N-m_1,1} - v^{N-m_1,2}\}, N\}, \\ \tau_2^* &= \min\{\min\{m_2 : \tau_1^* < m_2 \le N, \xi_{m_2} \ge v^{N-m_2,1}\}, \{N+1 : \tau_1^* = N\}\}. \end{aligned}$$

This completes the proof. □

5.1.3 An optimal stopping rule for a random walk

Following Sofronov (2016), let us consider a sequence $\{\xi_n\}_{n=1}^N$ with no trend, $f_n = 0$, and the noise component $\{\varepsilon_n\}$ is a sequence of dependent random variables, namely, $\xi_1 = \eta_1, \xi_2 = \eta_1 + \eta_2, \dots, \xi_N = \eta_1 + \dots + \eta_N$, where $\eta_1, \eta_2, \dots, \eta_N$ are i.i.d. random variables with a finite expectation.

Here we assume that we can stop twice and after two stops we obtain a gain $Z_{m_1,m_2} = \xi_{m_2} - \xi_{m_1}, m_1, m_2, 1 \leq m_1 < m_2 \leq N$. If we buy and sell nothing until time N, we will obtain the gain $Z_{m_1,m_2} = 0$, so $m_1 = N$, $m_2 = N+1$. Therefore,

$$1 \leq m_1 \leq N_1, \quad N_1 = N,$$
$$m_1 < m_2 \leq N_2(m_1), \quad N_2(m_1) = \begin{cases} N & \text{if } m_1 < N, \\ N+1 & \text{if } m_1 = N. \end{cases}$$

We have

$$\begin{aligned} V_{m_1,m_2} &= \max\{X_{m_1,m_2}, \mathbf{E}_{m_1,m_2} V_{m_1,m_2+1}\} \\ &= \max\{\xi_{m_2} - \xi_{m_1}, \mathbf{E}_{m_1,m_2} V_{m_1,m_2+1}\} \\ &= \max\{\xi_{m_2}, \boldsymbol{E}_{m_1,m_2} V_{m_1,m_2+1} + \xi_{m_1}\} - \xi_{m_1} \\ &= \max\{\xi_{m_2}, v^{N-m_2,1}\} - \xi_{m_1}, \end{aligned}$$

where

$$\begin{aligned} v^{N-m_2+1,1} &= \boldsymbol{E}_{m_1,m_2-1}\big(\max\{\xi_{m_2}, v^{N-m_2,1}\}\big) \\ &= \boldsymbol{E}\big(\max\{\eta_{m_2}, v^{N-m_2,1} - \eta_1 - \cdots - \eta_{m_2-1}\}\big) + \eta_1 + \cdots + \eta_{m_2-1} \end{aligned}$$

for $m_1 < m_2 < N$, $v^{1,1} = \boldsymbol{E}\big(\max\{\eta_N, v^{0,1} - \eta_1 - \cdots - \eta_{N-1}\}\big) + \eta_1 + \cdots + \eta_{N-1} = \boldsymbol{E}\eta_N + \eta_1 + \cdots + \eta_{N-1}$ since $v^{0,1} = -\infty$. Then

$$\begin{aligned} v^{2,1} &= \boldsymbol{E}\big(\max\{\eta_{N-1}, v^{1,1} - \eta_1 - \cdots - \eta_{N-2}\}\big) + \eta_1 + \cdots + \eta_{N-2} \\ &= \boldsymbol{E}\big(\max\{\eta_{N-1}, \boldsymbol{E}\eta_N + \eta_1 + \cdots + \eta_{N-1} - \eta_1 - \cdots - \eta_{N-2}\}\big) \\ &+ \eta_1 + \cdots + \eta_{N-2} \\ &= \boldsymbol{E}\big(\max\{\eta_{N-1}, \boldsymbol{E}\eta_N + \eta_{N-1}\}\big) + \eta_1 + \cdots + \eta_{N-2}. \end{aligned}$$

This means that the decision-maker should sell if $\eta_{N-1} \geq \boldsymbol{E}\eta_N + \eta_{N-1}$, that is, $\boldsymbol{E}\eta_N \leq 0$, and the decision-maker should continue if $\eta_{N-1} < \boldsymbol{E}\eta_N + \eta_{N-1}$, that is, $\boldsymbol{E}\eta_N > 0$. Hence,

$$v^{2,1} = \begin{cases} \boldsymbol{E}\eta_{N-1} + \eta_1 + \cdots + \eta_{N-2}, & \text{if } \boldsymbol{E}\eta_N \leq 0; \\ \boldsymbol{E}\eta_N + \boldsymbol{E}\eta_{N-1} + \eta_1 + \cdots + \eta_{N-2}, & \text{if } \boldsymbol{E}\eta_N > 0. \end{cases}$$

Similarly, if $\boldsymbol{E}\eta_N = \boldsymbol{E}\eta_{N-1} \leq 0$,

$$\begin{aligned} v^{3,1} &= \boldsymbol{E}\big(\max\{\eta_{N-2}, v^{2,1} - \eta_1 - \cdots - \eta_{N-3}\}\big) + \eta_1 + \cdots + \eta_{N-3} \\ &= \boldsymbol{E}\big(\max\{\eta_{N-2}, \boldsymbol{E}\eta_{N-1} + \eta_1 + \cdots + \eta_{N-2} - \eta_1 - \cdots - \eta_{N-3}\}\big) \\ &+ \eta_1 + \cdots + \eta_{N-3} \\ &= \boldsymbol{E}\big(\max\{\eta_{N-2}, \boldsymbol{E}\eta_{N-1} + \eta_{N-2}\}\big) + \eta_1 + \cdots + \eta_{N-3}. \end{aligned}$$

This means that the decision-maker should stop since $\eta_{N-2} \geq \boldsymbol{E}\eta_{N-1} + \eta_{N-2}$.

If $\boldsymbol{E}\eta_N = \boldsymbol{E}\eta_{N-1} > 0$,

$$\begin{aligned} v^{3,1} =& \boldsymbol{E}\big(\max\{\eta_{N-2}, v^{2,1} - \eta_1 - \cdots - \eta_{N-3}\}\big) + \eta_1 + \cdots + \eta_{N-3} \\ =& \boldsymbol{E}\big(\max\{\eta_{N-2}, \boldsymbol{E}\eta_N + \boldsymbol{E}\eta_{N-1} + \eta_1 + \cdots + \eta_{N-2} - \eta_1 - \cdots - \eta_{N-3}\}\big) \\ & + \eta_1 + \cdots + \eta_{N-3} \\ =& \boldsymbol{E}\big(\max\{\eta_{N-2}, \boldsymbol{E}\eta_N + \boldsymbol{E}\eta_{N-1} + \eta_{N-2}\}\big) + \eta_1 + \cdots + \eta_{N-3}. \end{aligned}$$

This indicates that the decision-maker should continue since $\eta_{N-2} < \boldsymbol{E}\eta_N + \boldsymbol{E}\eta_{N-1} + \eta_{N-2}$.

Following this process, we can see that the rule depends on $\boldsymbol{E}\eta_1, \ldots, \boldsymbol{E}\eta_N$, which are already known, therefore, the optimal rule will be trivial: if the expectations are positive, buy in the beginning and sell in the end of the time interval, otherwise do not buy anything.

5.1.4 An optimal stopping rule for a geometric random walk

Following Sofronov (2018), let us consider a sequence $\xi_1, \xi_2, \ldots, \xi_N$,

$$\xi_n = \xi_0 \lambda^{\varepsilon_1 + \cdots + \varepsilon_n}, \quad \xi_0 \in \{\lambda^i, i = 0, \pm 1, \ldots\}, \quad \lambda > 1,$$

where ε_i are independent and identically distributed random variables with $P(\varepsilon_i = 1) = p$ and $P(\varepsilon_i = -1) = q$, $p + q = 1$, $0 < p < 1$. This sequence, which plays an important role in stochastic financial modeling, is called a geometric random walk [Shiryaev (1999) Chapter II, §1e].

Using Theorem 5.3, we will find an optimal double stopping rule for the sequence $\xi_1, \xi_2, \ldots, \xi_N$. We have, for $m_1 < m_2 < N$,

$$v^{N-m_2+1,1} = \boldsymbol{E}_{m_2-1}\big(\max\{\xi_{m_2}, v^{N-m_2,1}, \}\big)$$

where $\boldsymbol{E}_m$ is the conditional expectation with respect to $\mathcal{F}_m = \sigma(\xi_1, \ldots, \xi_m)$, σ-algebra generated by random variables $\xi_1, \ldots, \xi_m$. For $m_2 = N$, taking into account that $v^{0,1} = -\infty$, we have

$$v^{1,1} = \boldsymbol{E}_{N-1}\xi_N = r\xi_0\lambda^{\varepsilon_1 + \cdots + \varepsilon_{N-1}}, \quad r = \lambda p + \lambda^{-1} q.$$

Then

$$\begin{aligned} v^{2,1} &= \boldsymbol{E}_{N-2}\big(\max\{\xi_{N-1}, v^{1,1}\}\big) \\ &= \boldsymbol{E}_{N-2}\big(\max\{\xi_0\lambda^{\varepsilon_1 + \cdots + \varepsilon_{N-1}}, r\xi_0\lambda^{\varepsilon_1 + \cdots + \varepsilon_{N-1}}\}\big) \\ &= \boldsymbol{E}_{N-2}\big(\max\{\lambda^{\varepsilon_{N-1}}, r\lambda^{\varepsilon_{N-1}}\}\big)\xi_0\lambda^{\varepsilon_1 + \cdots + \varepsilon_{N-2}} \\ &= \begin{cases} r\xi_0\lambda^{\varepsilon_1 + \cdots + \varepsilon_{N-2}}, & r \le 1 \\ r^2\xi_0\lambda^{\varepsilon_1 + \cdots + \varepsilon_{N-2}}, & r > 1. \end{cases} \end{aligned}$$

This means that the decision-maker should stop if $r \leq 1$, and the decision-maker should continue if $r > 1$. In the same way,

$$\begin{aligned} v^{3,1} &= \boldsymbol{E}_{N-3}\big(\max\{\xi_{N-2}, v^{2,1}\}\big) \\ &= \begin{cases} \boldsymbol{E}_{N-3}\big(\max\{\lambda^{\varepsilon_{N-2}}, r\lambda^{\varepsilon_{N-2}}\}\big)\xi_0\lambda^{\varepsilon_1+\cdots+\varepsilon_{N-3}}, & r \leq 1 \\ \boldsymbol{E}_{N-3}\big(\max\{\lambda^{\varepsilon_{N-2}}, r^2\lambda^{\varepsilon_{N-2}}\}\big)\xi_0\lambda^{\varepsilon_1+\cdots+\varepsilon_{N-3}}, & r > 1 \end{cases} \\ &= \begin{cases} r\xi_0\lambda^{\varepsilon_1+\cdots+\varepsilon_{N-3}}, & r \leq 1 \\ r^3\xi_0\lambda^{\varepsilon_1+\cdots+\varepsilon_{N-3}}, & r > 1. \end{cases} \end{aligned}$$

Continuing this process, we can show that

$$v^{n,1} = \begin{cases} r\xi_{N-n}, & r \leq 1 \\ r^n\xi_{N-n}, & r > 1, \end{cases} \qquad 1 \leq n \leq N-1.$$

Similarly,

$$\begin{aligned} v^{2,2} &= \boldsymbol{E}_{N-2}\big(\max\{v^{1,1} - \xi_{N-1}, 0\}\big) \\ &= \boldsymbol{E}_{N-2}\big(\max\{(r-1)\xi_0\lambda^{\varepsilon_1+\cdots+\varepsilon_{N-1}}, 0\}\big) \\ &= \begin{cases} 0, & r \leq 1 \\ (r^2 - r)\xi_0\lambda^{\varepsilon_1+\cdots+\varepsilon_{N-2}}, & r > 1. \end{cases} \end{aligned}$$

Then

$$\begin{aligned} v^{3,2} &= \boldsymbol{E}_{N-3}\big(\max\{v^{2,1} - \xi_{N-2}, v^{2,2}\}\big) \\ &= \begin{cases} \boldsymbol{E}_{N-3}\big(\max\{(r-1)\lambda^{\varepsilon_{N-2}}, 0\}\big)\xi_0\lambda^{\varepsilon_1+\cdots+\varepsilon_{N-3}}, & r \leq 1 \\ \boldsymbol{E}_{N-3}\big(\max\{(r^2-1)\lambda^{\varepsilon_{N-2}}, (r^2-r)\lambda^{\varepsilon_{N-2}}\}\big)\xi_0\lambda^{\varepsilon_1+\cdots+\varepsilon_{N-3}}, & r > 1 \end{cases} \\ &= \begin{cases} 0, & r \leq 1 \\ (r^3 - r)\xi_0\lambda^{\varepsilon_1+\cdots+\varepsilon_{N-3}}, & r > 1. \end{cases} \end{aligned}$$

It can be proved by induction that

$$v^{n,2} = \begin{cases} 0, & r \leq 1 \\ (r^n - r)\xi_{N-n}, & r > 1, \end{cases} \qquad 2 \leq n \leq N.$$

with the value $v = v^{N,2}$.

This implies that the optimal decision rule depends solely on r, which is a known value. Therefore, the rule will be trivial: if $r \leq 1$, do not buy anything; otherwise, if $r > 1$, buy in the beginning and sell in the end of the time interval.

5.1.5 An optimal stopping rule for an AR(1) process

Here we consider an AR(1) sequence $\xi_1, \xi_2, \ldots, \xi_N$ (see Sofronov (2018)),

$$\xi_n = \alpha\xi_{n-1} + \varepsilon_n, \quad \xi_0 = 0,$$

where $\varepsilon_1, \ldots, \varepsilon_N$ are independent random variables having normal distribution $N(\mu, \sigma^2)$. Assume that $\alpha \in (-1, 1)$, then the AR(1) model is stationary.

Following the proof of Theorem 5.3, we have, for $m_1 < m_2 < N$,

$$v^{N-m_2+1,1} = \boldsymbol{E}_{m_2-1}\big(\max\{\xi_{m_2}, v^{N-m_2,1}\big)\},$$

where $\boldsymbol{E}_m$ is the conditional expectation with respect to $\mathcal{F}_m = \sigma(\xi_1, \ldots, \xi_m)$, σ-algebra generated by random variables $\xi_1, \ldots, \xi_m$. If $m_2 = N$, $v^{N-m_2+1,1} = v^{1,1} = \boldsymbol{E}_{N-1}\xi_N = \mu + \alpha\xi_{N-1}$ since $v^{0,1} = -\infty$. Note that $v^{N-m+1,i}$, $i = 1, 2$, is a function of ξ_{m-1}, $2 \le m \le N$.

Similarly, for $m_1 < N - 1$,

$$v^{N-m_1+1,2} = \boldsymbol{E}_{m_1-1}\big(\max\{v^{N-m_1,1} - \xi_{m_1}, v^{N-m_1,2}\}\big).$$

If $m_1 = N - 1$, taking into account that $v^{1,2} = 0$, we have

$$\begin{aligned} v^{2,2} &= \boldsymbol{E}_{N-2}\big(\max\{v^{1,1} - \xi_{N-1}, 0\}\big) \\ &= \boldsymbol{E}_{N-2}\big(\max\{\mu + \alpha\xi_{N-1} - \xi_{N-1}, 0\}\big), \end{aligned}$$

where the conditional distribution of ξ_{N-1} given ξ_{N-2} is Normal with mean $\mu + \xi_{N-2}$ and variance σ^2.

We see that in this case the $v^{n,1}$ and $v^{n,2}$ are functions of the previous observations. Note that for a buying-selling problem with independent observations the corresponding values do not depend on the past observations and, therefore, can be calculated in advance; see Sofronov et al. (2006).

In order to find the functions $v^{n,1} = v^{n,1}(t)$ and $v^{n,2} = v^{n,2}(t)$, we propose to use numerical integration methods. To do this, for each n, we evaluate the functions at a finite set of points, $t_1, t_2, \ldots, t_K$. Since in our case the integration is on the whole real line, we will use the Gauss-Hermite quadrature (see, for example, [Abramowitz and Stegun (1972) p. 890, eq. 25.4.46]):

$$\int_{-\infty}^{\infty} g(x)e^{-x^2}\,dx = \sum_{i=1}^{K} w_i g(x_i) + R_K,$$

where x_i is the i-th zero of the Hermite polynomial $H_K(x)$, $i = 1, 2, \ldots, K$, the weights w_i are given by

$$w_i = \frac{2^{K-1}K!\sqrt{\pi}}{K^2(H_{K-1}(x_i))^2},$$

and the remainder is

$$R_K = \frac{K!\sqrt{\pi}}{2^K(2K)!}g^{(2K)}(\eta), \quad -\infty < \eta < \infty.$$

Let

$$f_{t_k}(x) = \frac{1}{\sigma\sqrt{2\pi}}e^{-(x-\mu-\alpha t_k)^2/(2\sigma^2)}$$

be the probability density function of the normal distribution $N(\mu + \alpha t_k, \sigma^2)$. Then it can easily be shown that

$$\int_{-\infty}^{\infty} g(x)f_{t_k}(x)\,dx = \frac{1}{\sqrt{\pi}}\sum_{i=1}^{K} w_i g(\sqrt{2}\sigma x_i + \mu + \alpha t_k) + R_K.$$

Denote

$$\begin{aligned} g_{t_k}^{n,1}(x) &= \max\{x, v^{n,1}(x)\}, \\ g_{t_k}^{n,2}(x) &= \max\{v^{n,1}(x) - x, v^{n,2}(x)\}. \end{aligned}$$

Then, for $k = 1, 2, \dots, K$,

$$\begin{aligned} v^{1,1}(t_k) &= \mu + \alpha t_k, \\ v^{n,1}(t_k) &= \int_{-\infty}^{\infty} g_{t_k}^{n-1,1}(x) f_{t_k}(x)\, dx \\ &= \frac{1}{\sqrt{\pi}} \sum_{i=1}^{K} w_i g_{t_k}^{n-1,1}(\sqrt{2}\sigma x_i + \mu + \alpha t_k), \qquad (5.13) \\ & n = 2, \dots, N-1. \end{aligned}$$

Similarly, for $k = 1, 2, \dots, K$,

$$\begin{aligned} v^{1,2}(t_k) &= 0, \\ v^{n,2}(t_k) &= \int_{-\infty}^{\infty} g_{t_k}^{n-1,2}(x) f_{t_k}(x)\, dx \\ &= \frac{1}{\sqrt{\pi}} \sum_{i=1}^{K} w_i g_{t_k}^{n-1,2}(\sqrt{2}\sigma x_i + \mu + \alpha t_k), \qquad (5.14) \\ & n = 2, \dots, N. \end{aligned}$$

Though it is possible to choose the set of points $t_1, t_2, \dots, t_K$ arbitrarily, we can choose the points in such a way that the set is identical for all values of n. In this case, $t_k = \sqrt{2}\sigma x_k + \mu + \alpha t_k$, $k = 1, 2, \dots, K$. Then, for $\alpha \neq 1$,

$$t_k = \frac{\sqrt{2}\sigma x_k + \mu}{1 - \alpha}, \qquad k = 1, 2, \dots, K.$$

This allows us to sequentially evaluate the sets of values $v^{n,1}(t_k)$ and $v^{n,2}(t_k)$, $k = 1, 2, \dots, K$, using the previous layer of values $v^{n-1,1}(t_k)$ and $v^{n-1,2}(t_k)$; see (5.13) and (5.14). If $\alpha = 1$, similar to the case of the random walk considered in Section 5.1.3, the optimal rule is trivial (see also Section 5.1.4).

In the examples below, in order to compute the quadrature nodes x_i and the weights w_i, we use the function `gaussHermiteData` Blocker (2014). The function `approxfun` R Core Team (2018) is used to perform linear interpolation of the pairs $(t_k, v^{n,1}(t_k))$ and $(t_k, v^{n,2}(t_k))$, $k = 1, 2, \dots, K$.

Example 5.5. *Table 5.10 shows the values $v^{n,1}$ and $v^{n,2}$, $n = 1, 2, \dots, 7$, for a sequence with parameters $\alpha = 0$ (no dependence), $\mu = 0$, and $\sigma^2 = 1$. Since we observe a sequence of independent random variables, we can use Theorem 5.1 to calculate the $v^{n,1}$ and $v^{n,2}$ (for further details, see [Sofronov et al. (2006) Example 3]) and compare these values with the approximate values $v_g^{n,1}$ and $v_g^{n,2}$ obtained by the Gauss-Hermite quadrature (5.13) and (5.14) with $K = 500$. The small values of the errors in the table illustrate a good accuracy of the numerical procedure.*

TABLE 5.10
Comparison of $v^{n,1}$ and $v^{n,2}$ with $v_g^{n,1}$ and $v_g^{n,2}$ for $\mathsf{Normal}(0,1)$.

n	$v^{n,1}$	$v_g^{n,1}$	Error	$v^{n,2}$	$v_g^{n,2}$	Error
1	0.0000	0.0000	0.0000	0.0000	0.0000	0.0000
2	0.3989	0.3992	0.0003	0.3989	0.3992	−0.0003
3	0.6297	0.6301	−0.0004	0.7979	0.7985	−0.0006
4	0.7904	0.7902	0.0002	1.1184	1.1191	−0.0007
5	0.9126	0.9123	0.0003	1.3746	1.3749	−0.0003
6	1.0108	1.0101	0.0007	1.5844	1.5845	−0.0001
7	1.0924	1.0919	0.0005	1.7604	1.7605	−0.0001

Example 5.6. *Using the values of the parameters α, μ, and σ^2 specified above, we generated 10,000 sequences of length $N = 4$ with $\xi_0 = 0$ and applied the optimal double stopping rule. The estimated average gain of 1.030 is similar to the expected gain $v^{4,2} = 1.015$.*

For example, if we observe the sequence

$$\{\xi_n : n = 1, 2, 3, 4\} : -0.1206, -0.1193, 2.2498, -0.1196,$$

then $m_1 = 2$ (because $\xi_2 = -0.1193 \leq v^{2,1}(\xi_2) - v^{2,2}(\xi_2) = 0.3489 - 0.3609 = -0.0120$) and $m_2 = 3$ (because $\xi_3 = 2.2498 \geq v^{1,1}(\xi_3) = 0.2250$). This yields the gain $2.2498 - (-0.1193) = 2.3691$.

Example 5.7. *In Nagaraja et al. (2011), it was shown that AR(1) processes can successfully be applied to house log prices modeling. Assume $\alpha = 0.99$, $\mu = 0$, and $\sigma^2 = 0.002$. Using these values, we generated 10,000 sequences of length $N = 5$ with $\xi_0 = 0$ and applied the rules defined as follows:*

Rule 1 (optimal): *the optimal double stopping rule with the Gauss-Hermite quadrature ($K = 500$);*

Rule 2 (deterministic): *always buy at the beginning ($m_1 = 1$) and sell at the end ($m_2 = 5$) of the time interval;*

Rule 3 (deterministic): *do not buy anything.*

Rule 4 (random): *stop randomly at m_1 and m_2, subject to $1 \leq m_1 < m_2 \leq 5$.*

Table 5.11 displays the average gains of the rules indicating that Rule 1 (the optimal rule) outperforms the other three rules.

5.2 Multiple Selling Problems

We consider selling problems of selling several identical objects with a finite horizon.

TABLE 5.11
The average gains for the four different rules.

	Rule 1	Rule 2	Rule 3	Rule 4
Average gain	0.0007	−0.0020	0.0000	−0.0018

5.2.1 One offer per time period

Let $\xi_1, \xi_2, \ldots, \xi_N$ be a sequence of independent random variables. We observe these random variables sequentially and have to decide when we must stop. Our decision to stop depends on the observations already made, but does not depend on the future which is not yet known. After k ($k \geq 2$) stoppings at times $m_1, m_2, \ldots, m_k, 1 \leq m_1 < m_2 < \cdots < m_k \leq N$ we get a gain

$$Z_{m_1,m_2,\ldots,m_k} = \xi_{m_1} + \xi_{m_2} + \cdots + \xi_{m_k}. \tag{5.15}$$

The problem consists of finding a procedure for maximizing the expected gain.

The random variable ξ_n can be interpreted as the value of an asset (for example, a house) at time n. So we consider the problem of selling k identical objects with finite horizon N, with one offer per period and no recall of past offers. Note that within behavioral ecology we can consider the random variable ξ_n as quality of item (potential mate or place of foraging) which appears at time n; see Section 1.6.11 for further details.

Nikolaev (1980) considered an example in which he has obtained an optimal multiple-stopping rule τ^* and the value of a game v in case when $\xi_1, \xi_2, \ldots, \xi_N$ are independent identically distributed random variables with uniform distribution $U(0,1)$. The following theorem generalizes this result, see Nikolaev and Sofronov (2007).

Theorem 5.4. *Let $\xi_1, \xi_2, \ldots, \xi_N$ be a sequence of independent random variables with known distribution functions $F_1, F_2, \ldots, F_N$, Z_{m_k} be as in (5.15). Let $v^{L,l}$ be the value of a game with l ($l \leq k$) stoppings and L ($L \leq N$) steps. If there exist $\boldsymbol{E}\xi_1, \boldsymbol{E}\xi_2, \ldots, \boldsymbol{E}\xi_N$; then the value $v = v^{N,k}$, where*

$$\begin{aligned} v^{n,1} &= \boldsymbol{E}\big(\max\{\xi_{N-n+1}, v^{n-1,1}\}\big), \ 1 \leq n \leq N, \ v^{0,1} = -\infty, \\ v^{n,k-i+1} &= \boldsymbol{E}\big(\max\{v^{n-1,k-i} + \xi_{N-n+1}, v^{n-1,k-i+1}\}\big), \ k-i+1 \leq n \leq N, \\ & \quad v^{k-i,k-i+1} = -\infty, \ i = k-1, \ldots, 1. \end{aligned}$$

We put

$$\begin{aligned} \tau_1^* &= \min\{m_1 : 1 \leq m_1 \leq N-k+1, \xi_{m_1} \geq v^{N-m_1,k} - v^{N-m_1,k-1}\}, \\ \tau_i^* &= \min\{m_i : \tau_{i-1}^* < m_i \leq N-k+i, \xi_{m_i} \geq v^{N-m_i,k-i+1} - v^{N-m_i,k-i}\}, \\ & \qquad\qquad i = 2, \ldots, k-1, \\ \tau_k^* &= \min\{m_k : \tau_{k-1}^* < m_k \leq N, \xi_{m_k} \geq v^{N-m_k,1}\}, \end{aligned}$$

then $\tau^* = (\tau_1^*, \ldots, \tau_k^*)$ is the optimal multiple-stopping rule.

TABLE 5.12
The values for the uniform distribution $U(0,1)$.

L	$v^{L,1}$	$v^{L,2}$	$v^{L,3}$	$v^{L,4}$	$v^{L,5}$	$v^{L,6}$	$v^{L,7}$
1	0.5000						
2	0.6250	1.0000					
3	0.6953	1.1953	1.5000				
4	0.7417	1.3203	1.7417	2.0000			
5	0.7751	1.4091	1.9091	2.2751	2.5000		
6	0.8004	1.4761	2.0341	2.4761	2.8004	3.0000	
7	0.8203	1.5287	2.1318	2.6318	3.0287	3.3203	3.5000
8	0.8364	1.5712	2.2105	2.7568	3.2105	3.5712	3.8364
9	0.8498	1.6064	2.2756	2.8597	3.3597	3.7756	4.1064
10	0.8611	1.6360	2.3303	2.9462	3.4847	3.9462	4.3303

The proof of the theorem is similar to the proofs of Theorem 5.1 and Theorem 5.2.

We discuss here two examples in which we specify the distributions F_1, $F_2,\dots$, F_N. Let $\xi_1, \xi_2, \dots, \xi_N$ be a sequence of independent random variables, where $\xi_n = \eta_n - c_n$, $\{c_n, n = 1, 2, \dots, N\}$ is a known nonnegative deterministic function (a cost per observation), $\{\eta_n, n = 1, 2, \dots, N\}$ is a sequence of independent identically distributed random variables. We assume that the distribution of η_n is either uniform or normal (for comparison, see Example 5.1 and Example 5.3). In each case, we present a solution of the multiple optimal stopping problem.

Example 5.8 (Uniform distribution). *Let $\eta_1, \dots, \eta_N$ be a sequence of independent random variable having uniform distribution $U(a, b)$, a, b are fixed numbers. From Theorem 5.4, for $1 \le n \le N$, we have*

$$\begin{aligned} v^{n,1} &= \boldsymbol{E}\big(\max\{\xi_{N-n+1}, v^{n-1,1}\}\big) \\ &= (v^{n-1,1} + c_{N-n+1} - a)^2/(2(b-a)) + (a+b)/2 - c_{N-n+1}, \end{aligned}$$

where $v^{0,1} = a - c_N$.

Similarly, for $k - i + 1 \le n \le N$,

$$\begin{aligned} v^{n,k-i+1} &= \boldsymbol{E}\big(\max\{v^{n-1,k-i} + \xi_{N-n+1}, v^{n-1,k-i+1}\}\big) \\ &= (v^{n-1,k-i+1} - v^{n-1,k-i} + c_{N-n+1} - a)^2/(2(b-a)) \\ &\quad + (a+b)/2 + v^{n-1,k-i} - c_{N-n+1}, \end{aligned}$$

where $v^{k-i,k-i+1} = a + (k-i)(a+b)/2 - (c_{N-k+i} + \dots + c_N)$, $i = k-1, \dots, 1$.

Table 5.12 shows the values of the game $v^{L,l}$ with l ($l \le 7$) stops and L ($L \le 10$) steps for the choice $c_n = 0$, $a = 0$, $b = 1$.

If $N = 10$, $k = 3$; *then the expected gain* $v = v^{10,3} = 2.3303$. *We have the optimal multiple-stopping rule* $\tau^* = (\tau_1^*, \tau_2^*, \tau_3^*)$:

$$\begin{aligned}
\tau_1^* &= \min\{m_1 : 1 \le m_1 \le 8, \xi_{m_1} \ge v^{10-m_1,3} - v^{10-m_1,2}\},\\
\tau_2^* &= \min\{m_2 : \tau_1^* < m_2 \le 9, \xi_{m_2} \ge v^{10-m_2,2} - v^{10-m_2,1}\},\\
\tau_3^* &= \min\{m_3 : \tau_2^* < m_3 \le 10, \xi_{m_3} \ge v^{10-m_3,1}\}.
\end{aligned}$$

Assume, for example, the following sequence $\{\xi_n, n = 1, \ldots, 10\}$ *was observed: 0.5335, 0.4892, 0.8619, 0.9711, 0.2289, 0.5518, 0.2496, 0.8356, 0.6542, 0.2138. Then*

- $m_1 = 3$ *because* $\xi_3 = 0.8619 \ge v^{7,3} - v^{7,2} = 2.1318 - 1.5287 = 0.6031$,
- $m_2 = 4$ *because* $\xi_4 = 0.9711 \ge v^{6,2} - v^{6,1} = 1.4761 - 0.8004 = 0.6757$, *and*
- $m_3 = 8$ *because* $\xi_8 = 0.8356 \ge v^{2,1} = 0.6250$.

In this case, we would receive the gain $0.8619 + 0.9711 + 0.8356 = 2.6686$.

Example 5.9 (Normal distribution). *Let* $\eta_1, \ldots, \eta_N$ *be a sequence of independent random variable having normal distribution* $N(\mu, \sigma^2)$ *with probability density function*

$$g(x) = \frac{1}{\sqrt{2\pi}\sigma} \exp\left\{-\frac{(x-\mu)^2}{2\sigma^2}\right\}, \quad x \in (-\infty, \infty),$$

where $\mu \in (-\infty, \infty)$, $\sigma > 0$ *are fixed numbers. As in Example 5.8, we can show that, for* $1 \le n \le N$,

$$\begin{aligned}
v^{n,1} &= \boldsymbol{E}\big(\max\{\xi_{N-n+1}, v^{n-1,1}\}\big)\\
&= \sigma\psi\left(\frac{v^{n-1,1} + c_{N-n+1} - \mu}{\sigma}\right) + \mu - c_{N-n+1},
\end{aligned}$$

where $\psi(x) = \varphi(x) + x\Phi(x)$, $\varphi(x)$ *is the density function of the standard normal distribution,* $\Phi(x)$ *is the distribution function of the standard normal distribution,* $v^{0,1} = -\infty$.

Likewise, for $k - i + 1 \le n \le N$,

$$\begin{aligned}
v^{n,k-i+1} &= \boldsymbol{E}\big(\max\{v^{n-1,k-i} + \xi_{N-n+1}, v^{n-1,k-i+1}\}\big)\\
&= \sigma\psi\left(\frac{v^{n-1,k-i+1} - v^{n-1,k-i} + c_{N-n+1} - \mu}{\sigma}\right)\\
&\quad + \mu + v^{n-1,k-i} - c_{N-n+1},
\end{aligned}$$

where $v^{k-i,k-i+1} = -\infty$, $i = k-1, \ldots, 1$.

Table 5.13 presents the values of the game $v^{L,l}$ *with* l ($l \le 7$) *stops and* L ($L \le 10$) *steps for the choice* $c_n = 0$, $\mu = 0$, $\sigma = 1$.

TABLE 5.13
The values for the normal distribution $N(0,1)$.

L	$v^{L,1}$	$v^{L,2}$	$v^{L,3}$	$v^{L,4}$	$v^{L,5}$	$v^{L,6}$	$v^{L,7}$
1	0.0000						
2	0.3989	0.0000					
3	0.6297	0.6297	0.0000				
4	0.7904	1.0287	0.7904	0.0000			
5	0.9127	1.3198	1.3198	0.9127	0.0000		
6	1.0108	1.5478	1.7187	1.5478	1.0108	0.0000	
7	1.0924	1.7344	2.0380	2.0380	1.7344	1.0924	0.0000
8	1.6121	1.8918	2.3034	2.4369	2.3034	1.8918	1.1621
9	1.2227	2.0276	2.5299	2.7727	2.7727	2.5299	2.0276
10	1.2762	2.1468	2.7270	3.0619	3.1716	3.0619	2.7270

If $N = 9$, $k = 2$*; then the expected gain* $v = v^{9,2} = 2.0276$*. We have the optimal multiple-stopping rule* $\tau^* = (\tau_1^*, \tau_2^*)$*:*

$$\tau_1^* = \min\{m_1 : 1 \le m_1 \le 8, \xi_{m_1} \ge v^{9-m_1,2} - v^{9-m_1,1}\},$$
$$\tau_2^* = \min\{m_2 : \tau_1^* < m_2 \le 9, \xi_{m_2} \ge v^{9-m_2,1}\}.$$

Suppose we observed the following sequence $\{\xi_n, n = 1, \dots, 9\}$*: 1.9022, −0.0272, −0.1018, 0.0394, −0.7181, 0.9786, 1.5822, 1.5154, 2.5732. Then*

- $m_1 = 1$ *because* $\xi_1 = 1.9022 \ge v^{8,2} - v^{8,1} = 1.8918 - 1.6121 = 0.2797$*, and*
- $m_2 = 6$ *since* $\xi_6 = 0.9786 \ge v^{3,1} = 0.6297$.

This would give us the gain $1.9022 + 0.9786 = 2.8808$.

5.2.2 A fixed number of offers per time period

We will consider a generalization of the problem with one offer per time period described in the previous section.

Let $\xi_1, \xi_2, \dots, \xi_N$ be a sequence of independent identically distributed random variables. We observe these random variables sequentially and have to decide when we must stop, given that there is no recall allowed, that is, a random variable once rejected cannot be chosen later on. Our decision to stop depends on the observations already made, but does not depend on the future which is not yet known. After k, $1 \le k \le K$, stoppings at times $m_1, m_2, \dots, m_k$, $1 \le m_1 < m_2 < \dots < m_k \le N$ we get a gain

$$Z_{m_1,m_2,\dots,m_k} = c_1\xi_{m_1} + c_2\xi_{m_2} + \dots + c_k\xi_{m_k}, \tag{5.16}$$

where $c_1 + c_2 + \dots + c_k = K$, $1 \le c_n \le C$. The problem consists of finding a procedure for maximizing the expected gain.

The random variable ξ_n can be interpreted as a value of asset (for example, a house) at time n, c_n is a number of the objects sold at time n. We consider the problem of selling K identical objects with finite horizon N, with a fixed rate of offers C, that is, a number of offers per time period, and no recall of past offers. If a decision-maker stops at time n, he or she can sell $1, 2, \dots, C$ objects. Clearly, this decision may affect the decision-maker's further strategy to sell the remaining objects. In this paper, we derive in an explicit view a decision rule for identifying the number of stoppings k and a corresponding optimal procedure for this selling problem.

Theorem 5.4 gives a multiple optimal stopping rule and the expected gain for a problem with a fixed number of stoppings k. However, in our problem, we have a certain degree of freedom to choose the k. Namely, if we stop at time n, we can sell 1,2 or C objects. Obviously our decision may affect the remaining number of stopping required to sell all of the K objects. The following theorem compares the relative efficiencies $v^{n,k}/k$, the ratio of the value of a game to the number of stoppings, of optimal multiple-stopping rules with different numbers of stoppings. In other words, $v^{n,k}/k$ is the expected gain of selling one object if we use the optimal stopping rule with k stoppings.

Theorem 5.5 (Sofronov (2013)). *Let $\xi_1, \xi_2, \dots, \xi_N$ be a sequence of independent random variables with a known distribution functions $F_1, F_2, \dots, F_N$, $Z_{\boldsymbol{m}_k}$ be as in* (5.16). *Let $v^{L,l}$ be the value of a game with l, $l \le L$, stoppings and L, $L \le N$, steps. If there exist $\boldsymbol{E}\xi_1, \boldsymbol{E}\xi_2, \dots, \boldsymbol{E}\xi_N$, then*

$$\frac{v^{n,k}}{k} \ge \frac{v^{n,k+1}}{k+1}, \quad 1 \le k \le n, \quad 1 \le n \le N,$$

where $v^{k,k+1} = -\infty$.

Proof: The proof is by induction on n.

For $n = 1$, the inequality is trivial. Indeed, it follows easily from $v^{1,1} = \boldsymbol{E}\xi_N$ and the fact that $v^{1,2} = -\infty$. In the same way, we can show that

$$\frac{v^{n,n}}{n} \ge \frac{v^{n,n+1}}{n+1}, \quad 1 \le n \le N.$$

Assume that the following inequality holds

$$v^{n-1,1} \ge \frac{v^{n-1,2}}{2}. \tag{5.17}$$

Now we show that

$$v^{n,1} \ge \frac{v^{n,2}}{2}, \quad 2 \le n \le N.$$

From Theorem 5.4, we have

$$\begin{aligned} v^{n,1} &= \boldsymbol{E}(\max\{\xi_{N-n+1}, v^{n-1,1}\}), \\ v^{n,2} &= \boldsymbol{E}(\max\{v^{n-1,1} + \xi_{N-n+1}, v^{n-1,2}\}). \end{aligned}$$

Put

$$\begin{aligned} g(x,1) &= \max\{x, v^{n-1,1}\} \\ &= \begin{cases} v^{n-1,1}, & x < v^{n-1,1}, \\ x, & x \geq v^{n-1,1}; \end{cases} \\ g(x,2) &= \frac{1}{2}\max\{v^{n-1,1}+x, v^{n-1,2}\} \\ &= \begin{cases} v^{n-1,2}/2, & x < v^{n-1,2}-v^{n-1,1}, \\ (v^{n-1,1}+x)/2, & x \geq v^{n-1,2}-v^{n-1,1}. \end{cases} \end{aligned}$$

It is not hard to prove that

$$g(x,1) \geq g(x,2), \quad -\infty < x < \infty. \tag{5.18}$$

Indeed, it is easily shown that the node on the curve $g(x,1)$ is also located on the curve $g(x,2)$. Note that the slope of the linear part of the function $g(x,1)$ is greater than the analogous slope of the $g(x,2)$ as $1 > 1/2$. By inductive assumption (5.17), it is clear that the function $g(x,1)$ is greater than the horizontal part of the function $g(x,2)$. This proves (5.18). Thus, $\boldsymbol{E}g(\xi_{N-n+1},1) \geq \boldsymbol{E}g(\xi_{N-n+1},2)$, that is, $v^{n,1} \geq v^{n,2}/2$.

Assume that the following two inequalities hold:

$$\frac{v^{n-1,k}}{k} \geq \frac{v^{n-1,k+1}}{k+1}, \tag{5.19}$$

$$\frac{v^{n-1,k-1}}{k-1} \geq \frac{v^{n-1,k}}{k}. \tag{5.20}$$

Let us show that

$$\frac{v^{n,k}}{k} \geq \frac{v^{n,k+1}}{k+1}.$$

Using Theorem 5.4, we obtain

$$\begin{aligned} v^{n,k} &= \boldsymbol{E}(\max\{v^{n-1,k-1}+\xi_{N-n+1}, v^{n-1,k}\}), \\ v^{n,k+1} &= \boldsymbol{E}(\max\{v^{n-1,k}+\xi_{N-n+1}, v^{n-1,k+1}\}). \end{aligned}$$

Put

$$\begin{aligned} g(x,k) &= \frac{1}{k}\max\{v^{n-1,k-1}+x, v^{n-1,k}\} \\ &= \begin{cases} v^{n-1,k}/k, & x < v^{n-1,k}-v^{n-1,k-1}, \\ (v^{n-1,k-1}+x)/k, & x \geq v^{n-1,k}-v^{n-1,k-1}; \end{cases} \\ g(x,k+1) &= \frac{1}{k+1}\max\{v^{n-1,k}+x, v^{n-1,k+1}\} \\ &= \begin{cases} v^{n-1,k+1}/(k+1), & x < v^{n-1,k+1}-v^{n-1,k}, \\ (v^{n-1,k}+x)/(k+1), & x \geq v^{n-1,k+1}-v^{n-1,k}. \end{cases} \end{aligned}$$

It can easily be checked that

$$g(x,k) \geq g(x,k+1), \quad -\infty < x < \infty. \tag{5.21}$$

$$(v^{n-1,k} - v^{n-1,k-1}, v^{n-1,k}/k),$$
$$(v^{n-1,k}/k, v^{n-1,k}/k),$$

respectively. By inductive assumption (5.19), it is clear that point A is located above the horizontal part of the function $g(x,k+1)$. By inductive assumption (5.20), we have

$$v^{n-1,k} - v^{n-1,k-1} \leq v^{n-1,k} - \frac{k-1}{k}v^{n-1,k} = \frac{v^{n-1,k}}{k}.$$

This proves (5.21). Thus, $\boldsymbol{E}g(\xi_{N-n+1},k) \geq \boldsymbol{E}g(\xi_{N-n+1},k+1)$, that is,

$$\frac{v^{n,k}}{k} \geq \frac{v^{n,k+1}}{k+1}.$$

This completes the proof. □

Corollary 5.22. *The optimal multiple-stopping rule with a smaller number of stoppings is more efficient, that is,*

$$v^{N,1} \geq \frac{v^{N,2}}{2} \geq \cdots \geq \frac{v^{N,k}}{k}.$$

Let us consider our initial problem (5.16). If $C \geq K$ and we use the optimal stopping rule with 1 stopping, then we obtain the expected gain $Kv^{N,1}$ such that

$$Kv^{N,1} \geq K\frac{v^{N,2}}{2} \geq \cdots \geq K\frac{v^{N,K}}{K},$$

where $Kv^{N,k}/k$ is the expected gain of selling K objects for the optimal stopping rule with k stoppings. Thus, the problem (5.16) is reduced to a problem with 1 stopping and a gain $Z^r_{m_1} = \xi_{m_1}$, $Z_{m_1} = KZ^r_{m_1}$. In other words, we get a higher expected gain if we stop once and sell all of the K objects than we stop more than once and sell the objects in parts.

If $K = kC$, where C, K, k are natural numbers, then we can consider a reduced problem with a gain

$$Z^r_{m_1,m_2,\ldots,m_k} = \xi_{m_1} + \xi_{m_2} + \cdots + \xi_{m_k},$$

where $Z_{m_1,m_2,\ldots,m_k} = CZ^r_{m_1,m_2,\ldots,m_k}$ is the initial gain.

Generally, it is readily seen that the number of stoppings k for the optimal sequential procedure given by Theorem 5.4 is as follows

$$k = \left\lceil \frac{K}{C} \right\rceil,$$

where $\lceil \cdot \rceil$ is the ceiling function, the smallest following integer number.

We discuss here two examples in which we specify the distribution function F. In Section 5.2.1, two examples were considered in which optimal multiple-stopping rules τ^* and the values v were obtained in the case where $\xi_1, \xi_2, \ldots, \xi_N$ are independent distributed random variables having either uniform or normal distribution. Using these examples, we illustrate the usefulness of the result obtained in Theorem 5.5.

Example 5.10 (Uniform distribution). *Let $\xi_1, \xi_2, \ldots, \xi_N$ be a sequence of independent random variable having uniform distribution $U(a, b)$, a, b are fixed numbers. From Example 5.8, we have*

$$\begin{aligned} v^{n,1} &= (v^{n-1,1} - a)^2/(2(b-a)) + (a+b)/2, \\ v^{n,k} &= (v^{n-1,k} - v^{n-1,k-1} - a)^2/(2(b-a)) + (a+b)/2 + v^{n-1,k-1}, \end{aligned}$$

where $v^{0,1} = a$, $v^{k,k+1} = a + k(a+b)/2$, $1 \le n \le N$.

Recall that in Table 5.12 we presented the values of the game $v^{L,l}$ with l ($l \le 7$) stoppings and L ($L \le 10$) steps for the choice $a = 0$, $b = 1$.

If $K = 2$, $C = 3$, $N = 10$, then $k = 1$ and the value of the game of the reduced problem

$$v = v^{10,1} = 0.8611.$$

This yields the expected gain for the initial problem $2 \cdot 0.8611 = 1.7222$, which is higher than $v^{10,2} = 1.6360$ if we had used the double optimal stopping rule. We have the optimal stopping rule $\tau^* = (\tau_1^*)$, where

$$\tau_1^* = \min\{m_1 : 1 \le m_1 \le 10, \xi_{m_1} \ge v^{10-m_1,1}\}.$$

Example 5.11 (Normal distribution). *Let $\xi_1, \xi_2, \ldots, \xi_N$ be a sequence of independent random variable having normal distribution $N(\mu, \sigma^2)$, here $\mu \in (-\infty, \infty)$, $\sigma > 0$ are fixed numbers. From Example 5.9, we obtain*

$$\begin{aligned} v^{n,1} &= \sigma\psi\left(\frac{v^{n-1,1} - \mu}{\sigma}\right) + \mu, \\ v^{n,k} &= \sigma\psi\left(\frac{v^{n-1,k} - v^{n-1,k-1} - \mu}{\sigma}\right) + \mu + v^{n-1,k-1}, \end{aligned}$$

where $\psi(x) = \varphi(x) + x\Phi(x)$, $\varphi(x)$ is the density function of the standard normal distribution, $\Phi(x)$ is the distribution function of the standard normal distribution, $v^{0,1} = -\infty$, $v^{k,k+1} = -\infty$, $1 \le n \le N$.

In Table 5.13, we presented the values of the game $v^{L,l}$ with l ($l \le 7$) stoppings and L ($L \le 10$) steps for the choice $\mu = 0$, $\sigma = 1$.

If $K = 6$, $C = 2$, $N = 8$, then $k = 3$ and the value of the game of the reduced problem

$$v = v^{8,3} = 2.3034.$$

This yields the expected gain for the initial problem $2 \cdot 2.3034 = 4.6068$, which is higher than $v^{8,6} = 1.8918$ if we had used the optimal stopping rule with 6 stoppings. We have the optimal stopping rule $\tau^* = (\tau_1^*, \tau_2^*, \tau_3^*)$:

$$\begin{aligned} \tau_1^* &= \min\{m_1 : 1 \le m_1 \le 6, \xi_{m_1} \ge v^{8-m_1,3} - v^{8-m_1,2}\}, \\ \tau_2^* &= \min\{m_2 : \tau_1^* < m_2 \le 7, \xi_{m_2} \ge v^{8-m_2,2} - v^{8-m_2,1}\}, \\ \tau_3^* &= \min\{m_3 : \tau_2^* < m_3 \le 8, \xi_{m_3} \ge v^{8-m_3,1}\}. \end{aligned}$$

5.2.3 A variable rate of offers per time period

Assume we sequentially observe independent identically distributed random variables $\xi_1, \xi_2, \ldots, \xi_N$, where the value ξ_n can be interpreted as the value of an asset (for example, a house) at time n. We have K identical objects, which we want to sell within the finite time horizon N, and at each time n we may receive no more than C_n offers, and we are not allowed to recall past offers. Our decision to sell c_n, $0 \le c_n \le C_n$, objects at time n solely depends on the information we know already and it does not depend on the observations we have not seen yet. If we decide to sell at times $m_1, m_2, \ldots, m_k$, $1 \le m_1 < m_2 < \cdots < m_k \le N$, $1 \le k \le K$, we get a gain

$$Z_{m_1,m_2,\ldots,m_k} = c_{m_1}\xi_{m_1} + c_{m_2}\xi_{m_2} + \cdots + c_{m_k}\xi_{m_k}, \tag{5.22}$$

where $c_{m_1} + c_{m_2} + \cdots + c_{m_k} = K$, $0 \le c_n \le C_n$, C_n are known positive constants, $n = 1, 2, \ldots, N$. In the other words, C_n is the demand and ξ_n is the price of the asset at time n. The problem consists of finding a sequential procedure for maximizing the expected gain. Since c_n are not fixed, the problem we consider in this paper becomes much more complicated compare to the problem with a fixed rate of offers $C_n = C$, $n = 1, 2, \ldots, N$, considered in the previous section.

Theorem 5.4 gives a multiple optimal stopping rule and the expected gain for a problem with a fixed number of stops k. However, in our initial problem, the k is not fixed since at time n we can sell $0, 1 \ldots$ or C_n objects, C_n are known positive constants, $n = 1, 2, \ldots, N$. Obviously, our decision may affect the remaining number of stops required to sell all of the K objects. In this case, c_n are $\mathcal{F}_n$-measurable random variables taking values from $\{0, 1, \ldots, C_n\}$, where $\mathcal{F}_n = \sigma(\xi_1, \ldots, \xi_n)$.

Let us consider an extended version of the problem. We observe a sequence of random variables $\tilde{\xi}_1, \tilde{\xi}_2, \ldots, \tilde{\xi}_{S_N}$, $S_N = C_1 + \cdots + C_N$, where

$$\tilde{\xi}_j = \begin{cases} \xi_1, & 1 \le j \le S_1, \\ \xi_2, & S_1 + 1 \le j \le S_2, \\ \vdots & \\ \xi_N, & S_{N-1} + 1 \le j \le S_N, \end{cases} \tag{5.23}$$

$\{S_n\}_{n=1}^N$ is a cumulative sum, $S_n = C_1 + \cdots + C_n$. In the other words, we observe the following sequence

$$\underbrace{\xi_1, \ldots, \xi_1}_{C_1}, \underbrace{\xi_2, \ldots, \xi_2}_{C_2}, \ldots, \underbrace{\xi_N, \ldots, \xi_N}_{C_N}.$$

The following theorem gives a multiple optimal stopping rule and the expected gain for the stopping problem of the extended sequence when we are allowed to stop K times and we receive the gain

$$Z_{m_1,m_2,\ldots,m_K} = \tilde{\xi}_{m_1} + \tilde{\xi}_{m_2} + \cdots + \tilde{\xi}_{m_K}. \tag{5.24}$$

Theorem 5.6. *Let $\tilde{\xi}_1, \tilde{\xi}_2, \ldots, \tilde{\xi}_{S_N}$ be a sequence of random variables as in (5.23) with the gain function (5.24). Let $v^{L,l}$ be the expected gain with l, $l \le K$, stops and*

L, $L \le S_N$, steps. If there exist $\boldsymbol{E}\tilde{\xi}_1, \boldsymbol{E}\tilde{\xi}_2, \ldots, \boldsymbol{E}\tilde{\xi}_{S_N}$, then the value $v = v^{S_N,K}$, where

$$
\begin{aligned}
v^{S_N-n+1,1} &= \begin{cases} \boldsymbol{E}\big(\max\{\tilde{\xi}_n, v^{S_N-n,1}\}\big), & n-1 \in \{S_1, \ldots, S_{N-1}\}, \\ \max\{\tilde{\xi}_n, v^{S_N-n,1}\}, & \textit{otherwise}. \end{cases} \\
& \quad 1 \le n \le S_N,\ v^{0,1} = -\infty, \\
v^{S_N-n+1,K-i+1} &= \begin{cases} \boldsymbol{E}\big(\max\{v^{S_N-n,K-i} + \tilde{\xi}_n, \\ \qquad v^{S_N-n,K-i+1}\}\big), & n-1 \in \{S_1, \ldots, S_{N-1}\}, \\ \max\{v^{S_N-n,K-i} + \tilde{\xi}_n, \\ \qquad v^{S_N-n,K-i+1}\}, & \textit{otherwise} \end{cases} \\
& \quad 1 \le n \le S_N - K + i,\ v^{K-i,K-i+1} = -\infty, \\
& \quad i = K-1, \ldots, 1.
\end{aligned}
$$

We put

$$
\begin{aligned}
\tau_1^* &= \min\{m_1 : 1 \le m_1 \le S_N - K + 1, \tilde{\xi}_{m_1} \ge v^{S_N-m_1,K} - v^{S_N-m_1,K-1}\}, \\
\tau_i^* &= \min\{m_i : \tau_{i-1}^* < m_i \le S_N - K + i, \tilde{\xi}_{m_i} \ge v^{S_N-m_i,K-i+1} - v^{S_N-m_i,K-i}\}, \\
& \qquad\qquad i = 2, \ldots, K-1, \\
\tau_K^* &= \min\{m_K : \tau_{K-1}^* < m_K \le S_N, \tilde{\xi}_{m_K} \ge v^{S_N-m_K,1}\}.
\end{aligned}
$$

then $\tau^* = (\tau_1^*, \ldots, \tau_K^*)$ is the optimal multiple-stopping rule.

Proof: From (2.20) we obtain

$$
\begin{aligned}
V_{m_1,\ldots,m_K} &= \max\{X_{m_1,\ldots,m_K}, \mathbf{E}_{m_K} V_{m_1,\ldots,m_{K-1},m_K+1}\} \\
&= \max\{\tilde{\xi}_{m_1} + \cdots + \tilde{\xi}_{m_K}, \mathbf{E}_{m_K} V_{m_1,\ldots,m_{K-1},m_K+1}\} \\
&= \max\{\tilde{\xi}_{m_K}, v^{S_N-m_K,1}\} + \tilde{\xi}_{m_1} + \cdots + \tilde{\xi}_{m_{K-1}}. \qquad (5.25)
\end{aligned}
$$

For $m_{K-1} < m_K < S_N$,

$$
\begin{aligned}
v^{S_N-m_K+1,1} &= \mathbf{E}_{m_K-1}\big(\max\{\tilde{\xi}_{m_K}, v^{S_N-m_K,1}\}\big) \\
&= \begin{cases} \boldsymbol{E}\big(\max\{\tilde{\xi}_{m_K}, v^{S_N-m_K,1}\}\big), & m_K - 1 \in \{S_1, \ldots, S_{N-1}\}, \\ \max\{\tilde{\xi}_{m_K}, v^{S_N-m_K,1}\}, & \text{otherwise.} \end{cases}
\end{aligned}
$$

For $m_K = S_N$,

$$
v^{S_N-m_K+1,1} = v^{1,1} = \begin{cases} \boldsymbol{E}\tilde{\xi}_{S_N}, & \text{if } m_K = S_{N-1}+1, \\ \tilde{\xi}_{S_N}, & \text{otherwise,} \end{cases}
$$

since $v^{0,1} = -\infty$.
In the other words, if $C_N = S_N - S_{N-1} = 1$,

$$
v^{C_N,1} = v^{1,1} = \boldsymbol{E}\tilde{\xi}_{S_N} = \boldsymbol{E}\xi_N.
$$

If $C_N = S_N - S_{N-1} > 1$,

$$v^{1,1} = v^{2,1} = \cdots = v^{C_N-1,1} = \xi_N,$$

whereas

$$v^{C_N,1} = \boldsymbol{E}\tilde{\xi}_{S_{N-1}+1} = \boldsymbol{E}\xi_N.$$

According to Theorem 5.4 if $\tilde{\xi}_{m_K} \geq v^{S_N-m_K,1}$, then we need to stop at time m_K. Therefore, should we decide to stop at time $S_{n-1}+1$, that is, in the beginning of the series

$$\underbrace{\xi_n, \ldots, \xi_n}_{C_n},$$

we will also stop at times $S_{n-1}+2, \ldots, S_n$. This implies that we will be able to sell C_n objects.

From (2.19) and (5.25), we have

$$\begin{aligned} X_{m_1,\ldots,m_{K-1}} &= \boldsymbol{E}_{m_{K-1}} V_{m_1,\ldots,m_{K-1},m_{K-1}+1} \\ &= \boldsymbol{E}_{m_{K-1}}\big(\max\{\tilde{\xi}_{m_{K-1}+1}, v^{S_N-m_{K-1}-1,1}\} + \tilde{\xi}_{m_1} + \cdots + \tilde{\xi}_{m_{K-1}}\big) \\ &= v^{S_N-m_{K-1},1} + \tilde{\xi}_{m_1} + \cdots + \tilde{\xi}_{m_{K-1}} \end{aligned} \tag{5.26}$$

since

$$\begin{aligned} &\boldsymbol{E}_{m_{K-1}}\big(\max\{\tilde{\xi}_{m_{K-1}+1}, v^{S_N-m_{K-1}-1,1}\} \\ &= \begin{cases} \boldsymbol{E}\big(\max\{\tilde{\xi}_{m_{K-1}+1}, v^{S_N-m_{K-1}-1,1}\}\big), & m_{K-1} \in \{S_1, \ldots, S_{N-1}\}, \\ \max\{\tilde{\xi}_{m_{K-1}+1}, v^{S_N-m_{K-1}-1,1}\}, & \text{otherwise.} \end{cases} \end{aligned}$$

In the same way,

$$\begin{aligned} V_{m_1,\ldots,m_i} &= \max\{X_{m_1,\ldots,m_i}, \boldsymbol{E}_{m_i} V_{m_1,\ldots,m_{i-1},m_i+1}\} \\ &= \max\{v^{S_N-m_i,K-i} + \tilde{\xi}_{m_1} + \cdots + \tilde{\xi}_{m_i}, \boldsymbol{E}_{m_i} V_{m_1,\ldots,m_{i-1},m_i+1}\} \\ &= \max\{v^{S_N-m_i,K-i} + \tilde{\xi}_{m_i},\ v^{S_N-m_i,K-i+1}\} \\ &+ \ \tilde{\xi}_{m_1} + \cdots + \tilde{\xi}_{m_{i-1}}, \end{aligned} \tag{5.27}$$

where, for $m_{i-1} < m_i < S_N - K + i$,

$$\begin{aligned} &v^{S_N-m_i+1,K-i+1} = \boldsymbol{E}_{m_i-1}\big(\max\{v^{S_N-m_i,K-i} + \tilde{\xi}_{m_i},\ v^{S_N-m_i,K-i+1}\}\big) \\ &= \begin{cases} \boldsymbol{E}\big(\max\{v^{S_N-m_i,K-i} + \tilde{\xi}_{m_i}, \\ v^{S_N-m_i,K-i+1}\}\big), & m_i - 1 \in \{S_1, \ldots, S_{N-1}\}, \\ \max\{v^{S_N-m_i,K-i} + \tilde{\xi}_{m_i}, \\ v^{S_N-m_i,K-i+1}\}, & \text{otherwise} \end{cases} \end{aligned}$$

with $v^{K-i,K-i+1} = -\infty$, $i = K-1, K-2, \ldots, 2$.

Similarly, for $i = K-1, K-2, \dots, 2$,

$$\begin{aligned} X_{m_1,\dots,m_{i-1}} &= \boldsymbol{E}_{m_{i-1}} V_{m_1,\dots,m_{i-1},m_{i-1}+1} \\ &= \boldsymbol{E}_{m_{i-1}} \big(\max\{v^{S_N-m_{i-1}-1,K-i} + \tilde{\xi}_{m_{i-1}+1},\, v^{S_N-m_{i-1}-1,K-i+1}\} \\ &\quad + \tilde{\xi}_{m_1} + \cdots + \tilde{\xi}_{m_{i-1}}\big) \\ &= v^{S_N-m_{i-1},K-i+1} + \tilde{\xi}_{m_1} + \cdots + \tilde{\xi}_{m_{i-1}} \end{aligned}$$

since

$$\begin{aligned} &\boldsymbol{E}_{m_{i-1}} \big(\max\{v^{S_N-m_{i-1}-1,K-i} + \tilde{\xi}_{m_{i-1}+1},\, v^{S_N-m_{i-1}-1,K-i+1}\}\big) \\ &= \begin{cases} \boldsymbol{E}\big(\max\{v^{S_N-m_{i-1}-1,K-i} + \tilde{\xi}_{m_{i-1}+1}, \\ v^{S_N-m_{i-1}-1,K-i+1}\}\big), & m_{i-1} \in \{S_1, \dots, S_{N-1}\}, \\ \max\{v^{S_N-m_{i-1}-1,K-i} + \tilde{\xi}_{m_{i-1}+1}, \\ v^{S_N-m_{i-1}-1,K-i+1}\}, & \text{otherwise.} \end{cases} \end{aligned}$$

Finally,

$$\begin{aligned} V_{m_1} &= \max\{X_{m_1}, \boldsymbol{E}_{m_1} V_{m_1+1}\} \\ &= \max\{v^{S_N-m_1,K-1} + \tilde{\xi}_{m_1}, \boldsymbol{E}_{m_1} V_{m_1+1}\} \\ &= \max\{v^{S_N-m_1,K-1} + \tilde{\xi}_{m_1},\, v^{S_N-m_1,K}\}, \end{aligned} \tag{5.28}$$

where, for $1 \le m_1 < S_N - K + 1$,

$$\begin{aligned} v^{S_N-m_1+1,K} &= \boldsymbol{E}_{m_1-1}(\max\{v^{S_N-m_1,K-1} + \tilde{\xi}_{m_1},\, v^{S_N-m_1,K}\}) \\ &= \begin{cases} \boldsymbol{E}(\max\{v^{S_N-m_1,K-1} + \tilde{\xi}_{m_1},\, v^{S_N-m_1,K}\}), & m_1 - 1 \in \{S_1, \dots, S_{N-1}\}, \\ \max\{v^{S_N-m_1,K-1} + \tilde{\xi}_{m_1},\, v^{S_N-m_1,K}\}, & \text{otherwise} \end{cases} \end{aligned}$$

with $v^{K-1,K} = -\infty$.

Taking into account Theorem 5.4, we obtain $v = v^{S_N,K}$.

Now, using Theorem 5.4, (5.25), (5.26), (5.27), and (5.28) we obtain the optimal multiple-stopping rule $\tau^* = (\tau_1^*, \dots, \tau_K^*)$:

$$\begin{aligned} \tau_1^* &= \min\{m_1 : 1 \le m_1 \le S_N - K + 1, \tilde{\xi}_{m_1} \ge v^{S_N-m_1,K} - v^{S_N-m_1,K-1}\}, \\ \tau_i^* &= \min\{m_i : \tau_{i-1}^* < m_i \le S_N - K + i, \tilde{\xi}_{m_i} \ge v^{S_N-m_i,K-i+1} - v^{S_N-m_i,K-i}\}, \\ &\qquad\qquad i = 2, \dots, K-1, \\ \tau_K^* &= \min\{m_K : \tau_{K-1}^* < m_K \le S_N, \tilde{\xi}_{m_K} \ge v^{S_N-m_K,1}\}. \end{aligned}$$

This completes the proof. □

Theorem 5.6 allows us to write down c_n in an explicit form, For example,

$$c_1 = \begin{cases} \min\{C_1, K\}, & \xi_1 \geq v^{S_N-1,K} - v^{S_N-1,K-1}, \\ 0, & \xi_1 < v^{S_N-1,K} - v^{S_N-1,K-1}; \end{cases}$$

$$c_2 = \begin{cases} \min\{C_2, K\}, & \xi_1 < v^{S_N-1,K} - v^{S_N-1,K-1}, \\ & \xi_2 \geq v^{S_N-C_1-1,K} - v^{S_N-C_1-1,K-1}, \\ \min\{C_2, K - c_1\}, & \xi_1 \geq v^{S_N-1,K} - v^{S_N-1,K-1}, \\ & \xi_2 \geq v^{S_N-C_1-1,K-C_1} - v^{S_N-C_1-1,K-C_1-1}, \\ 0, & \text{otherwise.} \end{cases}$$

We can see that c_1 depends on ξ_1; c_2 depends on ξ_1, ξ_2 and c_1. Similarly, we can obtain expressions for $c_3, c_4, \ldots, c_N$ with $c_n = \min\{C_n, K - s\}$ on $\{c_1 + \cdots + c_{n-1} = s\}, s \in \{0, 1, \ldots, K\}$. Note that the actual number of stops $k = \sum_{n=1}^{N} I(c_n > 0)$.

Corollary 5.29. *If $C_1 = \cdots = C_N = 1$, the extended sequence $\tilde{\xi}_1, \tilde{\xi}_2, \ldots, \tilde{\xi}_{S_N}$ coincides with the original sequence $\xi_1, \xi_2, \ldots, \xi_N$. So we consider the stopping problem of the original sequence with K stops allowed. Since $m_i \in \{S_1, \ldots, S_N\} = \{1, 2, \ldots, N\}$, $i = 1, 2, \ldots, K$, the values $v^{L,l}$, $l \leq K$, $L \leq N$, are calculated in the same way as in Theorem 5.4, which gives the multiple optimal stopping rule $\tau^* = (\tau_1^*, \ldots, \tau_K^*)$. If we stop at time n, then $c_n = 1$.*

Corollary 5.30. *If $C_1 = \cdots = C_N = C \geq 2$, $K = kC$, where C, K, k are natural numbers, then, according to the optimal rule, if we decide to stop at some time, we should sell all of the C objects (see Sofronov (2013)). In the other words, we can consider a reduced problem with a gain*

$$Z^r_{m_1,m_2,\ldots,m_k} = \xi_{m_1} + \xi_{m_2} + \cdots + \xi_{m_k},$$

where $Z_{m_1,m_2,\ldots,m_k} = CZ^r_{m_1,m_2,\ldots,m_k}$ is the initial gain (5.22). If we stop at time n, then $c_n = C$.

Corollary 5.31. *If $\min\{C_1, \ldots, C_N\} \geq K$, the problem (5.22) is reduced to a problem with 1 stop and a gain $Z^r_{m_1} = \xi_{m_1}$, $Z_{m_1} = KZ^r_{m_1}$. This means that the optimal rule will require us to stop once and sell all of the K objects (see Sofronov 2013)), that is, $c_n = K$ if we decide to stop at time n.*

5.3 Elfving's Problem

WE CONSIDER an optimization problem formulated based on the observation of the piecewise deterministic processes. Let sequence of random variables $\eta_1, \eta_2, \ldots$ appears at moment of jumps of the Poisson process. The Poisson process is defined by the stream of *i.i.d.* random variables ξ_n with the exponential distribution having parameter λ and let the moments of jumps be $\tau_0 = 0$, $\tau_n = \sum_{j=0}^{n} \xi_j$. Define nonincreasing function $r : \Re^+ \to \Re$. The problem states that we need to find the stopping moment σ^* that is optimal for the functional $\boldsymbol{E}(\eta_n r(\tau_n))$. Under general assumption $\int_0^\infty r(t)dt < \infty$, by making use of the continuous aspect of the problem, Elfving (1967) showed that he can derive the differential equation for the boundary $y(\cdot)$ of the optimal stopping region. The stopping rule σ is defined by the boundary function $y(\cdot)$ if $\sigma = \inf\{n \in \bar{\mathbb{N}} : \eta_n \geq y(\tau_n)\}$. This model shows that for an important class of processes with continuous time, optimization tasks come down to the optimal stopping of processes with discrete time, and the stopping region is determined by the border line, which can be determined analytically.

Contribution to Elfving's problem.

1. Krasnosielska-Kobos (2015) considered an optimal multiple-stopping problem with a random horizon and a general structure of rewards. A decision-maker has m commodities of the same type for sale (cf. Elfving (1967)). Therefore, he can accept at most m offers, which he observes sequentially. The decision on acceptance or rejection of the offer must be made based on past and current observations. Each accepted offer Y_n brings in a profit of G_n if the offer is accepted before a random horizon M, and 0 otherwise. This means that after time M the commodities cannot be sold. The decision-maker aims to maximize the expected total profit. An optimal selection strategy is obtained by applying a general theorem which shows that a model with random horizon can be reduced to a model with a discount function. A case with a random horizon independent of time and magnitude of offers is analyzed in detail. As applications, two problems are solved: a multiple-stopping problem with a random horizon, a discount function and a Poisson stream of offers, and a "timeless" multiple-stopping problem with a random number of offers and a discount factor.

2. It should be noted here that the problem analyzed in this paper can be treated as a question about the behavior of the auction participant (his strategy) when he plans to successfully complete the auction and purchase the desired object without overpaying (cf. Krishna (2010)).

3. Stadje (1987) presented at the 6-th Pannonian Symposium on Mathematical Statistics following problem. k commodities are for sale. The offers for one of the commodities arrive at times $0 < \tau_1 < \tau_2 < \ldots$ corresponding to the jumps of a homogeneous Poisson process, the offer at τ_i being Y_i, where the Y_i are *i.i.d.* with a piecewise continuous cumulative distribution function and are independent of the τ_i. The reward for accepting offers at $\{\tau_j, j = n_1, n_2, \ldots, n_k\}$ is $x_{n_1,\ldots,n_k} = \sum_{i=1}^{k} y_{n_i} r(\tau_{n_i})$, where $r(\cdot)$ is a discount function subject to certain conditions. The theory of finding optimal k stopping rules, relative to maximizing expected payoff, is developed in general, and such rules are explicitly derived in some special cases.

4. Bayraktar and Ludkovski (2011) studied the optimal trade execution strategies problem considering a financial market with discrete order flow. The agent has a finite liquidation horizon and must minimize price impact given a random number of incoming trade counterparties. In the first step, a full analysis of the properties and computation of the optimal dynamic execution strategy is provided, when the order flow is given by a Poisson process. Next, it is extended to the more general case where the order flow is a Markov-modulated compound Poisson process. Properties of the optimal actions and the value function are obtained. Numerical and graphical illustrations of the results with a 3-state regime-switching model conclude the paper. The two extensions proposed by the authors are interesting: one is to include time-dependent parameters (such as price impact, order intensity, and size distribution) or further constraints on the optimal strategy, and the other is to introduce a price dimension (cf. also Stadje (1990), Zabczyk (1984)).

5. Faller and Rüschendorf (2013) considered a sequence of independent random variables $X_1, \ldots, X_n$, for some $n \in \mathbb{N}$, and fix $m \in \mathbb{N}$ with $m \leq n$. The problem of identifying the optimal sequence of m stopping times $1 \leq \tau_1 \leq \ldots \leq \tau_m \leq n$ such that
$$E\Big(\sum_{i=1}^{m} X_{\tau_i}\Big) = \sup_{\tau_1,\ldots,\tau_m} E\Big(\sum_{i=1}^{m} X_{\tau_i}\Big),$$
where $1 \leq \tau_1 \ldots \leq \tau_m \leq n$ is any sequence of m stopping times. This is a *discrete-time optimal multistopping problem with sum payoff.*

5.4 The Optimal MSP Under Knightian Uncertainty

We present a theory of optimal stopping under Knightian uncertainty.

Riedel (2009) developed a theory of optimal stopping under Knightian uncertainty (v. Appendix A.2). A suitable martingale theory for multiple priors was derived that extends the classical dynamic programming or Snell envelope approach to multiple priors. The multiple prior theory to the classical setup was implemented via a minimax theorem. In a multiple prior version of the classical model of independent and identically distributed random variables, it is possible to discuss several examples from microeconomics, operation research, and finance. For monotone payoffs, the worst-case prior can be identified quite easily with the help of stochastic dominance arguments. For more complex payoff structures, such as barrier options, model ambiguity leads to stochastic changes in the worst-case beliefs.

Li (2022) has studied the optimal multiple-stopping problem under Knightian uncertainty. It is shown that the value function of the multiple-stopping problem coincides with the one corresponding to a new reward sequence or process. For the discrete-time case, this problem can be solved by an induction method, which is a straightforward generalization of the single stopping theory.

5.5 Insurance Strategies

Here we describe an application of multiple-stopping rules in insurance. We consider insurance products in which the policyholder has the option of insuring k of its annual losses for N years. This involves a choice of k out of N years in which to apply the insurance policy coverage by making claims against losses in the given year. The general insurance product described here can accommodate any form of insurance policy, but we will focus on two basic generic "building block" policies (see Definitions 5.1 and 5.2), which can be combined to create more complex types of products.

Let us assume that throughout a year a company incurs a random number of losses L with severities (loss amounts) $\zeta_1, \ldots, \zeta_L$. Next, suppose that the company holds an insurance product that lasts for N years and grants the company the right to mitigate k of its N annual losses using its insurance claims. Consider, for example, a given year n, $n \leq N$, where the company will incur L_n losses that add up to $\eta_n = \sum_{i=1}^{L_n} \zeta_i(n)$. Assuming that it has not yet utilized all its insurance mitigations k, it has the choice to make an insurance claim or not. If it utilizes the insurance claim in this year, the resulting annual loss will be denoted by $\widetilde{\eta}_n$. In this context, the company's aim is to choose k distinct years out of the N to minimize its expected loss over N years. From a mathematical point of view, the problem is to find a multiple optimal stopping rule to make the k sets of insurance claims.

As mentioned above, the insurance policies presented here should be treated as building blocks for more complex ones, leading to mitigation of more complex sources of risk. It also worth noticing that the policies presented are just a mathematical model of the actual policies that would be sold in practice, and although some characteristics, such as deductibles, can be incorporated in the model, they are not presented at this stage. Here, we present basic insurance policies the company can use in the insurance product.

Definition 5.1 (Individual Loss Policy (ILP)). *This policy applies a constant threshold to the loss process in year n in which individual losses experience a Top Cover Limit (TCL) as specified by*

$$\widetilde{\eta} = \sum_{i=1}^{L} \max\left(\zeta_i - TCL,\ 0\right).$$

Definition 5.2 (Accumulated Loss Policy (ALP)). *The ALP provides a specified maximum compensation on losses experienced over a year. If this maximum compensation is denoted by ALP, then the annual insured process is defined as*

$$\widetilde{\eta} = \left(\sum_{i=1}^{L} \zeta_i - ALP \right) \mathbb{I}_{\left(\sum_{i=1}^{L} \zeta_i > ALP\right)}.$$

Targino et al. (2017) discuss two different gain (or, to be precise, loss) functions:

1. Global Risk Transfer Strategy: the expected total loss over the period with the finite horizon N is minimized;
2. Local Risk Transfer Strategy: the expected sum of the losses at the insurance times (i.e. stopping times) is minimized.

For the first loss function, the formal objective is to minimize

$$\sum_{\substack{n=1 \\ n \notin \{m_1,\dots,m_k\}}}^{N} \eta_n + \sum_{j=1}^{k} \widetilde{\eta}_{m_j} = \sum_{n=1}^{N} \eta_n - \sum_{\substack{n=1 \\ n \in \{m_1,\dots,m_k\}}}^{N} \left\{ \eta_n - \widetilde{\eta}_n \right\}.$$

Since $\sum_{n=1}^{N} \eta_n$ does not depend on the choice of $m_1, \dots, m_k$, this problem is equivalent to maximizing

$$\sum_{j=1}^{k} \xi_{m_j} = \sum_{j=1}^{k} \left\{ \eta_{m_j} - \widetilde{\eta}_{m_j} \right\},$$

where the process ξ is defined as $\xi_n = \eta_n - \widetilde{\eta}_n$.

For the second objective function, the company aims to minimize the total loss, not over N years, but instead only at times at which the decisions are taken to apply for insurance and therefore claim against losses in the given year,

$$\sum_{j=1}^{k} \widetilde{\eta}_{m_j}$$

and, in this case, the process ξ should be viewed as $\xi_n = -\widetilde{\eta}_n$.

Example 5.12 (Pareto distribution). *Assume that the insured losses $\widetilde{\eta}_1, \dots, \widetilde{\eta}_N$ form a sequence of i.i.d. random variables such that $\widetilde{\eta} \sim \mathsf{Pareto}(\alpha, \beta)$ with $f(y) = \alpha\beta^{\alpha} y^{-\alpha-1}$, $y \in [\beta, \infty)$, $\alpha > 1$, $\beta > 0$. The multiple optimal rule that minimizes the expected loss can be calculated using $\xi = -\widetilde{\eta}$. Then the values of expected losses are as follows.*

$$v^{n,1} = \frac{\beta}{\alpha - 1} \left(\alpha - \left(\frac{\beta}{v^{n-1,1}} \right)^{\alpha-1} \right),$$

$$v^{n,k-i+1} = \frac{\beta}{\alpha - 1} \left(\alpha - \left(\frac{\beta}{v^{n-1,k-i+1} - v^{n-1,k-i}} \right)^{\alpha-1} \right) + v^{n-1,k-i},$$

TABLE 5.14
The values of expected losses for Pareto$(2,1)$.

n	$v^{n,1}$	$v^{n,2}$	$v^{n,3}$	$v^{n,4}$	$v^{n,5}$	$v^{n,6}$
1	2.0000					
2	1.5000	4.0000				
3	1.3333	3.1000	6.0000			
4	1.2500	2.7673	4.7552	8.0000		
5	1.2000	2.5909	4.2642	6.4470	10.0000	
6	1.1667	2.4811	3.9933	5.8061	8.1655	12.0000
7	1.1429	2.4059	3.8198	5.4417	7.3823	9.9047
8	1.1250	2.3511	3.6986	5.2032	6.9264	8.9858
9	1.1111	2.3094	3.6090	5.0340	6.6229	8.4408
10	1.1000	2.2766	3.5399	4.9072	6.4046	8.0728

$v^{k-i,k-i+1} = \infty$, $i = k, \ldots, 1$. *Table 5.14 presents the values of the expected losses* $v^{n,j}$ *with* $n \leq 10$ *steps and* $j \leq 6$ *stops left for* Pareto$(\alpha = 2, \beta = 1)$.

For example, if $k = 3$ *and* $N = 9$*, then the expected loss*

$$v = v^{9,3} = 3.6090.$$

We have the optimal stopping rule $\tau^* = (\tau_1^*, \tau_2^*, \tau_3^*)$*, where*

$$\tau_1^* = \min\{m_1 : 1 \leq m_1 \leq 7, y_{m_1} \leq v^{9-m_1,3} - v^{9-m_1,2}\},$$
$$\tau_2^* = \min\{m_2 : \tau_1^* < m_2 \leq 8, y_{m_2} \leq v^{9-m_2,2} - v^{9-m_2,1}\},$$
$$\tau_3^* = \min\{m_3 : \tau_2^* < m_3 \leq 9, y_{m_3} \leq v^{9-m_3,1}\}.$$

Targino et al. (2017) present some models in which the optimal rules can be calculated explicitly. For a Poisson–Inverse Gaussian model, where $\zeta_i \sim$ Inverse Gaussian(λ, μ) and $L \sim$ Poisson(λ_L), the optimal times (years) to exercise or make claims on the insurance policy for the Accumulated Loss Policy (ALP) can be calculated analytically regardless of where the global or local gain (objective) functions are considered. For the Individual Loss Policy (ILP), when using the gain function as the local objective function given by the sum of the losses at the stopping times (insurance claim years), it is proposed to model the losses after the insurance policy is applied and, in this case, analytical solutions for the stopping rules are presented. On the contrary, the ILP total loss case given by the global objective function does not produce a closed form solution. Still, a simple MCM can be used to accurately estimate the results.

6

The Best Choice Problems

> **Johannes Kepler** spent 2 years investigating 11 candidates for marriage during 1611 – 1613 after the death of his first wife which motivated mathematicians to formulate *the fiancée problem*.
>
> Ferguson (1989).

A version of the secretary problem is considered. Items ranked from 1 to N are randomly selected without replacement, one at a time, and to win is to stop at any item whose overall (absolute) rank belongs to the given set of ranks, given only the relative ranks of the items drawn so far. The methods of analysis are based on the existence of an embedded Markov chain and use the technique of backward induction. The requirement of choosing one of the item with the prescribed value of the absolute rank can lead to more complicated than threshold strategies. The approach can be used to give exact results for any set of absolute ranks. The exact results for the optimal strategy and the probability of success are given for a few sets. These examples are chosen to illustrate the variety of characters of optimal stopping sets. The asymptotic behavior is also investigated (cf. Suchwalko and Szajowski (2002)).

6.1 Introduction and Summary

Although a version of the secretary problem (the beauty contest problem, the dowry problem or the marriage problem) was first solved by Cayley (1875), it was not until six decades ago that there had been a sudden resurgence of interest in this problem (v. Problem 1.2). Since the articles by Gardner (1960) the secretary problem has been extended and generalized in many different directions. It is worth to emphasize that Lindley (1961) gives a first deeper analysis of the problem based on application of the dynamic programming method (v. Bellman (1957) for the introductory monograph on this methodology). Excellent reviews of the development of this colorful problem and its extensions have been given by Rose (1982), Freeman (1983), Samuels (1991) and Ferguson (1989). The formulation of the classical secretary problem in its simplest form can be formulated following Ferguson (1989). He defined the secretary problem in its standard form as having the following features:

1. There is only one secretarial position available.

DOI: 10.1201/9781003407102-6

2. The number of applicants, N, is known in advance.
3. The applicants are interviewed sequentially in a random order.
4. All applicants can be ranked from the best to the worst without any tie. Furthermore, the decision to accept or reject an applicant must be based solely on the relative ranks of the interviewed applicants.
5. Once rejected, an applicant cannot be recalled later.
6. The employer is satisfied with nothing but the very best. The reward is 1 if the best of the N applicants is chosen and 0 otherwise.

In our consideration, we change the assumption 6 and will be happy to accept the candidate who has the rank belonging to a fixed set A. In the literature on the original "secretary problem", i.e. when $A = \{1\}$, and its extension (see e.g. Ferguson (1989) for a comprehensive bibliography), the exact optimal strategy for the more general secretary problem is not given. In this paper, the derivation of the exact optimal strategy for the more general secretary problem is based on backward induction and using the existence of an embedded Markov chain. These techniques have been used by several authors (see for instance Ferguson (1989) for an extensive bibliography), and the difficulties appear when trying to extend existing accounts to derive exact results for more sophisticated sets A. In special cases, when $A = \{1, 2, \ldots, s\}$, the statement of the optimal strategy for $s = 2$ has been given by Gilbert and Mosteller (1966). Dynkin and Yushkevich (1967) outline a proof. For the case $s = 3$, the paper by Quine and Law (1996) has been devoted. For $s \geq 3$, authors such as Guseĭn-Zade (1966) and Frank and Samuels (1980) provide asymptotic results for the optimal strategy. In all these papers, the character of the set A is such that it contains all the ranks from 1 to some s. The more complicated, when the sequence of ranks has *holes*, has been considered by Rose (1982), Móri (1988) and Szajowski (1982). In these papers, the set A contains some nonextremal ranks. In Szajowski (1982) the set A consists of only one element s (one relative rank). The exact results have been given for $s = 1, 2$ and the asymptotic solution has been obtained for $s = 3, 4, 5$. In this paper, the results of Szajowski (1982) are extended to some subsets of $\{1, 2, \ldots, s\}$.

In Section 6.2 related to the secretary problem, the Markov chain is formulated. This section is based mainly on the suggestion from Dynkin and Yushkevich (1967) and the results of Szajowski (1982). In the next sections, the solutions to the secretary problem with A being a subset of $\{1, 2, \ldots, s\}$ for $s = 3, 4$ are given. We provide an exact and asymptotic solution for all cases. In Section 6.5, we present the optimal strategy for $A = \{6, 7\}$. In this special case, the strategy is no more a threshold.

In the last section, we give a comparison of the results obtained.

6.2 The Embedded Markov Model of the Secretary Problem

Let $\mathbb{S} = \{1, 2, \ldots, N\}$ be the set of ranks of items and $\{x_1, x_2, \ldots, x_N\}$ their permutation. We assume that all of them are equally likely. If X_k is the rank of the

k-th candidate, we define

$$Y_k = \{1 \le i \le k : X_i \le X_k\}.$$

The random variable Y_k is called *relative rank* of k-th candidate with respect to items investigated to the moment k.

We observe sequentially the permutation of items in set $\mathbb{S}$. The mathematical model of such an experiment is the probability space $(\Omega, \mathcal{F}, \boldsymbol{P})$. The elementary events are permutations of elements from $\mathbb{S}$, and the probability measure $\boldsymbol{P}$ is a uniform distribution on Ω. The observation of random variables Y_k, $k = 1, 2, \ldots, N$, generates the sequence of σ-fields $\mathcal{F}_k = \sigma\{Y_1, Y_2, \ldots, Y_k\}$, $k = 1, 2, \ldots, N$. The random variables Y_k are independent and $\boldsymbol{P}\{Y_k = i\} = \frac{1}{k}$.

Denote by $\mathfrak{M}^N$ the set of all Markov moments τ with respect to σ-fields $\{\mathcal{F}_k\}_{k=1}^N$. Let $q : \mathbb{S} \to \Re^+$ be the gain function. Define

$$v_N = \sup_{\tau \in \mathfrak{M}^N} \boldsymbol{E} q(X_\tau). \tag{6.1}$$

We are looking for $\tau^* \in \mathfrak{M}^N$ such that $\boldsymbol{E} q(X_{\tau^*}) = v_N$. Since $\mathfrak{M}^N$ is finite, then such τ^* exists and v_N is finite. In this paper, we consider the payoff function of the form

$$q(x) = \begin{cases} 1, & \text{if } x \in A, \\ 0, & \text{otherwise}, \end{cases} \tag{6.2}$$

where $A \subset \mathbb{S}$. By (6.1), we have $v_N = \boldsymbol{P}\{X_{\tau^*} \in A\} = \sup_{\tau \in \mathfrak{M}^N} \boldsymbol{P}\{X_\tau \in A\}$.

Such problems have been investigated, as stated in Section 6.1, by Gilbert and Mosteller (1966) and others. They constructed the optimal strategy for $A = \{1\}$, $A = \{1, 2\}$. Frank and Samuels (1980) and Guseĭn-Zade (1966) have considered $A = \{1, 2, \ldots, s\}$. In these papers, the gain functions are monotone. We consider the problems with the gain functions, which are not monotone.

6.2.1 The probability of success

Let $q(\cdot)$ be defined by (6.2). We have

$$\begin{aligned} \boldsymbol{P}\{X_\tau \in A\} &= \boldsymbol{E} q(X_\tau) = \sum_{r=1}^{N} \int_{\{\tau = r\}} q(X_\tau)\, d\boldsymbol{P} \\ &= \sum_{r=1}^{N} \int_{\{\tau = r\}} \sum_{a \in A} \boldsymbol{P}\{X_r = a | Y_r\}\, d\boldsymbol{P} \\ &= \boldsymbol{E} \sum_{a \in A} g_a(\tau, Y_\tau) = \boldsymbol{E} g_A(\tau, Y_\tau), \end{aligned}$$

where

$$g_a(r, l) = \boldsymbol{P}\{X_r = a | Y_r = l\} = \frac{\binom{a-1}{l-1}\binom{N-a}{r-l}}{\binom{N}{r}} \tag{6.3}$$

for $a = 1, 2, \ldots, N$, $l = 1, 2, \ldots, \min(a, r)$, $r = 1, 2, \ldots, N$ (see Gilbert and Mosteller (1966)).

6.2.2 Solution by recursive algorithm

Let $\mathfrak{M}_r^N = \{\tau \in \mathfrak{M}^N : r \leq \tau \leq N\}$ and $\tilde{v}_N(r) = \sup_{\tau \in \mathfrak{M}_r^N} \boldsymbol{E}q(X_\tau)$. The following algorithm allows to construct the value of the problem v_N.

$$\tilde{v}_N(N) = \boldsymbol{E}q(X_N) = \frac{\text{card}(A)}{N}. \tag{6.4}$$

Let

$$\begin{aligned} w_N(N,l) &= q(l) = \begin{cases} 1, & \text{if } l \in A, \\ 0, & \text{otherwise,} \end{cases} & (6.5)\\ w_N(r,l) &= \max\{g_A(r,l), \boldsymbol{E}w_N(r+1, Y_{r+1})\}, & (6.6)\\ \tilde{v}_N(r) &= \boldsymbol{E}w_N(r, Y_r) = \frac{1}{r}\sum_{l=1}^{r} w_N(r,l). & (6.7) \end{aligned}$$

We have then $v_N = \tilde{v}_N(1)$. The optimal stopping time τ^* is defined as follows: one have to stop at the first moment r when $Y_r = l$, unless $w_N(r,l) > g_A(r,l)$. We can define the stopping set $\Gamma = \{(r,l) : g_A(r,l) \geq w_N(r+1)\}$.

6.2.3 The embedded Markov chain

Let $a = \max(A)$. The function $g_a(r,l)$ defined in (6.3) is equal to 0 for $l > \min(a,r)$ and non-negative for $l \leq \min(a,r)$. It means that we can choose the required item at moments r with state (r,l) such that $l \leq \min(a,r)$.

Define $W_0 = (1, Y_1) = (1,1)$, $\gamma_t = \inf\{r > \gamma_{t-1} : Y_r \leq \min(a,r)\}$ ($\inf \emptyset = \infty$) and $W_t = (\gamma_t, Y_{\gamma_t})$. If $\gamma_t = \infty$ then define $W_t = (\infty, \infty)$. W_t is the Markov chain with following one step transition probabilities (see Szajowski (1982)

$$\begin{aligned} p(r,s) &= \boldsymbol{P}\{W_{t+1} = (s, l_s) | W_t = (r, l_r)\} \\ &= \begin{cases} \frac{1}{s}, & \text{if } r < a,\ s = r+1, \\ \frac{(r)_a}{(s)_{a+1}}, & \text{if } a \leq r < s, \\ 0, & \text{if } r \geq s \text{ or } r < a,\ s \neq r+1, \end{cases} \end{aligned} \tag{6.8}$$

with

$$p(\infty,\infty) = 1, \quad p(r,\infty) = 1 - a \sum_{s=r+1}^{N} p(r,s),$$

where $(s)_a = s(s-1)(s-2)\dots(s-a+1)$, $(s)_0 = 1$. Let $\mathcal{G}_t = \sigma\{W_1, W_2, \dots, W_t\}$ and $\tilde{\mathfrak{M}}^N$ be the set of stopping times with respect to $\{\mathcal{G}_t\}_{t=1}^N$. Since γ_t is increasing, then we can define $\tilde{\mathfrak{M}}_{r+1}^N = \{\sigma \in \tilde{\mathfrak{M}}^N : \gamma_\sigma > r\}$.

Let $\boldsymbol{P}_{(r,l)}(\cdot)$ be probability measure related to the Markov chain W_t, with trajectory starting in state (r,l) and $\boldsymbol{E}_{(r,l)}(\cdot)$ the expected value with respect to $\boldsymbol{P}_{(r,l)}(\cdot)$. From (6.8) we can see that the transition probabilities do not depend on relative ranks, but only on moments r where items with relative rank $l \leq \min(a,r)$ appear. Based on the following lemma we can solve the problem (6.1) with gain function (6.2) using the embedded Markov chain $\{W_t\}$ and the gain function given by (6.3).

Lemma 6.1. *(see Szajowski (1982))*

$$\boldsymbol{E}w_N(s+1, Y_{s+1}) = \boldsymbol{E}_{(s,l)}w_N(W_1) \text{ for every } l \leq \min(a, r). \tag{6.9}$$

6.2.4 Distribution of stopped Markov process

Let $\Gamma_r = \{(s,l) : s > r, g_A(s,l) \geq \boldsymbol{E}_{(s,l)}w_N(W_1)\}$ and $\sigma_r = \inf\{t : W_t \in \Gamma_r\}$. For $\sigma_r \in \tilde{\mathfrak{M}}^N_{r+1}$, we have $\tau^*_{r+1} = \inf\{s > r : (s, Y_s) \in \Gamma_r\}$. Moment τ^*_{r+1} is optimal Markov time in $\mathfrak{M}^N_{r+1}$ by definition of Γ_r. From (6.7) and (6.9), we have $\tilde{v}_N(r+1) = \boldsymbol{E}_{(r,l)}w_N(W_1)$, and by (6.6) and optimality τ^*_{r+1} in $\mathfrak{M}^N_{r+1}$ we get $\tilde{v}_N(r+1) = \boldsymbol{E}_{(r,l)}g_A(\tau^*_{r+1}, Y_{\tau^*_{r+1}}) = \boldsymbol{E}_{(r,l)}g_A(W_{\sigma_r})$. We have $\boldsymbol{E}_{(r,l)}w_N(W_1) = \boldsymbol{E}_{(r,l)}g_A(W_{\sigma_r})$. We need the distribution of random variables W_ν, where $\nu = \inf\{t : W_t \in \Gamma\}$, where Γ is subset of $\mathbb{E} = \{(r,l) : 1 \leq r \leq N,\ l = 1, 2, \ldots, a\}$.

Let $A = \{m_1, m_2, \ldots, m_k\}$. Denote $\Gamma_{r,s}(A) = \{(u,l) : r < u \leq s,\ l \in A\}$ for $k \leq a$, $m_i \leq a$, $i = 1, 2, \ldots, k$, $1 \leq r < s \leq N$, $\Gamma = \bigcup_{i=1}^N \Gamma_{r_{i-1}, r_i}(A_i)$ and $\Gamma^i = \bigcup_{j=i}^N \Gamma_{r_{j-1}, r_j}(A_j)$, where A_i is some set of relative ranks. Let $\nu_{r,s}(A) = \inf\{t : W_t \in \Gamma_{r,s}(A)\}$ and $\nu^i = \inf\{t : W_t \in \Gamma^i\}$. For $' \leq r$, $1 \leq l' \leq a$, we have

$$\begin{aligned} &\boldsymbol{P}_{(r',l')}\{W_{\nu_{r,s}(A)} = (u, m_i), m_i \in A\} \\ &= \frac{(r)_k}{(u)_{k+1}} \text{ for } r < u \leq s, i = 1, 2, \ldots, k, k = \text{card}(A), \end{aligned} \tag{6.10}$$

$$\begin{aligned} &\boldsymbol{P}_{(r',l')}\{W_{\nu^i} = (u,l)\} \\ &= \begin{cases} \frac{(r_{i-1})_{k_i}}{(r_i)_{k_i}} \boldsymbol{P}_{(r',l')}\{W_{\nu^{i+1}} = (u,l)\} \text{ for } (u,l) \in \Gamma^{i+1}, \\ \frac{(r_{i-1})_{k_i}}{(u)_{k_i+1}} \text{ for } (u,l) \in \Gamma_{r_{i-1}, r_i}(A_i), k_i = \text{card}(A_i). \end{cases} \end{aligned} \tag{6.11}$$

Based on the formulae (6.10) and (6.11) we can get the distribution of W_ν, $\nu = \inf\{t : W_t \in \Gamma\}$ recursively.

6.2.5 Construction of optimal stopping set

We get the recursive algorithm for determining the optimal strategy and the value of the problem for optimal choosing the item with absolute rank from A. This is justified by backward induction.

(i) For each l we assume $(N, l) \in \Gamma$. Let $S_0 = \{l : g_A(N-1, l) \geq \boldsymbol{E}_{(N-1,l)}g_A(W_1)\}$. The set $\{(N-1, l) : l \in S_0\} \subset \Gamma$.

(ii) Let $(s,l) \in \Gamma$ for $l \in S_0$ and $s > r$. Denote $\Gamma_r = \{(s,l) : l \in S_0, s > r\}$ and $\sigma_r = \inf\{t : W_t \in \Gamma_r\}$. It means that

$$g_A(s,l) \geq \boldsymbol{E}_{(s,l)}g_A(W_{\sigma_r}) \text{ for } s > r, l \in S_0 \tag{6.12}$$

and

$$g_A(s,l) < \boldsymbol{E}_{(s,l)}g_A(W_{\sigma_r}) \text{ for } s > r, l \notin S_0. \tag{6.13}$$

(iii) Let r_1 be highest r such that condition (6.12) or (6.13) is not valid.

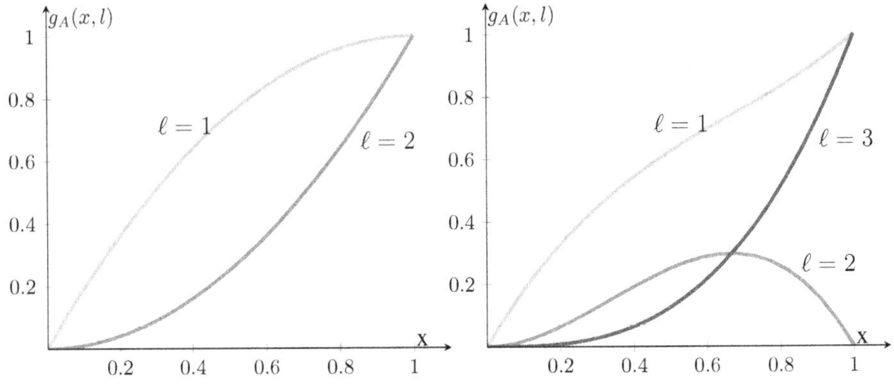

(a) For $A = \{1, 2\}$. **(b)** For $A = \{1, 3\}$.

FIGURE 6.1
Asymptotic payoff function when success means various quality of selected candidate.

(a) Let (6.12) be not valid at $r = r_1$ and $l \in S^{'} = \{m_1^{'}, \ldots, m_k^{'}\}$. The subset of stopping set Γ, for induction assumption, will be $\Gamma_{r_1-1} = \Gamma_{r_1-1,r_1}(S_1) \cup \Gamma_{r_1,N}(S_0)$, where $S_1 = S_0 \setminus S^{'}$.

(b) Let (6.13) be not valid at $r = r_1$ and $l \in S^{'} = \{m_1^{'}, \ldots, m_k^{'}\}$. The subset of stopping set Γ, for induction assumption, will be $\Gamma_{r_1-1} = \Gamma_{r_1-1,r_1}(S_1) \cup \Gamma_{r_1,N}(S_0)$, where $S_1 = S_0 \cup S^{'}$.

(c) If both conditions are breaking at r_1, (6.12) for $l \in S_1^{'} = \{m_1^{'}, \ldots, m_{k'}^{'}\}$ and (6.13) for $l \in S_2^{'} = \{m_1^{''}, \ldots, m_{k''}^{''}\}$ then the subset of stopping set Γ, for the induction assumption, will be $\Gamma_{r_1-1} = \Gamma_{r_1-1,r_1}(S_1) \cup \Gamma_{r_1,N}(S_0)$, where $S_1 = (S_0 \setminus S_1^{'}) \cup S_2^{'}$.

6.2.6 Asymptotic solution

Let the number of candidates goes to infinity. For such large number of candidates we can find optimal solution based on the following consideration. As $N \to \infty$ such that $\frac{r}{N} \to x \in (0,1]$ the embedded Markov chain $(W_t, \mathcal{F}_t, \boldsymbol{P}_{(1,1)})$ with state space $\mathbb{E} = \{1,2,\ldots,N\} \times \{1,2,\ldots,\max(A)\}$ can be treated as Markov chain $(W_t^{'}, \mathcal{F}_t, \boldsymbol{P}_{(\frac{1}{N},1)})$ on $\{\frac{1}{N}, \frac{2}{N}, \ldots, 1\} \times \{1,2,\ldots,\max(A)\}$. The gain function $g_A([Nx], l)$ has limit

$$g_A(x,l) = \sum_{a \in A} \binom{a-1}{l-1} x^l (1-x)^{a-l},\ l = 1,2,\ldots,\max(A).$$

We get $\lim_{N\to\infty} \boldsymbol{E}_{(\frac{r}{N},l)} g_A(W_1) = \boldsymbol{E}_{(x,l)} g_A(W_1^{''})$, where $(W_t^{''}, \mathcal{F}_t, \boldsymbol{P}_{(x,1)})$ is Markov chain with the state space $(0,1] \times \{1,2,\ldots,\max(A)\}$ and the transition density function

$$p(x,y)=\begin{cases}\frac{x^a}{y^{a+1}}, & 0<x<y\le 1,\\ 0, & x\ge y.\end{cases}\qquad(6.14)$$

The expected value with respect to the conditional distribution given in (6.14) is following

$$\boldsymbol{E}_{(x,l')}g_A(W_1^{''})=\sum_{l=1}^{\max(A)}\int_x^1 p(x,y)g_A(y,l)\,dy.\qquad(6.15)$$

The recursive formulae (6.4)–(6.7) in asymptotic case have the form

$$\begin{aligned} v(1) &= 0, &(6.16)\\ w(x,l) &= \max\{g_A(x,l),\boldsymbol{E}_{(x,l)}w(W_1^{''})\}, &(6.17)\\ v(x) &= \boldsymbol{E}_{(x,l)}w(W_1^{''}), &(6.18)\end{aligned}$$

where $w(x,l)$ is the limit of $w_N(r,l)$, when $\frac{r}{N}\to x\in(0,1]$, i.e. $\lim_{N\to\infty}w_N([Nx],l)=w(x,l)$. The asymptotic solution we get by "recursive" method based on (6.16)–(6.18). The distribution of stopped Markov process, given by (6.10)–(6.11) in finite case, in the asymptotic case has the form

$$f_{(x',l')}((y,m_i)) = \frac{x^k}{y^{k+1}},\ x'\le x<y\le 1,\qquad(6.19)$$

$$f_{(x',l')}((y,m_i)) = \begin{cases}\frac{x_{i-1}^{k_i}}{x_i^{k_i}}f_{(x',l')}((y,l)), & (y,l)\in\Gamma'(i+1),\\ \frac{x_{i-1}^{k_i}}{y^{k_i+1}}, & (y,l)\in\Gamma_{x_{i-1},x_i}(k_i),\end{cases}\qquad(6.20)$$

where $\Gamma_{x,z}(k)=\{(y,l):x<y\le z, l=m_1,m_2,\ldots,m_k\}$, $\Gamma'(i+1)=\bigcup_{j=i}^n\Gamma_{x_{j-1},x_j}(k_j)$.

The algorithm of optimal stopping set construction (analogous to the one introduced in Section 6.2.5) will be presented on examples in Section 6.3.1.

6.2.7 Modified BCP–alternative payoffs

The basic model of the secretary problem (v. Section 6.1) is formulated for the sequential observation of permutations of different numbers, assuming that all permutations are equally likely. Some authors also assumed that these numbers are randomly selected, and for the sake of consideration, they are practically sequentially added observations of independent random variables $X_1,X_2,\ldots,X_n$ with a uniform distribution on $[0,1]$. This allows them to formulate a selection problem with a criterion that refers to the value of this sequence, but limits the knowledge of the decision-maker to the relative ranks of them (cf. Kubicka et al. (2023)). For example of such topics, the problem of the search for the best candidate (**BCP**) will be presented, where the additional objective of the DM is to have the candidate with a value lower than a given level value a of the choice label of the candidate X_τ. τ is the moment when the accepted candidate appears. DM maximizes the probability of the event that the candidate appearing at τ is relatively the first, and $\boldsymbol{P}\{X_\tau \le a, Y_\tau = 1\}$ (v. Section 6.2 for definition of the relative and absolute ranks).

In this section, **DM** is equipped with stopping times strategy set $\mathfrak{S}$ with respect to filtering $\mathcal{F}_k = \sigma\{Y_1, \ldots, Y_k\}$, $k = 1, 2, \ldots, n$, looking for a stopping moment that maximizes $\boldsymbol{P}(X_\tau < a, Y_\tau = 1)$, where $a = \frac{1}{n}$ (unlike the task considered by Kubicka et al. (2023), where the relative rank of the retained candidate is not important). The problem can be reduced to the optimal stopping of an auxiliary Markov chain defined in Section 6.2, 6.2.3, with the payoff function $g(r, k, a) = F_{r,k}(a) = \boldsymbol{P}(X_k \le a, Y_k = r)$–the value of the distribution function of the rth order statistic for the set of k–elements sample at a (cf. Stepanov (2021)), that is, to determine

$$\boldsymbol{P}\left(X_{\tau^\star} < \frac{1}{n}, Y_{\tau^\star} = 1\right) = \sup_\tau \boldsymbol{P}\left(X_\tau < \frac{1}{n}, Y_\tau = 1\right) = \sup_\tau \boldsymbol{E}_{(1,1)} g\left(1, \tau, \frac{1}{n}\right), \tag{6.21}$$

and the strategy $\tau^\star$. Construction of the value and the optimal strategy in the problem is a standard task, but the straight description for a general class of distributions is not known. It will be presented here for the labels of the uniform distribution on $[0, 1]$.

Taking this into account, we have $F_{rk}(x) = \mathbf{Beta}_{r,k-r+1}(x)$, where k-is the size of the sample, r- is the rank of the observation. We have

$$g\left(1, k, \frac{1}{n}\right) = \boldsymbol{P}\left(X_k \le \frac{1}{n} | Y_k = 1\right) = 1 - \left(1 - \frac{1}{n}\right)^k.$$

Let, by Lemma 6.1,

$$G\left(1, k, \frac{1}{n}\right) = \boldsymbol{E}_{(1,k)} g\left(W_1, 1, \frac{1}{n}\right) = \sum_{k=j+1}^{n} \frac{k}{j(j-1)} \left(1 - \left(1 - \frac{1}{n}\right)^k\right).$$

Based on the monotonic properties of g by k (it is increasing), and G in some neighborhood of n, we easily prove that the problem is monotone. The stopping region is of threshold type. The threshold is $k^* \cong [0.32303n]$ and the value of the problem is $v \cong 0.276867$ (v. Figures 6.2 and 6.3).

Value-added examines of the asymptotic of the strategy and the value of the problem were provided.

6.2.8 Modified BCP–alternative payoffs with 2 selections

Selecting more objects in one observation sequence leads to multiple stop tasks. A systematic presentation of this issue is given, for example, in Sections 1.4 and 2.3. The problem is similar to the problem considered by Nikolaev (1976), however, in any optimization problem, a small change in the model elements may lead to the need for different methods of analysis. In order to illustrate the complexity of such models, let us consider the tasks of selecting two objects so that their labels are smaller than a given number, and we will make the selection among the relatively first ones. So

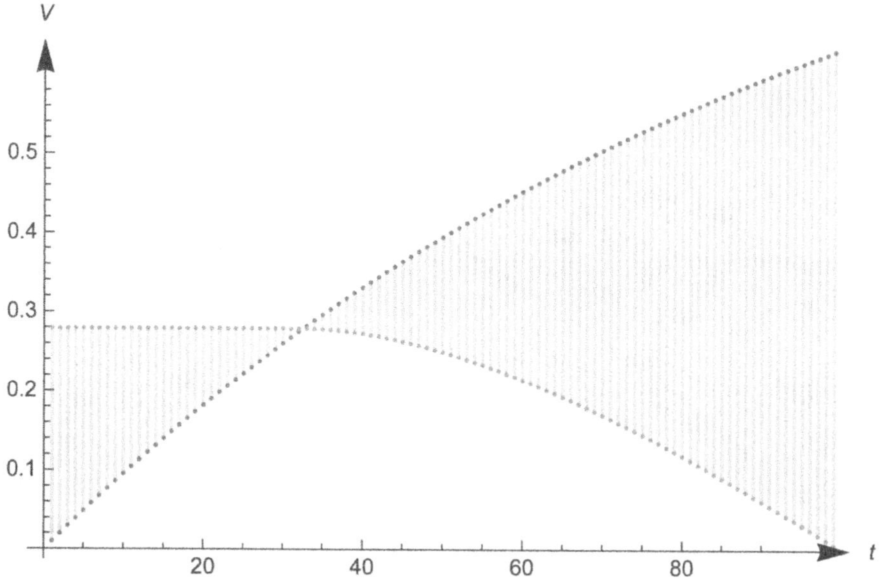

FIGURE 6.2
The payoff (gray blue) and the value (pink) of the OS problem for stopping at **BCP** with an absolute value lower than $\frac{1}{n}$. The maximum number of candidates is $M = 100$.

we search for the optimal pair of stopping moments $\tau^\star < \sigma^\star$ among the stopping moments in the set $\mathfrak{S}$ such that

$$\boldsymbol{P}\left(X_{\tau^\star} \vee X_{\sigma^\star} < \frac{1}{n}, Y_{\tau^\star} = 1, Y_{\sigma^\star} = 1\right) = \sup_{\tau<\sigma} \boldsymbol{P}\left(X_\tau \vee X_\sigma < \frac{1}{n}, Y_\tau = 1, Y_\sigma = 1\right).$$

An analysis of the distribution functions of the order statistics for the sample from the uniform distribution shows that the selected objects should appear successively with a relative rank of 1. Let us say that they will be observations from moments r and s, respectively, where $r < s$. Then the reward after selecting such two objects is the value of the distribution function of the 2nd order statistic from the s-element sample: $g(s, 2, \frac{1}{n}) = F_{s,2}(\frac{1}{n})$. Therefore, we reduce the issue to the problem of optimal stopping of a Markov chain describing a random walk through the moments of occurrence of observations with a relative rank of 1 and the payoff function:

$$H_1\left(s, M, \frac{1}{n}\right) = \boldsymbol{E}_{s,1} g\left(W_1, 2, \frac{1}{n}\right).$$

The $\{W_k\}_{k=1}^M$ process assumes values equal to the moments of appearance of the next relatively first observations. The payoff related to the first stop and the value function (the horizon of the problem $n = 100$) is shown in Figure 6.4.

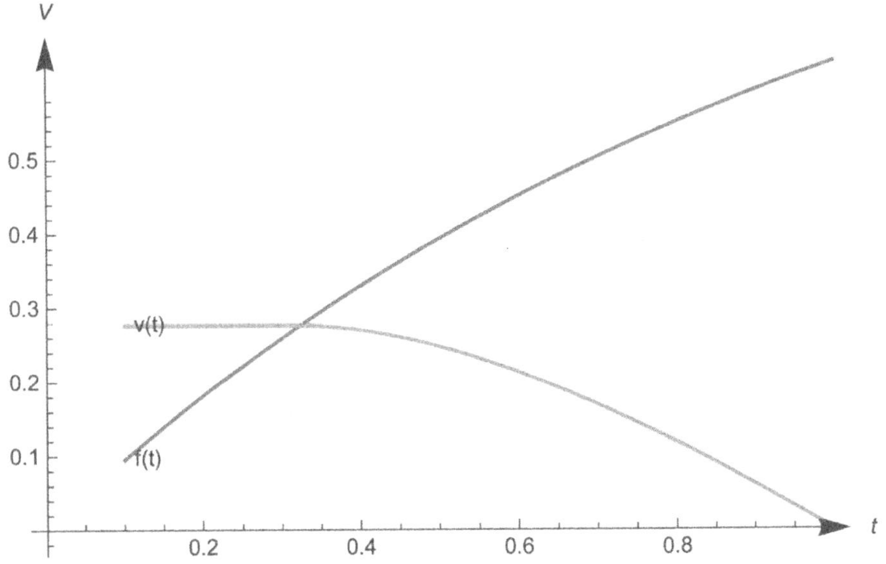

FIGURE 6.3
The asymptotic of the payoff and its expected payoff for stopping at **BCP** with absolute value lower than $\frac{1}{n}$.

6.3 Optimal Selection 2 Items having Absolute Rank Lower than 3

Now let's examine the task of choosing two objects. We will devise the best strategy for selecting the item that has an absolute rank within a subset of two ranks that are both less than 3. In simpler terms, we will consider the selection a success if the absolute rank of the chosen objects is not higher than 3. In practice, such a need may arise when we want to add two employees to a three-person team with diversified competencies that justify different salaries, and we already have one employee, whose skills we consider average. The better of the chosen ones will receive a salary slightly higher than the average in this group, and the second one will receive a salary slightly lower - in accordance with their rank.

6.3.1 Optimal stopping on the best or the third absolute rank item

Basing on results of the Section 6.2 we construct the optimal strategy for choosing the item with absolute rank 1 or 3. For a finite horizon N we can give the numerical solution. The results of the calculation for some N are in Table 6.1.

Let the requirement of decision-maker be to choose the first- or the third-rank applicant. Taking into account the consideration of the Section 6.2 we solve the

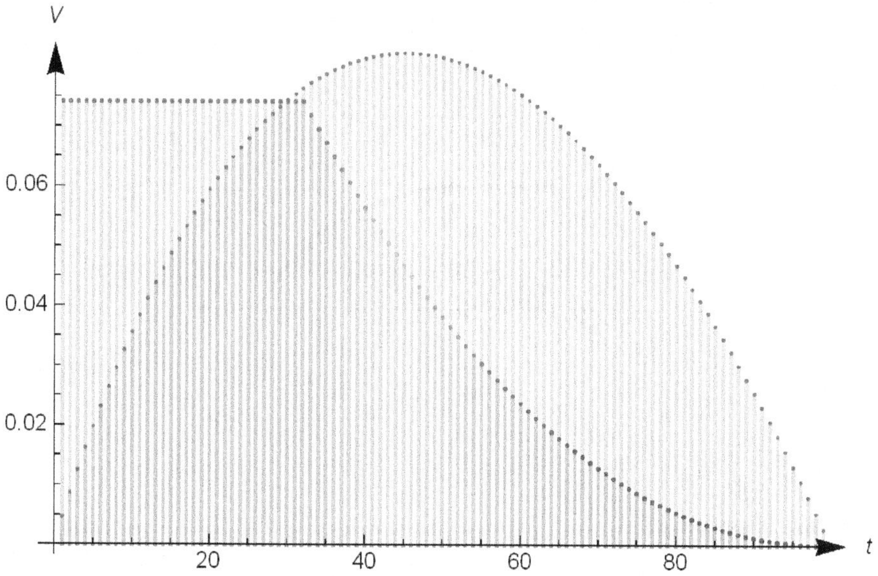

FIGURE 6.4
The payoff (gray blue) and the value (pink) of the OS problem for double stopping at relative best with an absolute value lower than $\frac{1}{n}$. The maximum number of candidates is $M = 100$.

TABLE 6.1
Optimal choosing an item from $A = \{1, 3\}$.

N	Strategies-relative ranks						Probability
	1		2		3		
4	2	4	3	4	4	4	0.6250
5	2	5	5	5	5	5	0.5833
6	3	6	5	6	5	6	0.5722
7	3	7	6	7	6	7	0.5619
8	3	8	7	8	7	8	0.5464
9	4	9	8	9	7	9	0.5421
10	4	10	9	10	8	10	0.5379
20	8	20	19	20	15	20	0.5107
100	35	100	99	100	72	100	0.4917
200	69	200	199	200	143	200	0.4894
∞	[0.339N]	N	N	N	[0.710N]	N	0.4870

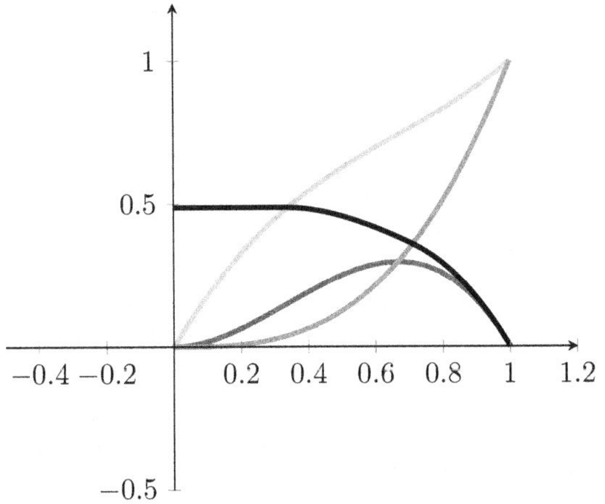

FIGURE 6.5
Take 1 or 3 to win.

optimal stopping problem for Markov chain with transition probability function given by (6.8), with $a = \max(A)$, where $A = \{1, 3\}$, and the gain function

$$g_A(r,l) = \begin{cases} g_1(r,1) + g_3(r,1), & l = 1, \\ g_3(r,2), & l = 2, \\ g_3(r,3), & l = 3. \end{cases}$$

Based on the algorithm given in Section 6.2.5 we get the results of Table 6.1. We show in an example for $N = 10$ how to obtain the strategy from Table 6.1. The stopping set for this horizon is

$$\begin{aligned} \Gamma &= \{(r,l) : 4 \le r \le 10, l = 1\} \cup \{(r,l) : 9 \le r \le 10, l = 2\} \\ &\quad \cup \{(r,l) : 8 \le r \le 10, l = 3\}, \end{aligned}$$

and the maximal probability of realization of the requirement is $v \cong 0.5379$.

For asymptotic solution we use the gain function

$$g_A(x,l) = \begin{cases} x + x(1-x)^2, & l = 1, \\ 2x^2(1-x), & l = 2, \\ x^3, & l = 3. \end{cases} \tag{6.22}$$

We get the asymptotically optimal stopping time by constructing the asymptotic stopping set. Let us assume that $\Gamma_x(1) = \Gamma_{x,1}(\{1,2,3\}) = \{(x,l) : x \in (1-\epsilon, 1], l = 1, 2, 3\} \subset \Gamma$ for the sufficiently small $\epsilon > 0$, where Γ is the asymptotically optimal stopping set. From (6.22), (6.14) and (6.15) we get

$$w_0(x) = \sum_{l=1}^{3} \int_x^1 p(x,y) g_A(y,l)\, dy = x(1-x^2).$$

Solving the inequality $w_1(x) - g_A(x, l) \leq 0$ for $l = 1, 2, 3$ and $x \in (1 - \epsilon, 1]$ we get that there are no $\epsilon > 0$ for which this inequality holds when $l = 2$. Then our assumption is false. Suppose now, that we change the definition of the stopping set in the neighborhood of 1 and set $\Gamma_x(1) = \Gamma_{x,1}(\{1, 3\}) = \{(x, l) : x \in (1 - \epsilon, 1], l = 1, 3\}$ and $\nu_1 = \inf\{t : W_t^{''} \in \Gamma^2_{x,1}, x \in (1 - \epsilon, 1]\}$. We have

$$w_1(x) = \boldsymbol{E}_{(x,l)} g_A\left(W_{\nu_1}^{''}\right) = \int_x^1 \frac{x^2}{y^3}\left(x + x(1-x)^2 + x^3\right) dy = 2x(1 + x\log(x) - x^2).$$

For $x \in (1 - \epsilon, 1]$ and $\epsilon > 0$ enough small are the inequalities

$$w_1(x) - g_A(x, l) \leq 0 \tag{6.23}$$

for $l = 1, 3$ hold and $w_0(x) \leq w_1(x)$. The nearest point on the left-hand side of 1, at which the (6.23) does not hold is $\alpha \cong 0.7105$. This point fulfills the equation $w_1(x) = g_A(x, 3)$. Define $v(x) = w_1(x)$ for $x \in (\alpha, 1]$.

Let $\Gamma_x(2) = \Gamma_{x,\alpha}(\{1\}) \cup \Gamma_{\alpha,1}(\{1, 3\})$ and $\nu_2 = \inf\{t : W_t^{''} \in \Gamma_x(2)\}$. We have by using (6.19) and (6.20)

$$\begin{aligned} w_2(x) &= \boldsymbol{E}_{(x,l)} g_A\left(W_{\nu_2}^{''}\right) = \int_x^\alpha \frac{x}{y^2}\left(x + x(1-x)^2\right) dy + \frac{x}{\alpha} w_1(\alpha) \\ &= x\left(2\log\left(\frac{\alpha}{x}\right) - 2(\alpha - x) + \frac{1}{2}\left(\alpha^2 - x^2\right)\right) + \frac{x}{\alpha} w_1(\alpha). \end{aligned}$$

The recursive procedure gives $w_2(x) \leq g_A(x, 1)$ and the next change of stopping set is in point $\beta \cong 0.3389$ which is a solution of equation $w_2(x) = g_A(x, 1)$ in $(0, \alpha]$. Define $v(x) = w_2(x)$ for $x \in (\beta, \alpha]$ and $v(x) = w_2(\beta)$ for $x \in (0, \beta]$.

We have got the optimal stopping set of the form

$$\Gamma = \Gamma_{\beta,\alpha}(\{1\}) \cup \Gamma_{\alpha,1}(\{1, 3\})$$

and

$$v(x) = w_2(\beta)\mathbb{I}_{\{x<\beta\}} + w_2(x)\mathbb{I}_{\{\beta\leq x\leq\alpha\}} + w_1(x)\mathbb{I}_{\{\alpha<x\leq 1\}}.$$

The value of the problem is then $v = v(\beta) \cong 0.4870$.
The last row of the Table 6.1 contains the form of asymptotic optimal strategy when it is applied to the big number of candidates N.

This method of determining the asymptotic optimal stopping set is a consequence of dynamic programming.

6.3.2 Optimal stopping on the second or the third

Analogous to the solution presented in details in the Section 6.3.1, we will show the solution of the problem $A = \{2, 3\}$, with $a = \max(A)$. We have the gain function

$$g_A(r, l) = \begin{cases} g_2(r, 1) + g_3(r, 1), & l = 1, \\ g_2(r, 2) + g_3(r, 2), & l = 2, \\ g_3(r, 3), & l = 3. \end{cases}$$

TABLE 6.2
Optimal choosing an item from $A = \{2, 3\}$.

N	Strategies-relative ranks						Probability
	1		2		3		
4	4	4	3	4	4	4	0.6667
5	5	5	3	5	5	5	0.6000
6	6	6	4	6	5	6	0.5750
7	7	7	4	7	6	7	0.5571
8	8	8	5	8	7	8	0.5357
9	9	9	5	9	7	9	0.5278
10	10	10	6	10	8	10	0.5179
20	20	20	10	20	15	20	0.4830
100	100	100	48	100	73	100	0.4575
200	200	200	94	200	144	200	0.4544
∞	N	N	[0.468N]	N	[0.716N]	N	0.4514

For asymptotic solution we use the gain function

$$g_A(x, l) = \begin{cases} x(1-x) + x(1-x)^2, & l = 1, \\ x^2 + 2x^2(1-x), & l = 2, \\ x^3, & l = 3. \end{cases} \tag{6.24}$$

Taking into account the following as in Section 6.3.1, we get the form of the value function and the optimal stopping set (see Table 6.2 for numerical values of the thresholds). We have

$$v(x) = w_2(\beta)\mathbb{I}_{\{x<\beta\}} + w_2(x)\mathbb{I}_{\{\beta \le x \le \alpha\}} + w_1(x)\mathbb{I}_{\{\alpha < x \le 1\}},$$

where

$$\begin{aligned} w_1(x) &= x^2(x - 3\log(x) - 1) \\ w_2(x) &= x^3 - 3x^2 + 3xe^{-\frac{1}{3}} + \frac{x}{\alpha} w_1(\alpha), \end{aligned}$$

and the constants α and β are determined similar to the Section 6.3.1. The constant $\alpha = e^{-\frac{1}{3}} \cong 0.7165$ is the closest 1 to the solution on the left side of equation $w_1(x) - g_A(x, 3) = 0$. The constant $\beta = 1 - \sqrt{1 - e^{-\frac{1}{3}}} \cong 0.4676$ is the closest α to the solution on the left side of equation $w_2(x) - g_A(x, 2) = 0$.
The optimal stopping set has the form

$$\Gamma = \Gamma_{\beta,\alpha}(\{2\}) \cup \Gamma_{\alpha,1}(\{2, 3\})$$

The value of the problem is $v = v(\beta) \cong 0.4514$.

TABLE 6.3
The solution of the $A = \{1, 4\}$.

N	Strategies-relative ranks								Probability
	1		2		3		4		
4	3	4	4	4	3	4	4	4	0.5667
5	3	5	5	5	4	5	5	5	0.5278
6	3	6	6	6	5	6	5	6	0.5278
7	4	7	7	7	5	7	6	7	0.5147
8	4	8	8	8	6	8	7	8	0.5107
9	4	9	9	9	7	9	8	9	0.5024
10	4	10	10	10	8	10	8	10	0.4926
20	8	20	20	20	15	20	16	20	0.4716
100	36	100	100	100	75	100	76	100	0.4531
200	71	200	200	200	151	200	151	200	0.4509
∞	[0.353N]	N	N	N	[0.751N]	N	[0.753N]	N	0.4487

6.4 Choosing Elements with a Rank that does not Exceed 4

We construct the optimal strategy for choosing the item with absolute rank belonging to the two-element subset of ranks less than 4.

6.4.1 Optimal stopping on the best or the fourth

Analogous to the solution presented in details in Section 6.3.1, we will show the solution of the problem $A = \{1, 4\}$, with $a = \max(A)$. We have the gain function

$$g_A(r, l) = \begin{cases} g_1(r, 1) + g_4(r, 1), & l = 1, \\ g_4(r, 2), & l = 2, \\ g_4(r, 3), & l = 3, \\ g_4(r, 4), & l = 4. \end{cases}$$

For asymptotic solution we use the gain function

$$g_A(x, l) = \begin{cases} x + x(1 - x)^3, & l = 1, \\ 3x^2(1 - x)^2, & l = 2, \\ 3x^3(1 - x), & l = 3, \\ x^4, & l = 4. \end{cases} \tag{6.25}$$

By consideration as in Section 6.3.1 we get the form of value function and the optimal stopping set (see Table 6.3 for the numerical values of the thresholds). We have

$$\begin{aligned} v(x) &= w_3(\gamma)\mathbb{I}_{\{x<\gamma\}} + w_3(x)\mathbb{I}_{\{\gamma\le x\le\beta\}} + w_2(x)\mathbb{I}_{\{\beta<x\le\alpha\}} + w_1(x)\mathbb{I}_{\{\alpha<x\le 1\}} \\ &= w_3(\gamma)\mathbb{I}_{[0,\gamma)}(x) + w_3(x)\mathbb{I}_{[\gamma,\beta]}(x) + w_2(x)\mathbb{I}_{(\beta,\alpha]}(x) + w_1(x)\mathbb{I}_{(\alpha,1]}(x), \end{aligned}$$

where

$$w_1(x) = 3x^4 - x^3 - 3x^2 + x - 6x^3\log(x)$$
$$w_2(x) = -3x^3 + x^2\left(3\alpha - 3\log(\alpha) - \frac{2}{\alpha}\right) + 2x + 3x^2\log(x) + \frac{x^2}{\alpha^2}w_1(\alpha)$$
$$w_3(x) = \frac{1}{3}x^4 - \frac{3}{2}x^3 + 3x^2 + x\left(-\frac{1}{3}\beta^3 + \frac{3}{2}\beta^2 - 3\beta + 2\log(\beta)\right)$$
$$- 2x\log(x) + \frac{x}{\beta}w_2(\beta),$$

and the constants α and β are determined similar to the Section 6.3.1. Then $\alpha \doteq 0.7528$ is such that $w_1(\alpha)-g_A(\alpha,3) = 0$, $\beta \cong 0.7507$ is such that $w_2(\beta)-g_A(\beta,4) = 0$, $\gamma \cong 0.3531$ is such that $w_3(\gamma) - g_A(\gamma,1) = 0$.
The optimal stopping set has the form

$$\Gamma = \Gamma_{\gamma,\beta}(\{1\}) \cup \Gamma_{\beta,\alpha}(\{1,4\}) \cup \Gamma_{\alpha,1}(\{1,3,4\})$$

The value of the problem is $v = v(\gamma) \cong 0.4487$.

6.4.2 Optimal stopping on the second or the fourth

Analogous to the solution presented in details in the Section 6.3.1, we will show the solution of the problem $A = \{2,4\}$, with $a = \max(A)$. We have the gain function

$$g_A(r,l) = \begin{cases} g_2(r,1) + g_4(r,1), & l = 1, \\ g_2(r,2) + g_4(r,2), & l = 2, \\ g_4(r,3), & l = 3, \\ g_4(r,4), & l = 4. \end{cases}$$

For asymptotic solution we use the gain function

$$g_A(x,l) = \begin{cases} x(1-x) + x(1-x)^3, & l = 1, \\ x^2 + 3x^2(1-x)^2, & l = 2, \\ 3x^3(1-x), & l = 3, \\ x^4, & l = 4. \end{cases}$$

By consideration as in Section 6.3.1 we get the form of value function and the optimal stopping set (see Table 6.4 for the numerical values of the thresholds). We have

$$v(x) = w_3(\gamma)\mathbb{I}_{\{x<\gamma\}} + w_3(x)\mathbb{I}_{\{\gamma\le x\le\beta\}} + w_2(x)\mathbb{I}_{\{\beta<x\le\alpha\}} + w_1(x)\mathbb{I}_{\{\alpha<x\le 1\}},$$

where

$$\begin{aligned} w_1(x) &= -x^4 - 3x^3 + 4x^2 + 3x^2\log(x), \\ w_2(x) &= 3x^3 + x^2(4\log(\alpha) - 3\alpha) - 4x^2\log(x) + \frac{x^2}{\alpha^2}w_1(\alpha), \\ w_3(x) &= -x^4 + 3x^3 - 4x^2 + x(\beta^3 - 3\beta^2 + 4\beta) + \frac{x}{\beta}w_2(\beta), \end{aligned}$$

TABLE 6.4
The solution of the $A = \{2, 4\}$.

N	Strategies-relative ranks								Probability
	1		2		3		4		
4	4	4	2	4	3	4	4	4	0.6250
5	5	5	3	5	4	5	5	5	0.5333
6	6	6	3	6	5	6	5	6	0.4833
7	7	7	4	7	5	7	6	7	0.4750
8	8	8	4	8	6	8	7	8	0.4607
9	9	9	5	9	7	9	8	9	0.4511
10	10	10	5	10	8	10	8	10	0.4405
20	20	20	10	20	15	20	16	20	0.4123
100	100	100	46	100	73	100	75	100	0.3904
200	200	200	91	200	146	200	150	200	0.3878
∞	N	N	[0.449N]	N	[0.727N]	N	[0.744N]	N	0.3853

and $\alpha \cong 0.7442$ is such that $w_1(\alpha) - g_A(\alpha, 4) = 0$, $\beta \cong 0.7274$ is such that $w_2(\beta) - g_A(\beta, 3) = 0$, $\gamma \cong 0.4491$ is such that $w_3(\gamma) - g_A(\gamma, 2) = 0$. For the details of method see Sections 6.3.1 and 6.3.2.
The optimal stopping set has the form

$$\Gamma = \Gamma_{\gamma,\beta}(\{2\}) \cup \Gamma_{\beta,\alpha}(\{2,3\}) \cup \Gamma_{\alpha,1}(\{2,3,4\}).$$

The value of the problem is $v = v(\gamma) \cong 0.3853$.

6.4.3 Optimal stopping on the third or the fourth

Analogous to the solution presented in details in the Section 6.3.1, we will show the solution of the problem $A = \{3, 4\}$, with $a = \max(A)$. We have the gain function

$$g_A(r,l) = \begin{cases} g_3(r,1) + g_4(r,1), & l = 1, \\ g_3(r,2) + g_4(r,2), & l = 2, \\ g_3(r,3) + g_4(r,3), & l = 3, \\ g_4(r,4), & l = 4. \end{cases}$$

For asymptotic solution we use the gain function

$$g_A(x,l) = \begin{cases} x(1-x)^2 + x(1-x)^3, & l = 1, \\ 2x^2(1-x) + 3x^2(1-x)^2, & l = 2, \\ x^3 + 3x^3(1-x), & l = 3, \\ x^4, & l = 4. \end{cases}$$

By consideration as in Section 6.3.1 we get the form of value function and the optimal stopping set (see Table 6.5 for the numerical values of the thresholds). We have

$$v(x) = w_3(\gamma)\mathbb{I}_{\{x<\gamma\}} + w_3(x)\mathbb{I}_{\{\gamma \le x \le \beta\}} + w_2(x)\mathbb{I}_{\{\beta < x \le \alpha\}} + w_1(x)\mathbb{I}_{\{\alpha < x \le 1\}},$$

TABLE 6.5
The solution of the $A = \{3, 4\}$.

N	Strategies-relative ranks								Probability
	1		2		3		4		
4	4	4	2	4	3	4	4	4	0.7500
5	5	5	3	5	4	5	5	5	0.6000
6	6	6	3	6	4	6	5	6	0.5556
7	7	7	4	7	5	7	6	7	0.5286
8	8	8	4	8	5	8	7	8	0.5086
9	9	9	5	9	6	9	8	9	0.4947
10	10	10	5	10	7	10	8	10	0.4804
20	20	20	10	20	12	20	16	20	0.4421
100	100	100	46	100	57	100	76	100	0.4134
200	200	200	91	200	112	200	152	200	0.4101
∞	N	N	[0.450N]	N	[0.556N]	N	[0.753N]	N	0.4069

where

$$\begin{aligned} w_1(x) &= -x^4 - 4x^3 + 5x^2 + 4x^3\log(x), \\ w_2(x) &= 4x^3 + x^2(-4\alpha + 5\log(\alpha)) - 5x^2\log(x) + \frac{x^2}{\alpha^2}w_1(\alpha), \\ w_3(x) &= -x^4 + 4x^3 - 5x^2 + x(\beta^3 - 4\beta^2 + 5\beta) + \frac{x}{\beta}w_2(\beta), \end{aligned}$$

and $\alpha \cong 0.7529$ is such that $w_1(\alpha) - g_A(\alpha, 4) = 0$, $\beta \cong 0.5557$ is such that $w_2(\beta) - g_A(\beta, 3) = 0$, $\gamma \cong 0.4505$ is such that $w_3(\gamma) - g_A(\gamma, 2) = 0$. For the details of the method, see Sections 6.3.1 and 6.3.2.
The optimal stopping set has the form

$$\Gamma = \Gamma_{\gamma,\beta}(\{2\}) \cup \Gamma_{\beta,\alpha}(\{2,3\}) \cup \Gamma_{\alpha,1}(\{2,3,4\}).$$

The value of the problem is $v = v(\gamma) \cong 0.4069$.

6.5 Choosing the Item with Rank 6 or 7

Analogously to the solution presented in detail in Section 6.3.1, we will show the solution of the problem $A = \{6, 7\}$, with $a = \max(A)$. We have the gain function

$$g_A(r,l) = \begin{cases} g_6(r,1) + g_7(r,1), & l = 1, \\ g_6(r,2) + g_7(r,2), & l = 2, \\ g_6(r,3) + g_7(r,3), & l = 3, \\ g_6(r,4) + g_7(r,4), & l = 4, \\ g_6(r,5) + g_7(r,5), & l = 5, \\ g_6(r,6) + g_7(r,6), & l = 6, \\ g_7(r,7), & l = 7. \end{cases}$$

TABLE 6.6
The solution of the $A = \{6, 7\}$.

N	Strategies-relative ranks										Prob.
	4		4		5		6		7		
7	4	5	7	7	5	7	6	7	7	7	0.6667
8	5	5	8	8	6	8	7	8	8	8	0.5357
9	5	6	9	9	6	9	7	9	8	9	0.4947
10	6	7	10	10	7	10	8	10	9	10	0.4755
20	11	15	20	20	13	20	15	20	17	20	0.3904
100	54	74	100	100	61	100	71	100	83	100	0.3478
200	107	149	200	200	120	200	140	200	165	200	0.3433
∞	$[\epsilon$ N]	$[\beta$ N]	N	N	$[\delta$ N]	N	$[\gamma$ N]	N	$[\alpha$ N]	N	0.3389

The results of Table 6.6 presented in a bit different form because of occurrence of the island strategy. We show on example for $N = 20$ how to retrieve the strategy from the Table 6.6. The stopping set for this horizon is

$$\begin{aligned}\Gamma &= \{(r,l) : r = 20, l = 1,2,3,4\} \cup \{(r,l) : 11 \le r \le 17, l = 4\}\\ &\quad \cup\{(r,l) : 13 \le r \le 20, l = 5\} \cup \{(r,l) : 15 \le r \le 20, l = 6\}\\ &\quad \cup\{(r,l) : 17 \le r \le 20, l = 7\},\end{aligned}$$

and the maximal probability of realization of the requirement is $v \cong 0.3904$. Because in all cases we stop on the ranks ≤ 3 only in the last moment (an exception is only the case $N = 7$, in which we stop at the relative third item in moment $r = 3$), we have omitted in Table 6.6 these ranks to simplify the notation.

For asymptotic solution we use the gain function

$$g_A(x,l) = \begin{cases} x(1-x)^5 + x(1-x)^6, & l = 1,\\ 5x^2(1-x)^4 + 6x^2(1-x)^5, & l = 2,\\ 10x^3(1-x)^3 + 15x^3(1-x)^4, & l = 3,\\ 10x^4(1-x)^2 + 20x^4(1-x)^3, & l = 4,\\ 5x^5(1-x) + 15x^5(1-x)^2, & l = 5,\\ x^6 + 6x^6(1-x), & l = 6,\\ x^7, & l = 7. \end{cases}$$

Taking into account the following as in Section 6.3.1, we get the form of the value function and the optimal stopping set. We have

$$\begin{aligned} v(x) &= w_5(\epsilon)\mathbb{I}_{\{x<\epsilon\}} + w_5(x)\mathbb{I}_{\{\epsilon\le x\le\delta\}} + w_4(x)\mathbb{I}_{\{\delta<x\le\gamma\}}\\ &\quad + w_3(x)\mathbb{I}_{\{\gamma<x\le\beta\}} + w_2(x)\mathbb{I}_{\{\beta<x\le\alpha\}} + w_1(x)\mathbb{I}_{\{\alpha<x\le1\}}, \end{aligned}$$

where

$$
\begin{aligned}
w_1(x) &= \int_x^1 \frac{x^3}{y^4}(g_A(y,5)+g_A(y,6)+g_A(y,7))\,dy,\\
w_2(x) &= \int_x^{\alpha} \frac{x^2}{y^3}(g_A(y,5)+g_A(y,6))\,dy+\frac{x^2}{\alpha^2}w_1(\alpha),\\
w_3(x) &= \int_x^{\beta} \frac{x^3}{y^4}(g_A(y,4)+g_A(y,5)+g_A(y,6))\,dy+\frac{x^3}{\beta^3}w_2(\beta),\\
w_4(x) &= \int_x^{\gamma} \frac{x^2}{y^3}(g_A(y,4)+g_A(y,5))\,dy+\frac{x^2}{\gamma^2}w_3(\gamma),\\
w_5(x) &= \int_x^{\delta} \frac{x}{y^2}g_A(y,4)\,dy+\frac{x}{\delta}w_4(\delta),
\end{aligned}
$$

and $\alpha \cong 0.8212$ is such that $w_1(\alpha) - g_A(\alpha, 7) = 0$, $\beta \cong 0.7483$ is such that $w_2(\beta) - g_A(\beta, 4) = 0$, $\gamma \cong 0.6950$ is such that $w_3(\gamma) - g_A(\gamma, 6) = 0$, $\delta \cong 0.5963$ is such that $w_4(\delta) - g_A(\delta, 5) = 0$, $\epsilon \cong 0.5310$ is such that $w_5(\epsilon) - g_A(\epsilon, 4) = 0$. For the details of method see Sections 6.3.1 and 6.3.2.
The optimal stopping set has the form

$$\Gamma = \Gamma_{\epsilon,\delta}(\{4\}) \cup \Gamma_{\delta,\gamma}(\{4,5\}) \cup \Gamma_{\gamma,\beta}(\{4,5,6\}) \cup \Gamma_{\beta,\alpha}(\{5,6\}) \cup \Gamma_{\alpha,1}(\{5,6,7\}).$$

The value of the problem is $v = v(\epsilon) \cong 0.3389$.

The main points of the analysis of the problem.

We have examined the most effective approaches for addressing problems with gain functions that are not monotonic. The recursive algorithm that was introduced to identify the optimal strategy (which is equivalent to defining the optimal stopping set) relies on the distribution of the stopped process (refer to Section 6.2.4).

We have provided various examples of optimal stopping sets, showcasing a wide range of possibilities. We have thoroughly analyzed the simplest sets A in Sections 6.3 and 6.4, presenting our findings. We have obtained interesting results from our analysis. For instance, in the case of $A = \{1, 3\}$, we have observed that stopping at the second relative rank is never optimal. Similarly, in the problem with $A = \{2, 3\}$, stopping at the first relative rank is not recommended. In the case of $A = \{1, 4\}$, we have determined that stopping at the second rank is not the optimal choice. Additionally, we have found that the moments of the first possibility of optimal stopping in the third and fourth relative ranks are nearly equal.

Let's compare the values of the problems. Luckily, it's easier to choose a candidate with an absolute rank of $A = \{1, 3\}$ rather than $A = \{2, 3\}$. A similar situation occurs in cases $A = \{1, 4\}$ and $A = \{3, 4\}$. Moreover, the probability of winning according to the optimal strategy doesn't decrease for sets $A = \{a, 4\}$, where $1 \le a \le 3$. The closed form strategy can only be given in the case $A = \{2, 3\}$.

In problem $A = \{5\}$, we have island asymptotic optimal strategy (for details see Szajowski (1982)). If we add one absolute rank < 5 more, the optimal strategy is threshold. We have tried to get an answer for a question "what sets A generate island optimal strategies?".

Island strategies can arise in two simplest cases: if we try to hit a small group of big absolute ranks or if we try to stop at long far apart groups of ranks. We provided a numerical analysis of optimal strategies for the sets A consisting of two ranks and the large number ($N = 300$) of candidates in these two classes. If we have found the island strategy, we prove the related asymptotic result. The problem of the first class (remote groups of ranks) were inspired by result from Szajowski (1982). We had to study how distant the ranks must be to get an island strategy. Thus, we have investigated sets of the form $\{1, a\}$ for the possible small a. In the second class are sets of type $\{a, a+1\}$ for small numbers a (large ranks, but close to each other). The island strategies appear in cases $A = \{1, 10\}$ and $A = \{6, 7\}$, respectively. To get these results in the introduced manner is arduous, but using the presented algorithm, we can obtain them numerically for any number of applicants. In Section 6.5, we have presented a result for $A = \{6, 7\}$, but in this way the result for $A = \{1, 10\}$ can also be obtained.

Consider the $A = \{6, 7\}$ asymptotic case. We can see that if we did not encounter a relatively fourth item between the moments $[0.531N]$ and $[0.748N]$, stopping later at the relatively fourth is no longer optimal.

? Alternative selection problems of candidates based of their ranks.

The fundamental element of the sequential selection model is the ability to relatively evaluate incoming candidates. The formulated objective is usually defined in reference to unknown global ratings. In addition to tasks involving the construction of optimal strategies, the investigation of heuristic tactics is also popular. For example, one might inquire about the actual benefits of implementing an optimal strategy (cf. Quine and Law (1996)).

For instance, Helmi and Panholzer (2013) provides a precise mathematical analysis of the behavior of "hiring above the median" strategies for a problem in the context of "online selection under uncertainty", known (at least in computer science-related literature) as the "hiring problem" (v. also Móri (1988)). A sequence of candidates is interviewed sequentially and based on the "score" of the current candidate an immediate decision whether to hire him or not has to be made. For "hiring above the median" selection rules, a new candidate will be hired if he has a score better than the median score of the *already recruited candidates*. Under the natural probabilistic model assuming that the ranks of the first n candidates are forming a random permutation, it is shown exact and asymptotic results for various quantities of interest to describe the dynamics of the hiring process and the quality of the hired staff. In particular one can characterize the limiting distribution of the number of hired candidates of a sequence of n candidates, which reveals the somewhat surprising effect, that slight changes in the selection rule, i.e., assuming the "lower" or the "upper" median as the threshold, have a strong influence on the asymptotic behavior. These allow extend considerably previous analysis of such selection rules proposed by Krieger et al. (2007), Broder et al. (2008), Archibald and Martínez (2009). There are some connections between the hiring process and the Chinese restaurant process introduced by Pitman (2006).

6.6 The Problem of Multiple Attempts to Choose the Best Items

In this section, we provide more details to the multiple best choice (MBC) problem introduced in Section 1.4.3.

Suppose that our gain is the probability of choosing k best objects. Let us remind the notations introduced earlier. Denote by $(a_1, a_2, \dots, a_N)$ any permutation of numbers $(1, 2, \dots, N)$. (1 corresponds to the best object, N corresponds to the worst one) If a_i is the n-th object in order on quality among $(a_1, a_2, \dots, a_i)$, we write $\xi_i = n$ for all $i = 1, 2, \dots, N$, a_i is called absolute rank, and ξ_i is called relative rank. It can be shown that

$$\boldsymbol{P}\{\xi_i = j\} = \frac{1}{i}, \quad j = 1, 2, \dots, i, \tag{6.26}$$

$$\boldsymbol{P}\{a_i = n \mid \xi_i = j\} = \frac{C_{n-1}^{j-1} C_{N-n}^{i-j}}{C_N^i}. \tag{6.27}$$

Let $(i_1, \dots, i_k)$ be any permutation of numbers $1, 2, \dots, k$. A rule $\tau^* = (\tau_1^*, \dots, \tau_k^*)$, $1 \le \tau_1^* < \tau_2^* < \dots < \tau_k^* \le N$ is an optimal rule if

$$\boldsymbol{P}\Big\{ \bigcup_{(i_1,\dots,i_k)} \{a_{\tau_1^*} = i_1, \dots, a_{\tau_k^*} = i_k\} \Big\}$$

$$= \sup_{\tau} \boldsymbol{P}\Big\{ \bigcup_{(i_1,\dots,i_k)} \{a_{\tau_1} = i_1, \dots, a_{\tau_k} = i_k\} \Big\} = \boldsymbol{P}_N^*,$$

where $\tau = (\tau_1, \dots, \tau_k)$. We are interested in finding the optimal rule $\tau^* = (\tau_1^*, \dots, \tau_k^*)$.

By $Z_{\boldsymbol{m}_k}^{(i)_k} = Z_{m_1,\dots,m_k}^{i_1,\dots,i_k}$ denote a conditional probability of event $\{a_{m_1} = i_1, \dots, a_{m_k} = i_k\}$ with respect to σ-algebra $\mathcal{F}_{\boldsymbol{m}_k} = \mathcal{F}_{m_k}$, generated by observations $(\xi_1, \dots, \xi_{m_k})$. There is one-to-one correspondence between $(a_{m_1}, a_{m_2}, \dots, a_{m_k})$ and $(\xi_{m_1}, \xi_{m_2}, \dots, \xi_{m_k})$ such that $\xi_{m_1} = 1, \xi_{m_2} = 1, 2, \dots, \xi_{m_k} = 1, 2, \dots, k$. Let us introduce the following random events

$$B_l = \{\omega : \xi_{m_l} = j_l, \xi_{m_l+1} > l, \dots, \xi_{m_{l+1}-1} > l\}, \quad l = 1, 2, \dots, k-1.$$

Then

$$Z_{\boldsymbol{m}_k}^{(i)_k} = \frac{m_k(m_k - 1)\dots(m_k - k + 1)}{N(N-1)\dots(N-k+1)} \mathbb{I}\left(\bigcap_{l=1}^{k-1} B_l \right) \mathbb{I}(\xi_{m_k} = j_k). \tag{6.28}$$

Put

$$Z_{m_k} = \sum_{(i_1,\dots,i_k)} Z^{(i)_k}_{m_k}.$$

Using (1.12), we get the value of the game v

$$\boldsymbol{P}^*_N = \boldsymbol{E}Z_{\tau^*} = \sup_{\tau\in\mathfrak{S}_1} \boldsymbol{E}Z_\tau = v.$$

This means that the multiple BC problem of choosing k best objects can be reduced to the problem of multiple-stopping of the random sequence Z_{m_k}.

The solution of this problem is the following optimal strategy: there exists a set $\pi^* = (\pi^*_1, \dots, \pi^*_k)$, $1 \le \pi^*_1 < \dots < \pi^*_k \le N$ such that

⇒it is necessary to skip first $\pi^*_1 - 1$ objects, and then we stop on the first object, which is better than all precursors, or on the $(N - k + 1)$-th object, if the best one does not appear by the moment $N - k + 1$;

⇒at second time we stop on the first object, which is better than all precursors, or worse than one object (if we already have observed $\pi^*_2 - 1$ objects), if any, or, otherwise, on $(N - k + 2)$-th object;

⇒the third choice should be made on the first object, which is better than all precursors, or worse than one object (if we already have observed $\pi^*_2 - 1$ objects) or worse than two objects (if we already have observed $\pi^*_3 - 1$ objects), if any, or on $(N - k + 3)$-th object etc.

More formally,

$$\begin{aligned}
\tau^*_1 &= \min\{m_1 \ge \pi^*_1 : \xi_{m_1} = 1\},\\
\tau^*_i &= \min\big[\min\{m_i > m_{i-1} : \xi_{m_i} = 1\},\\
&\quad \min\{m_i > m_{i-1} : m_i \ge \pi^*_2, \xi_{m_i} = 2\},\\
&\quad \dots, \min\{m_i > m_{i-1} : m_i \ge \pi^*_i, \xi_{m_i} = i\}\big]
\end{aligned}$$

on the set $F_{i-1} = \{\omega : \tau^*_1 = m_1, \dots, \tau^*_{i-1} = m_{i-1}\}$, $i = 2, \dots, k$, $F_0 = \Omega$.

We refer here to the papers of Nikolaev (1977, 1998), Tamaki (1979), Vanderbei (1980), Ano (1989), and Lehtinen (1993, 1997).

Table 6.7 displays the values of the game and the thresholds π^*.

Example 6.1. *If $N = 6$, $k = 2$ then $v = 0.3139$, $\pi^*_1 = 2$, $\pi^*_2 = 5$. So we have the following optimal rule:*

- *it is necessary to skip the first object, and then we stop on the first object, which is better than all precursors, or on the fifth object, if the best one does not appear by the moment 5;*
- *at second time we stop on the first object, which is better than all precursors, or worse than one object (if we already have observed 4 objects), if any, or, otherwise, on the sixth object.*

TABLE 6.7
The sets $\pi^* = (\pi_1^*, \ldots, \pi_k^*)$ and the values v.

$N \backslash k$	2	3	4
3	$\pi^* = (1,3)$ $v = 0.5000$		
4	$\pi^* = (1,3)$ $\pi^* = (1,4)$ $\pi^* = (2,3)$ $\pi^* = (2,4)$ $v = 0.3333$	$\pi^* = (1,2,4)$ $v = 0.4583$	
5	$\pi^* = (2,4)$ $v = 0.3333$	$\pi^* = (1,3,5)$ $v = 0.3333$	$\pi^* = (1,2,4,5)$ $v = 0.4333$
6	$\pi^* = (2,5)$ $v = 0.3139$	$\pi^* = (1,3,5)$ $\pi^* = (1,3,6)$ $\pi^* = (2,3,5)$ $\pi^* = (2,3,6)$ $\pi^* = (2,4,5)$ $\pi^* = (2,4,6)$ $v = 0.2500$	$\pi^* = (1,3,4,6)$ $v = 0.3139$

Consider the following random permutation $r_1, \ldots, r_6$: 3, 4, 2, 6, 1, 5. Then we observe the sequence of $\xi_1, \ldots, \xi_6$: 1, 2, 1, 4, 1, 5. So we stop on the third object ($m_1 = 3$, $\xi_3 = 1$) and on the fifth one ($m_2 = 5$, $\xi_5 = 1$). Thus we choose the best two objects.

The following asymptotic results were obtained:

- for the two-choice (or double) problem (see Nikolaev (1977) and Tamaki (1979)),

$$\lim_{N\to\infty} \frac{\pi_1^*}{N} = 0.2291, \quad \lim_{N\to\infty} \frac{\pi_2^*}{N} = 0.6065,$$
$$\lim_{N\to\infty} \boldsymbol{P}_N^* = 0.2254; \tag{6.29}$$

- for the three-choice problem (see Ano (1989)),

$$\lim_{N\to\infty} \frac{\pi_1^*}{N} = 0.1666, \quad \lim_{N\to\infty} \frac{\pi_2^*}{N} = 0.4370, \quad \lim_{N\to\infty} \frac{\pi_3^*}{N} = 0.7165,$$
$$\lim_{N\to\infty} \boldsymbol{P}_N^* = 0.1625; \tag{6.30}$$

- for the four-choice problem (see Lehtinen (1993)),

$$\lim_{N\to\infty} \frac{\pi_1^*}{N} = 0.1310, \quad \lim_{N\to\infty} \frac{\pi_2^*}{N} = 0.3418,$$
$$\lim_{N\to\infty} \frac{\pi_3^*}{N} = 0.5591, \quad \lim_{N\to\infty} \frac{\pi_4^*}{N} = 0.7788,$$
$$\lim_{N\to\infty} \boldsymbol{P}_N^* = 0.1271; \tag{6.31}$$

- for the five-choice problem (see Lehtinen (1997)),

$$\lim_{N\to\infty}\frac{\pi_1^*}{N}=0.1079,\quad \lim_{N\to\infty}\frac{\pi_2^*}{N}=0.2807,$$
$$\lim_{N\to\infty}\frac{\pi_3^*}{N}=0.4586,\quad \lim_{N\to\infty}\frac{\pi_4^*}{N}=0.6382,\quad \lim_{N\to\infty}\frac{\pi_5^*}{N}=0.8187,$$
$$\lim_{N\to\infty}\boldsymbol{P}_N^*=0.1043. \tag{6.32}$$

In the general case of k-choice problem, Vanderbei (1980) derived a lower bound for the probability of choosing k best objects,

$$\boldsymbol{P}_N^*\geq\frac{1}{e(k+1)},\qquad k\in\mathbb{N}.$$

This result was improved by Lehtinen (1997), who found an approximation for $\boldsymbol{P}_N^*$,

$$\frac{1}{e^{\gamma}(k+\frac{1}{2})}\leq\boldsymbol{P}_N^*\approx\frac{1}{(e-1)k+1},\qquad k\in\mathbb{N},k\geq 2,$$

where $\gamma\approx 0.5772$ is the Euler constant.

6.6.1 The double choice BCP

Let us consider the two-choice (or double) problem; for further details, see Nikolaev (1977) and Tamaki (1979).

Following (6.28), we have

$$Z_{m_1,m_2}^{(1,2)}=\frac{m_2(m_2-1)}{N(N-1)}\mathbb{I}(\xi_{m_1}=1,\xi_{m_1+1}>1,\ldots,\xi_{m_2-1}>1,\xi_{m_2}=2),\tag{6.33}$$
$$Z_{m_1,m_2}^{(2,1)}=\frac{m_2(m_2-1)}{N(N-1)}\mathbb{I}(\xi_{m_1}=1,\xi_{m_1+1}>1,\ldots,\xi_{m_2-1}>1,\xi_{m_2}=1),\tag{6.34}$$

where $Z_{m_1,m_2}^{(1,2)}$ and $Z_{m_1,m_2}^{(2,1)}$ are the conditional probabilities of $\{a_{m_1}=1,a_{m_2}=2\}$ and $\{a_{m_1}=2,a_{m_2}=1\}$ with respect to σ-algebra $\mathcal{F}_{m_1,m_2}=\mathcal{F}_{m_2}$, generated by observations $(\xi_1,\ldots,\xi_{m_2})$. Hence,

$$Z_{m_1,m_2}=Z_{m_1,m_2}^{(1,2)}+Z_{m_1,m_2}^{(2,1)}.$$

From (2.19) and (2.20), we can define the sequences $\{V_{m_1,m_2}^N\}$ and $\{V_{m_1}^N\}$:

$$V_{m_1,m_2}^N=\max\{Z_{m_1,m_2},\boldsymbol{E}_{m_2}V_{m_1,m_2+1}^N\},\quad N\geq m_2>m_1,\tag{6.35}$$
$$V_{m_1,N+1}^N=0,\quad m_1\geq 1,$$
$$V_{m_1}^N=\max\{X_{m_1}^N,\boldsymbol{E}_{m_1}V_{m_1+1}^N\},\quad N>m_1\geq 1,\tag{6.36}$$
$$V_N^N=0.$$

The double optimal stopping rule $\tau^* = (\tau_1^*, \tau_2^*)$ is given by:

$$\begin{aligned}\tau_1^* &= \min\{m_1 \geq 1 : X_{m_1}^N = V_{m_1}^N\},\\ \tau_2^* &= \min\{m_2 > m_1 : V_{m_1,m_2}^N = Z_{m_1,m_2}\} \text{ on the set } \{\tau_1^* = m_1\}.\end{aligned} \tag{6.37}$$

Denote $u_{m_1,m_2} = \boldsymbol{E}_{m_2-1}V_{m_1,m_2}$. From (6.35), we have

$$u_{m_1,m_2} = \boldsymbol{E}_{m_2-1}(\max\{Z_{m_1,m_2}^{(1,2)} + Z_{m_1,m_2}^{(2,1)}, u_{m,m_2+1}\}).$$

Since ξ_k, $k = 1, \ldots, N$, are independent and $X_{m_1}^N = u_{m_1,m_1+1}$, then $\boldsymbol{E}_{m_1}V_{m_1+1}^N = \boldsymbol{E}V_{m_1+1}^N$. From Theorem 2.3, $\boldsymbol{E}V_m^N = v_m$. Using (6.36), we have

$$v_{m_1}^N = \boldsymbol{E}(\max\{u_{m_1,m_1+1}, v_{m_1+1}^N\}). \tag{6.38}$$

Then we rewrite the double optimal stopping rule (6.37) in the following way:

$$\begin{aligned}\tau_1^* &= \min\{m_1 \geq 1 : u_{m_1,m_1+1} \geq v_{m_1+1}^N\}, \qquad (6.39)\\ \tau_2^* &= \min\{m_2 > m_1 : Z_{m_1,m_2}^{(1,2)} + Z_{m_1,m_2}^{(2,1)} \geq u_{m_1,m_2+1}\} \text{ on } \{\tau_1^* = m_1\}.\end{aligned}$$

Let $Z_{m_1,m_2}^{(1,2)}(\omega) = \overline{Z}_{m_1,m_2}^{(1,2)}$, $u_{m_1,m_2+1}(\omega) = \bar{u}_{m_1,m_2+1}$ on the set $\{\omega : \xi_{m_1} = 1, \xi_{m_1+1} > 1, \ldots, \xi_{m_2-1} > 1, \xi_{m_2} = 2\}$ and

$$\begin{aligned}\pi_1^* &= \min\{m_1 \geq 1 : \bar{u}_{m_1,m_1+1} \geq v_{m_1+1}^N\},\\ \pi_2(m_1) &= \min\{m_2 > m_1 : \overline{Z}_{m_1,m_2}^{(1,2)} \geq \bar{u}_{m_1,m_2+1}\}, \quad \pi_2(\pi_1^*) = \pi_2^*.\end{aligned}$$

Hence, the double optimal stopping rule (6.39) can be written as follows:

$$\begin{aligned}\tau_1^* &= \min\{m_1 \geq \pi_1^* : \xi_{m_1} = 1\},\\ \tau_2^* &= \min\big[\min\{m_2 > m_1 : \xi_{m_2} = 1\},\\ &\qquad \min\{m_2 > m_1 : m_2 \geq \pi_2^*, \xi_{m_2} = 2\}\big] \text{ on } \{\tau_1^* = m_1\}.\end{aligned} \tag{6.40}$$

Finally, we will outline how the results (6.29) can be derived. Let us consider all rules (τ_1, τ_2) defined by (6.40) with some (not necessarily optimal) thresholds (π_1, π_2), $1 \leq \pi_1 < \pi_2 \leq N$. Denote

$$\lim_{N\to\infty} \frac{\pi_1}{N} = \alpha, \quad \lim_{N\to\infty} \frac{\pi_2}{N} = \beta.$$

Then we can have one of the following three cases:

(1) $\pi_1 \leq m_1 < m_2 < \pi_2$,

(2) $\pi_2 \leq m_1 < m_2 \leq N$,

(3) $\pi_1 \leq m_1 < \pi_2 \leq m_2 \leq N$.

Let $\boldsymbol{P}_N^{(i)}$, $i = 1, 2, 3$, be the probability of selecting two best objects in the case (i) above. If we use the rule (τ_1, τ_2) with the thresholds (π_1, π_2), the total probability of selecting two best objects is $P_N = P_N^{(1)} + P_N^{(2)} + P_N^{(3)}$. Using the combinatorial argument, we can derive the expressions for $P_N^{(i)}$, $i = 1, 2, 3$.

In case (1), $\pi_1 \le m_1 < m_2 < \pi_2$, it is clear that $P(a_{m_1} = 1, a_{m_2} = 2) = 0$. If $\pi_1 > 1$, using the independence of relative ranks and (6.26), we obtain

$$\begin{aligned}
&P(a_{m_1} = 2, a_{m_2} = 1)\\
&= P(\xi_{\pi_1} > 1, \ldots, \xi_{m_1-1} > 1, \xi_{m_1} = 1, \xi_{m_1+1} > 1, \ldots, \xi_{m_2-1} > 1, \xi_{m_2} = 1) Z^{(2,1)}_{m_1,m_2}\\
&= \frac{\pi_1 - 1}{\pi_1} \cdot \ldots \cdot \frac{m_1 - 2}{m_1 - 1} \cdot \frac{1}{m_1} \cdot \frac{m_1}{m_1 + 1} \cdot \ldots \cdot \frac{m_2 - 2}{m_2 - 1} \cdot \frac{1}{m_2} Z^{(2,1)}_{m_1,m_2}\\
&= \frac{\pi_1 - 1}{(m_1 - 1)(m_2 - 1)m_2} Z^{(2,1)}_{m_1,m_2}.
\end{aligned}$$

If $\pi_1 = 1$, then $m_1 = 1$ since $\xi_1 = 1$.

$$\begin{aligned}
P(a_{m_1} = 2, a_{m_2} = 1) &= P(\xi_1 = 1, \xi_2 > 1, \ldots, \xi_{m_2-1} > 1, \xi_{m_2} = 1) Z^{(2,1)}_{m_1,m_2}\\
&= 1 \cdot \frac{1}{2} \cdot \ldots \cdot \frac{m_2 - 2}{m_2 - 1} \cdot \frac{1}{m_2} Z^{(2,1)}_{m_1,m_2}\\
&= \frac{1}{(m_2 - 1)m_2} Z^{(2,1)}_{m_1,m_2}.
\end{aligned}$$

Cases (2) and (3) can be considered in a similar way.

Summing up over all possible pairs of m_1 and m_2, we can show that if $\pi_1 > 1$,

$$\begin{aligned}
P_N^{(1)} &= \sum_{m_1=\pi_1}^{\pi_2-2} \sum_{m_2=m_1+1}^{\pi_2-1} \frac{\pi_1 - 1}{(m_1 - 1)(m_2 - 1)m_2} Z^{(2,1)}_{m_1,m_2},\\
P_N^{(2)} &= \sum_{m_1=\pi_2}^{N-1} \sum_{m_2=m_1+1}^{N} \frac{\pi_1 - 1}{(m_2 - 2)(m_2 - 1)m_2} \left(Z^{(1,2)}_{m_1,m_2} + Z^{(2,1)}_{m_1,m_2} \right),\\
P_N^{(3)} &= \sum_{m_1=\pi_1}^{\pi_2-1} \sum_{m_2=\pi_2}^{N} \frac{(\pi_1 - 1)(\pi_2 - 2)}{(m_1 - 1)(m_2 - 2)(m_2 - 1)m_2} \left(Z^{(1,2)}_{m_1,m_2} + Z^{(2,1)}_{m_1,m_2} \right).
\end{aligned}$$

Using (6.33) and (6.34), we have

$$\begin{aligned}
P_N = &\frac{(\pi_1 - 1)(\pi_2 - 2)}{N(N - 1)} \sum_{m_1=\pi_1}^{\pi_2-2} \frac{1}{m_1 - 1} - \frac{(\pi_1 - 1)(\pi_2 - \pi_1 - 1)}{N(N - 1)}\\
&+ \frac{2(\pi_1 - 1)}{N(N - 1)} \sum_{m_1=\pi_2}^{N-1} \sum_{m_2=m_1+1}^{N} \frac{1}{m_2 - 2}\\
&+ \frac{2(\pi_1 - 1)(\pi_2 - 2)}{N(N - 1)} \sum_{m_1=\pi_1}^{\pi_2-1} \sum_{m_2=\pi_2}^{N} \frac{1}{(m_1 - 1)(m_2 - 2)}.
\end{aligned}$$

Then, for large values of N,

$$\sum_{m_1=\pi_1}^{\pi_2-1} \frac{1}{m_1-1} \sim \ln \frac{\pi_2}{\pi_1},$$
$$\sum_{m_1=\pi_2}^{N-1} \sum_{m_2=m_1+1}^{N} \frac{1}{m_2-2} \sim N\left(\frac{\pi_2}{N} \ln \frac{\pi_2}{N} + 1 - \frac{\pi_2}{N}\right),$$
$$\sum_{m_1=\pi_1}^{\pi_2-1} \sum_{m_2=\pi_2}^{N} \frac{1}{(m_1-1)(m_2-2)} \sim \ln \frac{N}{\pi_2} \ln \frac{\pi_2}{\pi_1},$$

where $\lambda_N \sim \delta_N$ means that $\lim_{N\to\infty} \frac{\lambda_N}{\delta_N} = 1$.

Then

$$\lim_{N\to\infty} \boldsymbol{P}_N = \boldsymbol{P}(\alpha,\beta) = \alpha\beta(1 - 2\ln\beta)\ln\frac{\beta}{\alpha} + \alpha^2 + 2\alpha - 3\alpha\beta + 2\alpha\beta\ln\beta$$

with the maximum value of $\boldsymbol{P}(\alpha,\beta)$, $0 \le \alpha \le \beta \le 1$, being

$$\boldsymbol{P}_{\max} = \frac{\mu}{\sqrt{e}}(\sqrt{e}\mu - 2\ln\mu + 2\sqrt{e} - 5) \approx 0.2254,$$

where $\mu \approx 0.2291$ is the root of the equation $\sqrt{e}\mu - \ln\mu = \frac{7}{2} - \sqrt{e}$. Hence,

$$\lim_{N\to\infty} \frac{\pi_1^*}{N} = \mu, \quad \lim_{N\to\infty} \frac{\pi_2^*}{N} = e^{-1/2},$$
$$\lim_{N\to\infty} \boldsymbol{P}_N^* = \boldsymbol{P}_{\max}.$$

These results are illustrated in Figures 6.6a, 6.6b and 6.7.

The embedded Markov chain approach.

Adopting the notation for the payout obtained from the i-th stop (here $i \in \{1,2\}$) at time t and the relative rank of the accepted object $Y_t = 1$, denoted by $f_i(t)$, we have

$$v_1(t) = \sup_{\tau<\sigma} \boldsymbol{E}_t(f_1(\tau) + f_2(\sigma)) = \sup_{\tau} \boldsymbol{E}_t(f_1(\tau) + \sup_{\sigma} \boldsymbol{E}_\tau f_2(\sigma)),$$
$$v_2(t) = \sup_{\sigma} \boldsymbol{E}_{t,1} f_2(\sigma).$$

Auxiliary functions are

$$H_1(t) = \sum_{j=t+1}^{n} \frac{t}{j(j-1)} \frac{j}{n}, \tag{6.41}$$

$$H_2(t) = \sum_{j=t+1}^{n} \frac{t}{j(j-1)} v_2(j), \tag{6.42}$$

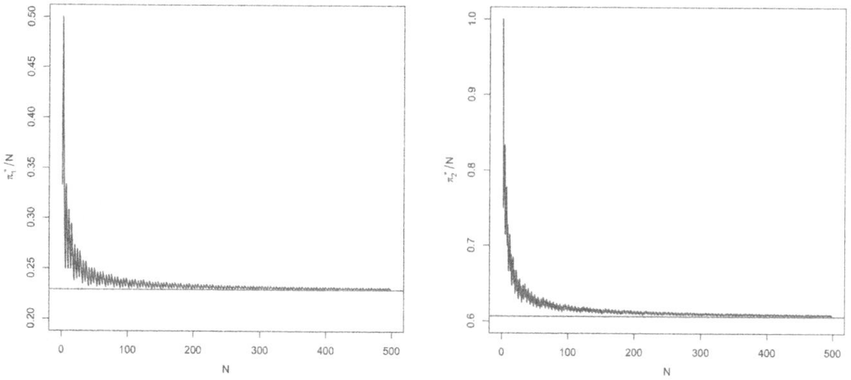

(a) The first threshold, π_1^*/N.

(b) The second threshold, π_2^*/N.

FIGURE 6.6
The first and the second threshold, π_1^*/N and π_2^*/N, in the double choice problem (blue line). The asymptotic values are shown in red.

where

$$v_2(t) = \max\left\{H_1(d_2^*), \frac{t}{n}\right\} \tag{6.43}$$

and d_2^* is the optimal threshold determining the optimal second stopping time.

The function $v_2(t) = \sup_{\sigma > t} \boldsymbol{E}_t f_2(\sigma)$. Solving the problem of selecting the best object when one stop is possible by backward induction, we have:

$$v_2(k) = \begin{cases} \frac{k}{n} & k \geq d_2^*, \\ \mathbf{T}v_2(k) & k < d_2^* \end{cases}$$
$$= \max\left\{H_1(d_2^*), \frac{k}{n}\right\},$$

where $d_2^* = \inf_k\{1 \leq k \leq n : \frac{k}{n} \geq \sum_{j=k+1}^{n} p(k,j) f_2(j)\}$.

Determining the optimal behavior with the first stop is based on the gain for it.

$$f_1(k) = \frac{k}{n} + \boldsymbol{E}_k f_2(\sigma^*) = \frac{k}{n} + \boldsymbol{E}_{(k,1)} v_2(W_1). \tag{6.44}$$

By the backward induction, we have:

$$v_1(k) = \begin{cases} \frac{k}{n} + H_2(k) & k \geq d_1^*, \\ \mathbf{T}v_1(k) & k < d_1^* \end{cases}$$
$$= \max\left\{H_1(d_1^*), \frac{k}{n}\right\},$$

where $d_2^* = \inf_k\{1 \leq k \leq n : \frac{k}{n} + H_2(k) \geq H_1(k) + \sum_{j=k+1}^{n} p(k,j) v_2(j)\}$.

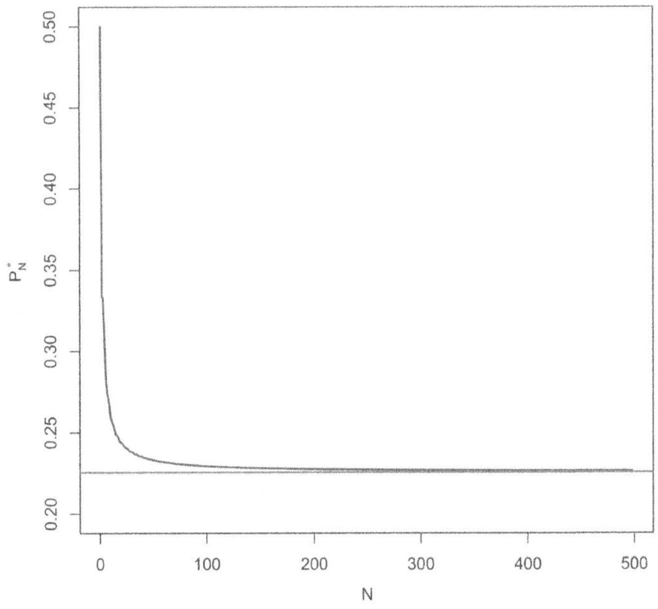

FIGURE 6.7
The probability of selecting two best objects, $\boldsymbol{P}^*_N$, in the double choice problem (blue line). The asymptotic value is shown in red.

6.7 Selection of Given Ranks Secretaries

We can generalize the problem of multiple best choices to a case when the gain is to maximize the probability of choosing k objects with given ranks $r_1, r_1, \ldots, r_k$, $1 \le r_1 < \ldots < r_k \le N$; see Nikolaev et al. (2007).

Let $\boldsymbol{l}_k = (l_1, \ldots, l_k)$ be a permutation of the integer numbers $r_1, r_2, \ldots, r_k$. A rule $\tau^* = (\tau_1^*, \ldots, \tau_k^*)$, $1 \le \tau_1^* < \tau_2^* < \cdots < \tau_k^* \le N$ is optimal if

$$\boldsymbol{P}\Big\{\bigcup_{\boldsymbol{l}_k}\{a_{\tau_1^*} = l_1, \ldots, a_{\tau_k^*} = l_k\}\Big\}$$
$$= \sup_{\tau} \boldsymbol{P}\Big\{\bigcup_{\boldsymbol{l}_k}\{a_{\tau_1} = l_1, \ldots, a_{\tau_k} = l_k\}\Big\} = \boldsymbol{P}^*_N, \qquad (6.45)$$

where $\tau = (\tau_1, \ldots, \tau_k)$. We are interested in finding the optimal rule $\tau^* = (\tau_1^*, \ldots, \tau_k^*)$. In the same way, we define the sequence $Z_{\boldsymbol{m}_k}$ and the value of the game v.

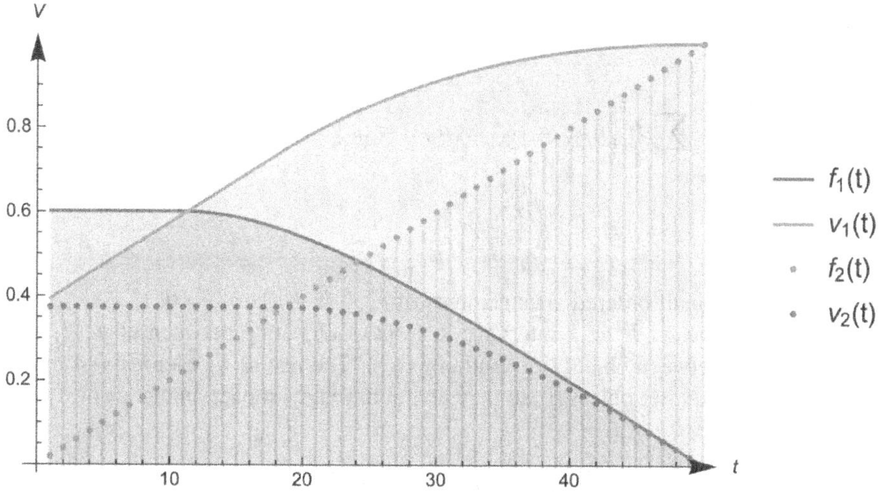

FIGURE 6.8
The gain functions for the best choice problem with 2 stops (for $n = 100$ offers).

By $Z^{\boldsymbol{l}_k}_{\boldsymbol{m}_k}$ denote a conditional probability of event $\{a_{m_1} = l_1, \ldots, a_{m_k} = l_k\}$ with respect to σ-algebra $\mathcal{F}_{\boldsymbol{m}_k}$, generated by observations $(\xi_1, \ldots, \xi_{m_k})$, and put

$$Z_{\boldsymbol{m}_k} = \sum_{\boldsymbol{l}_k} Z^{\boldsymbol{l}_k}_{\boldsymbol{m}_k}.$$

Using (6.45), we get the value of the game v: $\boldsymbol{P}^*_N = \boldsymbol{E}Z_{\tau^*} = \sup_\tau \boldsymbol{E}Z_\tau = v$.

As above, we can reduce the Best Choice Problem of k objects with ranks $r_1, \ldots, r_k$ to the problem of multiple-stopping of the random sequence $Z_{\boldsymbol{m}_k}$.

Let $R = \{r_1, \ldots, r_k\}$ be the set of ranks $r_1, \ldots, r_k$. It is clear that objects from $R' = \{1, 2, \ldots, r_k\} \setminus R$ have impact on the values of the relative ranks from the set R. Let us introduce the following random events

$$\begin{aligned}
B_0 &= \bigcap_{s=1}^{n_0} \{\omega : \xi_{t_{0,s}} = j_{0,s}, \xi_{t_{0,s}+1} > s, \ldots, \xi_{t_{0,s+1}-1} > s\}, \\
&\quad 1 \le t_{0,1} < \cdots < t_{0,n_0} < t_{0,n_0+1} = m_1, \\
B_i &= \{\omega : \xi_{m_i} = j_i, \xi_{m_i+1} > n_0 + \cdots + n_{i-1} + i - 1, \ldots, \\
&\quad \xi_{m_i+t_{i,1}-1} > n_0 + \cdots + n_{i-1} + i - 1\} \\
&\quad \bigcap_{s=1}^{n_i} \{\omega : \xi_{m_i+t_{i,s}} = j_{i,s}, \xi_{m_i+t_{i,s}+1} > n_0 + \cdots + n_{i-1} + i + s - 1, \ldots, \\
&\quad \xi_{m_i+t_{i,s+1}-1} > n_0 + \cdots + n_{i-1} + i + s - 1\}, \\
&\quad 1 \le t_{i,1} < \cdots < t_{i,n_i} < t_{i,n_i+1} = m_{i+1} - m_i, \quad i = 1, \ldots, k-1,
\end{aligned}$$

where $(j_1, \ldots, j_k)$, $1 \le j_i \le r_k - k + i$, is a set of the relative ranks from $R = \{r_1, \ldots, r_k\}$, $(j_{i,1}, \ldots, j_{i,n_i})$, $i = 0, 1, \ldots, k-1$, is a set of the relative ranks from

$R' = \{1, 2, \ldots, r_k\} \setminus R$.

Then

$$\begin{aligned} Z^{l_k}_{\boldsymbol{m}_k} &= \sum \boldsymbol{P}\Big\{a_{m_1} = l_1, \ldots, a_{m_k} = l_k \mid \{\omega : \xi_{m_k} = j_k\} \bigcap_{i=0}^{k-1} B_i\Big\} \\ &\times \quad \mathbb{I}(\xi_{m_k} = j_k)\mathbb{I}\Big(\bigcap_{i=0}^{k-1} B_i\Big), \end{aligned}$$

where we sum up all ordered arrangements of $r_k - k - n_k$, $n_k = 0, 1, \ldots, r_k - k$, elements from the set R' in k intervals: in front of object l_1, between object l_1 and object l_2, ..., between object l_{k-1} and object l_k. The rest of n_k elements from the set R' are located after object l_k, so they do not have any impact on the event

$$\{\omega : \xi_{m_k} = j_k\} \bigcap_{i=0}^{k-1} B_i.$$

Since $\xi_1 = 1$, $\xi_2 = 1$ or 2, ..., $\xi_i = 1, 2, \ldots, i$, then $\mathbb{I}(B_0) = 1$. Hence, using independence of the relative ranks $\xi_1, \ldots, \xi_N$, we obtain

$$\begin{aligned} Z^{l_k}_{\boldsymbol{m}_k} &= \sum C^{r_k-k}_{n_k} P^{N-m_k}_{n_k} \frac{m_k(m_k-1)\ldots(m_k - r_k + n_k + 1)}{N(N-1)\ldots(N - r_k + 1)} \\ &\times \quad \mathbb{I}(\xi_{m_k} = j_k)\mathbb{I}\Big(\bigcap_{i=1}^{k-1} B_i\Big), \end{aligned} \tag{6.46}$$

where $C^{r_k-k}_{n_k}$ is the number of n_k-combinations from a set with $r_k - k$ elements, $P^{N-m_k}_{n_k}$ is the number of n_k-permutations from a set with $N - m_k$ elements.

Using Theorem 2.2, we obtain the optimal multiple-stopping rule, which can be described as follows

$$\begin{aligned} &\tau_1^* = \min\{m_1 : \xi_{m_1} \in \Gamma_{1,m_1}\}, \quad \Gamma_1 = (\Gamma_{1,1}, \ldots, \Gamma_{1,N-k+1}), \\ &\Gamma_{1,s} \subseteq \{1, \ldots, s\} \cap \{1, \ldots, r_k - k + 1\}, \quad s = 1, \ldots, N - k, \\ &\Gamma_{1,N-k+1} = \{1, 2, \ldots, N - k + 1\}, \end{aligned}$$

where Γ are stopping sets derived beforehand.

Note that stopping at moment m_i depends on values ξ_{m_i}, and $\bigcap_{s=1}^{i-1} B_s$. Therefore, stopping set Γ_{i,m_i} depends on values $\xi_{m_1}, \xi_{m_1+1}, \ldots, \xi_{m_i-1}$. Hence,

$$\begin{aligned} &\tau_i^* = \min\{m_i > m_{i-1} : \xi_{m_i} \in \Gamma_{i,m_i}(\xi_{m_1}, \xi_{m_1+1}, \ldots, \xi_{m_i-1})\}, \\ &\Gamma_i = (\Gamma_{i,i}, \ldots, \Gamma_{i,N-k+i}), \quad \Gamma_{i,N-k+i} = \{1, 2, \ldots, N - k + i\}, \\ &\Gamma_{i,s} \subseteq \{1, \ldots, s\} \cap \{1, \ldots, r_k - k + i\}, \quad s = i, \ldots, N - k + i - 1. \end{aligned}$$

Stopping sets $\Gamma_1, \ldots, \Gamma_k$ can be defined by backward induction. Since the structure of the sets solely depends on the values $r_1, \ldots, r_k$, finding of the optimal stopping rules and the value of the game is a problem for each particular case.

Example 6.2. *Suppose we would like to select the best k objects. This means that $r_1 = 1, \dots, r_k = k$. Then $R = \{1, 2, \dots, k\}$, $R' = \emptyset$, $n_0 = n_1 = \cdots = n_k = 0$, and*

$$B_i = \{\omega : \xi_{m_i} = j_i, \xi_{m_i+1} > i, \dots, \xi_{m_{i+1}-1} > i\}, \quad i = 1, 2, \dots, k-1.$$

Since $C_{n_k}^{r_k - k} = C_0^0 = 1$, $P_{n_k}^{N-m_k} = P_0^{N-m_k} = 1$, then from (6.46), we have

$$Z_{m_k}^{l_k} = \frac{m_k(m_k-1)\dots(m_k-k+1)}{N(N-1)\dots(N-k+1)} \mathbb{I}(\xi_{m_k} = j_k)\mathbb{I}\Big(\bigcap_{i=1}^{k-1} B_i\Big),$$

which coincides with (6.28).

6.7.1 Choosing the best and the second best by 2 stops

Let us present the problem and its solution using the embedded Markov chain of Section 6.2 and the current relative ranks. The detailed presentation of the problem will be used to present the general model of the optimal multiple-stopping problem (**OMSP**) in Section 6.6.1. Let $\mathcal{F}_s = \sigma\{Y_1, Y_2, \dots, Y_s\}$ and $\mathfrak{S}$ the set of stopping times related to the filtration $\{\mathcal{F}_s\}_{s=0}^n$. The problem under consideration is to find a pair of stopping times $\tau^*, \sigma^* \in \mathfrak{S}$, $\tau^* < \sigma^*$ such that

$$\boldsymbol{P}(\omega : \{X_{\tau^*}, X_{\sigma^*}\} = \{1, 2\}) = \sup_{\substack{\tau, \sigma \in \mathfrak{S} \\ \tau < \sigma}} \boldsymbol{P}(\omega : \{X_{\tau^*}, X_{\sigma^*}\} = \{1, 2\}). \qquad (6.47)$$

The decision-maker's knowledge is restricted to the filtration $\{\mathcal{F}_s\}_{s=0}^n$, or even to sub-filtration $\{\mathcal{G}_t\}_{t=0}^n$, $\mathcal{G}_t = \sigma\{W_1, W_2, \dots, W_t\}$. The second stop should be optimal for any first stop. It will be discussed and shown below that the first stop is possible, for the positive output, on the relative first only. Consequently, the payoff of the second stop should take into account the moment of the first stop.

Definition 6.1. *Let us sequentially observe a sequence of continuous real random variables $X_1, X_2, \dots, X_n$. The random variable*

$$\rho_{kr}(\omega) = \sum_{j=1}^{r} \mathbb{I}_{\{\omega : X_j(\omega) \le X_k(\omega)\}}(\omega) \qquad (6.48)$$

is called the current relative rank of the observation k and moment r.

The current relative rank $\rho_{rr} = \rho_r$ as defined in Definition 1.8. The reward analysis of choosing items is based on the following facts:

1. successful potential candidates should be the relatively best or the relatively second best;

2. the selected first (a consequence of the first stop) should have the current relative rank belonging to $\{1, 2\}$ at the moment of the second stop (v. Preater (1994a)).

These circumstances make a ground for the construct of the payoffs. Let us define the sequence of positive decision points, pairs of stopping moments with a chance for success. The analysis begins by tracking the current relative rank of the selected item at the first stop (moment s) and the observational ranks at the potential second stop (moment r). Deliberation on the second stop at r should be taken into account to skip the current observation and look for the next good choice:

$$Y_s = 1 \quad (\rho_{sr}, Y_r) = (1,2) \qquad (\rho_{st}, Y_t) = (1,2) \tag{6.49a}$$

$$Y_s = 1 \quad (\rho_{sr}, Y_r) = (2,1) \qquad (\rho_{st}, Y_t) = (3,2) \tag{6.49b}$$

$$Y_s = 1 \quad (\rho_{sr}, Y_r) = (2,1) \qquad (\rho_{st}, Y_t) = (3,1). \tag{6.49c}$$

The process underlying the description of the item stream, limited to the valuable one, has been previously discussed and detailed in Section 6.2.3. The process $\{W_t\}_{t=0}^n$ should have the state space $\{1,2,\ldots,n\} \times \{1,2\}$. It is a Markov process having the transition probabilities (6.8) (see Szajowski (1982)).

$$\begin{aligned}
&\boldsymbol{P}(\{X_\tau, X_\sigma\} = \{1,2\}) \\
&= \boldsymbol{E}\mathbb{I}_{\{\omega:\{X_\tau,X_\sigma\}=\{1,2\}\}}(\omega) \\
&= \sum_{s=1}^{n}\sum_{r=s+1}^{n} \int_{\{\tau=s,\sigma=r\}} \mathbb{I}_{\{\omega:\{X_s,X_r\}=\{1,2\}\}}(\omega)\, d\boldsymbol{P} \\
&= \sum_{s=1}^{n}\sum_{r=s+1}^{n} \int_{\{\tau=s,\sigma=r\}} \boldsymbol{E}[\mathbb{I}_{\{\omega:\{X_s,X_r\}=\{1,2\}\}}(\omega) \mid \mathcal{F}_r]\, d\boldsymbol{P} \\
&= \sum_{s=1}^{n}\sum_{r=s+1}^{n} \int_{\{\tau=s,\sigma=r\}} \boldsymbol{E}[\mathbb{I}_{\{\omega:\{X_s,X_r\}=\{1,2\}\}}(\omega) \mid \rho_{sr}, Y_r]\, d\boldsymbol{P} \\
&= \sum_{s=1}^{n}\sum_{r=s+1}^{n} \int_{\{\tau=s,\sigma=r\}} \boldsymbol{P}(\omega : \{X_s, X_r\} = \{1,2\} \mid \rho_{sr}, Y_r)\, d\boldsymbol{P} \\
&= \boldsymbol{E} g_{\{1,2\}}(\tau, \rho_{(\tau,\sigma)}, \sigma, Y_\sigma),
\end{aligned}$$

where

$$g_{\{1,2\}}(s,k,r,l) = \boldsymbol{P}(\{X_s, X_r\} = \{1,2\} | s, \rho_{sr} = k, r, Y_k = l) \tag{6.50a}$$

$$= \begin{cases} \frac{r(r-1)}{n(n-1)} & k=1, l=2 \\ \frac{r(r-1)}{n(n-1)} & k=2, l=1 \quad 1 \le s < r < n, \\ 0 & \text{otherwise,} \end{cases} \tag{6.50b}$$

by straightforward calculation (v Gilbert and Mosteller (1966), Dynkin and Yushkevich (1967), Nikolaev (1976), Tamaki (1979), Szajowski (1982), Preater (1994a)). By observation (6.49), we have

$$\boldsymbol{E}_{(r,1)} g_{\{1,2\}}(s, 2, W_1) = 0; \tag{6.51a}$$

$$\boldsymbol{E}_{(r,2)} g_{\{1,2\}}(s, k, W_1) > 0 \text{ for } k = 1,2. \tag{6.51b}$$

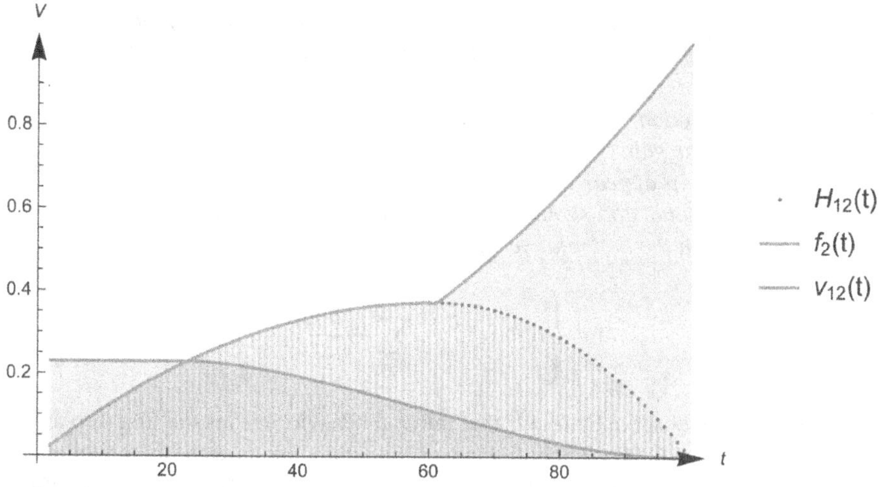

FIGURE 6.9
The gain functions for choosing the first and the second best with 2 stops (for $n = 100$ offers).

This means that after accepting the relatively first candidate at time s, the next relatively first candidate must be accepted if the goal is to select the best and the second best. However, if the next stage of the process $W_.$ at time $r > s$ is relatively second, there is a positive chance of finding a better candidate to pair with the one chosen at time s in the future as well. This is confirmed by the formal analysis of optimality equations. It is not difficult to notice that the optimal selection of the second element to pair with the one chosen at time s is a monotonic problem for candidates with relative rank 2, and the optimal second stopping time can be determined by the **OLA** rule in the state $(r, 2)$, and the appearance of the next state $(r, 1)$ of the process $W_.$ is an indication to stop this candidate immediately. Therefore, the solution to the auxiliary optimal stopping problem after accepting the state $(s, 1)$ is formulated below in Lemma 6.2. Its value is a function of the payout of the auxiliary optimization task for determining the first stopping time. Let us define

$$w_2(t, \rho_{(t,s)}, s, r) = \sup_{\sigma} \boldsymbol{E}_{(s,r)} g_{\{1,2\}}(t, \rho_{(t,\sigma)}, \sigma, Y_\sigma) \tag{6.52a}$$

If optimal solution $\sigma^\star$ of (6.52a) exists, we have

$$v_2(t, k) = \boldsymbol{E}_{(t,k)} g_{\{1,2\}}(t, \rho_{(t,\sigma^\star)}, \sigma^\star, Y_{\sigma^\star}) \tag{6.52b}$$

$$w_1(t, k) = \sup_{\tau<\sigma} \boldsymbol{E}_{(t,k)} g_{\{1,2\}}(\tau, \rho_{(\tau,\sigma)}, \sigma, Y_\sigma) = \sup_{\tau} \boldsymbol{E}_{(t,k)} v_2(\tau, Y_\tau). \tag{6.52c}$$

for $k = 1, 2, \ldots, t$.

Lemma 6.2. *[Tamaki (1979)] In the problem of constructing optimal stopping strategies in the selection problem of the best and second-best candidate of the no-information secretary problem, the construction of the first optimal stopping problem has the payoff function, which is the value of the auxiliary stopping problem of the process $W.$ with the gain function* (6.50).

Let the first stop appear at, $(s,1)$ *and the next state of* $W.$ *is* (r,k), $k=1,2$. *Optimally, the second stop should be made at the first relatively first, and after the moment* $r>d_2^*$, *on the first relatively first or relatively second observation, where*

$$d_2^* = \inf\left\{0<r\leq n : \sum_{j=r}^{n-1}\frac{1}{j-1}\leq\frac{1}{2}\right\}. \tag{6.53}$$

Proof: Let us observe by (6.49) that rational behavior means the first stop at the relative one. From (6.51) and the backward induction, we have the subset of states $\mathbf{B}_{(t,1)}$, where the second stop is at least as good as the continuation of observations:

$$\begin{aligned}
\mathbf{B}_{(t,1)} &= \{(s,1):s>t\}\\
&\quad\cup\{(s,2):(s>t)\ \&\ g_{\{1,2\}}(t,k,s,r)\geq \boldsymbol{E}_{(s,r)}w_2(t,\rho_{(t,W_{11})},W_1)\}\\
&=\left\{(s,1):s>t\right\}\\
&\quad\cup\left\{(s,2):(s>t)\ \&\ \frac{s(s-1)}{n(n-1)}\geq\sum_{j=s+1}^{n}\frac{s(s-1)}{j(j-1)(j-2)}\left[\frac{j(j-1)}{n(n-1)}\right.\right.\\
&\qquad\left.\left.+\,w_2(t,\rho_{(t,j)},j,2)\right]\right\}\\
&=\left\{(s,1):s>t\right\}\cup\left\{(s,2):(s>t)\ \&\ \frac{1}{2}\geq\sum_{j=s+1}^{n}\frac{1}{(j-2)}\right\}.
\end{aligned}$$

This concludes the proof of this lemma. □

Lemma 6.3. *If the first selection appears at the moment t the payoff of the problem is given by* (6.52b) *and it has form*

$$v_2(t,1)=\begin{cases}2\frac{t(t-1)}{n(n-1)}\sum_{j=t}^{n-1}\frac{1}{j-1} & \text{if } t\geq d_2^\star,\\[2ex] \frac{t(d_2^\star-t-1)}{n(n-1)}+2\frac{t(d_2^\star-2)}{n(n-1)}\sum_{j=d_2^\star-1}^{n-1}\frac{1}{j-1} & \text{if } t<d_2^\star.\end{cases} \tag{6.54}$$

Proof: From Lemma 6.2 and by the threshold form of the optimal second stop $\sigma^\star$ with the threshold given by (6.53) the gain of the first stop defined by (6.52b) for

$t \geq d_2^\star$ is equal

$$v_2(t,k) = \boldsymbol{E}_{(t,k)} g_{\{1,2\}}(t, \rho_{(t,\sigma^\star)}, \sigma^\star, Y_{\sigma^\star})$$
$$= \sum_{j=t+1}^{n} \frac{t(t-1)}{j(j-1)(j-2)} \frac{2j(j-1)}{n(n-1)} = 2\frac{t(t-1)}{n(n-1)} \sum_{j=t}^{n-1} \frac{1}{j-1}.$$

For $t < d_2^\star$, we have

$$v_2(t,k) = \sum_{j=t+1}^{d_2^\star - 1} \frac{t}{j(j-1} \frac{j(j-1)}{n(n-1)} + \frac{t}{d_2^\star - 1} 2\frac{(d_2^\star - 1)(d_2^\star - 2)}{n(n-1} \sum_{j=d_2^\star}^{n} \frac{1}{j-2}$$
$$= \frac{t(d_2^\star - t - 1)}{n(n-1)} + 2\frac{t(d_2^\star - 2)}{n(n-1)} \sum_{j=d_2^\star - 1}^{n-1} \frac{1}{j-1}.$$

This ends the proof of this lemma. □

6.8 The Multiple BCP with Minimal Summarized Rank

Let η_i be an absolute rank of selected object. That is, η_i is one plus the number of objects from $(a_1, a_2, \ldots, a_n) < a_i$. The objective is to find procedure such that the expected gain $\boldsymbol{E}(\eta_{\tau_1} + \cdots + \eta_{\tau_k})$, $k \geq 2$ is minimal (see Nikolaev (1976, 1998)).

Formally, let $(\eta_1, \eta_2, \ldots, \eta_N)$ denote a permutation of the integers $(1, 2, \ldots, N)$, all $N!$ permutations being equally likely. For any $i = 1, 2, \ldots, N$ let ξ_i be the number of objects $a_1, a_2, \ldots, a_i$ which are less than or equal to a_i. In the other words, ξ_i is the relative rank of the i-th object. By definition, put

$$v = \inf_{\tau} \mathbf{E}(\eta_{\tau_1} + \cdots + \eta_{\tau_k}), \quad \tau = (\tau_1, \ldots, \tau_k).$$

We want to find the optimal procedure $\tau^* = (\tau_1^*, \ldots, \tau_k^*)$ and the value of the game v.

Let $\mathcal{F}_{\boldsymbol{m}_i}$ be the σ-algebra, generated by $(\xi_1, \xi_2, \ldots, \xi_{m_i})$. If we put

$$Z_{\boldsymbol{m}_k} = \boldsymbol{E}(\eta_{m_1} + \cdots + \eta_{m_k} \mid \mathcal{F}_{\boldsymbol{m}_k}), \tag{6.55}$$

then

$$v = \inf_{\tau} \boldsymbol{E} Z_{\tau}, \quad \tau = (\tau_1, \ldots, \tau_k).$$

So we reduce our problem to the problem of multiple-stopping of the sequence $Z_{\boldsymbol{m}_k}$.

Using (6.27) and a variant of the Chu–Vandermonde identity (see, for example, Koepf (2nd ed. 2014)), we have

$$\begin{aligned} \boldsymbol{E}(\eta_i \mid \xi_i = j) &= \sum_{k=1}^{N} k\boldsymbol{P}(\eta_i = k \mid \xi_i = j) = \sum_{k=0}^{N} k\boldsymbol{P}(\eta_i = k \mid \xi_i = j) \\ &= \sum_{k=0}^{N} \frac{kC_{k-1}^{j-1}C_{N-k}^{i-j}}{C_N^i} = \frac{j}{C_N^i}\sum_{k=0}^{N} C_k^j C_{N-k}^{i-j} \\ &= \frac{jC_{N+1}^{i+1}}{C_N^i} = \frac{j(N+1)}{i+1}. \end{aligned} \tag{6.56}$$

Hence, from (6.55) and (6.56),

$$\begin{aligned} Z_{\boldsymbol{m}_k} &= \boldsymbol{E}_{\boldsymbol{m}_k}(\eta_{m_1} + \cdots + \eta_{m_k}) = \boldsymbol{E}_{\boldsymbol{m}_k}(\eta_{m_1} + \cdots + \eta_{m_{k-1}}) + \boldsymbol{E}_{\boldsymbol{m}_k}\eta_{m_k} \\ &= \sum_{i=1}^{k-1} \boldsymbol{E}_{\boldsymbol{m}_k}\eta_{m_i} + \boldsymbol{E}\eta_{m_k} = \sum_{i=1}^{k-1} g_{m_i,m_k} + \frac{N+1}{m_k+1}\xi_{m_k}, \end{aligned}$$

where $g_{m_i,m_k} = \boldsymbol{E}_{\boldsymbol{m}_k}\eta_{m_i}$.

Using (2.19), (2.20), and taking into account the independence of $\xi_1, \xi_2, \ldots, \xi_N$, we obtain

$$\begin{aligned} V_{\boldsymbol{m}_k} &= \min\left\{\sum_{i=1}^{k-1} g_{m_i,m_k} + \frac{N+1}{m_k+1}\xi_{m_k}, \boldsymbol{E}_{\boldsymbol{m}_k}V_{\boldsymbol{m}_{k-1},m_k+1}\right\} \\ &= \sum_{i=1}^{k-1} g_{m_i,m_k} + \min\left\{\frac{N+1}{m_k+1}\xi_{m_k}, u_{m_k+1}^{(1)}, \right\}, \end{aligned} \tag{6.57}$$

with $u_{N+1}^{(1)} = \infty$ and

$$\begin{aligned} u_{m_k}^{(1)} &= \boldsymbol{E}_{\boldsymbol{m}_{k-1},m_k-1}\left(\min\left\{\frac{N+1}{m_k+1}\xi_{m_k}, u_{m_k+1}^{(1)}, \right\}\right) \\ &\quad + \sum_{i=1}^{k-1} \boldsymbol{E}_{\boldsymbol{m}_{k-1},m_k-1} g_{m_i,m_k} - \sum_{i=1}^{k-1} g_{m_i,m_k-1} \\ &= \boldsymbol{E}\left(\min\left\{\frac{N+1}{m_k+1}\xi_{m_k}, u_{m_k+1}^{(1)}\right\}\right) \\ &= \frac{1}{m_k}\sum_{j=1}^{m_k} \min\left\{\frac{N+1}{m_k+1}j, u_{m_k+1}^{(1)}\right\}. \end{aligned} \tag{6.58}$$

Hence,

$$X_{\boldsymbol{m}_{k-1}} = \boldsymbol{E}_{\boldsymbol{m}_{k-1}}V_{\boldsymbol{m}_{k-1},m_{k-1}+1} = u_{m_{k-1}+1}^{(1)} + \sum_{i=1}^{k-1} g_{m_i,m_{k-1}}.$$

From (2.19) and (2.20),

$$V_{\boldsymbol{m}_{k-1}} = \min\left\{\sum_{i=1}^{k-2} g_{m_i,m_{k-1}} + u^{(1)}_{m_{k-1}+1} + \frac{N+1}{m_{k-1}+1}\xi_{m_{k-1}}, \boldsymbol{E}_{\boldsymbol{m}_{k-1}} V_{\boldsymbol{m}_{k-2},m_{k-1}+1}\right\}$$
$$= \sum_{i=1}^{k-2} g_{m_i,m_{k-1}} + \min\left\{\frac{N+1}{m_{k-1}+1}\xi_{m_{k-1}} + u^{(1)}_{m_{k-1}+1}, u^{(2)}_{m_{k-1}+1}\right\}, \tag{6.59}$$

where $u^{(2)}_N = \infty$ and

$$\begin{aligned} u^{(2)}_{m_{k-1}} &= \boldsymbol{E}\left(\min\left\{\frac{N+1}{m_{k-1}+1}\xi_{m_{k-1}} + u^{(1)}_{m_{k-1}+1}, u^{(2)}_{m_{k-1}+1}\right\}\right) \\ &= \frac{1}{m_{k-1}} \sum_{j=1}^{m_{k-1}} \min\left\{\frac{N+1}{m_{k-1}+1}j + u^{(1)}_{m_{k-1}+1}, u^{(2)}_{m_{k-1}+1}\right\} \\ &= u^{(1)}_{m_{k-1}+1} + \frac{1}{m_{k-1}} \sum_{j=1}^{m_{k-1}} \min\left\{\frac{N+1}{m_{k-1}+1}j, u^{(2)}_{m_{k-1}+1} - u^{(1)}_{m_{k-1}+1}\right\}. \end{aligned} \tag{6.60}$$

Continuing this process for $i = k-2, \dots, 2$ we obtain

$$V_{\boldsymbol{m}_i} = \sum_{l=1}^{i-1} g_{m_l,m_i} + \min\left\{\frac{N+1}{m_i+1}\xi_{m_i} + u^{(k-i)}_{m_i+1}, u^{(k-i+1)}_{m_i+1}\right\}, \tag{6.61}$$

where $m_{i-1} < m_i \leq N-k+i$, $u^{(k-i+1)}_{N-k+i+1} = \infty$ and

$$u^{(k-i+1)}_{m_i} = u^{(k-i)}_{m_i+1} + \frac{1}{m_i} \sum_{j=1}^{m_i} \left\{\frac{N+1}{m_i+1}j, u^{(k-i+1)}_{m_i+1} - u^{(k-i)}_{m_i+1}\right\}. \tag{6.62}$$

From Theorem 2.3, $\boldsymbol{E}V^N_m = v_m$. Then

$$\begin{aligned} V_{m_1} &= \min\left\{\frac{N+1}{m_1+1}\xi_{m_1} + u^{(k-1)}_{m_1+1}, \boldsymbol{E}_{m_1} V_{m_1+1}\right\} \\ &= \min\left\{\frac{N+1}{m_1+1}\xi_{m_1} + u^{(k-1)}_{m_1+1}, v^N_{m_1+1}\right\}, \end{aligned} \tag{6.63}$$

where $1 \leq m_1 \leq N-k+1$, $v^N_{N-k+2} = \infty$,

$$\begin{aligned} v^N_{m_1} &= \boldsymbol{E}\left(\min\left\{\frac{N+1}{m_1+1}\xi_{m_1} + u^{(k-1)}_{m_1+1}, v^N_{m_1+1}\right\}\right) \\ &= \frac{1}{m_1}\sum_{j=1}^{m_1} \min\left\{\frac{N+1}{m_1+1}j, v^N_{m_1+1} - u^{(k-1)}_{m_1+1}\right\} + u^{(k-1)}_{m_1+1}. \end{aligned} \tag{6.64}$$

Therefore, the value of the sequence is given by

$$v^N = v_1^N = u_2^{(k-1)} + \min\left\{\frac{N+1}{2}, v_2^N - u_2^{(k-1)}\right\}, \quad N > 2.$$

Note that $u_n^{(r)}$ means the optimal expected summarized rank of r objects selected after the moment n.

Using the notations introduced by Chow et al. (1964), the optimal rule can be described in the following way. Denote

$$\delta_i^{(1)} = \left\lfloor \frac{i+1}{N+1} u_{i+1}^{(1)} \right\rfloor, \quad i = k, k+1, \ldots, N-1,$$

$$\delta_l^{(j)} = \left\lfloor \frac{l+1}{N+1}\left(u_{l+1}^{(j)} - u_{l+1}^{(j-1)}\right) \right\rfloor, \quad j = 2, \ldots, k-1,$$

where $\lfloor d \rfloor$ is the ceiling function, that is, the greatest integer less than or equal to d. Then, from (6.58),

$$\begin{aligned} u_{m_k}^{(1)} &= \frac{1}{m_k}\left(\frac{N+1}{m_k+1}(1 + \cdots + \delta_{m_k}^{(1)}) + (m_k - \delta_{m_k}^{(1)}) u_{m_k+1}^{(1)}\right) \\ &= \frac{1}{m_k}\left(\frac{N+1}{m_k+1} \cdot \frac{\delta_{m_k}^{(1)}(1 + \delta_{m_k}^{(1)})}{2} + (m_k - \delta_{m_k}^{(1)}) u_{m_k+1}^{(1)}\right). \end{aligned} \tag{6.65}$$

From (6.60) and (6.62),

$$\begin{aligned} u_{m_i}^{(k-i+1)} = u_{m_i+1}^{(k-i)} + \frac{1}{m_i}\Bigg(&\frac{N+1}{m_i+1} \cdot \frac{\delta_{m_i}^{(k-i+1)}\left(1 + \delta_{m_i}^{(k-i+1)}\right)}{2} \\ &+ \left(m_i - \delta_{m_i}^{(k-i+1)}\right)\left(u_{m_i+1}^{(k-i+1)} - u_{m_i+1}^{(k-i)}\right)\Bigg), \end{aligned} \tag{6.66}$$

for $i = 2, \ldots, k-1$. Finally, from (6.64), we have

$$\begin{aligned} v_{m_1}^N = u_{m_1+1}^{(k-1)} + \frac{1}{m_1}\Bigg(&\frac{N+1}{m_1+1} \cdot \frac{\delta_{m_1}^{(k)}\left(1 + \delta_{m_1}^{(k)}\right)}{2} + \\ &+ \left(m_1 - \delta_{m_1}^{(k)}\right)\left(v_{m_1+1}^{(N)} - u_{m_1+1}^{(k-1)}\right)\Bigg), \end{aligned} \tag{6.67}$$

where $\delta_{m_1}^{(k)} = \left\lfloor \frac{m_1+1}{N+1}\left(v_{m_1+1}^N - u_{m_1+1}^{(k-1)}\right)\right\rfloor$, $1 \le m_1 \le N-k$.

Put $\delta_N^{(1)} = N$, $\delta_{N-1}^{(2)} = N, \ldots, \delta_{N-k+2}^{(k-1)} = N$, $\delta_{N-k+1}^{(k)} = N$. From Theorem 2.2, (6.57), (6.59), (6.61), and (6.63), the optimal rule τ^* can be written as follows:

$$\tau_i^* = \min\{m_i > m_{i-1} : \xi_{m_i} \le \delta_{m_i}^{(k-i+1)}\}$$

on the set $D_{i-1} = \{\omega : \tau_1^* = m_1, \ldots, \tau_{i-1}^* = m_{i-1}\}$, where $D_0 = \Omega$.

Thus, the solution of this problem is the following optimal strategy: there exist integer vectors

$$\begin{aligned}
&\boldsymbol{\delta}^{(k)} = (\delta_1^{(k)}, \dots, \delta_{N-k+1}^{(k)}), \quad 0 \le \delta_1^{(k)} \le \dots \le \delta_{N-k}^{(k)} < \delta_{N-k+1}^{(k)} = N, \\
&\vdots \\
&\boldsymbol{\delta}^{(2)} = (\delta_{k-1}^{(2)}, \dots, \delta_{N-1}^{(2)}), \quad 0 \le \delta_{k-1}^{(2)} \le \dots \le \delta_{N-2}^{(2)} < \delta_{N-1}^{(2)} = N, \qquad (6.68) \\
&\boldsymbol{\delta}^{(1)} = (\delta_k^{(1)}, \dots, \delta_N^{(1)}), \quad 0 \le \delta_k^{(1)} \le \dots \le \delta_{n-1}^{(1)} < \delta_N^{(1)} = N, \\
&\delta_j^{(i_1)} \le \delta_j^{(i_2)}, \quad 1 \le i_1 < i_2 \le k, \quad k - i_1 + 1 \le j \le N - i_2 + 1
\end{aligned}$$

such that

$$\begin{aligned}
\tau_1^* &= \min\{m_1 : \xi_{m_1} \le \delta_{m_1}^{(k)}\}, \\
\tau_i^* &= \min\{m_i > m_{i-1} : \xi_{m_i} \le \delta_{m_i}^{(k-i+1)}\},
\end{aligned}$$

on the set $F_{i-1} = \{\omega : \tau_1^* = m_1, \dots, \tau_{i-1}^* = m_{i-1}\}, i = 2, \dots, k, F_0 = \Omega$.

Nikolaev (1976) showed that in the case of double choice problem,

$$\lim_{N\to\infty} v = 2\prod_{j=1}^{\infty}\left(\frac{j+2}{j}\right)^{1/(j+1)} \approx 7.739.$$

Table 6.8 displays the vectors $\boldsymbol{\delta}^{(k)}, \dots, \boldsymbol{\delta}^{(1)}$, and the values v, where (l_5, l_6), $(l_7, l_8) \in \{(1,2), (1,3), (2,2), (2,3)\}$.

Example 6.3. *If* $N = 6$, $k = 3$ *then the value of the game* $v = 8.4583$. *We can choose any optimal rule mentioned in the proper cell of Table 6.8. In particular,* $\delta^{(3)} = (1,2,3,6), \delta^{(2)} = (1,1,2,6), \delta^{(1)} = (1,2,2,6)$. *We have the following optimal rule:*

$$\begin{aligned}
\tau_1^* &= \min\{m_1 : \xi_{m_1} \le \delta_{m_1}^{(3)}\}, \\
\tau_2^* &= \min\{m_2 > m_1 : \xi_{m_2} \le \delta_{m_2}^{(2)}\}, \\
\tau_3^* &= \min\{m_3 > m_2 : \xi_{m_3} \le \delta_{m_3}^{(1)}\}.
\end{aligned}$$

If we observe the following sequence $a_1, \dots, a_6$*: 3, 4, 2, 6, 1, 5; then* $\xi_1, \dots, \xi_6$*: 1, 2, 1, 4, 1, 5. So we get* $m_1 = 1$ *(*$\xi_1 = 1 \le \delta_1^{(3)} = 1$*),* $m_2 = 3$ *(*$\xi_3 = 1 \le \delta_3^{(2)} = 1$*),* $m_3 = 5$ *(*$\xi_5 = 1 \le \delta_5^{(3)} = 2$*). Hence we obtain three best objects with summarized rank* $3 + 2 + 1 = 6$.

6.8.1 The double choice problem with minimal summarized rank

Let us consider in detail the double choice problem with minimal summarized rank. First, note that

$$v_i^N = \inf_{s \in \mathfrak{S}_i^{N-k+1}} EX_s = \inf_{s \in \mathfrak{S}_i^{N-k+1}} \boldsymbol{E}\left(\frac{N+1}{s+1}\xi_s + u_{s+1}^{(k-1)}\right),$$

TABLE 6.8
The vectors $\boldsymbol{\delta}^{(k)}, \dots, \boldsymbol{\delta}^{(1)}$ and the values v.

$N\backslash k$	2	3	4	5
3	$\boldsymbol{\delta}^{(2)} = (1,3)$ $\boldsymbol{\delta}^{(1)} = (1,3)$ $v = 3.6667$			
4	$\boldsymbol{\delta}^{(2)} = (0,1,4)$ $\boldsymbol{\delta}^{(1)} = (0,l_1,4)$ or $\boldsymbol{\delta}^{(2)} = (1,l_2,4)$ $\boldsymbol{\delta}^{(1)} = (1,l_3,4)$ $l_1,l_2,l_3 \in \{1,2\}$ $v = 4.3750$	$\boldsymbol{\delta}^{(3)} = (1,4)$ $\boldsymbol{\delta}^{(2)} = (1,4)$ $\boldsymbol{\delta}^{(1)} = (l_4,4)$ $l_4 \in \{1,2\}$ $v = 6.8750$		
5	$\boldsymbol{\delta}^{(2)} = (0,1,2,5)$ $\boldsymbol{\delta}^{(1)} = (0,1,2,5)$ $v = 4.6000$	$\boldsymbol{\delta}^{(3)} = (1,2,5)$ $\boldsymbol{\delta}^{(2)} = (1,2,5)$ $\boldsymbol{\delta}^{(1)} = (1,2,5)$ $v = 7.6000$	$\boldsymbol{\delta}^{(4)} = (1,5)$ $\boldsymbol{\delta}^{(3)} = (1,5)$ $\boldsymbol{\delta}^{(2)} = (2,5)$ $\boldsymbol{\delta}^{(1)} = (2,5)$ $v = 11.0500$	
6	$\boldsymbol{\delta}^{(2)} = (0,1,l_5,l_6,6)$ $\boldsymbol{\delta}^{(1)} = (0,1,l_7,l_8,6)$ $v = 4.9583$	$\boldsymbol{\delta}^{(3)} = (0,1,2,6)$ $\boldsymbol{\delta}^{(2)} = (0,l_5,l_6,6)$ $\boldsymbol{\delta}^{(1)} = (0,l_7,l_8,6)$ or $\boldsymbol{\delta}^{(3)} = (1,2,3,6)$ $\boldsymbol{\delta}^{(2)} = (1,l_5,l_6,6)$ $\boldsymbol{\delta}^{(1)} = (1,l_7,l_8,6)$ $v = 8.4583$	$\boldsymbol{\delta}^{(4)} = (1,2,6)$ $\boldsymbol{\delta}^{(3)} = (1,2,6)$ $\boldsymbol{\delta}^{(2)} = (l_5,l_6,6)$ $\boldsymbol{\delta}^{(1)} = (l_7,l_8,6)$ $v = 11.9583$	$\boldsymbol{\delta}^{(5)} = (1,6)$ $\boldsymbol{\delta}^{(4)} = (2,6)$ $\boldsymbol{\delta}^{(3)} = (2,6)$ $\boldsymbol{\delta}^{(2)} = (l_9,6)$ $\boldsymbol{\delta}^{(1)} = (l_{10},6)$ $l_9,l_{10} \in \{2,3\}$ $v = 16.2167$

where $\mathfrak{S}_m^n = \{s \in \mathfrak{S}_m : \boldsymbol{P}(s \le n) = 1\}$. Suppose $k = r$, then from (6.63), we have

$$v_{i,r}^N = \boldsymbol{E}\left(\min\left\{\frac{N+1}{i+1}\xi_i + u_{i+1}^{(r-1)}, v_{i+1,r}^N\right\}\right),$$

where $v_{N-r+2,r}^N = \infty$. Using (6.58), (6.60), and (6.62), we obtain $v_{i,r}^N = u_i^{(r)}$. Therefore, for $r = 1, 2, \dots, k$,

$$u_i^{(r)} = u_i^{(r)}(N) = \inf_{s \in \mathfrak{S}_i^{N-r+1}} \boldsymbol{E}\left(\frac{N+1}{s+1}\xi_s + u_{s+1}^{(r-1)}(N)\right), \tag{6.69}$$

where $u_m^{(0)} = 0$, $u_i^{(k)}(N) = v_i^N$.

Now assume that $k = 2$. In this case, $u_i^{(1)} = u_i^{(1)}(N)$ is the minimal expected rank of one object selected after moment i, whereas $u_i^{(2)} = u_i^{(2)}(N)$ is the minimal expected summarized rank of two objects.

Lemma 6.4. *The following inequalities hold.*

(1) $u_i^{(2)} \geq u_i^{(1)} + u_{i+1}^{(1)} \geq 2u_i^{(1)}$, $i = 1, 2, \ldots, N-1$.

(2) $u_{i+1}^{(2)} - u_i^{(2)} \geq u_{i+1}^{(1)} - u_i^{(1)}$, $i = 1, 2, \ldots, N-2$.

Proof:

(1) From (6.69),

$$u_i^{(2)} \geq \inf_{s \in \mathfrak{S}_i^N} \boldsymbol{E}\left(\frac{N+1}{s+1}\xi_s\right) + u_{i+1}^{(1)} = u_i^{(1)} + u_{i+1}^{(1)} \geq 2u_i^{(1)}.$$

(2) From (6.64) and (6.69), we can see that $\tau_{1,i} = \min\{s \in \mathfrak{S}_i^{N-1} : \xi_s \leq \delta_s^{(2)}\}$ is the optimal stopping time, that is,

$$u_i^{(2)} = \boldsymbol{E}\left(\frac{N+1}{\tau_{1,i}+1} y_{\tau_{1,i}} + u_{\tau_{1,i}+1}^{(1)}\right).$$

Similarly, using (6.58) and (6.69), we obtain

$$u_i^{(1)} = \boldsymbol{E}\left(\frac{N+1}{\tau_{2,i}+1}\xi_{\tau_{2,i}}\right),$$

where $\tau_{2,i} = \min\{s \in \mathfrak{S}_i^N : \xi_s \leq \delta_s^{(1)}\}$. It is clear that $\boldsymbol{P}(\tau_{1,i} \leq \tau_{2,i}) = 1$ since $\delta_i^{(1)} \leq \delta_i^{(2)}$. Then we can have one of the following two cases.

(i) $\tau_{1,i} \geq i$, $\tau_{2,i} > i$. In this case, $u_{i+1}^{(1)} - u_i^{(1)} = 0$, and $u_{i+1}^{(2)} - u_i^{(2)} \geq 0$.

(ii) $\tau_{1,i} = i$, $\tau_{2,i} = i$. Then

$$u_{i+1}^{(2)} - u_i^{(2)} = u_{i+1}^{(2)} - \boldsymbol{E}\left(\frac{N+1}{i+1}\xi_i\right) - u_{i+1}^{(1)},$$

$$u_{i+1}^{(1)} - u_i^{(1)} = u_{i+1}^{(1)} - \boldsymbol{E}\left(\frac{N+1}{i+1}\xi_i\right).$$

Using the inequality proved in the first part of this lemma, we have

$$\left(u_{i+1}^{(2)} - u_i^{(2)}\right) - \left(u_{i+1}^{(1)} - u_i^{(1)}\right) = u_{i+1}^{(2)} - 2u_{i+1}^{(1)} \geq 0.$$

□

Denote

$$\alpha_i^{(1)} = \frac{i+1}{N+1} u_{i+1}^{(1)}, \quad i = 0, 1, \ldots, N-1, \tag{6.70}$$

$$\alpha_i^{(2)} = \frac{i+1}{N+1}\left(u_{i+1}^{(2)} - u_{i+1}^{(1)}\right), \quad i = 0, 1, \ldots, N-2. \tag{6.71}$$

Lemma 6.5. *The following inequalities hold*

(1) For $N > 2$,

$$\alpha_0^{(2)} < \alpha_1^{(2)} < \cdots < \alpha_{N-2}^{(2)} < \frac{5}{8}(N-1) + \frac{3}{4}.$$

(2) For $i = 0, 1, \ldots, N-2$,

$$\frac{3(i+1)}{2(N-i+1)} \leq \alpha_i^{(2)} \leq \frac{4N}{N-i+3}.$$

Proof:

(1) From Lemma 6.4(2), it follows that

$$u_{i+1}^{(2)} - u_{i+1}^{(1)} \geq u_i^{(2)} - u_i^{(1)}.$$

Therefore, $\alpha_{i-1}^{(2)} < \alpha_i^{(2)}$. To show the last inequality, notice that

$$u_N^{(1)} = \frac{N+1}{2}, \quad \delta_{N-1}^{(1)} = \left\lfloor \frac{N}{2} \right\rfloor, \quad u_{N-1}^{(2)} = N+1.$$

Then, using (6.65), we obtain

$$u_{N-1}^{(1)} \geq \frac{N+1}{2(N-1)} \left(\frac{\frac{N}{2}\left(\frac{N}{2}-1\right)}{N} + \left(\frac{N}{2}-1\right)\right) \geq \frac{3}{8}(N-1).$$

Hence,

$$\alpha_{N-2}^{(2)} \leq \frac{N-1}{N+1}\left(N+1-\frac{3}{8}N+\frac{3}{8}\right) < \frac{5}{8}(N-1)+\frac{3}{4}.$$

(2) The first inequality directly follows from Lemma 2 Chow et al. (1964), which states that

$$\alpha_i^{(1)} \geq \frac{3(i+1)}{2(N-i+2)}.$$

Then, using (6.70), (6.71) and Lemma 6.4(1), we have

$$\alpha_i^{(2)} \geq \frac{i+1}{N+1} u_{i+2}^{(1)} = \frac{i+1}{i+2} a_{i+1}^{(1)} \geq \frac{3(i+1)}{2(N-i+1)}.$$

Now let us prove the second inequality. Using (6.70) and (6.67), we can write that

$$\alpha_{i-1}^{(2)} = \frac{i}{i+1}\alpha_i^{(1)} - \alpha_{i-1}^{(1)} + \frac{\delta_i^{(2)}(\delta_i^{(2)}+1) + 2(i-\delta_i^{(2)})\alpha_i^{(2)}}{2(i+1)}. \tag{6.72}$$

Taking into account that $\delta_j^{(2)} = \left\lfloor \alpha_i^{(2)} \right\rfloor = \alpha_i^{(2)} - \alpha$, with $0 \leq \alpha < 1$,

$$\alpha_{i-1}^{(2)} = \frac{i}{i+1}\alpha_i^{(1)} - \alpha_{i-1}^{(1)} + \frac{\alpha_i^{(2)}(1+2i-\alpha_i^{(2)})}{2(i+1)} - \frac{\alpha(1-\alpha)}{2(i+1)}. \tag{6.73}$$

Let us show that

$$\frac{i}{i+1}\alpha_i^{(1)} - \alpha_{i-1}^{(1)} \le \frac{i}{i+1}\alpha_i^{(2)} - \alpha_{i-1}^{(2)}. \tag{6.74}$$

It is clear that

$$\frac{i}{i+1}\alpha_i^{(1)} - \alpha_{i-1}^{(1)} = \frac{i}{N+1}(u_{i+1}^{(1)} - u_i^{(1)}), \tag{6.75}$$

$$\frac{i}{i+1}\alpha_i^{(2)} - \alpha_{i-1}^{(2)} = \frac{i}{N+1}(u_{i+1}^{(2)} - u_{i+1}^{(1)} - u_i^{(2)} + u_i^{(1)}). \tag{6.76}$$

If $\tau_{1,i} \ge i$, $\tau_{2,i} > i$ (see case (i) in the proof of Lemma 6.4(2)),

$$\frac{i}{i+1}\alpha_i^{(1)} - \alpha_{i-1}^{(1)} = 0,$$

$$\frac{i}{i+1}\alpha_i^{(2)} - \alpha_{i-1}^{(2)} = \frac{i}{N+1}(u_{i+1}^{(2)} - u_i^{(2)}) \ge 0.$$

This means that (6.74) holds.

Assume now that $\tau_{1,i} = i$, $\tau_{2,i} = i$ (see case (ii) in the proof of Lemma 6.4(2)). Then

$$u_i^{(1)} = \boldsymbol{E}\left(\frac{N+1}{i+1}y_i\right),$$

$$u_i^{(2)} = \boldsymbol{E}\left(\frac{N+1}{i+1}y_i\right) + u_{i+1}^{(1)},$$

which gives us $u_i^{(2)} - u_i^{(1)} = u_{i+1}^{(1)}$. From Lemma 6.4(1), $u_{i+1}^{(2)} - u_{i+1}^{(1)} \ge u_{i+2}^{(1)}$. Using this, (6.76) gives the following inequality

$$\frac{i}{i+1}\alpha_i^{(2)} - \alpha_{i-1}^{(2)} \ge \frac{i}{N+1}(u_{i+2}^{(1)} - u_{i+1}^{(1)}).$$

Combining the obtained inequality and (6.75), we can see that (6.74) holds if

$$u_{i+2}^{(1)} - u_{i+1}^{(1)} \ge u_{i+1}^{(1)} - u_i^{(1)}.$$

From (6.65), we have

$$\begin{aligned}(u_{i+2}^{(1)} - u_{i+1}^{(1)}) - (u_{i+1}^{(1)} - u_i^{(1)}) &= \frac{\delta_{i+1}^{(1)}}{i+1}u_{i+2}^{(1)} - \frac{\delta_i^{(1)}}{i}u_{i+1}^{(1)} \\ &+ \frac{N+1}{i(i+1)}\frac{\delta_i^{(1)}(\delta_i^{(1)}+1)}{2} \\ &- \frac{N+1}{(i+1)(i+2)}\frac{\delta_{i+1}^{(1)}(\delta_{i+1}^{(1)}+1)}{2}.\end{aligned} \tag{6.77}$$

Denote $W = (u_{i+2}^{(1)} - u_{i+1}^{(1)}) - (u_{i+1}^{(1)} - u_i^{(1)})$. Let us show that $W \ge 0$.

First, assume $\delta_i^{(1)} = \delta_{i+1}^{(1)} > 0$. In case if $\delta_i^{(1)} = \delta_{i+1}^{(1)} = 0$, then (6.74) is trivial. From (6.70) and (6.77),

$$\begin{aligned} W &= \delta_i^{(1)} \left(\frac{u_{i+2}^{(1)}}{i+1} - \frac{u_{i+1}^{(1)}}{i} + \frac{N+1}{2(i+1)} \left(\frac{\delta_i^{(1)}+1}{i} - \frac{\delta_i^{(1)}+1}{i+2} \right) \right) \\ &\geq \delta_i^{(1)} \left(-\frac{u_{i+2}^{(1)}}{i(i+1)} + \frac{N+1}{i+1} \cdot \frac{\delta_i^{(1)}+1}{i(i+2)} \right) \\ &= \frac{\delta_i^{(1)}(N+1)}{i(i+1)(i+2)} \left(\delta_i^{(1)} + 1 - \frac{i+2}{N+1} u_{i+2}^{(1)} \right) \\ &= \frac{\delta_i^{(1)}(N+1)}{i(i+1)(i+2)} \left(\delta_i^{(1)} + 1 - \alpha_{i+1}^{(1)} \right). \end{aligned}$$

From the definition of $\alpha_{i+1}^{(1)}$, we can see that $\alpha_{i+1}^{(1)} < \delta_{i+1}^{(1)} + 1$. Therefore, $W > 0$. Using (6.70), (6.77) can be rewritten in the following way:

$$\begin{aligned} W &= \frac{\delta_{i+1}^{(1)}}{i+1} \left(u_{i+2}^{(1)} - \frac{N+1}{i+2} \cdot \frac{\delta_{i+1}^{(1)}+1}{2} \right) - \frac{\delta_i^{(1)}}{i} \left(u_{i+1}^{(1)} - \frac{N+1}{i+1} \cdot \frac{\delta_i^{(1)}+1}{2} \right) \\ &= \frac{N+1}{i+1} \left(\frac{\delta_{i+1}^{(1)}}{i+2} \left(\alpha_{i+1}^{(1)} - \frac{\delta_{i+1}^{(1)}+1}{2} \right) - \frac{\delta_i^{(1)}}{i} \left(\alpha_i^{(1)} - \frac{\delta_i^{(1)}+1}{2} \right) \right). \end{aligned} \tag{6.78}$$

Since $\delta_i^{(1)} = \delta_{i+1}^{(1)}$, then

$$W = \frac{\delta_i^{(1)}(N+1)}{i+1} \left(\frac{1}{i+2} \left(\alpha_{i+1}^{(1)} - \frac{\delta_i^{(1)}+1}{2} \right) - \frac{1}{i} \left(\alpha_i^{(1)} - \frac{\delta_i^{(1)}+1}{2} \right) \right).$$

We can see that $W > 0$ if $\delta_i^{(1)} = \delta_{i+1}^{(1)} > 0$. This means that

$$\frac{1}{i+2} \left(\alpha_{i+1}^{(1)} - \frac{\delta_i^{(1)}+1}{2} \right) - \frac{1}{i} \left(\alpha_i^{(1)} - \frac{\delta_i^{(1)}+1}{2} \right) > 0 \tag{6.79}$$

This inequality also holds in case if $\delta_i^{(1)} < \delta_{i+1}^{(1)}$.

Next, assume that $\delta_{i+1}^{(1)} = \delta_i^{(1)} + d$, $d \geq 1$. From (6.78),

$$\begin{aligned} W &= \frac{N+1}{i+1} \left(\frac{\delta_i^{(1)}+d}{i+2} \left(\alpha_{i+1}^{(1)} - \frac{\delta_i^{(1)}+1}{2} - \frac{d}{2} \right) - \frac{\delta_i^{(1)}}{i} \left(\alpha_i^{(1)} - \frac{\delta_i^{(1)}+1}{2} \right) \right) \\ &= \frac{N+1}{i+1} \left(\frac{\delta_i^{(1)}}{i+2} \left(\alpha_{i+1}^{(1)} - \frac{\delta_i^{(1)}+1}{2} \right) - \frac{\delta_i^{(1)}}{i} \left(\alpha_i^{(1)} - \frac{\delta_i^{(1)}+1}{2} \right) \right. \\ &\quad \left. + \frac{d}{i+2} \left(\alpha_{i+1}^{(1)} - \left(\delta_i^{(1)} + \frac{d+1}{2} \right) \right) \right). \end{aligned} \tag{6.80}$$

Since $\delta_{i+1}^{(1)} = \delta_i^{(1)} + d \le \alpha_{i+1}^{(1)}$,

$$\alpha_{i+1}^{(1)} - \left(\delta_i^{(1)} + \frac{d+1}{2}\right) \ge \alpha_{i+1}^{(1)} - (\delta_i + d) \ge 0, \; d \ge 1.$$

Using (6.79), we obtain that $W \ge 0$. This means that (6.74) is proved. From (6.73),

$$\alpha_{i-1}^{(2)} \le \frac{\alpha_i^{(2)}(1 + 4i - \alpha_i^{(2)})}{4(1+i)} - \frac{\alpha(1-\alpha)}{4(1+i)}.$$

We can easily show that

$$h(\alpha_i^{(2)}) = \frac{\alpha_i^{(2)}(1 + 4i - \alpha_i^{(2)})}{4(1+i)}$$

is an increasing function when $\alpha_i^{(2)} \le 2i + \frac{1}{2}$. Also,

$$\alpha_{i-1}^{(2)} \le h(\alpha_i^{(2)}). \tag{6.81}$$

Now the lemma can be proved by backward induction. If $i = N - 2$, then the lemma follows from Lemma 6.5(1) with $\alpha_{N-2}^{(2)} < \frac{5}{8}(N-1) + \frac{4}{5}N$. Assume now that the second inequality in Lemma 6.5(2) holds for $1 \le i \le N-2$. Let us show the inequality also holds for $i-1$. From Lemmas 6.4(2) and 6.5(1), we have

$$\alpha_{i-1}^{(2)} = \frac{i}{N+1}(u_i^{(2)} - u_i^{(1)}) \le \frac{i}{N+1}(u_{N-2}^{(2)} - u_{N-2}^{(1)}) < \frac{5}{8}i + \frac{3}{4}.$$

If $\frac{5}{8}i + \frac{3}{4} \le \frac{4N}{N-i+4}$, then $2i + \frac{1}{2} \ge \frac{4N}{N-i+3}$. Since $h(\alpha_i^{(2)})$ increases when $\alpha_i^{(2)} \le 2i + \frac{1}{2}$, then from (6.81), we have

$$\alpha_{i-1}^{(2)} \le h(\alpha_i^{(2)}) \le h\left(\frac{4N}{N-i+3}\right) \le \frac{4N}{N-i+4}.$$

This concludes the proof.

□

Lemma 6.6. *The value of the problem* $v_1^N < 20$.

Proof: From Lemmas 6.4(2) and 6.5(2), we have

$$v_1^N - v_1^{(1)} \le v_i^N - v_i^{(1)} = \frac{N+1}{i}\alpha_{i-1}^{(2)} \le \frac{N+1}{i}\frac{4N}{N-i+4} < 16,$$

where $i = \lfloor \frac{N}{2} \rfloor$. Hence, $v_1^N < v_1^{(1)} + 16 < 20$ since $v_1^{(1)} < 4$ (see Chow et al. (1964)).
□

For $r = 1, 2, \ldots, N$, denote

$$i_r^{(m)} = \min\{j \ge 1 : r \le \delta_j^{(m)}\}, \quad m = 1, 2. \tag{6.82}$$

Lemma 6.7. *The sequences* $i_r^{(1)}$ *and* $i_r^{(2)}$ *have the following properties.*

(1) $i_r^{(2)} \leq i_r^{(1)}$.

(2)

$$\frac{i_1^{(2)}}{N} > \frac{1}{32}, \quad N \geq 30. \tag{6.83}$$

$$1 - \frac{4}{r} < \frac{i_r^{(2)}}{N} \leq 1 - \frac{1}{2r}, \quad N \geq 12r, \tag{6.84}$$

(3)

$$\lim_{N\to\infty} \alpha^{(2)}_{i_r^{(2)}} = \lim_{N\to\infty} \alpha^{(2)}_{i_r^{(2)}-1} = r. \tag{6.85}$$

Proof:

(1) The inequality directly follows from the fact that $\delta_j^{(1)} \leq \delta_j^{(2)}, j = 1, 2, \ldots, N-1$.

(2) If $i_1^{(2)} \geq \lfloor \frac{N}{2} \rfloor$, then $\frac{i_1^{(2)}}{N} > \frac{1}{32}$. If $i_1^{(2)} < \lfloor \frac{N}{2} \rfloor$, using (6.82), we obtain

$$\begin{aligned} 1 \leq \delta^{(2)}_{i_1^{(2)}} \leq \alpha^{(2)}_{i_1^{(2)}} &= \frac{i_1^{(2)}+1}{N+1}\left(u^{(2)}_{i_1^{(2)}+1} - u^{(1)}_{i_1^{(2)}+1}\right) \\ &\leq \frac{i_1^{(2)}+1}{N+1}\left(u^{(2)}_{\lfloor \frac{N}{2} \rfloor} - u^{(1)}_{\lfloor \frac{N}{2} \rfloor}\right) < \frac{i_1^{(2)}+1}{N+1}16. \end{aligned}$$

This proves (6.83).

Let us now prove the double inequality (6.84). Since $\delta^{(2)}_{i_r^{(2)}} \geq r$, then $\alpha^{(2)}_{i_r^{(2)}} \geq r$ and $\frac{4N}{N-i_r^{(2)}+3} \geq r$. This yields the left inequality $\frac{i_r^{(2)}}{N} > 1 - \frac{4}{r}$. The right inequality follows from $\frac{i_r^{(1)}}{N} \leq 1 - \frac{1}{2r}$, $N \geq 12r$ (see Chow et al. (1964)) and the first part of this lemma, that is, $i_r^{(2)} \leq i_r^{(1)}$.

(3) From Chow et al. (1964), we have

$$\frac{\alpha_i^{(1)}(1 + 2i - \alpha_i^{(1)})}{2(i+1)} - \frac{\alpha_1(1-\alpha_1)}{2(i+1)} \geq \frac{i}{i+1}\alpha_i^{(1)}\left(1 - \frac{\alpha_i^{(1)}}{2(i+1)}\right),$$

where $\alpha_1 = \alpha_i^{(1)} - [\alpha_i^{(1)}]$. Using Lemma 6.5(1), (6.73) and (6.75), we obtain

$$\begin{aligned} 0 < \alpha^{(2)}_{i_r^{(2)}} - \alpha^{(2)}_{i_r^{(2)}-1} &\leq \alpha^{(2)}_{i_r^{(2)}} - \left(\frac{i_r^{(2)}}{i_r^{(2)}+1}\alpha^{(1)}_{i_r^{(2)}} - \alpha^{(1)}_{i_r^{(2)}-1}\right) \\ &- \frac{i_r^{(2)}}{i_r^{(2)}+1}\alpha^{(2)}_{i_r^{(2)}}\left(1 - \frac{\alpha^{(2)}_{i_r^{(2)}}}{2(i_r^{(2)}+1)}\right) \\ &\leq \frac{(1+\alpha^{(2)}_{i_r^{(2)}})^2}{2(i_r^{(2)}+1)} - \frac{i_r^{(2)}}{N+1}(u^{(1)}_{i_r^{(2)}+1} - u^{(1)}_{i_r^{(2)}}). \end{aligned}$$

From Lemma 6.5(2), (6.84) and (6.83), we have

$$0 < \alpha^{(2)}_{i_r^{(2)}} - \alpha^{(2)}_{i_r^{(2)}-1} \leq \frac{\left(1+\frac{4N}{N-i_r^{(2)}+3}\right)^2}{2(i_r^{(2)}+1)} < \frac{(1+8r)^2}{\frac{1}{16}}\frac{1}{N},$$

with $N \geq 30$. Then, using

$$\alpha^{(2)}_{i_r^{(2)}-1} < r \leq \alpha^{(2)}_{i_r^{(2)}},$$

gives us (6.85).

□

The following theorem gives the asymptotic value of the double choice problem with minimal summarized rank as $N \to \infty$.

Theorem 6.8. *Let v_1^N be the value of the double choice problem. Then*

$$\lim_{N\to\infty} v_1^N = 2 \lim_{N\to\infty} u_1^{(1)}(N) = 2\prod_{j=1}^{\infty}\left(\frac{j+2}{j}\right)^{\frac{1}{j+1}} \approx 7.739.$$

Proof: From (6.85) and the inequalities $r \leq \delta^{(2)}_{i_r^{(2)}} \leq \alpha^{(2)}_{i_r^{(2)}}$, we can see that $\delta^{(2)}_{i_r^{(2)}} = r$ for large enough values of N. It is clear that

$$\lim_{N\to\infty}\left(\alpha^{(2)}_{i_{r+1}^{(2)}} - \alpha^{(2)}_{i_r^{(2)}}\right) = 1,$$

$$\lim_{N\to\infty}\left(\alpha^{(2)}_{i_r^{(2)}} - \alpha^{(2)}_{i_r^{(2)}-\lambda}\right) = 0, \quad \lambda = 1, 2, \ldots .$$

Therefore,

$$\lim_{N\to\infty}\left(i_{r+1}^{(2)} - i_r^{(2)}\right) = \infty.$$

Let N be large enough so that $\delta^{(2)}_{i_r^{(2)}} = r$, $\delta^{(2)}_{i_{r+1}^{(2)}} = r+1$, and $i_{r+1}^{(2)} - r \geq i_r^{(2)} + 3$, where r is fixed. Denote

$$R(i) = \frac{i}{i+1}\alpha_i^{(1)} - \alpha_{i-1}^{(1)}.$$

From (6.72),

$$\alpha^{(2)}_{i_{r+1}^{(2)}-2} = R(i_{r+1}^{(2)}-1) + \frac{r(r+1) + 2(i_{r+1}^{(2)} - 1 - r)\alpha^{(2)}_{i_{r+1}^{(2)}-1}}{2i_{r+1}^{(2)}}.$$

Let $\gamma_i = \alpha_i^{(2)} - \frac{r}{2}$. Then

$$\begin{aligned}
\gamma_{i_{r+1}^{(2)}-1} &= \frac{i_{r+1}^{(2)}}{i_{r+1}^{(2)} - r - 1}\gamma_{i_{r+1}^{(2)}-2} - \frac{i_{r+1}^{(2)}}{i_{r+1}^{(2)} - r - 1} R(i_{r+1}^{(2)} - 1) \\
&= \frac{i_{r+1}^{(2)}}{i_{r+1}^{(2)} - r - 1} \cdot \frac{i_{r+1}^{(2)} - 1}{i_{r+1}^{(2)} - r - 2} \cdot \ldots \cdot \frac{i_r^{(2)} + 2}{i_r^{(2)} + 1 - r}\gamma_{i_r^{(2)}} \\
&- \frac{i_{r+1}^{(2)}}{i_{r+1}^{(2)} - r - 1} \cdot \frac{i_{r+1}^{(2)} - 1}{i_{r+1}^{(2)} - r - 2} \cdot \ldots \cdot \frac{i_r^{(2)} + 2}{i_r^{(2)} + 1 - r} R(i_r^{(2)} + 1) - \ldots \\
&- \frac{i_{r+1}^{(2)}}{i_{r+1}^{(2)} - r - 1} \cdot \frac{i_{r+1}^{(2)} - 1}{i_{r+1}^{(2)} - r - 2} R(i_{r+1}^{(2)} - 2) - \frac{i_{r+1}^{(2)}}{i_{r+1}^{(2)} - r - 1} R(i_{r+1}^{(2)} - 1) \\
&= \prod_{j=1}^{r+1} \frac{i_{r+1}^{(2)} + j - r - 1}{i_r^{(2)} + j - r}\gamma_{i_r^{(2)}} - \prod_{j=1}^{r+1} \frac{i_{r+1}^{(2)} + j - r - 1}{i_r^{(2)} + j - r} R(i_r^{(2)} + 1) \\
&- \frac{i_{r+1}^{(2)}}{i_{r+1}^{(2)} - r - 1} \cdot \frac{i_{r+1}^{(2)} - 1}{i_{r+1}^{(2)} - r - 2} \cdot \ldots \cdot \frac{i_r^{(2)} + 3}{i_r^{(2)} - r + 2} R(i_r^{(2)} + 2) - \ldots \\
&- \frac{i_{r+1}^{(2)}}{i_{r+1}^{(2)} - r - 1} R(i_{r+1}^{(2)} - 1).
\end{aligned}$$

For a fixed r,

$$\prod_{j=1}^{r+1} \lim_{N\to\infty} \frac{i_{r+1}^{(2)} + j - r - 1}{i_r^{(2)} + j - r} = \lim_{N\to\infty} \left(\frac{i_{r+1}^{(2)}}{i_r^{(2)}}\right)^{r+1} < \infty.$$

From Chow et al. (1964), we have

$$\lim_{N\to\infty} (\alpha_i^{(1)} - \alpha_{i-1}^{(1)}) = 0 \quad \text{on the set} \quad \left\{0 < a \le \frac{i}{N} \le b < 1\right\}.$$

Then, substituting $\alpha_i^{(2)} - \frac{r}{2}$ instead of γ_i, from (6.85), we obtain as $N \to \infty$,

$$\left(\frac{r+2}{r}\right)^{\frac{1}{r+1}} = \lim_{N\to\infty} \frac{i_{r+1}^{(2)}}{i_r^{(2)}}.$$

Then the proof of

$$\lim_{N\to\infty} \frac{i_1^{(2)}}{N} = \prod_{j=1}^{\infty} \left(\frac{j}{j+2}\right)^{\frac{1}{j+1}} \tag{6.86}$$

will follow the proof of

$$\lim_{N\to\infty} \frac{i_1^{(1)}}{N} = \prod_{j=1}^{\infty} \left(\frac{j}{j+2}\right)^{\frac{1}{j+1}}$$

in Chow et al. (1964). From (6.82),

$$\delta_1^{(2)} = \delta_2^{(2)} = \cdots = \delta_{i_1^{(2)}-1}^{(2)} = 0.$$

Then, using (6.62), we have

$$u_1^{(2)} = \cdots = u_{i_1^{(2)}}^{(2)},$$

and, similarly,

$$u_1^{(1)} = \cdots = u_{i_1^{(1)}}^{(1)}.$$

From Lemma 6.7(1), $i_1^{(2)} \leq i_1^{(1)}$, and, therefore,

$$\begin{aligned} 1 &= \lim_{N\to\infty} \alpha_{i_1^{(2)}-1}^{(2)} = \lim_{N\to\infty} \left(\frac{i_1^{(2)}}{N} \cdot \left(u_{i_1^{(2)}}^{(2)} - u_{i_1^{(2)}}^{(1)} \right) \right) \\ &= \lim_{N\to\infty} \frac{i_1^{(2)}}{N} \cdot \lim_{N\to\infty} \left(u_1^{(2)} - u_1^{(1)} \right). \end{aligned}$$

From (6.86), we have

$$\lim_{N\to\infty} \left(u_1^{(2)} - u_1^{(1)} \right) = \prod_{j=1}^{\infty} \left(\frac{j+2}{j} \right)^{\frac{1}{j+1}}.$$

Finally, using the result from Chow et al. (1964),

$$\lim_{N\to\infty} u_1^{(1)}(N) = \prod_{j=1}^{\infty} \left(\frac{j+2}{j} \right)^{\frac{1}{j+1}},$$

we obtain

$$\lim_{N\to\infty} v_1^N = \lim_{N\to\infty} u_1^{(2)}(N) = 2\prod_{j=1}^{\infty} \left(\frac{j+2}{j} \right)^{\frac{1}{j+1}}.$$

This concludes the proof. □

6.8.2 Application to online auction

Kleinberg (2005) proposed the following application of the generalized secretary problem. In the secretary problem, a set S of numbers is presented to *an on-line algorithm* in random order. At any time, the algorithm may stop and choose the current element, and the goal is to **maximize the probability of choosing the largest element in the set**. The generalization that is taken into account studies a variation in which the algorithm is allowed to choose k elements of S, and the goal is to maximize their sum. Based on the assumption that elements of S are different, it is equivalent to selecting k elements of S with the highest ranks. Let $A = \{1, 2, \ldots, k\}$, $\mathcal{P}_k$ the set of all permutation of A, and $\xi_k = \mathbf{card}\{1 \le j \le n : X_j > X_k\}$ the absolute rank of kth observation. The decision-maker is looking for k relative rank-based stopping times $\tau_j^\star$, $j = 1, 2, \ldots, k$ such that

$$\boldsymbol{P}\{\omega : (\xi_{\tau_1^\star}, \xi_{\tau_2^\star}, \ldots, \xi_{\tau_k^\star}) \in \mathcal{P}_k\} = \sup_{\tau_1 < \tau_2 < \ldots < \tau_k} \boldsymbol{P}\{\omega : (\xi_{\tau_1}, \xi_{\tau_2}, \ldots, \xi_{\tau_k}) \in \mathcal{P}_k\}. \tag{6.87}$$

The optimal strategy for the issue with given k has been constructed by Nikolaev (1976,1977), Platen (1980), Bruss and Louchard (2016), Preater (2006). It's worth emphasizing here the systematic analysis of the consequences of successive state choices in the generalized task of selecting the best object to the case of optimally assembling a set of predefined (desired) objects conducted by Preater (1994a, 1994b, 1991). A special case of the issue is discussed in the presentation of the solution selection two best by two stops (v. Sections 6.1, 6.6.1, 6.7). For analysis of the online auctions it was used the heuristic algorithm whose competitive ratio is $1 - O(\sqrt{1/k})$. To our knowledge, this is the first algorithm whose competitive ratio approaches 1 as $k \to \infty$. As an application we solve an open problem in the theory of online auction mechanisms.

6.9 Multilateral Selection of Secretaries

Almost thirty years have passed since Sakaguchi's attempt to organize his knowledge about multi-person tasks, the common feature of which is to meet individual needs, defined by a fixed deposit function, by selecting options from a sequential sequence. The considered models are not identical to Dynkin's games, although some of them lead to the game class with stopping stochastic processes. In his work, Sakaguchi (1995) distinguished the following features of decision problems that distinguish these problems and allow for a uniform description of them (he considered two person games only).

1. The decision problem is either two-person zero-sum game (2PZSG) or two-person non-zero-sum game (2PnZSG). *In further considerations, we will also take into account the problems of many* ***Decision-Makers (DM)*** *(NPG)).*
2. Decision-makers (**DM**) observe the components of the vector sequence, where we distinguish between cases, each gathers different knowledge, because it observes selected components of the incoming vectors (A) and tasks in which everyone has a full observation of the components (B). Sakaguchi's work focused on two specific models of the analyzed objects:

   2A: The observed objects are described by two features $\{(X_i, Y_i)\}_{i=1}^n$. The first **DM** focuses attention on X and the second on Y. In the analyses, the case of dependent features and independent features are examined separately. Item once accepted cannot be called later.

2B: The observed objects are characterized by one feature that is interesting for both decision-makers. An object can only be accepted (selected) by one decision-maker, who in the future cannot replace it with a "newer one". If both are not interested in the object and reject it, there will be no return to it in the future.

3. Each **DM** has his objective, according to which his payoff is given. The objectives can depend on both **DM**'s choice.

Analysis of possible cases, when the number of decision-makers is greater than 2, is wider. We will base the basic division on Sakaguchi's proposal. In the case of **2B** there is a problem of allocating the object indicated by both DMs simultaneously to one of them. In most of such situations, a mechanism is introduced to determine the priority of participants in the decision-making process. In our considerations, such a solution will be a special case of the proposed solution, although we believe that prioritizing, especially using a random mechanism, is sometimes more appropriate.

The considered models in Sakaguchi's works, selected from the work that were current at that time, have specific natural decision-making goals that allow for the definition of a natural solution to the problem. With many decision-makers, the goal is to establish criteria for their decisions and to make assumptions about their rationality. The rationality of players is usually the desire to maximize their own win or utility function and the ability to evaluate the effects of their own, and other decision-makers, decisions. In further analysis, we will use the game theory apparatus to address the issue of rationality of decision-makers. We will attach the priority to von Stackelberg's (1934) approach (cf. Colman and Stirk (1998), He et al. (2017)).

6.9.1 Research inspiring issue

In computer science, High-Throughput Computing (HTC), High Performance Computing (HPC) is the use of many computing resources over long periods of time to accomplish a computational task. Some **HTC** systems, such as **HT**Condor and PBS[1], can run tasks on opportunistic resources (alternative, sporadically available, dedicated to other tasks). However, operating in such an environment is a difficult problem.

The quasi-opportunistic management of supercomputers, unlike opportunistic computer networks, in which computing resources are used whenever they can be accessed, chooses the best resources to use. The premise for this approach is the observation that in various resource sharing networks, users often have access to opportunistic (available occasionally, dedicated to other tasks, but which can also perform other tasks) resources of dynamic quality. Since many user tasks are delay tolerant, this enables network users to expect an opportunistic resource and access it at the best quality so as to perform the task as favorably as possible. For such delay-resistant and opportunistic resource sharing networks, known resource access

[1] **Portable Batch System** (or simply **PBS**) is the name of computer software that performs job scheduling

strategies developed in the literature were developed with the following constraints (assumptions, rationale):

(i) focused mainly on one-user scenarios, while competing tasks reported by other users are ignored;

(ii) the possibility of interference by the provider (owner) of the resources used, who may take steps to implement its own resource sharing policy, is not taken into account;

(iii) the impact of the actions of both network users and the resource vendor on the dynamics of the resource quality is also not taken into account.

In order to overcome these limitations, the issue can be reduced to a multi-person decision problem in which we have a leader who has the ability to control access to resources, for which a game model taking into account different participant rights, based on a brand-controlled process and strategies that are stopping moments (leader-follower controlled Markov stopping game (LF-C-MSG)) seems to be perfect. In such a model, the solution is the Stackelberg equilibrium strategy in the LF-C-MSG, which can be used to control the behavior of both network users and the resource provider for better performance and resource utilization. Bearing in mind the model for such an issue, we will conduct an analysis of known models of selecting the best resources in the conditions of imperfect competition.

6.9.2 Mixed-type secretary games

An employer interviews a finite number n of applicants for a position (v. Sakaguchi and Szajowski (2000)). They are interviewed one by one sequentially in random order. As each applicant i is interviewed, two attributes are evaluated by the amounts X_i and Y_i, where X_i may be "talenta" (or quality), and Y_i may be the "look" (or degree of favorable impression) of the applicant. Suppose that $\{X_i\}_{i=1}^n$, $(\{Y_i\}_{i=1}^n)$ is under the condition of full (no)-information secretary problem and that X_i's and Y_i's are mutually independent. We consider the three kinds of the employer's object and for each of three cases the problem is formulated by dynamic programming, and the optimal policy is explicitly derived.

Problem 6.1. *[Description of the mixed-type secretary games.] An employer interviews a finite number n of applicants for a position. They are interviewed one by one sequentially in random order. Each applicant i has two attributes X_i, i.e. talent, ability or quality, and Y_i, i.e. the look or the degree of favorable impression. Suppose that $\{X_i\}_{i=1}^n$ is an iid sequence of r.v.s with common cdf $F(x)$, and $\{Y_i\}_{i=1}^n$ is the sequence of observable related ranks based on the second attribute. It is assumed that the two sequences $\{X_i\}_{i=1}^n$ and $\{Y_i\}_{i=1}^n$ are mutually independent. The employer observes (X_i, Y_i) sequentially one by one, as each applicant appears, and he must choose (=hire or stop) one applicant without recall (i.e. an applicant is once not hired, she is rejected and cannot be recalled later). There are three problems in which objective of the employer is to:*

1. *maximize the expected quality (i.e. X) of the applicant hired, with the condition that his (or her) look (i.e. Y) is the best among all applicants.*
2. *i.b.i.d., with the more generous condition that the look should be the best or the second best.*
3. *maximize the expected utility of the absolute rank in the look of the applicant hired, with the condition that his (or her) quality is highest among all applicants.*

The purpose of the section is to find the optimal hiring policy to achieve objectives: (1) $\sim$ (3) of Problem 6.1, by formulating and solving the suitable optimization problem.

Our assumption that X_i's and Y_i's are mutually independent is, of course, too much restrictive, and doesn't fit our real world. Looks and talent may be correlated. Dependence between the two attributes X and Y will introduce a new class of secretary problems different from that discussed in this paper (see Sakaguchi and Hamada (2000)).

Moreover the problems we consider here belong to a mixed-type of full-information (FI) times null-information (NI) secretary problems. We consider, also NI-times-NI and FI-times-FI types of secretary problems, in the similar context as discussed in the paper. The present work may be an alternative approach to the secretary problem for bivariate r.v.s than that which was tried in Sakaguchi and Saario (1994).

A very important and now classical literature in FI and NI secretary problems is Gilbert and Mosteller (1966). Recent look for the secretary problem and its extension can be found in Samuels (1991).

The above formulated tasks will be solved by the optimal stopping theory methods. Let us recall the formulation of such optimization problems. Let $(\xi_n, \mathcal{F}_n, \boldsymbol{P}_x)_{n=0}^N$ be a Markov process defined on the probability space $(\mathbb{E}, \mathcal{B})$. Define $g : \mathbb{E} \to \Re$ such that $\boldsymbol{E}_x|g(X_1)| < \infty$ and $\mathfrak{S}$ be the set of stopping times with respect to filtration $(\mathcal{F}_n)_{n=0}^N$. The optimal stopping problem can be formulated as determining of

$$v(x) = \sup_{\tau \in \mathfrak{S}} \boldsymbol{E}_x g(\xi_\tau)$$

and $\tau^* \in \mathfrak{S}$ such that $\boldsymbol{E}_x g(\xi_{\tau^*}) = v(x)$. The problems (1) $\sim$ (3) considered in this section can be reformulated as the optimal stopping problem of bidimensional Markov chain $\xi_n = (X_i, Y_i)$ with the gain function $g(x, y) = h_1(x)h_2(y)$. One of the coordinate is related to the no information best choice problem or its generalization. The description of the problem is as follows. Let $\mathbb{K} = \{x_1, x_2, \ldots, x_N\}$ be the set of characteristics, assuming that the values are different. The decision-maker observes a permutation $\{\eta_1, \eta_2, \ldots, \eta_N\}$ of the elements of $\mathbb{K}$ sequentially. All permutations are equally likely. Let R_k denote the absolute rank of the object with the characteristics η_k, i.e.

$$R_k = \min\{r : \eta_k = \bigwedge_{1 \le i_1 < \ldots < i_r \le N} \bigvee_{1 \le j \le r} \eta_{i_j}\},$$

($\wedge$ and $\vee$ denote minimum and maximum, respectively). The decisions at each time are based on relative ranks $Y_1, Y_2, \ldots, Y_N$ of the applicants, where

$$Y_k = \min\{r : \eta_k = \bigwedge_{1\le i_1<\ldots<i_r\le k} \bigvee_{1\le j\le r} \eta_{i_j}\}. \tag{6.88}$$

One can define the auxiliary payoff function

$$g_a(r,l) = \boldsymbol{P}\{R_r = a | Y_r = l\} = \frac{\binom{a-1}{l-1}\binom{N-a}{r-l}}{\binom{N}{r}}, \tag{6.89}$$

where $a = 1,2,\ldots,N$; $l = 1,2,\ldots,\min(a,r)$; $r = 1,2,\ldots,N$. The argument of the second coordinate is bi-dimensional. The first argument is the number of observation and the second one is the relative rank. The parameter a is the absolute rank of observation.

Let c_{R_i} be the payoff when we choose the object with absolute rank R_i. If we are using the strategies based on knowledge of the relative ranks then we have to construct the payoff measurable with respect to suitable filtration. Let $\xi_i = (X_i, Y_i)$, where Y_i is defined by (6.88). Denote $\mathcal{F}_i = \sigma(\xi_1, \xi_2, \ldots, \xi_i)$ and let $\tau \in \mathfrak{S}$. We have

$$\begin{aligned} \boldsymbol{E}c_{R_\tau} &= \sum_{i=1}^{n}\int_{\{\tau=i\}} c_{R_i}\, d\boldsymbol{P} = \sum_{i=1}^{n}\int_{\{\tau=i\}} \boldsymbol{E}[c_{R_i}|\mathcal{F}_i]\, d\boldsymbol{P} \\ &= \sum_{i=1}^{n}\int_{\{\tau=i\}} \boldsymbol{E}[c_{R_i}|Y_i]\, d\boldsymbol{P} = \sum_{i=1}^{n}\int_{\tau=i} \varphi(i,Y_i)\, d\boldsymbol{P} \\ &= \boldsymbol{E}\varphi(\tau, Y_\tau), \end{aligned} \tag{6.90}$$

where

$$\varphi(t,y) = \sum_{r=y}^{n} g_r(t,y)c_r. \tag{6.91}$$

Since the part of the gain function is related to the payoff in the no information secretary problem, it is convenient to consider the Markov chain with state space on which the gain function is strictly positive. Let us assume that the gain function is 0 for $l > \min(a,r)$ and positive for $l \le \min(a,r)$. It means the we are looking for the states (r,l) such that $l \le \min(a,r)$.

Let $W_1 = (1,Y_1) = (1,1)$ and for $t > 1$ let $\gamma_t = \inf\{r > \gamma_{t-1} : Y_r \le \min(a,r)\}$, ($\inf\emptyset = \infty$) and $W_t = (\gamma_t, Y_{\gamma_t})$. If $\gamma_t = \infty$, then we put $W_t = (\infty,\infty)$. The process W_t is the Markov chain with the following one step transition probabilities

$$\begin{aligned} p(r,s) &= P\{W_{t+1} = (s,l_s)|W_t = (r,l_r)\} \\ &= \begin{cases} \frac{1}{s}, & \text{for } r < a,\ s = r+1, \\ \frac{(r)_a}{(s)_{a+1}}, & \text{for } a \le r < s, \\ 0, & \text{for } r \ge s \text{ or } r < a,\ s \ne r+1, \end{cases} \end{aligned} \tag{6.92}$$

$$p(\infty,\infty) = 1,\ p(r,\infty) = 1 - a\sum_{s=r+1}^{N} p(r,s), \tag{6.93}$$

where $(s)_a = s(s-1)(s-2)\dots(s-a+1)$; $(s)_0 = 1$. Denote $\mathcal{G}_t = \sigma(X_{\gamma_1}, W_1, X_{\gamma_2}, W_2, \dots, X_{\gamma_t}, W_t)$ and $\mathfrak{M}$ the set of the stopping times with respect to the filtration $\{\mathcal{G}_t\}_{t=1}^n$.

6.9.3 Selecting good quality together with the best look

Let $\{(X_i, Y_i)\}_{i=1}^n$ be a sequence of independent bivariate r.v.s as given in the previous section. Observing the sequence (X_i, Y_i), $i = 1, 2, \dots, n$, one by one sequentially, we want to maximize $\boldsymbol{E}X_\tau c_{R_\tau}$, where τ is the stopping time belonging to $\mathfrak{S}$ and $c_1 = 1$ and $c_i = 0$ for $i = 2, 3, \dots, n$, i.e. we have $a = 1$ in (6.89)–(6.93). We have then by (6.90) and (6.91), for fixed horizon n, that for $\tau \in \mathfrak{S}$ there is $\sigma \in \mathfrak{M}$ such that

$$\boldsymbol{E}X_\tau c_{R_\tau} = \boldsymbol{E}\tilde{g}(X_{\gamma_\sigma}, W_\sigma) = \boldsymbol{E}g(X_\sigma, \gamma_\sigma). \tag{6.94}$$

We define state (x, i) which means that the i-th object has the first attribute $X_i = x$ and the second attribute $Y_i = 1$). The horizon is n and the stop reward at state (x, i) is, therefore,

$$g(x, i) = \frac{i}{n}x.$$

Denoting, by $v_n(x, i)$, the expected reward obtained by employing the optimal stopping rule for the n-object problem at state (x, i), we easily have, by (6.93) that

$$\boldsymbol{P}\{Y_{i+1}, \dots, Y_{j-1} \geq 2 \text{ and } Y_j = 1 | Y_i = 1\} = p(i, j) = \frac{i}{j(j-1)},$$

and

$$v_n(x, i) = \max\left\{\frac{i}{n}x, \sum_{j=i+1}^{n} \frac{i}{j(j-1)} \boldsymbol{E}_F v_n(X, j)\right\}, \tag{6.95}$$

$i = 1, 2, \dots, n$; $v_n(x, n) \equiv x$. Letting $V_{n,i} \equiv \frac{n}{i}\boldsymbol{E}_F v_n(X, i)$, and $d_{n,i} \equiv \sum_{j=i+1}^{n} \frac{V_{n,j}}{j-1}$, by equation (6.95) we have

$$V_{n,i} = \boldsymbol{E}_F(X \vee d_{n,i}) = S_F(d_{n,i}), \tag{6.96}$$

where $S_F(t)$ is defined by $S_F(t) \equiv \boldsymbol{E}_F(X \vee t)$.

Theorem 6.9. *The optimal stopping rule for the optimality equation (6.95) is : "Stop at the earliest object (X_i, Y_i) whose relative rank Y_i, when it appears, is unity and satisfies $X_i > d_{n,i}$". The sequence $\{d_{n,i}\}_{i=1}^n$ is determined by the recursion*

$$d_{n,i} = \frac{S_F(d_{n,i+1})}{i} + d_{n,i+1}, \tag{6.97}$$

$i = 1, 2, \dots, n-1$; $d_{n,n} = 0$.

The optimal expected reward is given by $\boldsymbol{E}_F v_n(X, 1)$, i.e. $S_F(d_{n,1})/n$.

Proof: The recursion (6.97) follows from the definition of $d_{n,i}$ and (6.96). □

TABLE 6.9
Decision points and values of the problem (Ex. 6.4).

i	$d_{10,i}$	$S_F(d_{10,i})$
10	0	0.5
9	0.0556	0.5015
8	0.1183	0.5070
7	0.1907	0.5182
6	0.2771	0.5384
5	0.3848	0.5740
4	0.5283	0.6396
3	0.7415	0.7749
2	1.1289	1.1289
1	2.2578	2.2578

Example 6.4. *Let $\{X_i\}_{i=1}^n$ be i.i.d. sequence of r.v.s with the uniform distribution on $[0,1]$ i.e. common cdf has a form $F(x) = x$ $(0 \leq x \leq 1)$ and $n = 10$. Then, since $S_F(t) = (1+t^2)/2$ for $0 \leq t \leq 1$; and $= t$ for $t > 1$, we have*

$$S_F(d_{10,i}) = \begin{cases} \frac{1+d_{10,i}^2}{2} & \text{if } 0 \leq d_{10,i} \leq 1, \\ d_{10,i} & \text{if } d_{10,i} > 1. \end{cases}$$

Hence, from (6.97), we get Table 6.9 which shows that the optimal rule is given by "Pass the first two objects and stop at the earliest object (X_i, Y_i), $3 \leq i \leq 10$, that satisfies $Y_i = 1$ and $X_i > d_{10,i}$". The expected reward obtained by employing this rule is $\frac{S_F(d_{10,1})}{10} \fallingdotseq 0.2258$.

Remark 18. *Consider the following two non-optimal stopping problems. The first one is to attempt to maximize $\boldsymbol{E}_F X_\tau$ and leave Ys in a random. This rule gives the expected reward $\frac{1}{n}v_n$, where v_n is given by Moser's sequence $v_n = (1+v_{n-1}^2)/2$, $(n \geq 1,\ v_0 = 0)$(see Remark 19 on p.254). The second one is to attempt to maximize the probability of selecting the best look and leave Xs in a random (see Gilbert and Mosteller (1966) Section 2a and Table 2). The optimal expected reward $\frac{1}{n}S_F(d_{n,1})$ in Theorem 6.9 is not smaller than $\max\{\frac{1}{n}v_n, E_F(X)\pi(s_n^*, n)\}$. We find that this is true since $\frac{1}{10}S_F(d_{10,1}) \fallingdotseq 0.2258$, $v_{10} \fallingdotseq 0.8611$ and $\pi(s_{10}^*, 10) \fallingdotseq 0.3987$.*

6.9.4 Selecting good quality together with one of the two bests in the look

We consider the same problem as in the previous section with a single difference that the required condition at the stopping time $\tau \in \mathfrak{S}$ is taken a little bit more generously. Let us denote by R_i, as in Section 6.9.3, the absolute rank of the i-th object in the second attribute and let $c_1 = 1$ for $i = 1, 2$ and $c_i = 0$ for $i = 3, 4, \ldots, n$. We can use formulae (6.89)–(6.93) with $a = 2$. The problem is to maximize $\boldsymbol{E}X_\tau c_{R_\tau}$, where

$\tau \in \mathfrak{S}$. We have that for $\tau \in \mathfrak{S}$ there is $\sigma \in \mathfrak{M}$ such that

$$\begin{aligned} \boldsymbol{E}X_\tau c_{R_\tau} &\equiv \boldsymbol{P}(R_\tau = 1 \text{ or } 2)\boldsymbol{E}[X_\tau | R_\tau = 1 \text{ or } 2] \\ &= \boldsymbol{E}\tilde{g}(X_{\gamma_\sigma}, W_\sigma) = \boldsymbol{E}g(X_{\gamma_\sigma}, \gamma_\sigma, Y_{\gamma_\sigma}). \end{aligned}$$

When horizon is n we are in state (x, i, k) when $X_i = x$ and the second attribute has relative rank $Y_i = k$. The stop reward is

$$g(x,i,k) = \begin{cases} \boldsymbol{P}(R_i = 1 \text{ or } 2 | Y_i = 1)x = \frac{i(2n-i-1)x}{n(n-1)}, & \text{for } k = 1, \\ \boldsymbol{P}(R_i = 2 | Y_i = 2)x = \frac{i(i-1)x}{n(n-1)}, & \text{for } k = 2, \end{cases}$$

which follows from (6.89).

Denote, by $v_n(x, i)$, the expected reward obtained by using the optimal stopping rule for the n-period problem at state $(x, i, 1)$ and denote by $u_n(x)$, the similar one at the state $(x, i, 2)$. We find that

$$v_n(x,i) = \max\left\{\frac{i(2n-i-1)}{n(n-1)}x, \Phi_n(i)\right\}, \tag{6.98}$$

and

$$u_n(x,i) = \max\left\{\frac{i(i-1)}{n(n-1)}x, \Phi_n(i)\right\}, \tag{6.99}$$

where

$$\Phi_n(i) \equiv \sum_{j=i+1}^{n} \frac{i(i-1)}{j(j-1)(j-2)} \boldsymbol{E}_F\{v_n(X,j) + u_n(X,j)\}$$

($i = 1, 2, \ldots, n$; $v_n(x, n) = u_n(x, n) = x$), since we have by (6.89)–(6.93) with $a = 2$

$$\boldsymbol{P}\{Y_{i+1}, \ldots, Y_{j-1} \geq 3 \text{ and } Y_j \leq 2 | Y_i = k\} = p(i,j) = \frac{i(i-1)}{j(j-1)(j-2)},$$

for $k = 1, 2$. Letting $V_{n,i} = \frac{n}{i}\boldsymbol{E}_F v_n(X, i)$ and $U_{n,i} = \frac{n}{i}\boldsymbol{E}_F u_n(X, i)$. Equations (6.98)–(6.99) becomes

$$V_{n,i} = \frac{2n-i-1}{n-1} S_F(d_{n,i}), \tag{6.100}$$

$$U_{n,i} = \frac{i-1}{n-1} S_F(e_{n,i}), \tag{6.101}$$

($i = 1, 2, \ldots, n$; $U_{n,n} = V_{n,n} = E_F(X)$), where

$$d_{n,i} \equiv \frac{n-1}{2n-i-1} \sum_{j=i+1}^{n} \frac{i-1}{(j-1)(j-2)} (V_{n,j} + U_{n,j}), \tag{6.102}$$

and

$$e_{n,i} = \frac{2n-i-1}{i-1} d_{n,i}. \tag{6.103}$$

Thus we can prove

Theorem 6.10. *The optimal stopping rule for the optimality equations (6.98)–(6.99) is: "Stop at either the earliest state* (X_i, i, Y_i) *satisfying* $Y_i = 1$ *and* $X_i > d_{n,i}$ *or* $Y_i = 2$ *and* $X_i > e_{n,i}$*, whichever occurs first". The sequence* $\{d_{n,i}\}$ *and* $\{e_{n,i}\}$ *are determined by the recursion*

$$d_{n,i} = \frac{1}{(2n-i-1)i}[(n-1)(V_{n,i+1}+U_{n,i+1})+(2n-i-2)(i-1)d_{n,i+1}], \quad (6.104)$$

$(i = 1, 2, \ldots, n-1;\ V_{n,n} = U_{n,n} = \boldsymbol{E}_F(X),\ d_{n,n} = e_{n,n} = 0)$*, where* $\{V_{n,i}\}$*,* $\{U_{n,i}\}$ *and* $\{e_{n,i}\}$ *satisfy (6.100)–(6.101) and (6.103). The optimal expected reward is given by* $E_F v_n(X, 1)$*, i.e.* $\frac{2}{n} S_F(d_{n,1})$.

Proof: The recursion (6.104) follows from (6.102). The rest is the consequence of easy calculations. □

Example 6.5. *Let* $F(x) = x,\ (0 \le x \le 1)$ *and* $n = 10$*. Since we have, from (6.104) and (6.103),*

$$\begin{aligned} d_{10,i} &= \frac{1}{(19-i)i}[9(V_{10,i+1}+U_{10,i+1}) + (18-i)(i-1)d_{10,i+1}] \\ e_{10,i} &= \frac{19-i}{i-1}d_{10,i} \end{aligned}$$

where

$$\begin{aligned} V_{10,i} &= \frac{19-i}{9}[\frac{1}{2}(1+d_{10,i}^2) \text{ or } d_{10,i}], \\ U_{10,i} &= \frac{i-1}{9}[\frac{1}{2}(1+e_{10,i}^2) \text{ or } e_{10,i}]. \end{aligned}$$

we get a table

This table shows that the optimal rule is : "Pass the first object and stop at either the earliest state (X_i, i, Y_i), $2 \le i \le 9$, satisfying $Y_i = 1$ and $X_i > d_{10,i}$ or $6 \le i \le 9$, satisfying $Y_i = 2$ and $X_i > e_{10,i}$, whichever occurs first. If none of the above event occurs, stop at (X_{10}, Y_{10})". The expected reward obtained by employing this rule is $\frac{2}{10} S_F(d_{10,1}) = 0.3731$ (v. Table 6.10).

Remark 19. *If the interviewer considers the second component* Y_i *as unimportant and neglects it, and wants to select one with* X_i *as large as possible, then the optimal stopping rule is: "Stop at the earliest one with* $X_i > v_{n-i}$*, where the sequence* $\{v_m\}_{m=1}^n$ *is determined by* $v_m = S_F(v_{m-1})$*,* $m = 1, 2, \ldots;\ v_0 \equiv 0$*", and the optimal expected reward is equal to* v_n*. We know, for example, that if* $F(x) = x$*,* $0 \le x \le 1$*, then* $v_m = \frac{1+v_{m-1}^2}{2}$ *(i.e. Moser's sequence, see Moser (1956), and* $v_{10} = 0.8611$*, see Gilbert and Mosteller (1966), Table 11 in Section 5). Also for this case* (6.5) *in the example 6.4 shows that the requirement that the selected object should be the best (the best or the second best) in the look diminishes the optimal value down to* 0.2258 (0.3731).

TABLE 6.10
Decision points and values of the problem (Ex. 6.10).

i	$d_{10,i}$	$V_{10,i}$	$e_{10,i}$	$U_{10,i}$
10	0	0.5	0	0.5
9	0.1	0.5611	0.125	0.4514
8	0.1831	0.6316	0.2877	0.4211
7	0.2567	0.7106	0.5134	0.4212
6	0.3281	0.8000	0.8531	0.4799
5	0.4083	0.9074	1.4291	0.6352
4	0.5172	1.0563	2.5860	0.8620
3	0.6829	1.3034	5.4632	1.2140
2	0.9877	1.8658	16.7910	1.8657
1	1.8657	3.7314	∞	∞

6.9.5 Selecting good look together with the best quality

Let $\{(X_i,Y_i)\}_{i=1}^n$ be an *iid* sequence of r.v.s as in Section 6.1 and assume that $\{X_i\}_{i=1}^n$ has uniform distribution over the unit interval $0 \le x \le 1$. Observing the sequence $\{(X_i,Y_i)\}_{i=1}^n$, one by one sequentially, we want to maximize $\boldsymbol{E} c_{R_\tau} \chi_{\{X_\tau=\max_{1\le i\le n} X_i\}}$, where the stopping time $\tau \in \mathfrak{S}$. One can reformulate the problem as the optimal stopping problem for some Markov chain. Denote $\xi_s = \chi_{\{X_s=\max_{1\le i\le s} X_i\}}$. We have

$$\begin{aligned} \boldsymbol{E} c_{R_\tau} \chi_{\{X_\tau=\max_{1\le i\le n} X_i\}} &= \sum_{s=1}^n \int_{\{\tau=s\}} c_{R_s} \chi_{\{X_s=\max_{1\le i\le n} X_i\}}\, d\boldsymbol{P} \qquad (6.105)\\ &= \sum_{s=1}^n \int_{\{\tau=s\}} \boldsymbol{E}[c_{R_s} \chi_{\{X_s=\max_{1\le i\le s} X_i\}} | \mathcal{F}_s]\, d\boldsymbol{P} \\ &= \sum_{s=1}^n \int_{\{\tau=s\}} \boldsymbol{E}[c_{R_s} \chi_{\{X_s=\max_{1\le i\le s} X_i\}} | \xi_s, Y_s]\, d\boldsymbol{P} \\ &= \sum_{s=1}^n \int_{\{\tau=s\}} \psi(s,\xi_s,Y_s)\, d\boldsymbol{P} = \boldsymbol{E}\psi(\tau,\xi_\tau,Y_\tau). \end{aligned}$$

Let $Z_1 = (1, X_1)$ and for $t > 1$ let $\gamma_t = \inf\{r > \gamma_{t-1} : \xi_r = 1\}$, $(\inf\emptyset = \infty)$ and $Z_t = (\gamma_t, X_{\gamma_t})$. If $\gamma_t = \infty$, then we put $Z_t = (\infty,\infty)$. The process Z_t is the Markov chain with the following one-step density of transition probabilities,

$$p((i,x),(j,z)) = \begin{cases} x^{j-i-1}, & \text{for } j > i,\ x < z \\ 0, & \text{otherwise.} \end{cases} \qquad (6.106)$$

For calculation of (6.105) it is enough to observe the sequence $\{(Z_t, Y_{\gamma_t})\}_{t=1}^n$. Let $\mathcal{G}_t = \sigma(Z_1, Y_1, Z_2, Y_{\gamma_2}, \ldots, Z_t, Y_{\gamma_t})$ and $\mathfrak{M}$ the set of stopping times with respect to $\{\mathcal{G}_t\}_{t=1}^n$. If we face the $\gamma_t = i$-th object (X_i, Y_i) with $X_i = \max(X_1, X_2, \ldots, X_i) = x$ and $Y_i = k$, we have Markov chain $\{(\gamma_t, X_{\gamma_t}, Y_{\gamma_t})\}_{t=1}^n$ in state (i, x, k). The stop reward at state (i, x, k) is based on the above consideration and from (6.89)

$$g(i, x, k) = \frac{x^{n-i}}{\binom{n}{i}} \sum_{r=k}^{n} \binom{r-1}{k-1}\binom{n-r}{i-k} c_r, \tag{6.107}$$

for $1 \leq k \leq i$, $k \leq r \leq n$, $1 \leq i \leq n$ and $0 \leq x \leq 1$. We assume that $c_1 \geq c_2 \geq \ldots \geq c_n \geq 0$. The gain function $g(i, x, k)$ is defined on $\{1, 2, \ldots, n\} \times [0, 1] \times \{1, 2, \ldots, n\}$. For every $\tau \in \mathfrak{S}$, we have $\sigma \in \mathfrak{M}$ such that

$$\boldsymbol{E} c_{R_\tau} \chi_{\{X_\tau = \max_{1\leq i\leq n} X_i\}} = \boldsymbol{E}\psi(\tau, \xi_\tau, Y_\tau) = \boldsymbol{E} g(\gamma_\sigma, X_{\gamma_\sigma}, Y_{\gamma_\sigma}).$$

Define

$$\begin{aligned} \mathbf{T}g(i, x, k) &= \boldsymbol{E}_{(i,x,k)} g(\gamma, X_\gamma, Y_\gamma) \qquad (6.108)\\ &= \sum_{r=i+1}^{n} \sum_{u=1}^{n} \frac{1}{r} \int_x^1 \frac{z^{n-r}}{\binom{n}{i}} \sum_{s=u}^{n} \binom{s-1}{u-1}\binom{n-s}{r-u} c_s x^{r-i-1} dz \\ &= \sum_{r=i+1}^{n} \frac{x^{r-i-1} - x^{n-i}}{r(n-r+1)} \sum_{u=1}^{r}\sum_{s=u}^{n} \frac{\binom{s-1}{u-1}\binom{n-s}{r-u}}{\binom{n}{i}} c_s. \end{aligned}$$

Denoting, by $v_n(i, x, k)$, the expected reward obtained by employing the optimal stopping rule for the n-period problem at state (i, x, k), we easily have

$$\begin{aligned} v_n(i, x, k) &= \max\left[\frac{x^{n-i}}{\binom{n}{i}} \sum_{r=k}^{n} \binom{r-1}{k-1}\binom{n-r}{i-k} c_r, \right. \qquad (6.109)\\ &\left. \sum_{j=i+1}^{n} \frac{x^{j-i-1}}{j} \sum_{w=1}^{j} \int_x^1 v_n(j, z, w) dz\right], \end{aligned}$$

where $1 \leq k \leq i$, $1 \leq i \leq n$, $0 \leq x \leq 1$, with $v_n(n, x, k) = c_k$.

The one-step look ahead stopping region (Ross (1970); pp. 137–139) corresponding to this optimality equation is

$$\begin{aligned} B &= \{(i, x, k) | g(i, x, k) \geq Tg(i, x, k)\} \qquad (6.110)\\ &= \{(i, x, k) | \frac{\sum_{r=k}^{n} \binom{r-1}{k-1}\binom{n-r}{i-k}}{\binom{n}{i}} c_r \\ &\geq \sum_{j=i+1}^{n} \left[\frac{x^{-n+j-1} - 1}{j(n-j+1)} \sum_{w=1}^{j}\sum_{r=w}^{n} \frac{\binom{r-1}{w-1}\binom{n-r}{j-w}}{\binom{n}{j}} a_r\right]\}. \end{aligned}$$

We discuss the following two simple cases.

Case 1. $a_1 = 1, a_2 = a_3 = \cdots = a_n = 0$.

For this case the stop reward (6.107) becomes

$$g(i, x, k|n) = \begin{cases} \frac{ix^{n-i}}{n}, & \text{if } k = 1 \\ 0, & \text{if } 2 \leq k \leq i \end{cases}$$

and the one-step stopping region (6.110) becomes, after simplification,

$$B = \{(i, x, k|n)|1 \geq \frac{1}{i}\sum_{l=1}^{n-i} \frac{x^{-l} - 1}{l}\}. \tag{6.111}$$

Lemma 6.11. *The region B given by (6.111) is "closed", i.e. if once a state enters B, the state never leaves B as the process goes on.*

Proof: For any $0 \leq x < z \leq 1$, we have

$$i^{-1}\sum_{l=1}^{n-i} \frac{x^{-l} - 1}{l} \geq i^{-1}\sum_{l=1}^{n-i} \frac{z^{-l} - 1}{l}$$

$$= i^{-1}\left[\sum_{l=1}^{n-i-1} \frac{z^{-l} - 1}{l} + \frac{z^{-(n-i)} - 1}{n-i}\right] \geq \frac{1}{i+1}\sum_{l=1}^{n-i-1} \frac{z^{-l} - 1}{l}.$$

Hence

$$1 \geq \frac{1}{i}\sum_{l=1}^{n-i} \frac{x^{-l} - 1}{l} \Rightarrow 1 \geq \frac{1}{i+1}\sum_{l=1}^{n-i-1} \frac{x^{-l} - 1}{l}.$$

□

Let $d_{n,i}$ ($i = 1, 2, \ldots, n-1$; $d_{n,n-1} = n^{-1}$, $d_{n,n} \equiv 0$) be a unique root in $[0, 1]$ of the equation

$$\frac{1}{i}\sum_{l=1}^{n-i} \frac{x^{-l} - 1}{l} = 1. \tag{6.112}$$

Evidently

$$1 \geq \frac{1}{i}\sum_{l=1}^{n-i} \frac{x^{-l} - 1}{l} \iff x > d_{n,i}.$$

For some small n, we find that $d_{2,1} = 1/2$, $d_{3,2} = 1/3$, $d_{3,1} = \frac{1+\sqrt{6}}{5} \fallingdotseq 0.6899$, $d_{4,i} = 1/4$, $\frac{1+\sqrt{8}}{7}(\fallingdotseq 0.5469)$, 0.7755 for $i = 3, 2, 1$, respectively. $d_{5,i} = 1/5$, $\frac{1+\sqrt{10}}{9}(\fallingdotseq 0.4625)$, 0.6591, 0.8246 for $i = 4, 3, 2, 1$, respectively and so on.

It is well known that if the one-step stopping region is realizable and "closed", it becomes the optimal stopping region. From Lemma 6.11 we thus have

Theorem 6.12. *The optimal stopping region for the optimality equation (6.109) in* **Case 1**, *is: "stop at the earliest* (X_i, Y_i) *that satisfies* $Y_i = 1$ *and* $X_i = \max_{1\le t\le i} X_t > d_{n,i}$, *where each* $d_{n,i}$ *is given by a unique root of the equation (6.112).*

Example 6.6. *For* $n = 5$ *the optimal stopping rule is as follows.*

If $y_1 = 1 \,\&\, x_1 \ge 0.8246$, *then* **STOP** *else* **observe** (x_2, y_2)

→ *If* $y_2 = 1 \,\&\, x_2 \ge x_1 \vee 0.6591$, *then* **STOP** *else* **observe** (x_3, y_3)

→ *If* $y_3 = 1 \,\&\, x_3 \ge x_1 \vee x_2 \vee 0.4625$, *then* **STOP** *else* **observe** (x_4, y_4)

→ *If* $y_4 = 1 \,\&\, x_4 \ge x_1 \vee x_2 \vee x_3 \vee 0.2$,
then **STOP** *else* **observe** (x_5, y_5)&**STOP**

Remark 20. *Suppose that the interviewer does not observe the second component* Y_i *or does observe but without considering it as important, and wants to only maximize the probability of selecting the applicant with the highest* X_i. *Then the well-known result (see Gilbert and Mosteller (1966), Table 8 in Section 3) in the theory of the secretary problem is that the optimal stopping rule is: "Stop at the earliest applicant with* $X_i = \max_{1\le t\le i} X_t > v_{n-i}$, *where each* v_m *in the sequence* $\{v_m\}_{m=1}^{n}$ *is determined by*

$$\sum_{j=1}^{m} \frac{v^{-j} - 1}{j} = 1, \ (m = 1, 2, \ldots). \tag{6.113}$$

We know that $v_i = 0.8246, 0.7758, 0.6899, 0.6392\ 1/2$, *for* $i = 5, 4, 3, 2, 1$, *respectively, and compare these values with* $d_{5,i}$*'s in Example 6.6.*

Remark 21. *(continuation to Remark 2). In Remark 2 it is known that the probability of selecting the maximum of X by following the optimal strategy is*

$$P_n = \frac{1}{n}\left[1 + \sum_{i=1}^{n-1}\sum_{y=1}^{i} \frac{v_i^{n-y}}{n-y}\right] \tag{6.114}$$

(see Gilbert and Mosteller (1966), Section 3 and Sakaguchi (1973)), and hence

$$\begin{aligned} P_5 &= \frac{1}{5}\left[1 + \sum_{i=1}^{4}\sum_{y=1}^{i} \frac{v_i^{5-y}}{5-y}\right] = \frac{1}{5}\left[1 + \frac{1}{4}v_1^4 + (\frac{1}{4}v_2^4 + \frac{1}{3}v_2^3) \right. \\ &\left. + (\frac{1}{4}v_3^4 + \frac{1}{3}v_3^3 + \frac{1}{2}v_3^2) + (\frac{1}{4}v_4^4 + \frac{1}{3}v_4^3 + \frac{1}{2}v_4^2 + v_4)\right] \fallingdotseq 0.6392 \end{aligned}$$

where v_i*'s are given by (6.113).*

The expected reward obtained by employing the optimal strategy in Theorem 6.12 is $W_n \equiv \int_0^1 v_n(1,x,1)dx$, and the explicit expression of W_n, like (6.114) is presently not known. It is clear, however, that by solving optimality equation (6.109) for Case 1, i.e.

$$v_n(i,x) = \max\left[\frac{i}{n}x^{n-i}, \sum_{j=i+1}^{n} \frac{x^{j-i-1}}{j} \int_x^1 v_n(j,z)dz\right]$$

($i = 1, 2, \ldots, n$; $v_n(n,x) \equiv 1$; $0 \le x \le 1$), (we need not consider states in which $2 \le y_i \le i$) recursively downward, we can derive $v_n(1,x)$ and hence obtain W_n. But it is a tediously lengthy job. For instance, we find, for $n = 5$

$$\begin{aligned}
v_5(5,x) &= 1 \\
v_5(4,x) &= \frac{4}{5}x \vee \frac{1}{5}\int_x^1 v_5(5,z)dz = \frac{4}{5}(x) \vee \frac{1-x}{5}, \\
v_5(3,x) &= (\frac{3}{5}x^2) \vee \sum_{j=4}^{5} \frac{x^{j-4}}{j}\int_x^1 v_5(j,z)dz \\
&= (\frac{3}{5}x^2) \vee \Phi_5(3,x).
\end{aligned}$$

where

$$\Phi_5(3,x) = \begin{cases} \frac{21}{200} + \frac{3}{20}x - \frac{7}{40}x^2, & \text{if } 0 \le x \le \frac{1}{5} \\ \frac{1}{10} + \frac{1}{5}x - \frac{3}{10}x^2, & \text{if } \frac{1}{5} \le x \le 1 \\ \frac{3}{5}x^2, & \text{if } d_{5,3} \le x \le 1 \end{cases}$$

and

$$v_5(2,x) = (\frac{2}{5}x^3) \vee \int_x^1 \left\{\frac{1}{3}v_5(3,z) + \frac{x}{4}v_5(4,z) + \frac{x^2}{5}v_5(5,z)\right\} dz$$

and so on. Note that the unique root of $\frac{3}{5}x^2 = \Phi_5(3,x)$ coincides with $d_{5,3} = (1+\sqrt{10})/9$ derived from (6.112).

Case 2. $a_1 = a_2 = 1, a_3 = a_4 = \cdots = a_n = 0$.

For this case the optimality equation (6.109) becomes

$$v_n(i,x) = \max\left\{\frac{i(2n-i-1)}{n(n-1)}x^{n-i}, \Psi_n(i,x)\right\} \quad \text{for states } (i,x,1) \tag{6.115}$$

$$u_n(i,x) = \max\left\{\frac{i(i-1)}{n(n-1)}x^{n-i}, \Psi_n(i,x)\right\} \quad \text{for states } (i,x,2).$$

where

$$\Psi_n(i,x) \equiv \sum_{j=i+1}^{n} \frac{x^{j-i-1}}{j}\int_x^1 (v_n(j,z) + u_n(j,z))dz,$$

and the one-step stopping region (6.110) becomes, after simplification, $B_1 \cup B_2$, where

$$B_1 = \{(i,x,1)|1 \geq \frac{n-1}{i(n-\frac{1}{2(i+1)})}\sum_{l=1}^{n-i}\frac{x^{-l}-1}{l}\} \tag{6.116}$$

$$B_2 = \{(i,x,2)|1 \geq \frac{2(n-1)}{i(i-1)}\sum_{l=1}^{n-i}\frac{x^{-l}-1}{l}\}.$$

Lemma 6.13. *The region $B_1 \cup B_2$ is "closed".*

Proof: For any $0 \leq x < z \leq 1$ we can easily prove that

$$\frac{1}{i(n-\frac{i+1}{2})}\sum_{l=1}^{n-i}\frac{x^{-l}-1}{l} \geq \frac{1}{(i+1)(n-\frac{i+2}{2})}\sum_{l=1}^{n-i-1}\frac{z^{-l}-1}{l},$$

and

$$\frac{1}{i(i-1)}\sum_{l=1}^{n-i}\frac{x^{-l}-1}{l} \geq \frac{1}{i(i+1)}\sum_{l=1}^{n-i-1}\frac{z^{-l}-1}{l}.$$

Therefore both of B_1 and B_2, and hence $B_1 \cup B_2$, are "closed". □

Let $e_{n,i}$ $(1 \leq i \leq n;\ e_{n,n-1} = \frac{2}{n+2},\ e_{n,n} \equiv 0)$ and $f_{n,i}$ $(2 \leq i \leq n;\ f_{n,n-1} = \frac{2}{n},\ f_{n,n} \equiv 0)$ be a unique root of the equation

$$\frac{n-1}{i(n-\frac{i+1}{2})}\sum_{l=1}^{n-i}\frac{x^{-l}-1}{l} = 1 \tag{6.117}$$

and

$$\frac{2(n-1)}{i(i-1)}\sum_{l=1}^{n-i}\frac{x^{-l}-1}{l} = 1, \tag{6.118}$$

respectively.

It is clear that $(i,x,1) \in B_1 \iff x > e_{n,i}$ and $(i,x,2) \in B_2 \iff x > f_{n,i}$.

For some small n, we find that

$$e_{2,1} = \tfrac{1}{2},\ e_{3,2} = \tfrac{2}{5},\ e_{3,1} = \tfrac{1+\sqrt{6}}{5} \risingdotseq 0.6899$$
$$e_{4,i} = \tfrac{1}{3},\ \tfrac{3+\sqrt{66}}{19} \risingdotseq 0.5855,\ 0.7755 \text{ for } i = 3,2,1, \text{ respectively}$$
$$e_{5,i} = \tfrac{2}{7},\ \tfrac{2+\sqrt{34}}{15} \risingdotseq 0.5221,\ 0.6834,\ 0.8248 \text{ for } i = 4,3,2,1, \text{ respectively}$$

and

$$f_{3,2} = \tfrac{2}{3},$$
$$f_{4,i} = \tfrac{1}{2},\ \tfrac{3+\sqrt{42}}{11}(\risingdotseq 0.8619), \text{ for } i = 3,2 \text{ respectively}$$
$$f_{5,i} = \tfrac{2}{5},\ \tfrac{2+\sqrt{22}}{9}(\risingdotseq 0.7434),\ 0.9246, \text{ for } i = 4,3,2, \text{ respectively.}$$

From Lemma 6.13 we obtain

Theorem 6.14. *The optimal stopping region for the optimal equation (6.109) in Case 2, is: "stop at either the earliest* (X_i, Y_i)*, with* $Y_i = 1$ *and* $X_i = \max_{1 \le t \le i} X_t > e_{n,i}$*, or the earliest* (X_i, Y_i)*, with* $Y_i = 2$ *and* $X_i = \max_{1 \le t \le i} X_t > f_{n,i}$*, whichever occurs first, where* $\{e_{n,i}\}$ *and* $\{f_{n,i}\}$ *are defined by the unique roots of the equations (6.117)–(6.118).*

Example 6.7. *For* $n = 5$ *the optimal stopping rule is as follows.*

If $y_1 = 1 \;\&\; x_1 \ge 0.8248$, *then* **STOP** *else* **observe** (x_2, y_2)

→ *If* $y_2{=}1 \;\&\; x_2 \ge x_1 \vee 0.6834$ *or* $y_2{=}2 \;\&\; x_2 \ge x_1 \vee 0.9246$, *then* **STOP** *else* **observe** (x_3, y_3)

→ *If* $y_3{=}1 \;\&\; x_3 \ge x_1 \vee x_2 \vee 0.5221$ *or* $y_3{=}2 \;\&\; x_3 \ge x_1 \vee x_2 \vee 0.7434$, *then* **STOP** *else* **observe** (x_4, y_4)

→ *If* $y_4{=}1 \;\&\; x_4 \ge x_1 \vee x_2 \vee x_3 \vee \frac{2}{7}$ *or* $y_4{=}2 \;\&\; x_4 \ge x_1 \vee x_2 \vee x_3 \vee \frac{2}{5}$, *then* **STOP** *else* **observe** (x_5, y_5) **&STOP**

6.9.6 Selecting good look together with the best quality

Rapoport et al. (2023) studied the performance of heuristics relative to the performance of optimal solutions in the rich domain of sequential search, where the decision to stop the search depends only on the applicant's relative rank. Considering multiple variants of the secretary problem, that vary from one another in their formulation and method of solution, they find that descriptive heuristics perform well only when the optimal solution prescribes a single threshold value. It is shown that a computational heuristic originally proposed as an approximate solution to a single variant of the secretary problem performs equally well in many other variants where the optimal solution prescribes multiple threshold values that gradually relax the criterion for stopping the search. In the paper a new heuristic with near optimal performance in a competitive or strategic variant of the secretary problem with multiple employers competing with one another to hire job applicants is proposed. Both heuristics share a simple computational component: the ratio of the number of interviewed applicants to the number of those remaining to be searched. They presented the subgame-perfect Nash equilibrium for this competitive variant and an algorithm for its computation.

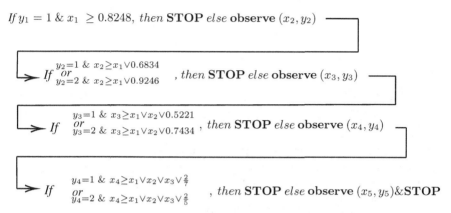

7

Numerical and Asymptotic Solutions for Optimal Stopping Problems

Motto:
Embracing Uncertainty, Harnessing Probability: Advancing Numerical Methods for Complex Realities.

OVER THE PAST CENTURY, there has been a remarkable surge in the utilization of probabilistic models within the realm of numerical methods. This surge is primarily attributed to the growing need for sophisticated tools to understand and address the complexities inherent in random phenomena and systems with stochastic parameters. At the turn of the 20th and 21st centuries, the proliferation of computational power played a pivotal role in unlocking the practical significance of probabilistic models. Early algorithms like the Monte Carlo (**MCM**) methods, developed in the 1940s, laid the groundwork for simulating and analyzing stochastic processes. However, during their inception, the computational resources available were inadequate to fully harness their potential.

With advancements in computing technology, particularly the exponential growth in processing power and the development of parallel computing architectures, the feasibility and efficiency of implementing probabilistic algorithms have vastly improved. This has led to their widespread adoption across various fields, including finance, engineering, physics, and biology, among others. Moreover, the scope of probabilistic models has expanded beyond mere statistical analyses. They are now integral to solving optimization problems that involve stochastic parameters. For instance, in optimal control tasks, such as optimal stopping of stochastic processes, probabilistic methods offer robust and effective solutions.

The significance of these methodologies is underscored by their ability to provide insights and solutions in scenarios where deterministic approaches fall short. By embracing uncertainty and randomness inherent in many real-world systems, probabilistic models empower researchers and practitioners to make informed decisions and optimize processes in an ever-evolving landscape of complex phenomena. Thus, the integration of probabilistic models into numerical methods represents a profound paradigm shift, shaping the way we approach and tackle a wide array of challenges across diverse domains. This is now a significant methodology also for optimal control tasks, including optimal stopping of stochastic processes (v. Pagès (2018)).

DOI: 10.1201/9781003407102-7

7.1 Dynamic Programming

Here we will review numerical approaches usually used in financial mathematics literature – with an emphasis on their application for valuating American options. In financial mathematics, and indeed in operations research, statistics and control theory, these methods refer to the use of *dynamic programming* or *approximate dynamic programming*. In the methods, the sample path is simulated, often by Monte Carlo simulation, according to stochastic models from financial theories. The problem of pricing the American option is then essentially the task of implementing or approximating the dynamic programming structure; either through a deterministic or stochastic method. In other fields, this concept is referred to as "reinforcement learning" – a subset of the field of machine learning.

The use of machine learning, deep learning and neural networks in problems with high complexity are increasingly more popular, and much of the work demonstrates that such machine learning methods can advance the current state-of-the-art algorithms in optimal stopping and financial research; see, for example, Becker et al. (2019) and Kohler et al. (2010). While numerous machine learning methods have been developed for single optimal stopping, there are fewer algorithms designed for multiple-stopping problems. A recent example includes a deep neural network algorithm for multiple-stopping proposed by Han and Li (2023).

7.1.1 Quantization Tree Algorithm

In Bally and Pagès (2003), the authors formulate a quantization method for solving multi-dimensional discrete-time optimal stopping problems. The general idea is to compute a large quantity of conditional expectations by projecting the stochastic process onto optimal "grids", which are chosen such that the mean square error of the projection is minimized. While in other similar methods, the grids are chosen a priori, the key difference with this method is that the grids are chosen optimally. This is achieved by simulating large samples of $\{\xi_n\}_{n=0,\ldots,N}$ in order to produce a grid Γ_n^* of size M_n which is optimally fitted to ξ_n (with respect to all grids Γ of the same size) using the closest neighbor projection $\text{Proj}_{\Gamma_n^*}(\xi_n)$, which is determined by least squares approximation. The best possible size-M_n grid can be used to approximate the d-dimensional vector ξ_n.

7.1.2 Stochastic Mesh Method

Letting $\{\xi_n\}_{n=0,\ldots,N}$ be a Markov process on $\Re^d$ with fixed initial state ξ_0. The idea of the method proposed by Broadie and Glasserman (2004) is to generate random vectors $\{R_n(i)\}_{n=1,\ldots,N}$ for $i = 1, \ldots, b$ and defining $R_0(1) = \xi_0$. We then compute some "weights" $W_n(R_t(i), R_{n+1}(j))$ for each arc joining $R_n(i)$ to $R_{n+1}(j)$.

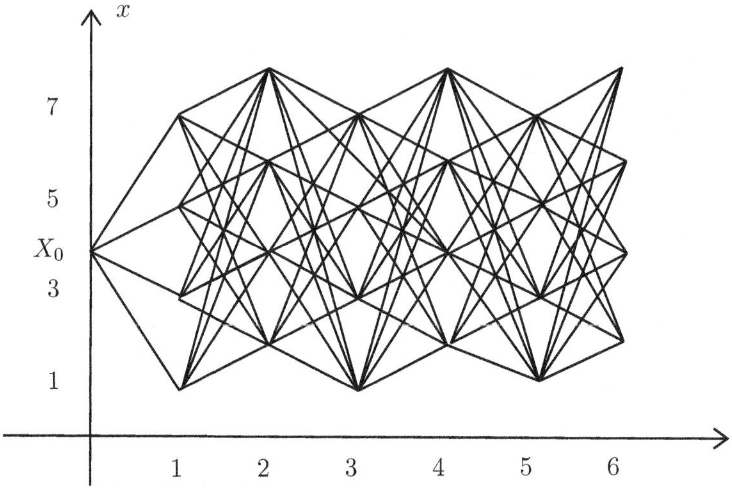

FIGURE 7.1
Illustration of stochastic mesh for a single asset ($N = 6$ and $b = 4$).

Figure 7.1 illustrates a mesh for $d = 1$, $N = 6$ and $b = 4$ but other examples, including $d = 2$, are provided in Broadie and Glasserman (2004) for the interested reader.

The random variables $R_n(i)$ are, in the simplest implementation of this method, generated by assuming they are independent identically distributed samples from some *carefully* chosen mesh density function $g_n(\cdot)$ (for $n = 1, \ldots, N$) through the average density method (see Broadie and Glasserman (2004) for more details), which is a function of the conditional distribution function of ξ_{n+1} conditioned on ξ_n and the marginal distribution of ξ_n. The choice of mesh density is vital to the performance of the algorithm.

7.2 Regression Based Methods

In the context of multiple-stopping, it is hard to evaluate sequences $\{V_{m_i}\}$ and $\{X_{m_i}\}$, $i = 1, 2, \ldots, k$, which are formulated in terms of conditional expectations. The algorithms discussed in this section were proposed for single optimal stopping but they can be extended to multiple-stopping problems. The algorithms aim to utilize Monte Carlo simulations to generate artificial samples, enabling the application of recursive regression to estimate the sequences. Tsitsiklis and Van Roy (2001) proposed an algorithm based on linear regression. Carriere (1996) employs various non-linear regression techniques, a concept further explored by Longstaff and Schwartz (2001). Although the approach closely resembles that of Tsitsiklis and Van Roy (2001), both methods were developed independently. We will give an brief overview of the Carriere and Longstaff-Schwartz algorithms.

7.2.1 Carriere Algorithm

Carriere's algorithm Carriere (1996) is one of the methods that involves the simulation of the stochastic processes. The algorithm relies on approximating the conditional expectations within the dynamic programming problem through the application of nonparametric regression. The initial step involves generating M sample paths from a Markovian process:

$$\xi_1 = (\xi_{1,0}, \ldots, \xi_{1,N}), \ldots, \xi_M = (\xi_{M,0}, \ldots, \xi_{M,N}),$$

where $\xi_{m,n}$ is the observed variable at time n in the m-th simulation. Nonparametric regression (q-splines and local polynomial smoothing are detailed in Carriere (1996)) is then used to estimate the conditional expectations with the simulated variables. In contrast with the stochastic mesh method, which approximates the Snell envelope for a fixed time point, Carriere's method estimates the Snell envelope for all points.

7.2.2 Longstaff-Schwartz Algorithm

The Longstaff-Schwartz algorithm Longstaff and Schwartz (2001) is considered to be a standard for the financial industry in the area of option pricing and has since been analyzed, verified and extended by many researchers (see, for example, Meinshausen and Hambly (2004)). It serves also as a useful benchmark for comparing other proposed methods in this area. The algorithm has a similar structure to Carriere's method in the sense that it also generates M sample paths as described above. However, the Snell envelope is not estimated but rather an approximation for the stopping region. Emphasis in this algorithm is placed on the approximation of the stopping time, for which for estimates of the stopping region are calculated backwards through least squares regression. The estimate of the value of the game is then the average of the estimated values across M generated sample paths, with respect to the optimal stopping rule.

7.3 Cross-Entropy Method

The Cross-Entropy method developed by Rubinstein(1997) is one of the versions of the Monte Carlo method developed for problems requiring the estimation of events with low probabilities. It is an alternative approach to combinatorial and continuous, multi-step, optimization tasks. In the field of rare event simulation, this method is used in conjunction with weighted sampling, a known technique of variance reduction, in which the system is simulated as part of another set of parameters, called reference parameters, to increase the likelihood of a rare event. The advantage of the relative entropy method is that it provides a simple and fast procedure for estimating optimal reference parameters in importance sampling (IS). The CE algorithm can be seen as a self-learning code covering the following two iterative phases:

1. Generating random data samples (trajectories, vectors, etc.) according to a specific random mechanism.
2. Updating parameters of the random mechanism based on data to obtain a "better" sample in the next iteration.

Now the main general algorithm behind the Cross-Entropy method will be presented on the example of applying the method to selected problems of optimal stopping. The CE method to determine the optimal set and game value using relative ranks and observation of extreme values. This problem was described and solved by the CE method by Polushina (2010) and Sofronov et al. (2006), and Sofronov and Polushina (2013). Its correctness and programming in a different environment was also considered by Stachowiak (2019) in her engineering thesis (v. also Stachowiak and Szajowski (2019)).

Consider the secretary problem (v. Ferguson(1989), Szajowski(1982) for multiple choice, i.e. the issue of selecting the best proposals from a finite set. The objective is to find such an optimal stopping rule that the sum of the ranks of all selected objects is the lowest (their value is then the highest), see Section 6.8.

7.3.1 Optimization by the method of Cross-Entropy

In stopping problems, for small N there is a possibility to calculate exact values by analytical methods whereas it may be quite hard for large values of N. However, it is much easier to obtain them through simulations.

Consider the following problem of minimizing the mean sum of ranks:

$$\min_{\boldsymbol{x}\in\mathcal{X}} \boldsymbol{E}\widehat{S}(x,\boldsymbol{R}), \tag{7.1}$$

where $\mathcal{X} = \{\boldsymbol{x} = (x^{(1)},\ldots,x^{(k)}) :$ conditions from (6.68) are preserved$\}$. $\boldsymbol{R} = (R_1,\ldots,R_N)$ is a random permutation of numbers $1,\ldots,N$. $\widehat{S}$ is an unbiased estimator of $\boldsymbol{E}\widehat{S}(\boldsymbol{x},\boldsymbol{R})$ with the following formula:

$$\widehat{S}(\boldsymbol{x}) = \frac{1}{N_1}\sum_{n=1}^{N_1}(R_{n\tau_1}+\cdots+R_{n\tau_k}),$$

where $(R_{n1},\ldots,R_{nN})$ is the n-th copy of a random permutation $\boldsymbol{R}$.
Now a Cross-Entropy algorithm can be applied Rubinstein (1997). Let us define the indicator collections $\{\mathbb{I}_{\{S(x)\le\gamma\}}\}$ on $\mathcal{X}$ for different levels γ. Let $\{f(\cdot\,;u)\}$ be a density family on $\mathcal{X}$ parameterized by the actual parameter value u. For a specific u, we can combine (7.1) with the problem of estimation

$$l(\gamma) = \boldsymbol{P}_u(S(X)\le\gamma) = \sum_x \mathbb{I}_{\{S(x)\le\gamma\}} f(x,u) = \boldsymbol{E}_u\mathbb{I}_{\{S(x)\le\gamma\}},$$

where $\boldsymbol{P}_u$ is a measure of the probability in which the random state X has a density $\{f(\cdot\,; u)\}$ and γ is a known or unknown parameter with l being estimated by the Kullback-Leibler distance minimization. First let us generate a pair $\{(\gamma_t, u_t)\}$, which we then update until the stopping criterion is met and the optimal pair is obtained $\{(\gamma^*, u^*)\}$. More precisely, arrange u_0 and choose not too small ϱ and we proceed as follows:

1) Updating γ_t

Generate a sample $X_1, \ldots, X_{N_2}$ from $\{f(\cdot\,; u_{t-1})\}$. Calculate $\widehat{S}(X_1), \ldots, \widehat{S}(X_{N_2})$ and sort in ascending order. For $\widehat{\gamma}_t$ choose

$$\widehat{\gamma}_t = \widehat{S}_{(\lceil \varrho N \rceil)}.$$

2) Updating u_t

$\widehat{u}_t$ is obtained from the Kullback-Leibler distance minimization, that is, from maximization

$$\max_u D(u) = \max_u \boldsymbol{E}_{u_{t-1}} \mathbb{I}_{\{\widehat{S}(x) \le \gamma_t\}} \log f(X; u), \tag{7.2}$$

so

$$\max_u \widehat{D}(u) = \max_u \frac{1}{N_2} \sum_{n=1}^{N_2} \mathbb{I}_{\{\widehat{S}(x) \le \widehat{\gamma}_t\}} \log f(X_n; u). \tag{7.3}$$

Instead of updating the parameter vector $\boldsymbol{u}$ we can use the following smoothed version

$$\widehat{\boldsymbol{u}}_t = \alpha \widehat{\boldsymbol{u}}_t + (1 - \alpha) \widehat{\boldsymbol{u}}_{t-1}, \tag{7.4}$$

where α is called the smoothing parameter, with $0.7 < \alpha \le 1$. Clearly, for $\alpha = 1$, we have our original updating rule.

3) Stopping Criterion

A stopping criterion should be specified. More precise transitions and the determination of the stopping criteria are described by Stachowiak and Szajowski (2020).

7.3.2 The Cross-Entropy Method for the multiple BCP

Recall that $\pi^* = (\pi_1^*, \ldots, \pi_k^*)$ is the set of thresholds that we wish to find. It is difficult to obtain the set π^* and the value v, but we can get them by simulation. So we consider the following maximization problem on $\mathcal{X} = \{\boldsymbol{x} = (x_1, \ldots, x_k) : 1 \le x_1 < \cdots < x_k \le N\}$

$$\max_{\boldsymbol{x} \in \mathcal{X}} \boldsymbol{E}\widehat{S}(\boldsymbol{x}, \boldsymbol{Y}),$$

where $\boldsymbol{Y} = (Y_1, \ldots, Y_N)$ is a random permutation of numbers $1, 2, \ldots, N$, $\widehat{S}(\boldsymbol{x})$ is an unbiased estimator of $\boldsymbol{E}\widehat{S}(\boldsymbol{x}, \boldsymbol{Y})$

$$\widehat{S}(\boldsymbol{x}) = \frac{1}{N_1} \sum_{n=1}^{N_1} \mathbb{I}_{\{Y_{n\tau_1} = i_1, \ldots, Y_{n\tau_k} = i_k\}},$$

where $(Y_{n1}, \dots, Y_{nN})$ is the n-th copy of random permutation $\boldsymbol{Y}$, $(i_1, \dots, i_k)$ is any permutation of numbers $1, 2, \dots, k$.

Therefore, we can apply the CE algorithm for noisy optimization (v. Rubinstein and Kroese (2004)). We consider a vector of parameters $\boldsymbol{u} = \{\boldsymbol{u}_l\}$,

$$\boldsymbol{u}_l = \boldsymbol{P}\{\boldsymbol{x} = \boldsymbol{b}_l\}, \quad l = 1, \dots, L, \quad L = C_N^k,$$

where $\boldsymbol{b}_l = (b_{l1}, \dots, b_{lk})$, $1 \le b_{l1} < \dots < b_{lk} \le N$, L is the number of all possible combinations without repetition. We can write the probability function of $\boldsymbol{x}$ as

$$f(\boldsymbol{x}; \boldsymbol{u}) = \sum_{l=1}^{L} u_l \mathbb{I}_{\{\boldsymbol{x}=\boldsymbol{b}_l\}}.$$

Using ibid, (3.34), we get $\widehat{\boldsymbol{u}}_t = \{\widehat{u}_l^{(t)}\}$:

$$\widehat{u}_l^{(t)} = \frac{\sum_{n=1}^{N_2} \mathbb{I}_{\{\widehat{S}(\boldsymbol{x}_n) \ge \widehat{\gamma}_t\}} W_n^{(t-1)} \mathbb{I}_{\{\boldsymbol{x}_n = \boldsymbol{b}_l\}}}{\sum_{n=1}^{N_2} \mathbb{I}_{\{\widehat{S}(\boldsymbol{x}_n) \ge \widehat{\gamma}_t\}} W_n^{(t-1)}}, \tag{7.5}$$

$$W_n^{(t-1)} = \frac{u_{l_1}^{(0)}}{u_{l_1}^{(t-1)}},$$

where l_1 such that $\boldsymbol{x}_n = \boldsymbol{b}_{l_1}$, $\boldsymbol{x}_n = (X_{n1}, \dots, X_{nk})$ is a discrete random vector from $f(\boldsymbol{x}; \widehat{\boldsymbol{u}}_{t-1})$. Another way of the parametrization of the vector $\boldsymbol{u} = \{\boldsymbol{u}_l\}$ was considered in Sofronov et al. (2006).

We propose to use the following stopping criterion; see, e.g., p.207 on Rubinstein and Kroese (2004).

To identify the stopping moment T, we consider the following moving average process

$$B_t(K) = \frac{1}{K} \sum_{s=t-K+1}^{t} \widehat{\gamma}_s, \quad t = s,\ s = 1, \dots, \quad s \ge K,$$

where K is fixed. Define

$$C_t(K) = \frac{\frac{1}{K-1}\{\sum_{s=t-K+1}^{t} (\widehat{\gamma}_s - B_t(K))^2\}}{B_t(K)^2}.$$

Define next

$$C_t^-(K, R) = \min_{j=1,\dots,R} C_{t+j}(K)$$

and

$$C_t^+(K, R) = \max_{j=1,\dots,R} C_{t+j}(K),$$

respectively, where R is fixed.

We define stopping criterion as

$$T = \min\left\{t : \frac{C_t^+(K,R) - C_t^-(K,R)}{C_t^-(K,R)} \le \varepsilon\right\}, \tag{7.6}$$

where K and R are fixed and ε is a small number, say $\varepsilon \le 0.01$.

The CE algorithm for the MBC problem can be described as the following procedure.

Algorithm (The CE Method for the MBC Problem)

1. Choose some $\widehat{\boldsymbol{u}}_0$. Set $t = 1$ (level counter).
2. Generate a sample $\boldsymbol{x}_1, \dots, \boldsymbol{x}_{N_2}$ from the density $f(\cdot; \boldsymbol{u}_{t-1})$ and compute the sample $(1-\rho)$-quantile $\widehat{\gamma}_t$ of the $\widehat{S}(\boldsymbol{x})$ under $\boldsymbol{u}_{t-1}$.
3. Using the same sample, update $\widehat{\boldsymbol{u}}_t$ according to (7.5) and then smooth it out using (7.4).
4. If convergence condition (7.6) is met, then stop; otherwise set $t = t+1$ and reiterate from step 2.

7.3.3 The CE method for the multiple BCP with minimal summarized rank

As was shown in Nikolaev (1976, 1998), the solution of this problem is the following optimal strategy: there exist integer vectors

$$\begin{aligned}
\boldsymbol{\delta}^{(k)} &= (\delta_1^{(k)}, \dots, \delta_{N-k+1}^{(k)}), \quad 0 \le \delta_1^{(k)} \le \dots \le \delta_{N-k}^{(k)} < \delta_{N-k+1}^{(k)} = N, \\
&\vdots \\
\boldsymbol{\delta}^{(2)} &= (\delta_{k-1}^{(2)}, \dots, \delta_{N-1}^{(2)}), \quad 0 \le \delta_{k-1}^{(2)} \le \dots \le \delta_{N-2}^{(2)} < \delta_{N-1}^{(2)} = N, \\
\boldsymbol{\delta}^{(1)} &= (\delta_k^{(1)}, \dots, \delta_N^{(1)}), \quad 0 \le \delta_k^{(1)} \le \dots \le \delta_{n-1}^{(1)} < \delta_N^{(1)} = N, \\
\delta_j^{(i_1)} &\le \delta_j^{(i_2)}, \quad 1 \le i_1 < i_2 \le k, \quad k - i_1 + 1 \le j \le N - i_2 + 1
\end{aligned} \tag{7.7}$$

such that

$$\begin{aligned}
\tau_1^* &= \min\{m_1 : \xi_{m_1} \le \delta_{m_1}^{(k)}\}, \\
\tau_i^* &= \min\{m_i > m_{i-1} : \xi_{m_i} \le \delta_{m_i}^{(k-i+1)}\},
\end{aligned}$$

on the set $D_{i-1} = \{\omega : \tau_1^* = m_1, \dots, \tau_{i-1}^* = m_{i-1}\}$, $i = 2, \dots, k$, $D_0 = \Omega$.

We consider the following problem

$$\min_{\boldsymbol{x} \in \mathcal{X}} \boldsymbol{E}\widehat{S}(\boldsymbol{x}, \boldsymbol{R}),$$

where

$$\mathcal{X} = \{\boldsymbol{x} = (x^{(1)}, \dots, x^{(k)}) : \text{conditions (7.7) are hold}\},$$

$\boldsymbol{R} = (R_1, \ldots, R_N)$ is a random permutation of numbers $1, 2, \ldots, N$, $\widehat{S}(\boldsymbol{x})$ is an unbiased estimator of $\boldsymbol{E}\widehat{S}(\boldsymbol{x}, \boldsymbol{R})$

$$\widehat{S}(\boldsymbol{x}) = \frac{1}{N_1} \sum_{n=1}^{N_1} (R_{n\tau_1} + \cdots + R_{n\tau_k}).$$

As before, we can use the CE algorithm for noisy optimization. We consider a 3-dimensional matrix of parameters $\boldsymbol{u} = \{u_{ijl}\}$

$$u_{ijl} = \boldsymbol{P}\{X_j^{(i)} = l\},\ i = 1, \ldots, k;\ j = k-i+1, \ldots, N-i+1;\ l = 0, \ldots, N-1.$$

It follows easily that

$$f(x_j^{(i)}; \boldsymbol{u}) = \sum_{l=0}^{N-1} u_{ijl} \mathbb{I}_{\{x_j^{(i)} = l\}}.$$

As above, we see that

$$\widehat{u}_{ijl}^{(t)} = \frac{\sum_{n=1}^{N_2} \mathbb{I}_{\{\widehat{S}(\boldsymbol{x}_n) \geq \widehat{\gamma}_t\}} W_{nij}^{(t-1)} \mathbb{I}_{\{X_{nij} = l\}}}{\sum_{n=1}^{N_2} \mathbb{I}_{\{\widehat{S}(\boldsymbol{x}_n) \geq \widehat{\gamma}_t\}} W_{nij}^{(t-1)}}, \quad W_{nij}^{(t-1)} = \frac{\widehat{u}_{ijX_{nij}}^{(0)}}{\widehat{u}_{ijX_{nij}}^{(t-1)}}, \tag{7.8}$$

where $\boldsymbol{x}_n = \{X_{nij}\}$, X_{nij} is a random variable from $f(x_j^{(i)}; \widehat{\boldsymbol{u}}_{t-1})$. Formula (7.3) becomes (7.8).

7.4 Asymptotic Methods

For stopping problems, where the distribution of the random variables is known, calculating the values of game explicitly may be computationally intensive. A theoretical approach can still be developed by computing the asymptotic behavior of the sequence so that the values need not be computed recursively. This is typically performed through the use of the asymptotic approximation of difference equations into differential equations.

Another interesting problem is the duration of the game. By the "duration" (sometimes referred to as "time") of the game or the stopping problem, we refer to how many observations the statistician observes before optimally stopping. It should be emphasized that knowing how long you would be waiting, on average, for potentially large sequences of observations is also useful to understand (see Ernst and Szajowski (2021)), which highlights the need for asymptotic analysis to address this question.

Less focus has been placed on understanding the asymptotic behavior of the stopping duration, with most pre-existing results focusing on secretary-type, so called "no-information", problems where the distribution of the observations is unknown.

The asymptotic expectation and variance of the stopping time for the secretary problem were studied in Mazalov and Peshkov (2004). Similar asymptotic analyses for other variants of no-information problems can be found in Demers (2021), Gilbert and Mosteller (1966), Yasuda (1984), Yeo (1997), where the techniques/formulations used differ depending on the particular variation or structure of the problem.

There is substantially less literature describing the asymptotic behavior for "full-information" problems, when the distribution of the variables is known a priori. A smaller subset addresses the multiple-stopping problem, which we focus on in this section (see Entwistle et al. (2024)).

Gilbert and Mosteller (1966) studied the optimal stopping strategy for the full-information problem in which the objective is to maximize the probability of attaining the best observation, known as the full-information best-choice problem Gnedin (1996). The optimal rule was shown to be a threshold strategy wherein the player would stop on ξ_m if it was the best observed so far after all other observations and its value exceeded a threshold that depended on m. The asymptotic behavior of this rule was also derived.

In scenarios where full information is available, the payoff can be expressed in terms of the actual values of the stopped variables. A notable instance of this is the stopping of the uniform random variables, as discussed in Section 5a of Gilbert and Mosteller's (1966) work, which shares a close connection with Cayley's problem (v. Moser (1956), Guttman (1960) which offer further insights into this area). In their paper Gilbert and Mosteller (1966) presented an asymptotic expression for the expected reward associated with a sequence of n independent and identically distributed (iid) random variables following the standard uniform distribution.

Additionally, Mazalov and Peshkov (2004) found the asymptotic behavior of the expected value and variance of the stopping time as $N/3$ and $N^2/18$, respectively, when the variables are from the uniform distribution. However, the methodologies employed were tailored to the distribution's specific characteristics, posing challenges for generalization to other distribution functions.

In works by Kennedy and Kertz (1990, 1991), the authors utilized extreme value theory to establish limit theorems for threshold-stopped random variables, deriving the asymptotic distribution of the reward sequence in optimal stopping scenarios with iid random variables. The analysis of the asymptotic payoff in multiple cases is briefly addressed in Section 5c of Gilbert and Mosteller's (1966) paper.

Entwistle et al. (2022) outlined a novel approach for a general asymptotic technique for calculating the asymptotic behavior of the payoff as well as $\boldsymbol{E}(\tau_N)$ and $\mathbf{Var}(\tau_N)$ in the single-stopping case, where τ_N is the single-stopping time, as $N \to \infty$ for general classes of probability distributions in the full-information problem where we wish to maximize the expected reward. The techniques in Entwistle et al. (2022), which are used in this section, employ the asymptotic analysis of difference and differential equations in order to establish and solve asymptotic differential equations for the quantities of interest. Differential equations were also used in Bayón et al. (2023), Pearce et al. (2012). In this section, we extend some of the results in Entwistle et al. (2022) to the multiple-stopping case Entwistle et al. (2024) through some inductive arguments and verify these results with simulations. For simplicity, we only analyze continuous distributions and reserve the notation $f(x), F(x)$, and $Q(y) = 1 - F(x)$ for the continuous probability density, cumulative distribution, and "survivor" function, respectively. We use the notation $f(x) \sim g(x)$ to establish the asymptotic relation that $\lim_{x\to\infty} \frac{f(x)}{g(x)} = 1$.

7.4.1 Computing v^{n,k} behavior

We derive a recurrence result akin to what was achieved in the case of single-stopping by Entwistle et al. (2022). It's worth noting that this relation for $v^{n,k}$ will now manifest as a second-order relation. To simplify the evaluation of the expectation, we leverage the property that for any continuously integrable random variable X with CDF $F(x)$, the expectation can be expressed as

$$\boldsymbol{E}(X) = \int_0^\infty (1 - F(x))\, dx - \int_{-\infty}^0 F(x)\, dx. \tag{7.9}$$

Theorem 7.1. *Consider Y as an integrable random variable with a well-defined expectation, drawn from a continuous PDF $f(y)$ with a survivor function $Q(y) = 1 - F(y)$. The value of a sequence with $n+1$ steps and k stops remaining can then be expressed as*

$$v^{n+1,k} = v^{n,k} + \int_{v^{n,k} - v^{n,k-1}}^\infty Q(y)\, dy. \tag{7.10}$$

Proof: For ease of notation, we let $v = v^{n-1,k} - v^{n-1,k-1}$. Then, by definition, we have

$$\begin{aligned} v^{n+1,k} &= \boldsymbol{E}(\max\{\xi_{N-n} + v^{n,k-1}, v^{n,k}\}) \\ &= v^{n,k-1} + \boldsymbol{E}(\max\{\xi_{N-n}, v^{n,k} - v^{n,k-1}\}) \\ &= v^{n,k-1} + \boldsymbol{E}(\max\{\xi_{N-n}, v\}) \end{aligned} \tag{7.11}$$

where the last expectation, using (7.9), is given by

$$\begin{aligned} -\int_{-\infty}^0 P(\max(Y, v \le y)\, dy + &\int_0^v P(\max(Y, v) > y)\, dy + \int_v^\infty P(\max(Y, v) > y)\, dy \\ &= 0 + \int_0^v 1\, dy + \int_v^\infty P(Y > y)\, dy \\ &= v + \int_v^\infty Q(y)\, dy = v^{n,k} - v^{n,k-1} + \int_{v^{n,k} - v^{n,k-1}}^\infty Q(y)\, dy. \end{aligned}$$

Substituting this result, noting that the $v^{n,k-1}$ terms cancel, we obtain the required result. □

We note that if $f(y)$ has bounded support in the positive direction such that $f(y) = 0$ for $y > y_{\max}$, it follows from above that

$$v^{n+1,k} = v^{n,k} + \int_{v^{n,k}-v^{n,k-1}}^{y_{\max}} Q(y)\,dy. \tag{7.12}$$

By the controlling factor method, we also have that $v^{n+1,k} - v^{n,k} \sim (v^{n,k})'$, which can be combined with the previous integral to establish the following:

$$(v^{n,k})' \sim \int_{v^{n,k}-v^{n,k-1}}^{\infty} Q(y)\,dy. \tag{7.13}$$

This may be differentiated on both sides to obtain the asymptotic relation for the second derivative:

$$(v^{n,k})'' \sim -Q(v^{n,k} - v^{n,k-1}) \cdot (v^{n,k} - v^{n,k-1})'. \tag{7.14}$$

This expression can now be rearranged for $Q(v^{n,k} - v^{n,k})$ to give

$$Q(v^{n,k} - v^{n,k-1}) \sim -\frac{(v^{n,k})''}{(v^{n,k} - v^{n,k-1})'}. \tag{7.15}$$

Depending on the asymptotic nature of $v^{n,k}$, $v^{n,k-1}$ and their derivatives, this result can be used through direct substitution. In other scenarios, the derivative expressions may not yield useful expressions and the behavior of $Q(v^{n,k} - v^{n,k-1})$ can be analyzed directly without this result. We will provide variations of this in the subsequent example calculations.

7.4.2 Example calculations

We illustrate the application of these ideas to some common distributions, such as the uniform and exponential distributions. However, the nature of the differential equations that arise from the multiple-stopping problem are much harder to solve—some of them have no solution.

Example 7.1. *The uniform distribution is given by $f(y) = \frac{1}{b-a}$ on $y \in [a, b]$, where $b > a$.*

For the differential equation for the large asymptotic behavior, we have that $Q(y) = \frac{b-y}{b-a}$. Defining $v := v^{n,2} - v^{n,1}$ and rearranging in the asymptotic relation for $(v^{n,2})'$ we obtain

$$v' \sim \int_v^b \frac{b-y}{b-a}\,dy + (v^{n,1})'.$$

From Entwistle et al. (2022), we have the asymptotic relation $v^{n,1} = v_n \sim b - \frac{2(b-a)}{n}$, which can be directly substituted into the above equation:

$$v' \sim \frac{(b-v)^2}{2(b-a)} - \frac{2(b-a)}{n^2}.$$

We can solve this formal differential equation to obtain

$$v \sim b + (a-b)\left(\frac{1+\sqrt{5}}{n} - \frac{2c}{n(n^{\sqrt{5}}+c)}\right) \sim b + \frac{(a-b)(1+\sqrt{5})}{n}$$

where c arises as a constant of integration but may be dropped since it is part of a sub-dominant term. We now replace v with $v^{n,2} - v^{n,1}$, substitute our asymptotic relation for $v^{n,1}$, and rearrange for $v^{n,2}$ to obtain

$$v^{n,2} \sim 2b - \frac{(3+\sqrt{5})(b-a)}{n} \quad \text{as} \quad n \to \infty.$$

We may notice, in general, that whenever v satisfies the relation

$$v' \sim \frac{(b-v)^2}{2(b-a)} - \frac{\Delta(b-a)}{n^2} \tag{7.16}$$

where Δ is some positive constant, that we obtain the asymptotic relation

$$v \sim b + \frac{(a-b)(1+\sqrt{1+2\Delta})}{n}. \tag{7.17}$$

This can be used to generalize the asymptotic behavior for $v^{n,k}$ for the uniform distribution.

Theorem 7.2. *Let $\xi_1, \xi_2, \ldots, \xi_N$ be independent and identically distributed uniform variables on the interval $[a, b]$, where $b > a$. The reward sequence $v^{n,k}$ follows the asymptotic relation*

$$v^{n,k} \sim kb - \frac{\Delta_k(b-a)}{n} \quad \text{as} \quad n \to \infty, \tag{7.18}$$

where $\Delta_{k+1} = 1 + \Delta_k + \sqrt{1+2\Delta_k}$ for $k \geq 0$ and $\Delta_0 = 0$.

Proof: We have shown that this is true for $k = 1$ and $k = 2$. Now, assume that

$$v^{n,k} \sim kb - \frac{\Delta_k(b-a)}{n} \quad \text{as} \quad n \to \infty.$$

Now define $v = v^{n,k+1} - v^{n,k}$. We have

$$v' \sim \frac{(b-v)^2}{2(b-a)} - \frac{\Delta_k(b-a)}{n^2}$$

which, from (7.16) and (7.17), has the asymptotic solution

$$v \sim b + \frac{(a-b)(1+\sqrt{1+2\Delta_k})}{n}.$$

To conclude the proof, we rearrange for $\upsilon^{n,k+1}$ to obtain

$$\upsilon^{n,k+1} \sim b(k+1) - \frac{(1+\Delta_k+\sqrt{1+2\Delta_k})(b-a)}{n} = b(k+1) - \frac{\Delta_{k+1}(b-a)}{n}.$$

□

We noted in Entwistle et al. (2022) that the behavior of $\upsilon^{n,1}$ was identical for the family of distributions with exponential tails. The next example will seek to unify such distributions that corresponded to $\alpha = 1$ in the multiple-stopping scenario.

Example 7.2. *A continuous probability density function $f(y)$ is given with a survival function that, for sufficiently large y, satisfies*

$$\left|Q(y) - \gamma e^{-(y/\beta)}\right| < \frac{e^{-(y/\beta)}}{y^{\Delta}} \tag{7.19}$$

for positive Δ, and where β and γ are positive constants. Assume each of the terms in the sequence of reward values $\upsilon^{n,1}, \ldots, \upsilon^{n,k}$ increases without bound.

We obtain the ordinary differential equation

$$\frac{d\upsilon^{n,k}}{dn} \sim \int_{\upsilon^{n,k}-\upsilon^{n,k-1}}^{\infty} Q(y)\,dy = \gamma\beta\Gamma\left(1, \frac{\upsilon^{n,k}-\upsilon^{n,k-1}}{\beta}\right) + g^*(n), \tag{7.20}$$

where Γ represents the upper incomplete gamma function, and

$$|g^*(n)| < \int_{\upsilon^{n,k}-\upsilon^{n,k-1}}^{\infty} \frac{e^{-y/\beta}}{y^{\Delta}}\,dy < \frac{1}{(\upsilon^{n,k}-\upsilon^{n,k-1})^{\Delta}} \int_{\upsilon^{n,k}-\upsilon^{n,k-1}}^{\infty} e^{-y/\beta}\,dy$$

which is sub-dominant in the asymptotic differential equation. The solution to the differential equation is thus approximated by

$$\frac{d\upsilon^{n,k}}{dn} \sim \gamma\beta\Gamma\left(1, \frac{\upsilon^{n,k}-\upsilon^{n,k-1}}{\beta}\right).$$

This gives

$$\frac{d\upsilon^{n,k}}{dn} \sim \gamma e^{-(\upsilon^{n,k}-\upsilon^{n,k-1})/\beta} \qquad \text{as} \qquad n \to \infty. \tag{7.21}$$

We have, from Entwistle et al. (2022), $\upsilon^{n,1} \sim \beta\log(n)$, and so the general case for $k > 2$ may be presented by mathematical induction.

Theorem 7.3. *Let $\xi_1, \xi_2, \ldots, \xi_N$ be random variables from a distribution $f(y)$ that, for sufficiently large y, satisfies*

$$\left|Q(y) - \gamma e^{-(y/\beta)}\right| < \frac{e^{-(y/\beta)}}{y^{\Delta}} \tag{7.22}$$

for positive Δ, and where β and γ are positive constants. Assume each of the terms in the sequence of reward values $\upsilon^{n,1}, \ldots, \upsilon^{n,k}$ increases without bound. Then, the asymptotic behavior of $\upsilon^{n,k}$ is given by

$$\upsilon^{n,k} \sim \beta\log(n^k) \qquad \text{as} \qquad n \to \infty. \tag{7.23}$$

Proof: From (7.21), we have that the behavior of $v^{n,k}$ is related by

$$\frac{dv^{n,k}}{dn} \sim \gamma e^{-v^{n,k}/\beta} e^{v^{n,k-1}/\beta}$$

We have verified the claim for $n = 1$ in Entwistle et al. (2022). Now we assume that $v^{n,k} \sim \beta \log(n^k)$ as $n \to \infty$ and use this to prove the same for $v^{n,k+1}$. We write the asymptotic differential equation $(v^{n,k+1})' \sim \gamma e^{-(v^{n,k+1}-v^{n,k})/\beta}$ and substitute for our assumed asymptotic equation for $v^{n,k}$ to obtain

$$(v^{n,k+1})' \sim \beta e^{-(v^{n,k+1}-\beta \log(n^k))/\beta} = \beta n^k e^{-v^{n,k+1}/\beta}$$

which is a separable differential equation with solution

$$v^{n,k+1} \sim \beta \log(n^{k+1})$$

as required. □

7.4.3 Calculating the Optimal Expectation

In this section, we extend the single-stopping calculation approach to handle multiple-stopping rules. We start by deriving the asymptotic for the expected value τ_1^*, the first optimal stopping time. Then, under certain conditions, we can recursively find the expectations for subsequent stopping times. We introduce notation to accommodate multiple reward sequences and k stopping variables. Let $\tau_1^\star, \tau_2^\star, \ldots, \tau_k^\star$ represent the consecutive optimal stopping times, respectively. Also, define $w_{i,j} = P(y < v_{i,j})$ and $v_{i,j} = v^{N-j,k-i+1} - v^{N-j,k-i}$, where $i = 1, \ldots, k$, and $j = 1, \ldots, N$.

7.4.4 An Asymptotic Equation for $\boldsymbol{E}(\tau_1^*)$

By recalling the notation from the beginning of this section, we obtain $\boldsymbol{E}(\tau_1^*)$ through

$$\begin{aligned}
\boldsymbol{E}(\tau_1^*) &= \sum_{n=1}^{N-k+1} nP(\xi_1 < v_{1,1}, \ldots, \xi_{n-1} < v_{1,n-1}, \xi_n \geq v_{1,n}) \\
&= (1 - w_{1,1}) + 2w_{1,1}(1 - w_{1,2}) + \cdots + Nw_{1,1}w_{1,2} \cdots w_{1,N-k} \\
&= 1 + \sum_{n=1}^{N-k} \prod_{j=n}^{N-k} w_{1,N-k+1-j},
\end{aligned}$$

We split the summation for $\boldsymbol{E}(\tau_1^*)$ on a value k^*, where $0 \ll k^* \ll N$:

$$\boldsymbol{E}(\tau_1^*) = 1 + \sum_{n=1}^{k^*-1} \prod_{j=n}^{N-k} w_{1,N-1-j} + \sum_{n=k^*}^{N-k} \prod_{j=n}^{N-k} w_{1,N-1-j}$$

where we apply the fact that $0 < w_{1,N-1-j} < 1$ to obtain a bound for the first summation term:

$$0 < \sum_{n=1}^{k^*-1} \prod_{j=n}^{N-k} w_{1,N-1-j} < k^* - 1.$$

For the second summation term, as k^* is large in the limit as $N \to \infty$, we may use the asymptotic approximations for $v^{n,k}$ obtained in the previous section.

$$\begin{aligned}\sum_{n=k^*}^{N-k} \prod_{j=n}^{N-k} w_{1,N-1-j} &= \sum_{n=k^*}^{N-k} \prod_{j=n}^{N-k} (1 - Q(v_{1,N-1-j})) \\ &\sim \sum_{n=k^*}^{N-k} \prod_{j=n}^{N-k} \left(1 + \frac{(v^{j+1,k})''}{(v^{j+1,k} - v^{j+1,k-1})'}\right).\end{aligned}$$

For many distributions, this can be simplified through using the large-n asymptotic for $v^{n,k}$ and its derivatives, or obtaining an asymptotic expression for $Q(v^{n,k} - v^{n,k-1})$ through other means. In the case where $Q(v_{1,N-1-k}) \sim \frac{\lambda}{j}$, we have from Entwistle et al. (2022) that

$$\boldsymbol{E}(\tau_1^*) \sim \frac{N}{\lambda + 1} \qquad \text{as} \qquad N \to \infty.$$

Example 7.3. *Uniform Distribution for k stops.*

For simplicity, we first consider the double stopping problem ($k = 2$). From Example 7.1, we have that

$$v^{n,1} \sim b - \frac{2(b-a)}{n} \qquad \text{and} \qquad v^{n,2} \sim 2b - \frac{(3+\sqrt{5})(b-a)}{n} \qquad \text{as} \qquad n \to \infty.$$

This obtains $v^{n,2} - v^{n,1} \sim b - \frac{(b-a)(1+\sqrt{5})}{n}$, and thus, $Q(v^{n,2} - v^{n,1}) \sim \frac{1+\sqrt{5}}{n}$.

We then have that

$$\boldsymbol{E}(\tau_1^*) \sim \sum_{n=k^*}^{N-2} \prod_{j=n}^{N-2} (1 - Q(v_{1,N-1-j})) \sim \sum_{n=k^*}^{N-2} \prod_{j=n}^{N-2} \left(1 - \frac{1+\sqrt{5}}{j}\right) \sim \frac{N}{2+\sqrt{5}}.$$

For the general result with $k > 2$ stops for the uniform distribution, we apply the asymptotic behavior of $Q(v^{n,k} - v^{n,k-1})$ to obtain

$$\boldsymbol{E}(\tau_1^*) \sim \sum_{n=k^*}^{N-k} \prod_{j=n}^{N-k} \left(1 - \frac{1+\sqrt{1+2\Delta_{k-1}}}{j}\right) \sim \frac{N}{2+\sqrt{1+2\Delta_{k-1}}}.$$

For $k = 1$, conveniently $\Delta_0 = 0$, this retrieves $\frac{N}{3}$. For $k = 2$, we have that $\Delta_1 = 2$, and so, this retrieves $\frac{N}{2+\sqrt{5}}$, consistent with our previous results.

Example 7.4 (Distributions with an Exponential Tail (k stops)). *We once again consider a probability distribution $f(y)$ is given with a survival function that, for sufficiently large y, satisfies*

$$\left|Q(y) - \gamma e^{-(y/\beta)}\right| < \frac{e^{-(y/\beta)}}{y^{\Delta}} \tag{7.24}$$

where the additional conditions are described in Example 7.2.

We found that the sequences $v^{n,k}$ satisfy the asymptotic relation $v^{n,k} \sim \beta \log(n^k)$ and so

$$v^{n,k} - v^{n,k-1} \sim \beta \log(n^k) - \beta \log(n^{k-1}) = \beta \log(n).$$

From this we obtain asymptotic relations for the derivatives:

$$(v^{n,k} - v^{n,k-1})'' \sim \frac{\beta}{n} \quad \text{and} \quad (v^{n,k})'' \sim -\frac{\beta k}{n^2} \quad \text{as} \quad n \to \infty.$$

Hence, we obtain

$$\boldsymbol{E}(\tau_1^*) \sim \sum_{n=k^*}^{N-k} \prod_{j=n}^{N-k} \left(1 + \frac{(v^{j+1,k})''}{(v^{j+1,k} - v^{j+1,k-1})'}\right) \sim \sum_{n=k^*}^{N-k} \prod_{j=n}^{N-k} \left(1 - \frac{k}{j}\right) \sim \frac{N}{k+1}.$$

Provided that $\boldsymbol{E}(\tau_1^*)$ is asymptotically of this form, we may make use of linearity of expectation to obtain convenient conditional formulae that are not as complicated as those encountered in the previous section.

7.4.5 An Inductive Approach for $\boldsymbol{E}(\tau_j^*), j > 1$

Due to the independence of observations, it starts to make physical sense to view the expectation of the $(j+1)$th stopping time as a function of only the previous stopping time. We, thus, investigate the properties of the $(j+1)$th stopping time when conditioned on the jth. We would expect then to only need to add the additional expected number of observations to stop one more time out of a "reduced" optimal stopping problem. We now introduce the notation $\tau_{N,j,k}^*$ to allow for more flexible interactions between stopping times.

Definition 7.1. *Let $\tau_{N,j,k}^*$ denote the jth stopping time (out of k) in the optimal stopping problem with N observations.*

Here, $\tau_{j,N,k}^*$ denotes the τ_j^* used in the previous section. $\tau_{1,N,1}^*$ corresponds to the τ^* in the single-stopping problem.

Theorem 7.4. *Let $\xi_1, \xi_2, \ldots, \xi_N$ be independent and identically distributed random variables where the expectations of the optimal stopping times $\{\tau_{j,N,k}^*\}$ exist. Suppose further that the first stopping time out of k stops has asymptotic expectation $\boldsymbol{E}(\tau_{1,N,k}^*) \sim \frac{N}{\lambda_{1,k}}$ for some constant $\lambda_{1,k}$. Then, the following relation holds:*

$$\boldsymbol{E}(\tau_{j+1,N,k}^*) \sim \boldsymbol{E}(\tau_{j,N,k}^*) + \frac{N - \boldsymbol{E}(\tau_{j,N,k}^*)}{\lambda_{1,k-j}} \tag{7.25}$$

where $\lambda_{1,k-j}$ is some other positive constant.

Proof: Let us denote

$$\boldsymbol{\tau}^*_{j,N,k}(m) = \boldsymbol{E}(\tau^*_{j,N,k}|\tau^*_{j-1,N,k} = m)$$

and

$$D_{m,j,k,N} = \{\omega : \xi_{m_j} \geq v^{N-m_j,k-j+1} - v^{N-m_j,k-j}\}$$

when $m < m_j \leq N - k + j$, $j = 1, 2, \ldots, k$. We have

$$\boldsymbol{E}\big[\min_{D_{m,j,k,N}} (m_{j+1})\big] = m + \boldsymbol{E}\big[\min_{D_{1,1,k-j,N-m}} (m_1)\big] = m + \boldsymbol{\tau}^*_{j+1,N,k}(1),$$

We first show that the conditional expectation, $\boldsymbol{E}(\tau^*_{j+1,N,k}|\tau^*_{j,N,k} = m)$, is given by

$$\begin{aligned}
&\boldsymbol{E}(\min\{m_{j+1} : m < m_{j+1} \leq N - k + j + 1,\\
&\qquad \xi_{m_{j+1}} \geq v^{N-m_{j+1},k-j} - v^{N-m_{j+1},k-j-1}\})\\
&= m + \boldsymbol{E}(\min\{m_1 : 1 \leq m_1 \leq (N-m) - (k-j) + 1,\\
&\qquad \xi_{m_1} \geq v^{N-m,k-j} - v^{N-m,(k-j)-1}\})\\
&= m + \boldsymbol{E}(\tau_{1,N-m,k-j}),
\end{aligned}$$

where the last equality follows directly from the definition of τ_1^* in the previous section. By the assumption of the form of the expectation, we thus have that

$$\boldsymbol{E}(\tau^*_{j+1,N,k}|\tau^*_{j,N,k}) \sim \tau^*_{j,N,k} + \frac{N - \tau^*_{j,N,k}}{\lambda_{1,k-j}} \tag{7.26}$$

and so through linearity properties of expectation, as well as the law of total expectation, we finally obtain

$$\boldsymbol{E}(\tau^*_{j+1,N,k}) \sim \boldsymbol{E}(\tau^*_{j,N,k}) + \frac{N - \boldsymbol{E}(\tau^*_{j,N,k})}{\lambda_{1,k-j}}. \tag{7.27}$$

□

This has a reasonable physical interpretation, since we would expect that the expectation of the $(j+1)$th stopping time will be the expectation of the jth stopping time, as well as the additional time it would take to stop one more time in the revised stopping problem to stop at the next observation. This revised stopping problem has a reduced $N - \boldsymbol{E}(\tau^*_{j,N,k})$ number of observations on average, as well as $k - j$ stops remaining, since we have already stopped j times out of the original k times. A consequence of this theorem is that the asymptotic expectations may all be considered linear.

7.4.6 Asymptotic Equations for $\boldsymbol{E}(\tau^*_{j,N,k})$

We now demonstrate how the ideas discussed in the previous section can be applied to obtain all of the multiple-stopping times for some classes of distribution. Simulations

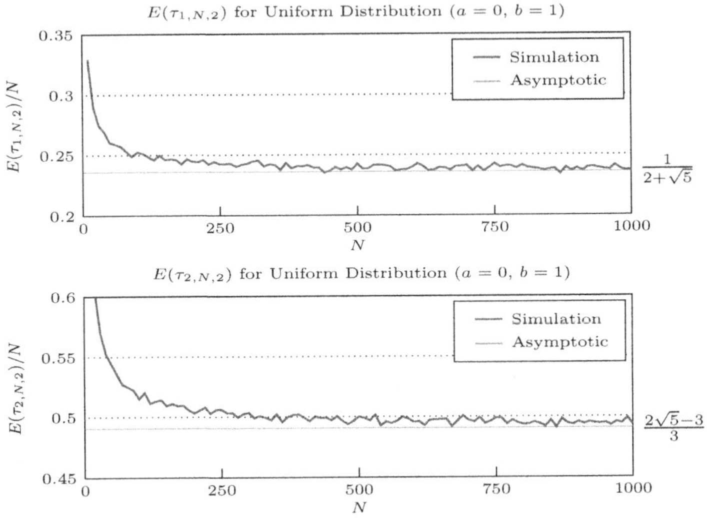

FIGURE 7.2
Large-N asymptotic predictions for the expectation of the optimal double stopping rule for the standard uniform distribution.

were also conducted to support the asymptotic calculations in this section. Figure 7.2 compares the large-N asymptotic predictions for the expected duration for the standard uniform distribution under the optimal double stopping rule and Figures 7.3 and 7.4 shows a similar comparison under a triple stopping rule for the exponential and the gamma distribution with the rows corresponding to the first, second, and third stops, respectively. In this section, we will now adopt the previous notation for the jth stopping time out of k, for N observations as $\tau^*_{j,N,k}$.

These results were verified by simulation (see Figure 7.2) which show the realized expectation converging to the asymptotic prediction. This procedure can be extended to calculate further results for more stopping times permitted. Simulation results (see Figures 7.3 and 7.4) for the exponential and gamma distribution further support these asymptotic calculations.

Example 7.5. *Double Stopping on the Uniform Distribution*
For the uniform distribution, we obtain an asymptotic expression for the first stopping time out of k stops. It is given by

$$\boldsymbol{E}(\tau^*_{1,N,k}) \sim \frac{N}{2 + \sqrt{1 + 2\Delta_{k-1}}},$$

where Δ_k is defined through the recursive scheme

$$\begin{cases} \Delta_{k+1} = 1 + \Delta_k + \sqrt{1 + 2\Delta_k} & k = 0, 1, 2, 3, \ldots \\ \Delta_0 = 0 \end{cases}$$

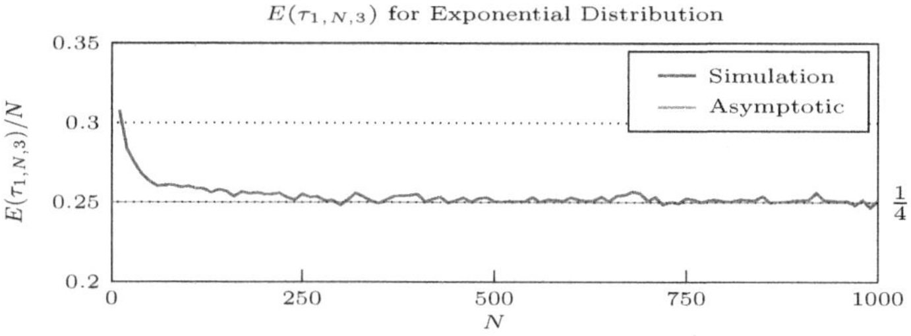

(a) $\boldsymbol{E}(\tau_1, N, 3)$ for the Exponential Distribution.

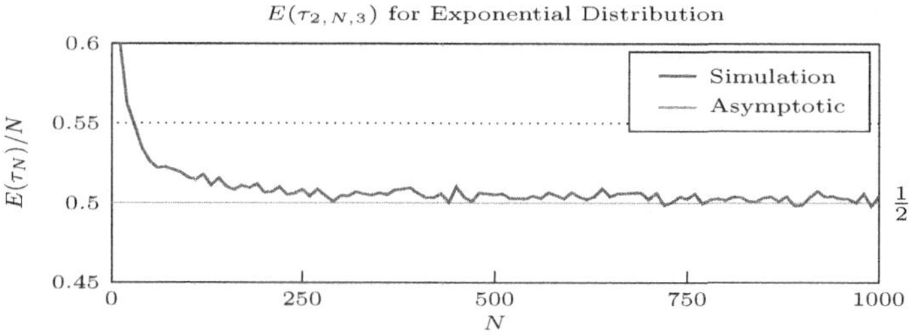

(b) $\boldsymbol{E}(\tau_2, N, 3)$ for the Exponential Distribution.

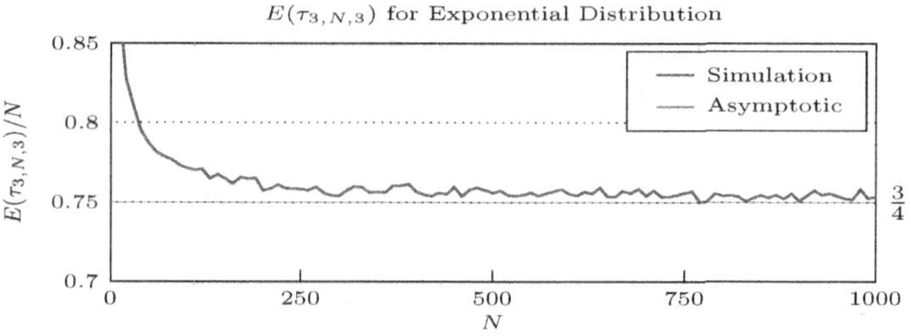

(c) $\boldsymbol{E}(\tau_3, N, 3)$ for the Exponential Distribution.

FIGURE 7.3
Large-N asymptotic predictions for the expectation of the optimal triple stopping rule for the standard exponential distribution.

We can then apply Equation (7.27), for the example of the double stopping case, recursively to obtain $\boldsymbol{E}(\tau_{1,N,2}) \sim \frac{1}{2+\sqrt{5}}N$ and $\boldsymbol{E}(\tau_{2,N,2}) \sim \frac{2\sqrt{5}-3}{3}N$.

Example 7.6. *Multiple-Stopping on Distributions with Exponential Tails*
We saw that the expressions can get somewhat out of hand for the uniform distribution. However, the structure can sometimes behave nicely to lead to a closed form result that need not be recursively evaluated. This is true for the class of distributions outlined in Example 7.2. We prove the following theorem for the general asymptotic expectation:

Theorem 7.5. *Let $\xi_1, \xi_2, \ldots, \xi_N$ be independent and identically distributed random variables. Define $\tau^*_{j,N,k}$ to be the jth stopping time out of k stops for the sequence of*

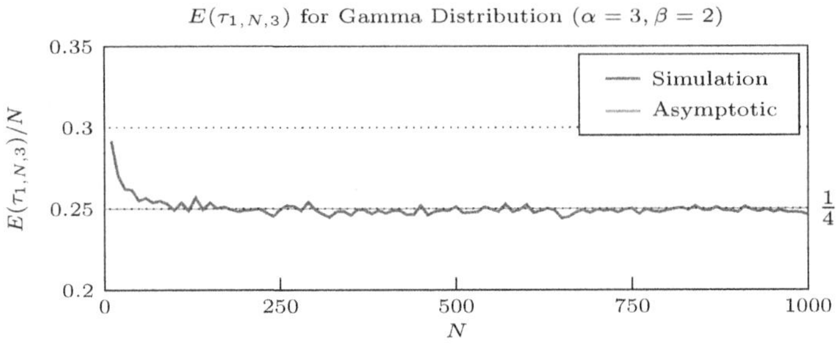

(a) $\boldsymbol{E}(\tau_1, N, 3)$ for the Gamma Distribution ($\alpha = 3, \beta = 2$).

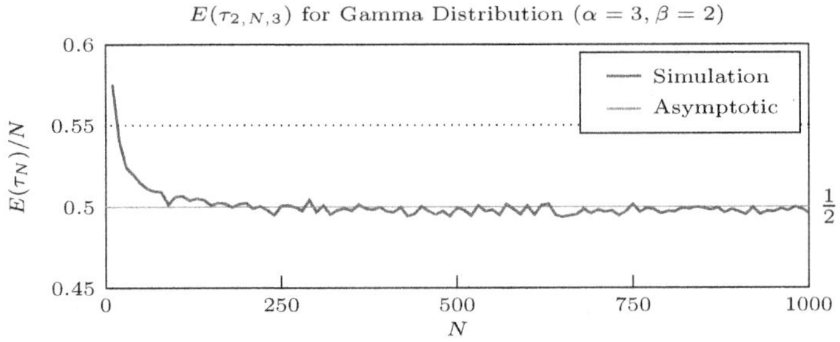

(b) $\boldsymbol{E}(\tau_2, N, 3)$ for the Gamma Distribution ($\alpha = 3, \beta = 2$).

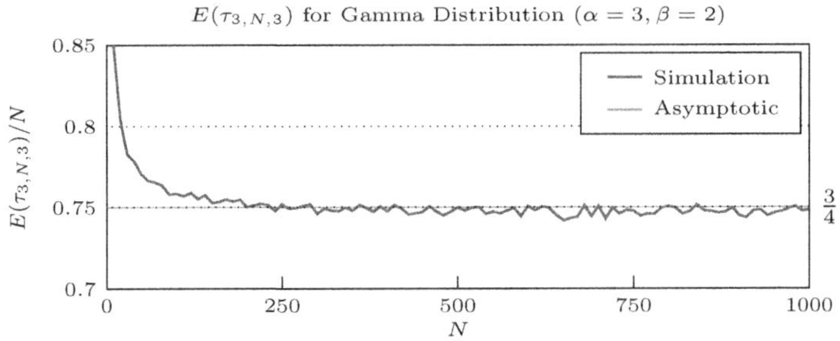

(c) $\boldsymbol{E}(\tau_3, N, 3)$ for the Gamma Distribution ($\alpha = 3, \beta = 2$).

FIGURE 7.4
Large-N asymptotic predictions for the expectation of the optimal triple stopping rule for the Gamma Distribution ($\alpha = 3$, $\beta = 2$).

N observations. If $\boldsymbol{E}(\tau^*_{1,N,k}) \sim \frac{N}{k+1}$, then we have that

$$\boldsymbol{E}(\tau^*_{j,N,k}) \sim \frac{jN}{k+1} \qquad \textit{for} \qquad j = 1, 2 \ldots, k. \tag{7.28}$$

Proof: We see that the base case is in agreement; we now proceed with the inductive argument that

$$\begin{aligned}\boldsymbol{E}(\tau^*_{j+1,N,k}) &\sim \boldsymbol{E}(\tau^*_{j,N,k}) + \frac{N - \boldsymbol{E}(\tau^*_{j,N,k})}{k-j+1} \qquad \text{(from Theorem 7.4)}\\ &= \frac{jN}{k+1} + \frac{N - \frac{jN}{k+1}}{k-j+1} = \frac{(j+1)N}{k+1}.\end{aligned}$$

□

Part IV

Auxiliary and Supplementary Material

Multiple optimal stopping problems have received considerable attention in recent years. We will present such a discrete-time stochastic model in which the moments of stopping constitute the strategies of one or more decision-makers. The one-decision-maker model of multiple-stopping was first formulated by Haggstrom (1967), and the two-decision-maker model by Dynkin (1969). Later work expanded the classes of model-defining processes (v. Nikolaev (1977, 1998), Nikolaev and Sofronov (2007), Sofronov (2013, 2016, 2018)), and rationalization methods with more decision-makers (v. Yasuda et al. (1980, 1982), Stadje (1985), Nakai (1997), Sakaguchi (1995), Szajowski (1993)).

Spectacular applications of **MSP**s appear in financial mathematics (cf. Chandramouli and Haugh (2012)). They appear in the context of pricing swing options in electricity markets (see, e.g., Keppo (2004)) and exotic options such as chooser caps (see, e.g., Meinshausen and Hambly (2004)) in the fixed income derivatives markets. Standard approaches to solving these problems are based on dynamic programming, but the curse of dimensionality implies that only small problems can be solved exactly; see, e.g., Jaillet et al. (2004). As a result, considerable effort has been spent to obtain good approximate solutions to multiple optimal stopping problems. These approximate solution techniques have been inspired by earlier work on the pricing of American options. In the case of American options, the cross-path regression approach of Longstaff and Schwartz (2001) and Tsitsiklis and Van Roy (2001) has been fundamental. They used simple linear regressions to obtain a very good approximation to the value function. Moreover, the resulting approximate value function could then be used to obtain a good feasible exercise policy. The value of this policy therefore constituted a lower, i.e., primal, bound on the true value of the option. To determine just how good such a policy was, Haugh and Kogan (2004) and Rogers (2002) independently developed *dual representations* that could be used to generate upper bounds, that is, dual, on the optimal value function of the optimal stopping problem. If the primal and dual bounds were sufficiently close to one another, then we could conclude that the suboptimal exercise policy corresponding to the lower bound was close to optimal. Since these dual representations were developed, there have been other significant developments, including, for example, by Andersen and Broadie (2004), Jamshidian (2007), Chen and Glasserman (2007), Desai et al. (2012).

A

Uncertainty, Risk and Simulations

Motto:
All's well that ends well

Uncertainty and risk are concepts that are often associated with decision-making and forecasting. While they are related, they have distinct characteristics.

Uncertainty refers to the lack of knowledge or information about future events or outcomes. It can arise from various sources such as incomplete data, limited understanding of the situation, or unpredictable external factors. In uncertain situations, the probabilities or likelihoods of different outcomes may not be known or accurately estimated. Uncertainty is often depicted with a variety of possible outcomes or scenarios and can make decision-making challenging due to the lack of clear guidance.

Probabilistic modeling is often used as an adequate method to address uncertainty. Probabilistic modeling involves representing uncertain factors through probability distributions and using them to make predictions or evaluate different scenarios.

Probabilistic models allow decision-makers to explicitly capture the uncertainty surrounding certain variables or events. By incorporating probabilistic distributions, such as normal distributions or Bayesian probability models, probabilistic modeling helps quantify the range of possible outcomes and their associated probabilities. This enables decision-makers to assess the likelihood of different outcomes and make informed decisions in the face of uncertainty.

Probabilistic modeling can be particularly helpful in situations where there are limited data, incomplete information, or unpredictable variables. It allows decision-makers to understand the risk associated with different choices and helps in evaluating trade-offs and potential consequences. Additionally, probabilistic modeling can provide insights into the sensitivity of outcomes to various inputs, enabling decision-makers to identify critical factors or sources of uncertainty.

Although probabilistic modeling is a valuable tool for addressing uncertainty, it is important to acknowledge that it relies on assumptions and simplifications. The accuracy and reliability of probabilistic models depend on the quality of the data, the appropriateness of chosen probability distributions, and the validity of underlying assumptions. Therefore, it is crucial to carefully consider the limitations and uncertainties associated with the model itself.

In general, probabilistic modeling is a widely used and effective method of managing uncertainty by providing a structured framework for quantifying and analyzing risks.

DOI: 10.1201/9781003407102-A

On the other hand, risk is a measure of the potential negative consequences or losses associated with a specific decision or action. Unlike uncertainty, risk assumes that the probabilities or likelihoods of different outcomes are known or can be estimated. Risk is typically quantified and expressed in terms of probability and impact. Capture the concept of the potential downside or harm that may result from a decision or event.

In summary, uncertainty refers to the lack of knowledge or information about future events, while risk quantifies the potential negative consequences associated with a specific decision or action. Both uncertainty and risk can affect decision-making and require strategies to address them effectively. A simulation is the imitation of the operation of a real-world process or system over time. Whether done by hand or on a computer, simulation involves the generation of an artificial history of a system and the observation of that artificial history to draw inferences concerning the operating characteristics of the real system (v. Banks et al. (2010)).

Reasons for introducing simulation come down to cost savings. Thinking in terms of costs is a simplification. A better expression of the philosophy of simulation is the creation and observation of a digital twin for the analyzed system or process.

In practice, simulations allow testing and experimentation in a more cost-effective or safer environment. Using data collected from the real world, we can create a virtual testbed to implement ideas without wasting physical resources or risking human and animal health.

The option of mathematical or digital model allows us to run tests whilst mitigating risk. No actual systems are being tampered with and if the model fails you can start again with no impact beyond the virtual realm. In practice, such experiments based on the mathematical model are the main simulation examples. Another reason for simulation would be the speed and ease of results[1].

A.1 Introduction to Probability Theory

Kolmogoroff (1933) formulated the idea that a normalized measurement space can serve to construct a theory of probability that meets all the customary demands of rigor in mathematics. Lebesgue's original investigations of the measure and integration concepts were purely geometrically motivated, and geometric ideas were in the foreground during their development. It was Émile Borel who first demonstrated the utility of their theory for solving probabilistic problems, in connection with the law of large numbers (cf. Borel (1909)).

[1]The Role of Simulation in Decision-Making.

A.1.1 The basic concepts of probability theory

The basic context for defining probability-theoretic concepts is always a normalized measure space $(\Omega, \mathcal{B}, \boldsymbol{P})$, where

- Ω is a set of elementary events;
- $\mathcal{B}$ is the σ-algebra of subsets in Ω;
- $\boldsymbol{P}$ is a measure on $\mathcal{B}$ normalized by $\boldsymbol{P}(\Omega) = 1$.

Kolmogoroff's axioms

The basic assumption of probability theory is to take the elementary event as read. The set of axioms was formulated by Kolmogoroff (1933).

1. The family of subsets $\mathcal{B}$ of the set of the elementary event Ω fulfills requirements
   (a) $\Omega \in \mathcal{B}$;
   (b) If the subset $A \in \mathcal{B}$, then $\bar{A} \in \mathcal{B}$, where $\bar{A}$ is the complementary event.
   (c) If $A_i \in \mathcal{B}$, $i = 1, 2, \ldots$, then $\bigcup_{i=1}^{\infty} A_i \in \mathcal{B}$.
2. The probability measure $\boldsymbol{P}$ on $(\Omega, \mathcal{B})$ fulfills requirements
   (a) For every $A \in \mathcal{B}$, we have $0 \leq \boldsymbol{P}(A) \leq 1$;
   (b) $\boldsymbol{P}(\Omega) = 1$;
   (c) If $A_i \in \mathcal{B}$, $i = 1, 2, \ldots$ such that $A_i \cap A_j = \emptyset$, then $\boldsymbol{P}(\bigcup_{i=1}^{\infty} A_i) = \sum_{i=1}^{\infty} \boldsymbol{P}(A_i)$.

Remark 22 (σ-algebra). *The family $\mathcal{B}$ has the following properties:*

- $\emptyset \in \mathcal{B}$;
- $A \cap B = \overline{\bar{A} \cup \bar{B}}$;
- $\cup, \cap$ – *are commutative, associative, distributive, idempotent*

Remark 23 (Further properties of $(\Omega, \mathcal{B}, \boldsymbol{P})$). *Let us present some important and useful properties of probabilities.*

- *For any $A, B \in \mathcal{B}$, we have $\boldsymbol{P}(A \cup B) = \boldsymbol{P}(A) + \boldsymbol{P}(B) - \boldsymbol{P}(A \cap B)$;*
- *For any $A \in \mathcal{B}$, we have $\boldsymbol{P}(\bar{A}) = 1 - \boldsymbol{P}(A)$;*
- *For any $A, B \in \mathcal{B}$, we have $A \subset B$ implies $\boldsymbol{P}(A) \leq \boldsymbol{P}(B)$;*
- *If $A \subset B$, $A, B \in \mathcal{B}$, then $\boldsymbol{P}(B \setminus A) = \boldsymbol{P}(B) - \boldsymbol{P}(A)$;*

Independence.

Definition A.1. *In probability theory, the independence of events refers to a situation where the occurrence or non-occurrence of one event does not affect the probability of another event happening. In simpler terms, two events, A and B, are considered independent if the probability of A occurring is the same whether or not B has occurred, and vice versa.*

- *The conditional probability of A given B for any two sets* $A, B \in \mathcal{B}$ *is defined as a number* $\boldsymbol{P}(A|B)$ *such that* $\boldsymbol{P}(A|B)\boldsymbol{P}(B) = \boldsymbol{P}(A \cap B)$. *If* $\boldsymbol{P}(B) > 0$ *then this is equivalent to the formula:*

$$\boldsymbol{P}(A|B) = \frac{\boldsymbol{P}(A \cap B)}{\boldsymbol{P}(B)};$$

  A is independent of B if $\boldsymbol{P}(A|B) = \boldsymbol{P}(A)$.
  If this equation holds, it indicates that knowledge about the occurrence or non-occurrence of one event provides no information about the likelihood of the other event happening.
- *Let* $\mathcal{A}$ *and* $\mathcal{B}$ *be families of events. We say that* $\mathcal{A}$ *and* $\mathcal{B}$ *are independent when for every* $A \in \mathcal{A}$ *and every* $B \in \mathcal{B}$, *we have* $\boldsymbol{P}(A \cap B) = \boldsymbol{P}(A) \cdot \boldsymbol{P}(B)$.

Independence is a fundamental concept in probability theory and is crucial in various statistical and decision-making applications.

Theorem A.1 (Formula of total probability). *On a given probability space* $(\Omega, \mathcal{F}, \boldsymbol{P})$ *let* $\{A_i\}_{i=1}^m \subset \mathcal{F}$ *be a family of disjoint events, where* $\boldsymbol{P}(A_i) > 0$ *for every* $i = 1, 2, \ldots, m$, $m \in \overline{\mathbb{N}}$ *and* $B \subset \bigcap_{i=1}^m A_i$, $B \in \mathcal{F}$. *We then have the following.*

$$\boldsymbol{P}(B) = \sum_{i=1}^{m} \boldsymbol{P}(B|A_i)\boldsymbol{P}(A_i). \tag{A.1}$$

Theorem A.2 (Bayes theorem). *Under the description of event like in Theorem A.1, we have*

$$\boldsymbol{P}(A_j \mid B) = \frac{\boldsymbol{P}(B|A_j)\boldsymbol{P}(A_j)}{\sum_{i=1}^{m} \boldsymbol{P}(B|A_i)\boldsymbol{P}(A_i)}. \tag{A.2}$$

Remark 24 (Terminology). *For an epistemological interpretation the following terminology for proposition* A_j *and evidence or background* B *is used.*

a priori $\boldsymbol{P}(A_j)$ *is called the prior probability or the initial degree of belief in* A_j *(v.* (A.1)*).*

a posteriori $\boldsymbol{P}(A_j|B)$ *is called the posterior probability, that is, the probability of* A_j *after taking into account* B *(cf.* (A.2)*).*

likelihood $\boldsymbol{P}(B|A_j)$ *is the conditional probability or likelihood, the degree of belief in* B *given that the proposition* A_j *is true.*

Random variables.

In order to provide analytical tools within well-known frameworks, random events are mapped to real numbers. This mapping is called a random variable when it meets the measurability conditions.

Definition A.2 (Random variable). *A random variable X is a mapping from the sample space Ω with the class of events $\mathcal{F}$ to the real line $\Re$ with the Borel σ field $\mathcal{B}$*[a] *that has properties for every $B \in \mathcal{B}$, we have $X^{-1}(B) \in \mathcal{F}$. That is, $X : \Omega \to \Re$ is a random variable if*

$$X^{-1}(B) \in \mathcal{F}, \text{ for every } B \in \mathcal{B}.$$

[a]The Borel σ-field in $\Re$ is the smolest σ-field that contains all intervals on $\Re$.

The measurability of X assures that probability can be assigned to all Borel subsets of $\Re$ via the assignment:

$$\boldsymbol{P}_X(B) = \boldsymbol{P}(\{\omega : X(\omega) \in B\} = \boldsymbol{P}(X^{-1}(B)), \text{ for every } B \in \mathcal{B}. \tag{A.3}$$

In this way, X induces a probability measure $\boldsymbol{P}_X$ on $(\Re, \mathcal{B})$. That is, $(\Re, \mathcal{B}, \boldsymbol{P}_X)$ is the probability space. Once $\boldsymbol{P}_X$ is determined, the structure of the underlying probability space $(\Omega, \mathcal{F}, \boldsymbol{P})$ is irrelevant in describing the probabilistic behavior of the random variable X.

The information contained in the probability measure $\boldsymbol{P}_X$ is more succinctly described in terms of the *cumulative distribution function* (cdf) of X, defined as

$$\boldsymbol{F}_X(x) = \boldsymbol{P}(\{\omega : X(\omega) \leq x\}) = \boldsymbol{P}_X((-\infty, x]), \text{ for every } x \in \Re.$$

Either of the two function $\boldsymbol{F}_X$ or $\boldsymbol{P}_X$ determines the other.

The family of all **cdf**'s $\mathcal{F}(x)$ is the set of all nondecreasing right-continuous functions, having left-hand limits with $\lim_{x\to-\infty} \boldsymbol{F}(x) = 0$ and $\lim_{x\to\infty} \boldsymbol{F}(x) = 1$. All **cdf**'s satisfy these properties. Given any function with these properties, a random variable can be constructed that has that function as its **cdf**. Random variables are classified according to the nature of their **cdf**'s. Two different types of interest are: *continuous random variables*, whose **cdf**'s are also absolutely continuous functions; and *discrete random variables*, whose **cdf**'s are piecewise constant.

Sometimes, it is of interest to generalize the notion of a random variable slightly to allow the values $\pm\infty$ to fall within the range of the random variable. To preserve measurability, the sets of outcomes in Ω for which the variable takes on the values $+\infty$ and $-\infty$ must be in $\mathcal{F}$. This generalization of a random variable is known as an *extended random variable*.

Expectation.

In simpler terms, the **cdf** provides a detailed account of how a random variable behaves probabilistically. However, for a more general overview, we can use the expected value of the random variable. When dealing with a simple random variable, which only has a finite number of possible values, the expected value is calculated by multiplying each value by its corresponding probability and summing up these products[2].

A simple random variable is one that takes on only finitely many values. The expected value of a simple random variable X taking on the values $x_1, x_2, \ldots, x_n$, is defined as

$$\boldsymbol{E}[X] = \sum_{i=1}^{n} x_i \boldsymbol{P}(\{\omega : X(\omega) = x_i\}). \tag{A.4}$$

The expected value of a general non–negative random variable X is defined as the (possibly infinitc) valuc

$$\boldsymbol{E}[X] = \sup_{\{\text{simple } Y \,:\, \boldsymbol{P}(Y \leq X)=1\}} \boldsymbol{E}(Y).$$

The expected value of an arbitrary random variable is defined if at least one of the non-negative random variables, $X^+ = \max\{0, X\}$ and $X^- = ({}^-X)^+$, has a finite expectation, in which case

$$\boldsymbol{E}[X] = \boldsymbol{E}[X^+] - \boldsymbol{E}[X^-].$$

Otherwise, $\boldsymbol{E}[X]$ is undefined (does not exist). The interpretation of $\boldsymbol{E}[X]$ is as the average value of X, where the average is taken over all values in the range of X weighted by the probabilities with which these values occur.

W hen $\boldsymbol{E}[X]$ exists, we write it as the integral

$$\boldsymbol{E}[X] = \int_\Omega X(\omega)\mathrm{d}\boldsymbol{P}(\omega) = \int_\Omega X\mathrm{d}\boldsymbol{P}. \tag{A.5}$$

If $\boldsymbol{E}[\mid X \mid] < \infty$ we say that X is *integrable*. Let $X = \mathbb{I}_A$. We have $\boldsymbol{E}[\mathbb{I}_A] = \boldsymbol{P}(A)$. The consequence is that knowledge of the expectation of all random variables is equivalent to knowledge of the probability distribution $\boldsymbol{P}$. For an event A and a random variable X whose expectation exists, we write

$$\boldsymbol{E}[X\mathbb{I}_A] = \int_A X(\omega)\mathrm{d}\boldsymbol{P}(\omega) = \int_A X\mathrm{d}\boldsymbol{P}.$$

The integral (A.5) is a Lebesgue-Stieltjes integral, and it equals the Lebesgue-Stieltjes integral

$$\int_\Omega X(\omega)\mathrm{d}\boldsymbol{P}(\omega) = \int_{\Re} x\boldsymbol{P}_X(dx),$$

which in turn equals the Rieman-Stieltjes integral

$$\int_{-\infty}^{\infty} xd\boldsymbol{F}_X(x).$$

[2]The use of the letter E to denote "expected value" goes back to Whitworth (1901))

If X is a random variable and $g : \Re \to \Re$ is a measurable function from $(\Re, \mathcal{B})$ to $(\Re, \mathcal{B})$, then the composite function $Y = g(X)$ is also a random variable and its expectation (assuming that it exists) is given by

$$\boldsymbol{E}[Y] = \int_{\Omega} g(X(\omega))\mathrm{d}\boldsymbol{P}(\omega) = \int_{\Re} g(x)\boldsymbol{P}_X(dx) = \int_{\Re} y\boldsymbol{P}_Y(dy). \tag{A.6}$$

Remark 25 (Related to expectation quantities of interest.). *Using the expectation of various functions of the random variables, we define a list of interesting quantities.*

i. The moments *of a random variable:*

$$m_n = \boldsymbol{E}[X^n] = \int_{\Re} x^n d\boldsymbol{F}(x), \quad n = 1, 2, \dots. \tag{A.7}$$

The first moment m_1 is called the mean of X.

ii. The central moments *of a random variable:*

$$\mu_n = \boldsymbol{E}[(X - \boldsymbol{E}[X])^n] = \int_{\Re} (x - m_1)^n d\boldsymbol{F}(x), \quad n = 1, 2, \dots. \tag{A.8}$$

The second central moment is the variance of X denoted σ^2.

iii. The function $\boldsymbol{M}_X(t) = \boldsymbol{E}[e^{tX}]$ for the complex t, which is known as the moment generating function *if $t \in \Re$, and the* characteristic function *if $t = iu$, with $i = \sqrt{-1}$ and $u \in \Re$. The characteristic function is sometimes written as $\varphi(u) = \boldsymbol{M}_X(iu)$. Recall that P_X and φ_X form a unique pair.*

Theorem A.3 (Jensen's inequality). *If the convex functions g and random variable X are such that $\boldsymbol{E}[g(Y)]$ exists, then*

$$\boldsymbol{E}[g(X)] \geq g(\boldsymbol{E}[X]). \tag{A.9}$$

If g is strictly convex, then the inequality in Jensen's inequality is strict unless X is almost surely constant.

Conditional expectation.

Following consideration of A.1 introducing the conditional probability, the conditional expectation is defined. Consider again a probability space $(\Omega, \mathcal{F}, \boldsymbol{P})$, an integrable random variable X, and an event A with $\boldsymbol{P}(A) > 0$.

Definition A.3 (Conditional expectation). *The conditional expectation of X given A is the constant*

$$\boldsymbol{E}[X \mid A] = \frac{\int_A X d\boldsymbol{P}}{\boldsymbol{P}(A)}. \tag{A.10}$$

Remark 26. *The equation of the definition A.3 can be rewritten that the constant $\boldsymbol{E}[X \mid A]$ satisfies the condition*

$$\int_A \boldsymbol{E}[X \mid A] d\boldsymbol{P} = \int_A X d\boldsymbol{P}. \tag{A.11}$$

This means that $\mathbf{E}[X \mid A]$ has the same $\mathbf{P}$-weighted integral that X does over the event A.

It is convenient to generalize this notion of *conditional expectation* to allow for conditioning on a family of events (in particular, σ-fields).

Definition A.4 (Fields of events vs. random variables). *The basic idea of the probability theory are following.*

i. A σ-field $\mathcal{G}$ is a sub-sigma field *of $\mathcal{F}$ if each element of $\mathcal{G}$ also belongs to $\mathcal{F}$, that is, $\mathcal{G} \subset \mathcal{F}$.*
ii. A random variable X is measurable with respect to the sub-σ-field $\mathcal{G}$ if $X^{-l}(B) \in \mathcal{G}$ for every $B \in \mathcal{B}$.
iii. The σ-field generated by a random variable X (denoted by $\sigma(X)$) is the smallest σ-field with respect to which X is measurable.

In a heuristic sense, $\sigma(X)$ can be conceptualized as the collection of events where the random variable X remains constant. Any additional refinement of $\mathcal{F}$ beyond $\sigma(X)$ cannot be observed through X. This interpretation holds precisely for simple random variables.

Definition A.5 (Generalized Conditional Expectation). *For an integrable random variable X and a sub-σ-field $\mathcal{G}$ the conditional expectation of X given $\mathcal{G}$ is defined as any random variable Z, measurable with respect to $\mathcal{G}$, such that*

$$\int_A \mathbf{E}[Z \mid \mathcal{G}] d\mathbf{P} = \int_A X d\mathbf{P}, \textit{ for every } A \in \mathcal{G}.$$

Such a random variable Z always exists, and we write $E[X \mid \mathcal{G}]$ to denote any such random variable. Any two versions of $E[X \mid \mathcal{G}]$ differ only on sets of probability zero. When X itself is cG-measurable, then $E[X \mid \mathcal{G}] = X$.

The interpretation of *conditional expectation* $\mathbf{E}[X \mid \mathcal{G}]$ is that it is a *random variable* that behaves like X does to the extent that is consistent with the constraint that it be $\mathcal{G}$-measurable. In this sense, when $\boldsymbol{E}(X^2) < \infty$, we have $\boldsymbol{E}[X \mid \mathcal{G}]$ is a projection of X onto $\mathcal{G}$. It is in fact minimization problem of $\boldsymbol{E}(Z - X)^2$ over all $\mathcal{G}$-measurable random variables Z. Again appealing to a heuristic interpretation, $\boldsymbol{E}[X \mid \mathcal{G}]$ is a random variable formed by replacing X by its centroid (with respect to $\boldsymbol{P}$) on each set in $\mathcal{G}$.

Example A.1. *Let us choose an event A with $0 < \boldsymbol{P}(A) < 1$, and consider the sub-$\sigma$-field $\mathcal{G} = \{A, A^c, \Omega, \emptyset\}$. Then, for any random variable X we have:*

$$\boldsymbol{E}[Z \mid \mathcal{G}] = \begin{cases} \boldsymbol{E}[Z \mid A] & \textit{for } \omega \in A, \\ \boldsymbol{E}[Z \mid A^c] & \textit{for } \omega \in A^c. \end{cases}$$

The existence of a conditional expectation is a consequence of the Radon-Nikodym theorem.

Radon-Nikodym derivatives.

Suppose $\mathbf{Q}$ is a probability measure in a measurable space $(\Omega, \mathcal{F})$. Then we have the following theorems.

Theorem A.4 (Lebesgue decomposition theorem). *There is a probability measure $\boldsymbol{P}$ and a random variable f (unique up to sets of $\boldsymbol{P}$-probability zero), and an event $H \in \mathcal{F}$ satisfying $\boldsymbol{P}(H) = 0$ such that*

$$\mathbf{Q}(A) = \int_A f d\boldsymbol{P} + \mathbf{Q}(H \cap A), \text{ for every } A \in \mathcal{F}.$$

Definition A.6. *The measure $\mathbf{Q}$ is* absolutely continuous with respect *to $\boldsymbol{P}$ (we say $\boldsymbol{P}$ dominates $\mathbf{Q}$ and we write $\mathbf{Q} \ll \boldsymbol{P}$) if $\boldsymbol{P}(H) = 0$ implies $\mathbf{Q}(H) = 0$.*
If $\mathbf{Q} \ll \boldsymbol{P}$ and $\mathbf{Q} \gg \boldsymbol{P}$ then we say that $\mathbf{Q}$ and $\boldsymbol{P}$ are equivalent *(i.e. $\mathbf{Q} \equiv \boldsymbol{P}$).*

Corollary A.12 (Radon-Nikodym theorem). *If $\mathbf{Q} \ll \boldsymbol{P}$, then there exists a random variable f such that*

$$\mathbf{Q}(A) = \int_A f d\boldsymbol{P}, \text{ for every } A \in \mathcal{F}.$$

The function f in claim of the corollary A.12 is called the *radon-Nikodym derivative* of $\mathbf{Q}$ with respect to $\boldsymbol{P}$. Usually we write

$$f(\omega) = \frac{\mathrm{d}\mathbf{Q}}{\mathrm{d}\boldsymbol{P}}(\omega), \text{ for every } \omega \in \Omega.$$

Properties of conditional expectation.

The existence of *conditional expectation* follows as a corollary of the Radon-Nikodym theorem. Suppose first that X is nonnegative. Consider the measurable space $(\Omega, \mathcal{G})$ and define two measures $\mathbf{Q}$ and $\boldsymbol{P}'$ on $(\Omega, \mathcal{G})$ by

$$\mathbf{Q}(A) = \int_A X \mathrm{d}\boldsymbol{P}, \text{ and } \boldsymbol{P}'(A) = \boldsymbol{P}(A) \text{ for every } A \in \mathcal{G}.$$

$\mathbf{Q}$ is a finite measure, but it is not a probability measure unless $\boldsymbol{E}[X] = 1$. The probability measure $\boldsymbol{P}'$ is the restriction of $\boldsymbol{P}$ to $\mathcal{G}$ and $\mathbf{Q} \ll \boldsymbol{P}'$. We have

$$\mathbf{Q}(A) = \int_A f \mathrm{d}\boldsymbol{P}', \text{ for every } A \in \mathcal{G},$$

with $f = \frac{\mathrm{d}\mathbf{Q}}{\mathrm{d}\boldsymbol{P}'}$. Comparing this with the definition of $\boldsymbol{E}[X \mid \mathcal{G}]$ we see that $\boldsymbol{E}[X \mid \mathcal{G}] = \frac{\mathrm{d}\mathbf{Q}}{\mathrm{d}\boldsymbol{P}'}$. If X is not nonnegative, then the existence of $\boldsymbol{E}[X \mid \mathcal{G}]$ follows by decomposing $X = X^+ - X^-$, and proceeding as above.

Corollary A.13. *If $\mathcal{H}$ is a sub-σ-field of $\mathcal{G}$, then it is easy to see that*

$$\boldsymbol{E}\left[\boldsymbol{E}[X \mid \mathcal{G}] \mid \mathcal{H}\right] = \boldsymbol{E}[X \mid \mathcal{H}].$$

The interpretation of this property is that the projection of X onto the $sigma$-field $\mathcal{H}$ can be found by first projecting onto the finer σ-field $\mathcal{G}$ and then projecting the result onto $\mathcal{H}$. So, for example, since $E[X]$ is the conditional expectation of X given the trivial σ field $\{\Omega, \emptyset\}$, we have

$$\boldsymbol{E}\left[\boldsymbol{E}[X \mid \mathcal{G}]\right] = \boldsymbol{E}[X], \text{ for any } \mathcal{G} \subset \mathcal{F}.$$

If Y is another random variable, then the conditional expectation of X given Y, is defined by

$$\boldsymbol{E}\left[X \mid Y\right] = \boldsymbol{E}[X \mid \sigma(Y)].$$

The quantity $\boldsymbol{E}\left[X \mid Y\right]$ has the interpretation of being the (probabilistically weighted) average value of X, given that Y is fixed. It can be shown that X is measurable $\sigma(Y)$ if and only if there is a measurable function $g : (\Re, \mathcal{B}) \to (\Re, \mathcal{B})$ such that $X = g(Y)$. Consequently, $\boldsymbol{E}\left[X \mid Y\right]$ can be considered a measurable function of Y. Thus, we can think of a measurable function $g : \Re \to \Re$ such that $\boldsymbol{E}\left[X \mid Y\right] = g(Y)$. It can also be written as $g(y) = \boldsymbol{E}\left[X \mid Y = y\right]$.

Conditional expectations vs. statistical independence.

Conditional expectations are closely related to the notion of (statistical) independence, defined in various circumstances as follows.

Definition A.7. *Following the consideration of A.1, we have:*

- *Two events A and B are said to be independent if $\boldsymbol{P}(A \cap B) = \boldsymbol{P}(A) \cdot \boldsymbol{P}(B)$.*
- *Two σ fields $\mathcal{H} \subset \mathcal{F}$ and $\mathcal{G} \subset \mathcal{F}$ are said to be independent if all elements of $\mathcal{H}$ are independent of all elements of $\mathcal{G}$.*
- *A random variable X is said to be independent of a σ-field $\mathcal{G}$ if $\sigma(X)$ is independent of $\mathcal{G}$.*
- *Two random variables X and Y are said to be independent if $\sigma(X)$ and $\sigma(F)$ are independent.*

If X is independent of $\mathcal{G}$, then $\boldsymbol{E}\left[X \mid \mathcal{G}\right] = \boldsymbol{E}\left[X\right]$; and so if X and Y are independent, then $\boldsymbol{E}\left[X \mid Y\right] = \boldsymbol{E}\left[X\right]$. This condition is the opposite extreme of the condition that X be $\sigma(Y)$-measurable, which means that X (and $\boldsymbol{E}\left[X \mid Y\right]$) is a function of Y.

The plane equipped with its Borel σ-field.

Consider the case in which the measurable space of interest is $(\Re^2, \mathcal{B} \times \mathcal{B})$–the plane equipped with its Borel σ-field; i.e. $(\Omega, \mathcal{F}) = (\Re^2, \mathcal{B}^2)$. Let $\mathcal{G}$ be the sub-σ-field consisting of those sets of the form

$$\{\bar{\omega} = (\omega_1, \omega_2) \in \Re^2 \mid \omega^2 \in B\}, \text{for } B \in \mathcal{B},$$

where $\mathcal{B}$ is the Borel σ-field of $\Re$. The set of $\mathcal{G}$ measurable random variables is the set of functions on Ω that is the mapping of the second coordinate of $\bar{\omega} \in \Omega$ only.

Based on this, let X be a random variable. The random variable $\boldsymbol{E}[X \mid \mathcal{G}]$ is any measurable function of ω_2 that satisfies the equation.

$$\int_{\Re\times B} \boldsymbol{E}[X \mid \mathcal{G}](\omega_2)\mathrm{d}\boldsymbol{P}(\bar{\omega}) = \int_{\Re\times B} X(\omega_1,\omega_2)\mathrm{d}\boldsymbol{P}(\bar{\omega}), \text{for every } B \in \mathcal{B}.$$

Assume that the probability measure $\boldsymbol{P}$ assigns probabilities via the integral

$$\boldsymbol{P}(A) = \int_A p(\omega_1,\omega_2)\mathrm{d}\omega_1\mathrm{d}\omega_2, \text{for every } A \in \mathcal{F}.$$

The function $p(\omega_1,\omega_2)$ is an integrable, nonnegative function of $\Re^2$, with total integral 1. Then

$$\int_B \boldsymbol{E}[X \mid \mathcal{G}](\omega_2)p_2(\omega_2)\mathrm{d}\omega_2 = \int_B\int_\Re X(\omega_1,\omega_2)p(\omega_1,\omega_2)\mathrm{d}\omega_1\mathrm{d}\omega_2,$$

where $p_2(\omega_2) = \int_\Re p(\omega_1,\omega_2)\mathrm{d}\omega_1$.

A.1.2 Simulation

To create the possibility of using the simulation approach to study a system, process (phenomenon), such as a jet engine, a new house or a vaccine, we have a number of options to choose from:

Experimentation with the real thing: i.e. a real engine, someone's home, or a human being.

Experimentation with a model using either a:

- **Physical model** , such as a Lego model of a house, a small scale of a ship or an aircraft;
- **Mathematical** or **Digital model** which describes the important parameters of the system and, by this way, give us possibility to observe them as in the real-time running object, such as a simulation of a jet engine or a *Digital Twin* of the *Human body*.

A simulation imitates how something has behaved previously, presently, and how it could behave in the future. Different processes have their own mechanisms of operation, creating unique systems. Therefore, simulations of different systems are based on different heuristics. Everything in the world has a unique way of operating, therefore different types of simulation are required to allow for various processes and systems to be modeled. Simulations fall into one of the following simulation taxonomy categories:

Discrete Simulation is when different parts of the simulation happen at specific points in time. Consequently, this type of simulation can be either:

→**Event-driven simulation**, describing a series of events or activities that happen or take effect at a particular point in time, sometimes with a duration.

→**Time-stepped simulation**, described by time steps or clock ticks, where things happen at each tick.

→**Continuous Model Simulation** is in the name: the simulation runs continuously with no gaps in between time stamps, unlike discrete simulation. It uses **ODE**s and **PDE**s such as **CFD**s and **FEA**.

Monte Carlo Simulation is where you are running a lot of simulations at the same time and in random orders until you find the solution.

A.1.3 Discrete-time Markov chain simulations

The dynamics of a Markov chain allows its simulation by generating subsequent states according to the discrete distribution determined by the row of the transition matrix, which is related to the current state of the chain. For example, in the MatLab Econometrics Toolbox (v. The MathWorks Inc., (2022)) there are functions that facilitate the definition of such a chain and its simulation.

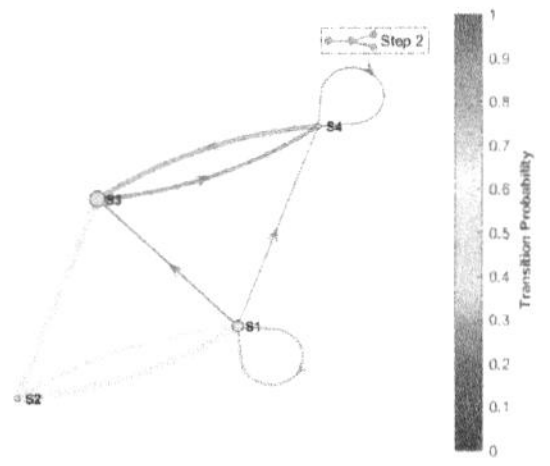

FIGURE A.1
Transition in simulation of Markov chain.

Sánchez-Salas et al. (2012) used Markov switching model to present processes in wireless communication.

A.2 Knightian Uncertainty

The economic crisis of the first decade of the XXI century has revived an old philosophical idea about risk and uncertainty. It has stemmed, in part, from the inability of financial institutions to effectively judge the riskiness of their investments. For this reason, the crisis has raised new attention to an idea of risk from decades past: "*Knightian uncertainty*".

Knight (2015)[3] formalized a distinction between risk and uncertainty in this book. As Knight saw it, an ever-changing world brings new opportunities for businesses to make profits, but also means that we have imperfect knowledge of future events. Therefore, according to Knight, risk applies to situations where we do not know the outcome of a given situation, but can accurately measure the odds. Uncertainty, on the other hand, applies to situations where we cannot know all the information we need in order to set accurate odds in the first place.[4]

Knight wrote

> There is a fundamental distinction between the reward for taking a known risk and that for assuming a risk whose value itself is not known. A known risk is *easily converted into an effective certainty*, while *true uncertainty is not susceptible to measurement*. An airline might forecast that the risk of an accident involving one of its planes is exactly one per 20 million take-offs. But the economic outlook for airlines 30 years from now involves so many unknown factors that it is incalculable.

Some economists have argued that this distinction is overblown. In the real business world, this objection says, all events are so complex that forecasting is always a matter of grappling with "true uncertainty", not risk; past data used to forecast risk may not reflect current conditions, anyway. In this view, "risk" would be best applied to a highly controlled environment, like a pure game of chance in a casino, and "uncertainty" would apply to nearly everything else.

However, Knight's distinction about risk and uncertainty may still help us analyze the recent behavior of, say, financial firms and other investors. Investment banks that in recent years regarded their own apparently precise risk assessments as trustworthy may have thought they were operating in conditions of Knightian risk, where they could judge the odds of future outcomes. Once the banks recognized those assessments were inadequate, however, they understood that they were operating in conditions of Knightian uncertainty — and may have held back from making trades or providing capital, further slowing the economy as a result.

When investors realize that their assumptions about risk are no longer valid and that Knightian uncertainty conditions apply, markets can witness "destructive flights to quality" in which participants rid their portfolios of everything but the safest of investments, such as US Treasury bonds.

One solution offered by some economists to stem these moments of panic is government-issued investment insurance for large financial institutions. In this sense, the existence of Knightian uncertainty is not just a quasiphilosophical dispute; the subjective perception of Knightian uncertainty among businesses is a pressing practical problem.

---

[3]Knight is best known as the author of the book Risk, Uncertainty and Profit (1921), based on his Ph.D. dissertation at Cornell University. In that book, he carefully distinguished between economic risk and uncertainty. Situations with risk were those where the outcomes were unknown but governed by probability distributions known at the outset. He argued that these situations, where decision-making rules such as maximizing expected utility can be applied, differ in a deep way from "uncertain" ones, in which not only the outcomes, but even the probability models that governed them, were unknown. Knight argued that uncertainty gave rise to economic profits that perfect competition could not eliminate.

[4]Peter Dizikes (2010) Explained: Knightian uncertainty. MIT News on campus and around the world. June 2, 2010.

B

Optimal Stopping

THE SEARCH for a rational choice of the observation sequence length for random variables should begin with statistical considerations. Until the mid-20th century, the role of statisticians was primarily to conclude the provided data. Even if the design and execution of an experiment were undertaken by the same person, the preparation of the experiment did not necessarily involve the intervention of a statistician. Therefore, during this time, statistical theory was limited to the design and analysis of random experiments in which the size and composition of the samples are entirely determined before the experiment began. Robbins (1952) succinctly explained the reasons for this state of affairs. He also observed that practical issues had emerged in quality control of production had emerged, creating the need for decisions on acceptance or rejection of production based on its quality, with varying sample sizes. Issues of this kind gave rise to the need for planning and analysis of sequential design of experiments, in which the size and composition of the samples are not fixed in advance but are functions of the observations themselves. The first works describing decision procedures that link the moment of decision-making with previous observations are attributed to Dodge and Romig (1929, 1941).

B.1 Snell Envelope

There are generally two approaches to solving optimal stopping problems. When the underlying process (or the gain process) is described by its unconditional finite-dimensional distributions, the appropriate solution technique is the martingale approach, so called because it uses martingale theory, the most important concept being the Snell envelope (v. Snell (1951, 1952)). In the discrete-time case, if the planning horizon T is finite, the problem can also be easily solved by dynamic programming.

DOI: 10.1201/9781003407102-B

Definition B.1. *Given a filtered probability space* $(\Omega, \mathcal{F}, \{\mathcal{F}_n\}_{n=1}^T, \boldsymbol{P})$, *and an absolutely continuous probability measure* $\mathbf{Q}$ *with respect to* $\boldsymbol{P}$, *then an adapted process* $\xi = \{\xi_n\}_{n\in[0,T]}$ *is the* Snell envelope *with respect to* $\mathbf{Q}$ *of the process* $\mathbf{X} = \{X_n\}_{n=0}^T$ *if*

1. ξ *is a* $\mathbf{Q}$*-supermartingale;*
2. ξ *dominates* $\mathbf{X}$, *i.e.* $\xi_n \geq X_n$, $\mathbf{Q}$*-a.e. for all times* $n \in \{0, 1, \ldots, T\}$*;*
3. *If* $\eta = \{\eta_n\}_{n=0}^T$ *is a* $\mathbf{Q}$*-supermartingale which dominates* $\mathbf{X}$, *then* η *dominates* ξ.

The application of the Snell envelope to the construction of the solution to the optimal stopping problem was the subject of research by Chow and Robbins (1961, 1963, 1967). The first monograph on optimal stopping by Chow et al. (1964) collected the main facts in the topic for the general setting of the problem.

B.2 Elimination Algorithm

In the optimal stopping problem for processes with a countable state space, the optimal rule can be characterized by a suitable set of states after reaching which the search process should be stopped. Analysis of expected payouts and potential payouts from further observation of the process shows that the optimal stopping set can be built by eliminating states in which stopping is not favorable at the time of analysis and will not be favorable in the future (v. Sonin (1999)). Let us assume that a Markov model $\mathcal{M}_1 = (\mathbb{E}_1, \boldsymbol{P}_1)$ is given and let (ξ_n), $n = 1, 2, \ldots$, be a Markov chain specified by the model $\mathcal{M}_1$. Let $D \subset \mathbb{E}_1$, and let $\tau_1, \tau_2, \ldots$, be the sequence of Markov times of the first, second, and so on, visits of the set $\mathbb{E}_2 \equiv \mathbb{E}_1 \setminus D$ by (ξ_n), so that $\tau_1 = \min\{k \geq 0 : \xi_k \in \mathbb{E}_2\}$, $\tau_{n+1} = \min\{k : \tau_n < k, \xi_k \in \mathbb{E}_2\}$, $0 \leq \tau_1 < \tau_2 < \ldots$. Let $u_1^{\mathbb{E}_2}(x, \cdot)$ be the distribution of the Markov chain (ξ_n) for the initial model $\mathcal{M}_1$ at the moment τ_1 of first visit to the set $\mathbb{E}_2$ (first exit from D) starting at z, $z \in D$. Consider the random sequence $\eta_n = \xi_{\tau_n}$, $n = 1, 2, \ldots$

Proposition 1. *The following properties of the stopped process have place:*

(a) The random sequence (η_n) *is a Markov chain in a model* $\mathcal{M}_2 = (\mathbb{E}_2, \boldsymbol{P}_2)$*;*

(b) the transition matrix $\boldsymbol{P}_2 = \{p_2(x, y)\}$ *is given by the formula*

$$p_2(x, y) = p_1(x, y) + \sum_{z\in D} p_1(x, z) u_1^{\mathbb{E}_2}(z, y), \; x, y \in \mathbb{E}_2. \tag{B.1}$$

Theorem B.1. *Let* $M_1 = (\mathbb{E}_1, \boldsymbol{P}_1, g)$ *be an optimal stopping problem,* $D \subset \{z \in \mathbb{E}_1 : \mathbf{T}_1 g(z) > g(z)\}$ *and* $\boldsymbol{P}_{1,x}(\tau_{\mathbb{E}_1 \setminus D} < \infty) = 1$ *for all* $x \in D$. *Consider an optimal stopping problem* $M_2 = (\mathbb{E}_2, \boldsymbol{P}_2, g)$ *with* $\mathbb{E}_2 = \mathbb{E}_1 \setminus D$, $p_2(x, y)$ *defined by* (B.1). *Let* S *be the optimal stopping set in the problem* $\mathcal{M}_2$. *Then* S *is the optimal stopping set in* $\mathcal{M}_1$ *also and* $v_1(x) = v_2(x)$ *for all* $x \in \mathbb{E}_2$.

B.3 Odds Algorithm

Observe sequentially the independent indicator random variables[1], $\mathbb{I}_1, \mathbb{I}_2, \ldots, \mathbb{I}_n$. The aim is to determine an optimal stopping rule that maximizes the probability of stopping at the last $k \leq n$ such that $\mathbb{I}_k = 1$. An elegant unification of the solution to the class of such optimal stopping problems was given by Bruss (2000). The general solution is following (v. Samuels (2001)). Let $p_k = \boldsymbol{P}(\mathbb{I}_k = 1)$ and $r_k = \frac{p_k}{1-p_k}$[2]. Now add these odds, starting with r_n and working backward until the sum reaches one. If $r_{s+1} + \ldots + r_n < 1 \leq r_s + \ldots + r_n$, then the rule that stops at the first $k \geq s$ such that $\mathbb{I}_k = 1$ (if any), is optimal.

Extension of the algorithm to multiple-stopping is subject of consideration in Section 2.5. The application to the game model was given by Ferguson (2016). Extension of the algorithm to multiple-stopping is subject of consideration in Section 2.5. The application to the game model was given by Ferguson (2016).

[1]The dichotomous random variables.
[2]i.e ., r_k is the odds in favor of the event $\{\omega : \mathbb{I}_k(\omega) = 1\}$.

C

Duality Based on Information Relaxations

Let us continue a discussion of Section 1.4.1. We begin with a general finite-horizon discrete-time dynamic program with a probability space $(\Omega, \mathcal{F}, \boldsymbol{P})$. Let $\mathcal{T} := \{0, 1, \ldots, T\}$, and the evolution of information is described by filtration $\mathbb{F} = \{\mathcal{F}_0, \ldots, \mathcal{F}_T\}$, with $\mathcal{F} = \mathcal{F}_T$. We make the usual assumption that $\mathcal{F}_0 = \{\emptyset, \Omega\}$, so that the decision-maker starts out with no information regarding the outcome of the uncertainty. There is a state vector, $x_t \in \Re^n$[1], whose dynamics satisfy

$$X_{t+1} = f_t(X_t, u_t, \xi_{t+1}),\ t = 0, \ldots, T-1, \tag{C.1}$$

where $u_t \in \mathcal{U}_t(x_t) \subset \Re^m$ is the control taken at time t and ξ_{t+1} is an $\mathcal{F}_{t+1}$-measurable random disturbance. Define

$$J_k(x_k, \overrightarrow{u_k}) = \boldsymbol{E}_{x_k} \sum_{t=k}^{T} g_t(X_t, U_t) = \boldsymbol{E}\Big[\sum_{t=k}^{T} g_t(X_t, U_t) \mid X_k = x_k\Big],$$

where $\overrightarrow{u_k} = (u_k, \ldots, u_T)$, and $U_j = u_j(X_j)$.

A feasible strategy, $\overrightarrow{u} := (u_0, \ldots, u_T)$, is one in which each individual action satisfies $u_t \in \mathcal{U}_t(x_t)$ for all t. We let $\mathcal{U} = \bigcup_x \mathcal{U}_0(x)$ denote the set of such strategies. *A feasible adapted strategy* is a feasible strategy that is $\mathcal{F}_t$-adapted. We let $\mathcal{U}_{\mathbb{F}}$ denote the set of all such $\mathcal{F}_t$-adapted strategies. The objective is to select a feasible adapted strategy, $u^\star$, to maximize the expected total gain,

$$g(x_0, \overrightarrow{u_0}) = \sum_{t=0}^{T} g_t(x_t, u_t).$$

We assume that all $g_t(x_t, u_t)$ are $\mathcal{F}_t$-measurable. In particular, the decision-maker's problem is then given by

1. determine

$$J_0^\star(x_0) = \sup_{\overrightarrow{u} \in \mathcal{U}_{\mathbb{F}}} \boldsymbol{E}_{x_0} g(x_0, \overrightarrow{U}) = \sup_{\overrightarrow{u} \in \mathcal{U}_{\mathbb{F}}} \boldsymbol{E}_{x_0} \sum_{t=0}^{T} g_t(X_t, U_t)$$

2. select $\overrightarrow{u}^\star \in \mathcal{U}_{\mathbb{F}}$ such that

$$J_0(x_0, \overrightarrow{u}^\star) = J_0^\star(x_0).$$

$J_0^\star(x)$ is the deterministic value function of the initial state x_0.

[1]The state of the system is deterministic element of the space $\Re^n$, but there are disturbances which makes that the system is stochastic.

DOI: 10.1201/9781003407102-C

C.1 The Dual Formulation

We now briefly describe the dual formulation of Brown et al. (2010) that may be used to construct dual, i.e. upper, bounds, on $J_0^*(x_0)$. We will only consider perfect information relaxations, as these relaxations are typically most useful in practice and are all that we will require in later consideration of the section. Let $\mathcal{S}$ denote the space of real-valued measurable functions that are defined on the state space $\Re^n$ and the dynamics of the system is defined by formula (C.1). We now define an operator Δ that maps $\mathcal{S}$ to the space of real-valued measurable functions on $\Re^n \times \Re^n \times \Re^m$ according to

$$(\Delta V_t)(x_t, x_{t-1}, u_{t-1}) := V_t(x_t) - \boldsymbol{E}[V_t(X_t)|x_{t-1}, u_{t-1}]. \tag{C.2}$$

Roughly speaking, Δ is an operator on (approximate) value functions. In particular, we have $\boldsymbol{E}_{x_0}[(\Delta V_t)(X_t, X_{t-1}, U_{t-1})] = 0$ for all integrable V_t. Let $\mathcal{D}$ be the space of real-valued functions on $\Re^n \times \mathcal{T}$ such that, if $V \in \mathcal{D}$, then $V_t := V(\cdot, t)$ is measurable and $\boldsymbol{E}_{x_0}[|V_t(X_t)|] < \infty$ for all $t \in \mathcal{T}$ and all feasible strategies $\vec{u} \in \mathcal{U}$. We now define an operator $\boldsymbol{F} : \mathcal{D} \to \mathcal{S}$ according to

$$\begin{aligned} \boldsymbol{F}V(x) := \boldsymbol{E}[\sup_{\vec{u}\in\mathcal{U}} G(X_0, \vec{u}, \omega) \mid X_0 = x] = \boldsymbol{E}\Big[\sup_{\vec{u}\in\mathcal{U}} \big\{g_0(x, u_0) \\ + \sum_{t=1}^{T} \big(g_t(X_t, U_t) - (\Delta V_t)(X_t, X_{t-1}, U_{t-1})\big)\big\} \mid X_0 = x\Big] \end{aligned} \tag{C.3}$$

where, based on (C.1)

$$G(x, \vec{u}, \omega) = g_0(x, u_0) + \sum_{t=1}^{T} \big(g_t(x_t, u_t) - (\Delta V_t)(X_t, x_{t-1}, u_{t-1})\big).$$

Here the supremum in (C.3) is over the space $\mathcal{U}$ of feasible strategies[2]. Following Brown et al. (2010) (cf. Chandramouli and Haugh (2012)) it is known that there is weak duality

Theorem C.1 (Weak Duality). ,

$$J_0^\star(x_0) \leq \boldsymbol{F}V(x_0) \quad \textit{for all } V \in \mathcal{D}. \tag{C.4}$$

Remark 27 (Dual problem). *Theorem C.1 suggests that we can get an upper bound on $J_0^\star(x_0)$ by evaluating $\boldsymbol{F}V(x_0)$ for any $V \in \mathcal{D}$. The theorem does not give guidance how close to $J_0^\star$ this bound should be. It suggests the following dual problem,*

1. *determine* $\inf_{V\in\mathcal{D}} \boldsymbol{F}V(x_0)$;
2. *find* $V^\star$ *such that* $\boldsymbol{F}V^\star(x_0) = \inf_{V\in\mathcal{D}} \boldsymbol{F}V(x_0)$.

[2]It is not the space $\mathcal{U}_{\mathcal{F}}$ of *feasible adapted strategies*.

Theorem C.2 (Strong Duality). *For all x,*

$$\boldsymbol{F}J_0^\star(x) = \inf_{V\in\mathcal{D}} \boldsymbol{F}V(x). \tag{C.5}$$

Theorem C.2 suggests that we might be able to obtain good upper bounds on the optimal value function, $J_0^\star(x)$, if we can find $V \cong J^\star$ and then compute $\boldsymbol{F}V(x)$. The following corollary, which follows immediately from the proof of Theorem C.2, has some significance for practical applications.

Corollary C.6. *When $V = J^\star$, we have*

$$J_0^\star(x) = \sup_{u\in\mathcal{U}} g_0(x, u_0) + \sum_{t=1}^{T} \left(g_t(x_t, u_t) - (\Delta J_t^\star)(x_t, x_{t-1}, u_{t-1})\right)$$

almost everywhere.

C.2 Conditional Expectation

The global economic crisis of the last two years has stemmed, in part, from the inability of financial institutions to effectively judge the riskiness of their investments. For this reason, the crisis has cast new attention on an idea about risk from decades past: "Knightian uncertainty".

Frank Knight was an idiosyncratic economist who formalized a distinction between risk and uncertainty in his 1921 book, Risk, Uncertainty, and Profit. As Knight saw it, an ever-changing world brings new opportunities for businesses to make profits, but also means we have imperfect knowledge of future events. Therefore, according to Knight, risk applies to situations where we do not know the outcome of a given situation, but can accurately measure the odds. Uncertainty, on the other hand, applies to situations where we cannot know all the information we need in order to set accurate odds in the first place.

"There is a fundamental distinction between the reward for taking a known risk and that for assuming a risk whose value itself is not known", Knight wrote. A known risk is "easily converted into an effective certainty", while "true uncertainty," as Knight called it, is "not susceptible to measurement". An airline might forecast that the risk of an accident involving one of its planes is exactly one per 20 million takeoffs. But the economic outlook for airlines 30 years from now involves so many unknown factors as to be incalculable.

Some economists have argued that this distinction is overblown. In the real business world, this objection goes, all events are so complex that forecasting is always a matter of grappling with "true uncertainty", not risk; past data used to forecast risk may not reflect current conditions, anyway. In this view, "risk" would be best applied to a highly controlled environment, like a pure game of chance in a casino, and "uncertainty" would apply to nearly everything else.

Even so, Knight's distinction about risk and uncertainty may still help us analyze the recent behavior of, say, financial firms and other investors. Investment banks that in recent years regarded their own apparently precise risk assessments as trustworthy may have thought they were operating in conditions of Knightian risk, where they could judge the odds of future outcomes. Once the banks recognized those assessments were inadequate, however, they understood that they were operating in conditions of Knightian uncertainty – and may have held back from making trades or providing capital, further slowing the economy as a result.

Ricardo Caballero, chair of MIT's Department of Economics and the Ford International Professor of Economics, Macroeconomics, and International Finance, is among those who have recently invoked Knightian uncertainty to explain the behavior of investors in times of financial panic. As Caballero stated in a lecture at the International Monetary Fund's research conference last November: When investors realize that their assumptions about risk are no longer valid and that conditions of Knightian uncertainty apply, markets can witness "destructive flights to quality" in which participants rid their portfolios of everything but the safest of investments, such as U.S. Treasury bonds.

One solution offered by Caballero to stem these moments of panic is government-issued investment insurance for large financial institutions. In this sense, the existence of Knightian uncertainty is not just a quasi-philosophical dispute; the subjective perception of Knightian uncertainty among businesses is a pressing practical problem.

C.3 Prophet Helps You Assess the Importance of Information

Continuing the considerations from Section 1.4.2, we will recall the conclusions that the introduction of an individual with exceptional abilities to predict the past provides for optimal detention.

Kleinberg et al. (2022) explore the implications of two central human biases studied in behavioral economics, reference points and loss aversion, in optimal stopping problems. In such problems, people evaluate a sequence of options in one pass, either accepting the option and stopping the search or giving up on the option forever. Authors assume that the best option seen so far sets a reference point that shifts as the search progresses, and a biased decision-maker's utility incurs an additional penalty when they accept a later option that is below this reference point. They results include tight bounds on the performance of a biased agent in this model relative to the best option obtainable in retrospect (a type of prophet inequality for biased agents), as well as tight bounds on the ratio between the performance of a biased agent and the performance of a rational one.

Bibliography

[1] M. Abramowitz and I. A. Stegun, editors. *Handbook of Mathematical Functions with Formulas, Graphs, and Mathematical Tables*. Dover, 9 edition, 1972. Cited on p. 174.

[2] M. Ajtai, N. Megiddo, and O. Waarts. Improved Algorithms and Analysis for Secretary Problems and Generalizations. *SIAM Journal on Discrete Mathematics*, 14(1):1–27, Jan. 2001. ISSN 0895-4801. doi: 10.1137/S0895480195290017. Cited on p. 82.

[3] M. Alario-Nazaret, J.-P. Lepeltier, and B. Marchal. Dynkin games. In *Stochastic differential systems (Bad Honnef, 1982)*, volume 43 of *Lect. Notes Control Inf. Sci.*, pages 23–32. Springer, Berlin, 1982. doi: 10.1007/BFb0044285. MR 814103. Cited on p. 29.

[4] T. Alpcan and T. Başar. *Network Security. A Decision and Game-theoretic Approach*. Cambridge University Press, Cambridge, 2011. ISBN 978-0-521-11932-0. MR 2757313. Cited on p. 107.

[5] L. Andersen and M. Broadie. Primal-dual simulation algorithm for pricing multidimensional American options. *Management Science*, 50(9):1222–1234, 2004. doi: 10.1287/mnsc.1040.0258. Cited on p. 287.

[6] K. Ano. Optimal selection problem with three stops. *J. Oper. Res. Soc. Japan*, 32:491–504, 1989. Cited on pp. 217 and 218.

[7] K. Ano. Bilateral secretary problem recognizing the maximum or the second maximum of a sequence. *J. Information & Optimization Sciences*, 11:177–188, 1990. Cited on p. 84.

[8] K. Ano, H. Kakinuma, and N. Miyoshi. Odds theorem with multiple selection chances. *J. Appl. Probab.*, 47(4):1093–1104, 2010. ISSN 0021-9002,1475-6072. doi: 10.1239/jap/1294170522. Cited on p. 18.

[9] M. Archibald and C. Martínez. The Hiring Problem and Permutations. *Discrete Mathematics & Theoretical Computer Science*, DMTCS Proceedings vol. AK, 21st International Conference on Formal Power Series and Algebraic Combinatorics (FPSAC 2009):63–76, Jan. 2009. doi: 10.46298/dmtcs.2731. Cited on p. 216.

[10] K. J. Arrow, D. Blackwell, and M. A. Girshick. Bayes and minimax solutions of sequentials decision problems. *Econometrica*, 17, 1949. ISSN 0012-9682. doi: 10.2307/1905525. Cited on p. 10.

[11] D. Assaf and E. Samuel-Cahn. The secretary problem: Minimizing the expected rank with i.i.d. random variables. *Adv. Appl. Probab.*, 28(3):828–852, 1996. ISSN 0001-8678. doi: 10.2307/1428183. Zbl 0857.60039. Cited on p. 16.

[12] D. Assaf and E. Samuel-Cahn. Simple ratio prophet inequalities for a mortal with multiple choices. *J. Appl. Probab.*, 37(4):1084–1091, 2000. ISSN 0021-9002. doi: 10.1239/jap/1014843085. MR 1808870. Cited on p. 23.

[13] V. Bally and G. Pagès. A quantization algorithm for solving multi-dimensional discrete-time optimal stopping problems. *Bernoulli*, 9(6):1003–1049, 2003. ISSN 1350-7265. doi: 10.3150/bj/1072215199. Cited on p. 263.

[14] J. Banks, J. Carson, B. Nelson, and D. Nicol. *Discrete-Event System Simulation*. Prentice Hall, 5 edition, 2010. ISBN 0136062121. Cited on p. 290.

[15] G. A. Barnard. Sequential tests in industrial statistics. *Supplement to the Journal of the Royal Statistical Society*, 8:1–21, 1946. ISSN 0035-9254. Cited on p. 10.

[16] M. Baron. Early detection of epidemics as a sequential change-point problem. In V. Antonov, C. Huber, M. Nikulin, and V. Polischook, editors, *Longevity, aging and degradation models in reliability, public health, medicine and biology, LAD 2004. Selected papers from the first French-Russian conference, St. Petersburg, Russia, June 7–9, 2004*, volume 2 of *IMS Lecture Notes-Monograph Series*, pages 31–43, St. Petersburg, Russia, 2004. St. Petersburg State Politechnical University. Cited on p. 38.

[17] J. Bartroff and T. L. Lai. Multistage tests of multiple hypotheses. *Commun. Stat., Theory Methods*, 39(8-9):1597–1607, 2010. ISSN 0361-0926. doi: 10.1080/03610920802592852. Zbl 1318.62248. Cited on p. 33.

[18] M. Basseville and A. Benveniste, editors. *Detection of abrupt changes in signals and dynamical systems. (Proceedings of a Conference on Detection of Abrupt Changes in Signals and Dynamical Systems, Paris, March 21-22, 1984).* Lecture Notes in Control and Information Sciences, 77. Berlin etc.: Springer-Verlag, 1986. Cited on pp. 37 and 38.

[19] N. Bäuerle and U. Rieder. *Markov Decision Processes with Applications to Finance*. Universitext. Berlin: Springer, 2011. ISBN 978-3-642-18323-2; 978-3-642-18324-9. doi: 10.1007/978-3-642-18324-9. Zbl 1236.90004. Cited on p. 108.

[20] L. Bayón, P. Fortuny, J. Grau, A. M. Oller-Marcén, and M. M. Ruiz. The best-or-worst and the postdoc problems with random number of candidates. *J. Comb. Optim.*, 38(1):86–110, 2019. ISSN 1382-6905. doi: 10.1007/s10878-018-0367-6. Zbl 1461.90115. Cited on p. 41.

[21] L. Bayón, P. Fortuny Ayuso, J. M. Grau, A. M. Oller-Marcén, and M. M. Ruiz. A new method for computing asymptotic results in optimal stopping problems. *Bulletin of the Malaysian Mathematical Sciences Society*, 46(1):46, 2023. Cited on p. 271.

[22] E. Bayraktar and M. Ludkovski. Optimal trade execution in illiquid markets. *Math. Finance*, 21(4):681–701, 2011. ISSN 0960-1627. doi: 10.1111/j.1467-9965.2010.00446.x. Cited on p. 191.

[23] E. Bayraktar, S. Dayanik, and I. Karatzas. The standard Poisson disorder problem revisited. *Stochastic Processes Appl.*, 115(9):1437–1450, 2005. doi: 10.1016/j.spa.2005.04.011. Cited on p. 122.

[24] S. Becker, P. Cheridito, and A. Jentzen. Deep optimal stopping. *Journal of Machine Learning Research*, 20:74, 2019. Cited on p. 263.

[25] X. Bei and S. Zhang. The Secretary Problem with Competing Employers on Random Edge Arrivals. In *Proceedings of the 36th AAAI Conference on Artificial Intelligence, AAAI 2022*, volume 36, pages 4818–4825, 2022. URL `https://www.scopus.com/inward/record.uri?eid=2-s2.0-85147539952&partnerID=40&md5=16d6b55157cea438e41596e95cb6c46c`. Cited on pp. 82 and 83.

[26] R. Bellman. Dynamic programming and a new formalism in the calculus of variations. *Proc. Natl. Acad. Sci. USA*, 40:231–235, 1954. ISSN 0027-8424. doi: 10.1073/pnas.40.4.231. Zbl 0058.36303. Cited on p. 108.

[27] R. Bellman. *Dynamic Programming*. Princeton University Press, Princeton, New Jersey, 1957. Zbl 0077.13605. Cited on pp. 108 and 195.

[28] R. Bellman, R. Kalaba, and D. Middleton. Dynamic programming, sequential estimation and sequential detection processes. *Proc. Nat. Acad. Sci. U.S.A.*, 47:338–341, 1961. ISSN 0027-8424. doi: 10.1073/pnas.47.3.338. Cited on p. 42.

[29] R. E. Bellman and S. E. Dreyfus. Applied dynamic programming, 1962. Zbl 0106.34901. Cited on p. 108.

[30] D. Belomestny and J. Schoenmakers. *Advanced Simulation-based Methods for Optimal Stopping and Control. With Applications in Finance*. London: Palgrave Macmillan, 2018. ISBN 978-1-137-03350-5; 978-1-137-03351-2. doi: 10.1057/978-1-137-03351-2. Zbl 6883074. Cited on p. 4.

[31] V. Belton and T. J. Stewart. *Multiple Criteria Decision Analysis: An Integrated Approach*. Springer US, Boston, MA, 2002. ISBN 978-1-4615-1495-4. doi: 10.1007/978-1-4615-1495-4_1. Cited on pp. 3, 7, and 47.

[32] A. Bensoussan and A. Friedman. Nonlinear variational inequalities and differential games with stopping times. *J. Funct. Anal.*, 16:305–352, 1974. Cited on p. 81.

[33] A. Bensoussan and A. Friedman. Nonzero-sum stochastic differential games with stopping times and free boundary problems. *Trans. Amer. Math. Soc.*, 231:275–327, 1977. Cited on pp. 80 and 81.

[34] B. A. Berezovskij, Y. M. Baryshnikov, and A. V. Gnedin. On a class of best-choice problems. *Inf. Sci.*, 39:111–127, 1986. ISSN 0020-0255. doi: 10.1016/0020-0255(86)90056-3. Cited on p. 24.

[35] J. O. Berger. Sequential Analysis. In *The New Palgrave Dictionary of Economics*, pages 1–3. Palgrave Macmillan UK, London, 2017. ISBN 978-1-349-95121-5. doi: 10.1057/978-1-349-95121-5_1295-1. [Access: 27 November 2016]. Cited on p. 10.

[36] J.-M. Bismut. Sur un Problème de Dynkin. *Z. Wahrsch. Ver. Gebite*, 39:31–53, 1977. Cited on p. 81.

[37] A. W. Blocker. *fastGHQuad: Fast Rcpp implementation of Gauss-Hermite quadrature*, 2014. URL `https://cran.r-project.org/package=fastGHQuad`. Cited on p. 175.

[38] T. Bojdecki. On optimal stopping of independent random variables appearing according to a renewal process, with random time horizon. *Bol. Soc. Mat. Mexicana (2)*, 22(1):35–40, 1977. Boletín de la Sociedad Matemática Mexicana. Segunda Serie. MR 651552. Cited on p. 20.

[39] T. Bojdecki. On optimal stopping of a sequence of independent random variables—probability maximizing approach. *Stochastic Process. Appl.*, 6(2): 153–163, 1977/78. ISSN 0304-4149. doi: 10.1016/0304-4149(78)90057-1. Cited on p. 16.

[40] T. Bojdecki. Probability maximizing approach to optimal stopping and its application to a disorder problem. *Stochastics*, 3(1):61–71, 1979. ISSN 0090-9491. doi: 10.1080/17442507908833137. Zbl 0432.60051. Cited on pp. 38, 122, 127, and 148.

[41] T. Bojdecki and J. Hosza. On a generalized disorder problem. *Stochastic Process. Appl.*, 18(2):349–359, 1984. ISSN 0304-4149. doi: 10.1016/0304-4149(84)90305-3. Cited on p. 122.

[42] E. Borel. Les probabilités dénombrables et leur applications arithmétiques. *Rendiconti Circolo mat. Palermo*, 27:247–270, 1909. Cited on p. 290.

[43] F. Boshuizen. Comparisons of threshold stopping rule and supremum expectations for independent random vectors. *Stochastic Anal. Appl.*, 8(4):389–396, 1990. ISSN 0736-2994. doi: 10.1080/07362999008809215. Zbl 0729.60038. Cited on p. 23.

[44] F. A. Boshuizen. *Prophet and Minimax Problems in Optimal Stopping Theory*. PhD thesis, Vrije Universiteit Te Amsterdam, Amsterdam, 1991. Cited on p. 23.

[45] M. Brenner and D. Galai. New financial instruments for hedging changes in volatility. *Financial Analysts Journal*, 45(4):61–65, 1989. ISSN 0015198X. URL `http://www.jstor.org/stable/4479241`. Cited on p. 72.

[46] M. Broadie and P. Glasserman. A stochastic mesh method for pricing high-dimensional american options. *Journal of Computational Finance*, 7:35–72, 2004. Cited on pp. 263 and 264.

[47] A. Z. Broder, A. Kirsch, R. Kumar, M. Mitzenmacher, and S. Vassilvitskii. The hiring problem and Lake Wobegon strategies. In *Proceedings of the nineteenth annual ACM-SIAM symposium on discrete algorithms, SODA 2008, San Francisco, CA, January 20–22, 2008*, pages 1184–1193. New York, NY: Association for Computing Machinery (ACM); Philadelphia, PA: Society for Industrial and Applied Mathematics (SIAM), 2008. ISBN 978-0-89871-647-4. Zbl 1192.90100. Cited on p. 216.

[48] B. Brodsky and B. Darkhovsky. *Nonparametric Methods in Change–Point Problems*, volume 243 of *Mathematics and its Applications*. Kluwer Academic Publishers, Dordrecht, 1993. Cited on p. 37.

[49] B. Brodsky and B. Darkhovsky. *Non-parametric Statistical Diagnosis. Problems and Methods.* Mathematics and its Applications (Dordrecht). 509. Dordrecht: Kluwer Academic Publishers. xvi, 452 p., 2000. Cited on p. 37.

[50] R. Bronfman-Nadas, N. Zincir-Heywood, and J. T. Jacobs. An Artificial Arms Race: Could it Improve Mobile Malware Detectors? In *2018 Network Traffic Measurement and Analysis Conference (TMA)*, pages 1–8. IEEE, jun 2018. ISBN 978-3-903176-09-6. doi: 10.23919/TMA.2018.8506545. Cited on p. 107.

[51] D. B. Brown, J. E. Smith, and P. Sun. Information relaxations and duality in stochastic dynamic programs. *Operations Research*, 58(4 PART 1):785–801, 2010. doi: 10.1287/opre.1090.0796. Cited on p. 306.

[52] F. T. Bruss. Sum the odds to one and stop. *Ann. Probab.*, 28(3):1384–1391, 2000. ISSN 0091-1798,2168-894X. doi: 10.1214/aop/1019160340. Cited on pp. 304 and 337.

[53] F. T. Bruss. What is known about Robbins' problem? *J. Appl. Probab.*, 42 (1):108–120, 2005. ISSN 0021-9002. doi: 10.1239/jap/1110381374. Zbl 1081.62059. Cited on pp. 14 and 16.

[54] F. T. Bruss and T. S. Ferguson. Multiple buying or selling with vector offers. *J. Appl. Probab.*, 34(4):959–973, 1997. ISSN 0021-9002. doi: 10.2307/3215010. Zbl 0905.60028. Cited on p. 42.

[55] F. T. Bruss and G. Louchard. Sequential selection of the κ best out of n rankable objects. *Discrete Math. Theor. Comput. Sci.*, 18(3):Paper No. 13, 12, 2016. MR 3601362. Cited on p. 246.

[56] F. T. Bruss and D. Paindaveine. Selecting a sequence of last successes in independent trials. *J. Appl. Probab.*, 37(2):389–399, 2000. ISSN 0021-9002. doi: 10.1017/s002190020001559x. Cited on p. 69.

[57] F. T. Bruss, M. Drmota, and G. Louchard. The complete solution of the competitive rank selection problem. *Algorithmica*, 22(4):413–447, 1998. ISSN 0178-4617. doi: 10.1007/PL00009232. Average-case analysis of algorithms. MR 1701621. Cited on p. 91.

[58] A. Bryson. Optimal control-1950 to 1985. *IEEE Control Systems Magazine*, 16(3):26–33, 1996. doi: 10.1109/37.506395. Cited on p. 6.

[59] R. Cairoli. Martingales à deux paramètres de carré intégrable. *C. R. Acad. Sci. Paris Sér. A-B*, 272:A1731–A1734, 1971. ISSN 0151-0509. MR 297007. Cited on p. 8.

[60] R. Cairoli and R. C. Dalang. *Sequential Stochastic Optimization*. Wiley Series in Probability and Statistics: Probability and Statistics. John Wiley & Sons, Inc., New York, 1996. ISBN 0-471-57754-5. doi: 10.1002/9781118164396. A Wiley-Interscience Publication. MR 1369770. Cited on p. 8.

[61] E. Carlstein, H. Múller, and D. Siegmund, editors. *Change-point Problems*, volume 23 of *IMS Lecture Notes-Monograph Series*, pages 78–92. Institute of Mathematical Statistics, Hayward, California, 1994. Cited on p. 37.

[62] J. F. Carriere. Valuation of the early-exercise price for options using simulations and nonparametric regression. *Insurance: Mathematics and Economics*, 19(1): 19–30, 1996. Cited on pp. 264 and 265.

[63] A. Cayley. Mathematical questions with their solutions. *The Educational Times*, 23:18–19, 1875. See The Collected Mathematical Papers of Arthur Cayley 10, 153–179 (1896). Cambridge University Press, Cambridge. Cited on pp. 4, 12, 90, and 195.

[64] S. S. Chandramouli and M. B. Haugh. A unified approach to multiple stopping and duality. *Oper. Res. Lett.*, 40(4):258–264, 2012. ISSN 0167-6377. doi: 10.1016/j.orl.2012.03.009. Cited on pp. 21, 287, and 306.

[65] N. Chen and P. Glasserman. Additive and multiplicative duals for American option pricing. *Finance and Stochastics*, 11(2):153–179, 2007. doi: 10.1007/s00780-006-0031-3. Cited on p. 287.

[66] R. W. Chen, B. Rosenberg, and L. A. Shepp. A secretary problem with two decision makers. *J. Appl. Probab.*, 34(4):1068–1074, 1997. doi: 10.2307/3215019. Cited on p. 84.

[67] S. Chen. *Multiple Testing and False Discovery Rate Control: Theory, Methods and Algorithms*. PhD thesis, University of California, San Diego, 2019. The dissertation supervisor: Ery Arias-Castro. Copyright - Database copyright ProQuest LLC; ProQuest does not claim copyright in the individual underlying works; Last updated - 2023-03-03. Cited on p. 33.

[68] R. Cheng, S. M. Seubert, and D. D. Wiegmann. Mate choice and the evolutionary stability of a fixed threshold in a sequential search strategy. *Computational and Structural Biotechnology Journal*, 10(16):8–11, 2014. Cited on p. 44.

[69] Y. S. Chow and H. Robbins. A martingale system theorem and applications. In J. Neyman, editor, *Proceedings of the Fourth Berkeley Symposium on Mathematical Statistics and Probability*, volume 4.1, pages 93–104, 1961. Zbl 0126.14002. Cited on pp. 23, 73, and 303.

[70] Y. S. Chow and H. Robbins. On optimal stopping rules. *Zeitschrift für Wahrscheinlichkeitstheorie und verwandte Gebiete*, 2(1):33–49, 1963. Cited on pp. 23 and 303.

[71] Y. S. Chow and H. Robbins. On values associated with a stochastic sequence. In L. M. Le Cam and J. Neyman, editors, *Proceedings of the Fifth Berkeley Symposium on Mathematical Statistics and Probability*, volume 5.1, pages 427–440. University of California Press, 1967. Cited on pp. 23 and 303.

[72] Y. S. Chow, S. Moriguti, H. Robbins, and S. M. Samuels. Optimal selection based on relative rank (the "Secretary problem"). *Israel J. Math.*, 2:81–90, 1964. ISSN 0021-2172. doi: 10.1007/BF02759948. MR 176583. Cited on pp. 16, 17, 26, 234, 238, 241, 242, 244, 245, and 303.

[73] Y. S. Chow, H. Robbins, and D. Siegmund. *Great Expectations: The Theory of Optimal Stopping*. Houghton Mifflin Co., Boston, MA, 1971. MR 331675. Cited on pp. 8, 17, 52, 59, 60, and 63.

[74] A. M. Colman and J. Stirk. Stackelberg reasoning in mixed-motive games: An experimental investigation. *Journal of Economic Psychology*, 19:279–293, 1998. doi: 10.1016/S0167-487098.00008-7. Cited on p. 247.

[75] R. Cowan and J. Zabczyk. A new version of the best choice problem. *Bull. Acad. Pol. Sci., Sér. Sci. Math. Astron. Phys.*, 24:773–778, 1976. ISSN 0001-4117. v. [76]. MR 428424, Zbl 0359.60040. Cited on p. 19.

[76] R. Cowan and J. Zabczyk. An optimal selection problem associated with the Poisson process. *Teor. Veroyatn. Primen.*, 23:606–614, 1978. ISSN 0040-361X. v. [77]. Zbl 0396.62063. Cited on pp. 315 and 316.

[77] R. Cowan and J. Zabczyk. An optimal selection problem associated with the Poisson process. *Theory Probab. Appl.*, 29:584–592, 1979. ISSN 0040-585X. doi: 10.1137/1123066. v.[76]. Zbl 0426.62058. Cited on pp. 19 and 315.

[78] D. Cownden and D. Steinsaltz. Effects of competition in a secretary problem. *Oper. Res.*, 62(1):104–113, 2014. ISSN 0030-364X. doi: 10.1287/opre.2013.1233. Cited on p. 40.

[79] M. H. A. Davis. *Markov Models and Optimization*. Chapman and Hall, New York, 1993. ISBN 9780203748039. doi: 10.1201/9780203748039. Cited on pp. 21 and 48.

[80] S. K. De. *Simultaneous Testing of Multiple Hypotheses in Sequential Experiments*. PhD thesis, The University of Texas at Dallas, 2012. The dissertation supervisor: Dr. Michael Baron. Copyright - Database copyright ProQuest LLC; ProQuest does not claim copyright in the individual underlying works; Last updated - 2023-03-04. Cited on p. 33.

[81] S. Demers. Expected duration of the no-information minimum rank problem. *Statistics & Probability Letters*, 168:108950, 2021. Cited on p. 271.

[82] C. Derman and J. Sacks. Replacement of periodically inspected equipment. (An optimal optional stopping rule.). *Nav. Res. Logist. Q.*, 7:597–607, 1960. ISSN 0028-1441. doi: 10.1002/nav.3800070429. Zbl 0173.46701. Cited on p. 73.

[83] V. V. Desai, V. F. Farias, and C. C. Moallemi. Pathwise Optimization for Optimal Stopping Problems. *Management Science*, 58(12):2292–2308, 2012. doi: 10.1287/mnsc.1120.1551. Cited on p. 287.

[84] R. L. Dobrushin, M. S. Pinsker, and A. N. Shiryaev. An application of the concept of entropy to signal detection problems with background noise. *Litov. Mat. Sb.*, 3(1):107–121, 1963. ISSN 0132-2818. Cited on p. 10.

[85] H. F. Dodge and H. G. Romig. A method of sampling inspection. *Bell System Technical Journal*, 8(4):613–631, 1929. doi: 10.1002/j.1538-7305.1929.tb01240.x. Cited on p. 302.

[86] H. F. Dodge and H. G. Romig. Single sampling and double sampling inspection tables. *Bell System Technical Journal*, 20(1):1–61, 1941. doi: 10.1002/j.1538-7305.1941.tb00851.x. Cited on p. 302.

[87] Y. Dombrovsky and N. Perrin. On adaptive search and optimal stopping mate choice. *The American Naturalist*, 144(2):355–361, 1994. Cited on p. 44.

[88] M. Dresher. *The Mathematics of Games of Strategy*. Dover Publications, Inc., New York, 1981. ISBN 0-486-64216-X. Cited on p. 92.

[89] P. Dube and R. R. Mazumdar. *A Framework for Quickest Detection of Traffic Anomalies in Networks*. Technical report, Electrical and Computer Engineering, Purdue University, West Lafayette, IN, 79406, USA, November 2001. citeseer.ist.psu.edu/506551.html. Cited on pp. 122 and 146.

[90] S. Dubuc. Solutions asymptotiques au problème des secrétaires. *Canadian J. Math.*, 25:495–505, 1973. ISSN 0008-414X. doi: 10.4153/CJM-1973-050-8. Cited on p. 16.

[91] E. B. Dynkin. The optimum choice of the instant for stopping a Markov process. *Soviet Mathematics*, 4:627–629, 1963. Zbl 0242.60018. Cited on p. 21.

[92] E. B. Dynkin. Game variant of a problem on optimal stopping. *Soviet Math. Dokl.*, 10:270–274, 1969. ISSN 0002-3264. translation from Е. Б. Дынкин, "Игровой вариант задачи об оптимальной остановке", Докл. АН СССР, 185:1 (1969), 16–19. Dokl. Akad. Nauk SSSR 185, 16–19 (1969). MR 0241121, Zbl 0186.25304. Cited on pp. 3, 28, 39, 48, 73, 78, 82, 91, and 287.

[93] E. B. Dynkin and A. A. Yushkevich. Теоремы и задачи о процессах Маркова. Izdat. "Nauka", Moscow, 1967. Title in English: *Theorems and problems in Markov processes*. MR 0222956. Cited on pp. 196 and 228.

[94] G. A. Edgar and L. Sucheston. *Stopping Times and Directed Processes*, volume 47 of *Encyclopedia of Mathematics and its Applications*. Cambridge University Press, Cambridge, 1992. ISBN 0-521-35023-9. doi: 10.1017/CBO9780511574740. MR 1191395. Cited on p. 49.

[95] Editorial Staff. The publications and writings of Herbert Robbins. *Ann. Statist.*, 31(2):407–413, 2003. ISSN 0090-5364. doi: 10.1214/aos/1051027874. Dedicated to the memory of Herbert E. Robbins. Cited on p. 14.

[96] N. V. Elbakidze. Construction of the cost and optimal policies in a game problem of stopping a Markov process. *Theory Probab. Appl.*, 21:163–168, 1976. Cited on p. 79.

[97] G. Elfving. A persistency problem connected with a point process. *J. Appl. Probability*, 4:77–89, 1967. ISSN 0021-9002. doi: 10.1017/s0021900200025237. Cited on pp. 7, 19, and 190.

[98] E. G. Enns and E. Ferenstein. The horse game. *J. Oper. Res. Soc. Japan*, 28 (1):51–62, 1985. ISSN 0453-4514. doi: 10.15807/jorsj.28.51. Cited on pp. 84 and 91.

[99] E. G. Enns and E. Z. Ferenstein. On a multiperson time-sequential game with priorities. *Sequential Anal.*, 6(3):239–256, 1987. ISSN 0747-4946. doi: 10.1080/07474948708836129. Cited on p. 84.

[100] H. N. Entwistle, C. J. Lustri, and G. Y. Sofronov. On asymptotics of optimal stopping times. *Mathematics*, 10(2):194, 2022. Cited on pp. 271, 272, 273, 275, 276, and 277.

[101] H. N. Entwistle, C. J. Lustri, and G. Y. Sofronov. Asymptotic duration for optimal multiple stopping problems. *Mathematics*, 12(5), 2024. ISSN 2227-7390. doi: 10.3390/math12050652. Cited on p. 271.

[102] M. Ernst and K. J. Szajowski. Average number of candidates surveyed by the headhunter in the recruitment. *Math. Appl. (Warsaw)*, 49(1):31–53, 2021. ISSN 1730-2668. doi: 10.14708/ma.v49i1.7082. Cited on p. 270.

[103] T. Ezra, M. Feldman, and R. Kupfer. On a competitive secretary problem with deferred selections. In *International Joint Conference on Artificial Intelligence*, 2020. Cited on p. 83.

[104] A. Faller and L. Rüschendorf. Optimal multiple stopping with sum-payoff. *Theory Probab. Appl.*, 57(2):325–336, 2013. ISSN 0040-585X. doi: 10.1137/S0040585X97986011. MR 3201659. Cited on pp. 42 and 191.

[105] A. D. Farooqui and M. A. Niazi. Game theory models for communication between agents: a review. *Complex Adaptive Systems Modeling*, 4(1):13, dec 2016. ISSN 2194-3206. doi: 10.1186/s40294-016-0026-7. Cited on p. 76.

[106] E. Z. Ferenstein. Two-person non-zero-sum games with priorities. In T. S. Ferguson and S. M. Samuels, editors, *Strategies for Sequential Search and Selection in Real Time, Proceedings of the AMS-IMS-SIAM Join Summer Research Conferences held June 21-27, 1990*, volume 125 of *Contemporary Mathematics*, pages 119–133, University of Massachusetts at Amherst, 1992. Cited on pp. 84, 85, and 91.

[107] E. Z. Ferenstein. A variation of the Dynkin's stopping game. *Math. Japonica*, 38(2):371–379, 1993. Cited on p. 81.

[108] E. Z. Ferenstein and A. Pasternak-Winiarski. Optimal stopping of a risk process with disruption and interest rates. In Michéle Breton and Krzysztof Szajowski, editor, *Advances in Dynamic Games. Theory, Applications, and Numerical Methods for Differential and Stochastic Games. Dedicated to the Memory of Arik A. Melikyan. Selected Papers Presented at the 13th International Symposium on Dynamic Games and Applications, Wrocław, Poland, Summer 2008*, pages 489–507. Boston, MA: Birkhäuser, 2011. ISBN 978-0-8176-8088-6/hbk; 978-0-8176-8089-3/ebook. doi: 10.1007/978-0-8176-8089-3_24. Zbl 1218.91078. Cited on pp. 38 and 146.

[109] T. S. Ferguson. Who Solved the Secretary Problem? *Statistical Science*, 4(3): 282–296, Aug 1989. ISSN 0883-4237. doi: 10.1214/ss/1177012493. Cited on pp. 4, 12, 14, 15, 24, 27, 82, 84, 90, 195, 196, and 266.

[110] T. S. Ferguson. Selection by committee. In *Advances in Dynamic Games. Applications to Economics, Finance, Optimization and Stochastic Control*, pages 203–209. Boston, MA: Birkhäuser, 2005. ISBN 0-8176-4362-1. doi: 10.1007/0-8176-4429-6_10. `https://www.math.ucla.edu/~tom/papers/committee.pdf`, Zbl 1123.91003. Cited on pp. 73 and 97.

[111] T. S. Ferguson. The sum-the-odds theorem with application to a stopping game of Sakaguchi. *Math. Appl. (Warsaw)*, 44(1):45–61, 2016. ISSN 1730-2668. doi: 10.14708/ma.v44i1.1192. Cited on p. 304.

[112] A. Frank and S. Samuels. On an optimal stopping of Gusein–Zade. *Stoch. Proc. Appl.*, 10:299–311, 1980. Cited on pp. 196 and 197.

[113] P. R. Freeman. The secretary problem and its extensions: a review. *Int. Statist. Rev.*, 51:189–206, 1983. Cited on pp. 14, 15, 82, 84, and 195.

[114] E. Frid. The optimal stopping for a two-person Markov chain with opposing interests. *Theory Probab. Appl.*, 14(4):713–716, 1969. Cited on p. 79.

[115] E. B. Frid. The optimal stopping rule for a Markov chain controlled by two persons with contradictory interests. *Teor. Veroyatn. Primen.*, 14:746–749, 1969. ISSN 0040-361X. English transl. in *Theory of Probability & Its Applications* vol.14(4), 713–716, 1969. Zbl 0194.20902. Cited on p. 28.

[116] A. Fuchsberger. Intrusion Detection Systems and Intrusion Prevention Systems. *Information Security Technical Report*, 10(3):134–139, jan 2005. ISSN 13634127. doi: 10.1016/j.istr.2005.08.001. Cited on p. 108.

[117] M. Fushimi. The secretary problem in a competitive situation. *J. Oper. Res. Soc. Japan*, 24:350–359, 1981. ISSN 0453-4514. doi: 10.15807/jorsj.24.350. ZBL0482.90090. Cited on pp. 82, 84, 85, and 86.

[118] C. Gabor and T. Halliday. Sequential mate choice by multiply mating smooth newts: females become more choosy. *Behav. Ecol.*, 8(2):162–166, 1997. Cited on p. 44.

[119] M. Gardner. Mathematical games. *Scientific American*, 202(1):150–156, 1960. The second part of the note: (3)(1960)172–182. Cited on pp. 15, 26, and 195.

[120] F. Gensbittel, P. Dana, and J. Renault. Competition and recall in selection problems. *Dyn. Games Appl.*, 13(Issue 4):40p., 2023. ISSN 2153-0785. doi: 10.1007/s13235-023-00539-2. Cited on p. 101.

[121] J. P. Gilbert and F. Mosteller. Recognizing the Maximum of a Sequence. *Journal of the American Statistical Association*, 61(313):35–73, 1966. doi: 10.1080/01621459.1966.10502008. MR 0198637. Cited on pp. 15, 16, 17, 18, 19, 20, 24, 84, 91, 196, 197, 228, 249, 252, 254, 258, and 271.

[122] M. Girshick and H. Rubin. A Bayes approach to a quality control model. *Ann. Math. Stat.*, 23:114–125, 1952. Cited on p. 37.

[123] K. S. Glasser. *Some generalizations of the Secretary Problem.* ProQuest LLC, Ann Arbor, MI, United States – District of Columbia, 1978. ISBN 979-8-204-74112-6. Advisor:Austin Barron. Cited on p. 19.

[124] K. S. Glasser, R. Holzsager, and A. Barron. The d choice Secretary Problem. *Communications in Statistics. C. Sequential Analysis*, 2(3):177–199, 1983. ISSN 0731-177X. doi: 10.1080/07474948308836035. MR 735911. Cited on p. 19.

[125] H. Glickman. A best-choice problem with multiple selectors. *J. Appl. Probab.*, 37(3):718–735, 2000. ISSN 0021-9002. doi: 10.1017/s0021900200015941. MR 1782448. Cited on p. 337.

[126] A. V. Gnedin. On the full information best-choice problem. *Journal of Applied Probability*, 33(3):678–687, 1996. Cited on p. 271.

[127] A. Goldenshluger, Y. Malinovsky, and A. Zeevi. A unified approach for solving sequential selection problems. *Probab. Surv.*, 17:214–256, 2020. ISSN 1549-5787. doi: 10.1214/19-PS333. Zbl 7210964. Cited on p. 26.

[128] S. M. Guseĭn-Zade. The problem of choice and the optimal stopping rule for a sequence of independent trials. *Teor. Verojatnost. i Primenen.*, 11:534–537, 1966. ISSN 0040-361x. MR 202256. Cited on p. 21.

[129] S. M. Guseĭn-Zade. The problem of choice and the optimal stopping rule for a sequence of independent trials. *Teor. Verojatnost. i Primenen.*, 11:534–537, 1966. ISSN 0040-361x. MR 202256. Cited on pp. 26, 196, and 197.

[130] I. Guttman. On a Problem of L. Moser. *Can. Math. Bull.*, 3(1):35–39, 1960. ISSN 0008-4395. doi: 10.4153/CMB-1960-008-8. ZBL0090.35802. Cited on p. 271.

[131] G. W. Haggstrom. Optimal stopping and experimental design. *Ann. Math. Statist.*, 37:7–29, 1966. ISSN 0003-4851. doi: 10.1214/aoms/1177699594. MR 195221. Cited on pp. 8, 31, and 60.

[132] G. W. Haggstrom. Optimal sequential procedures when more than one stop is required. *Ann. Math. Statist.*, 38:1618–1626, 1967. ISSN 0003-4851. doi: 10.1214/aoms/1177698595. MR 217946. Cited on pp. 3, 8, 9, 18, 47, 48, 50, 52, 53, and 287.

[133] J. R. Hall, S. A. Lippman, and J. J. McCall. Expected utility maximizing job search. In S. A. Lippman and J. J. McCall, editors, *Studies in the Economics of Search*, pages 133–135. North-Holland, Amsterdam, 1979. Cited on p. 42.

[134] K. Hammar and R. Stadler. Finding Effective Security Strategies through Reinforcement Learning and Self-Play. In *2020 16th International Conference on Network and Service Management (CNSM)*, pages 1–9. IEEE, nov 2020. ISBN 978-3-903176-31-7. doi: 10.23919/CNSM50824.2020.9269092. Cited on p. 108.

[135] K. Hammar and R. Stadler. Learning Intrusion Prevention Policies through Optimal Stopping. In *2021 17th International Conference on Network and Service Management (CNSM)*, pages 509–517. IEEE, oct 2021. ISBN 978-3-903176-36-2. doi: 10.23919/CNSM52442.2021.9615542. Cited on p. 108.

[136] K. Hammar and R. Stadler. Intrusion Prevention Through Optimal Stopping. *IEEE Transactions on Network and Service Management*, 19(3):2333–2348, sep 2022. ISSN 1932-4537. doi: 10.1109/TNSM.2022.3176781. Cited on p. 107.

[137] Y. Han and N. Li. A new deep neural network algorithm for multiple stopping with applications in options pricing. *Communications in Nonlinear Science and Numerical Simulation*, 117:106881, 2023. ISSN 1007-5704. Cited on p. 263.

[138] F. Harten, A. Meyerthole, and N. Schmitz. *Prophetentheorie. Prophetenungleichungen, Prophetenregionen, Spiele gegen einen Propheten*. Part I. The general and the independent case. Stuttgart: B. G. Teubner, 1997. ISBN 3-519-02737-2. Skripten zur Mathematische Statistik nr. 34. Cited on p. 22.

[139] M. B. Haugh and L. Kogan. Pricing American options: A duality approach. *Operations Research*, 52(2):258–270, 2004. doi: 10.1287/opre.1030.0070. Cited on p. 287.

[140] X. He, H. Dai, P. Ning, and R. Dutta. A Leader–Follower controlled Markov stopping game for delay tolerant and opportunistic resource sharing networks. *IEEE Journal on Selected Areas in Communications*, 35(3):615–627, 2017. doi: 10.1109/JSAC.2017.2659581. Cited on p. 247.

[141] A. Helmi and A. Panholzer. Analysis of the "hiring above the median" selection strategy for the hiring problem. *Algorithmica*, 66(4):762–803, 2013. ISSN 0178-4617. doi: 10.1007/s00453-012-9727-2. MR 3071846, Zbl 1307.62196. Cited on p. 215.

[142] T. Herberts and U. Jensen. Optimal detection of a change-point in a Poisson process for different observation schemes. *Scand. J. Stat.*, 31(3):347–366, 2004. doi: 10.1111/j.1467-9469.2004.02-102.x. Cited on pp. 38 and 146.

[143] T. P. Hill. Knowing When to Stop. *American Scientist*, 97(2):126–133, 2009. ISSN 1545-2786. doi: 10.1511/2009.77.126. Cited on pp. 4 and 12.

[144] T. P. Hill and D. P. Kennedy. Prophet inequalities for parallel processes. *J. Multivariate Anal.*, 31(2):236–243, 1989. ISSN 0047-259X. doi: 10.1016/0047-259X(89)90064-X. Zbl 0685.60048. Cited on p. 23.

[145] T. P. Hill and R. P. Kertz. Additive comparisons of stop rule and supremum expectations of uniformly bounded independent random variables. *Proc. Am. Math. Soc.*, 83:582–585, 1981. ISSN 0002-9939. doi: 10.2307/2044124. Zbl 0476.60044. Cited on p. 23.

[146] T. P. Hill and R. P. Kertz. Stop rule inequalities for uniformly bounded sequences of random variables. *Trans. Am. Math. Soc.*, 278:197–207, 1983. ISSN 0002-9947. doi: 10.2307/1999311. Cited on p. 23.

[147] T. P. Hill and R. P. Kertz. Multivariate prophet inequalities for negatively dependent random vectors. In *Strategies for Sequential Search and Selection in Real Time. Proceedings of the AMS-IMS-SIAM Joint Summer Research Conference, held June 21-27, 1990, at the University of Massachusetts, Amherst, MA, USA*, pages 183–190. Providence, RI: American Mathematical Society, 1992. ISBN 0-8218-5133-0. Zbl 0794.60040. Cited on p. 101.

[148] Y. Hochberg and A. C. Tamhane. *Multiple Comparison Procedures.* Wiley Ser. Probab. Math. Stat. John Wiley & Sons, Inc., New York, 1987. ISBN 0-471-82222-1. Zbl 0731.62125. Cited on p. 33.

[149] M. H. Hof, A. C. Ravelli, and A. H. Zwinderman. Adaptive list sequential sampling method for population-based observational studies. *BMC Medical Research Methodology*, 14(1):1–9, 2014. Cited on p. 111.

[150] S. Holm. A simple sequentially rejective multiple test procedure. *Scand. J. Stat.*, 6:65–70, 1979. ISSN 0303-6898. Zbl 0402.62058. Cited on p. 33.

[151] J. M. C. Hutchinson and K. Halupka. Mate choice when males are in patches: Optimal strategies and good rules of thumb. *Journal of Theoretical Biology*, 231(1):129–151, 2004. Cited on p. 44.

[152] N. Immorlica, R. Kleinberg, and M. Mahdian. Secretary Problems with Competing Employers. In P. Spirakis, M. Mavronicolas, and S. Kontogiannis, editors, *Internet and Network Economics. WINE 2006*, volume 4286 of *International Workshop on Internet and Network Economics*, pages 389–400, Berlin, Heidelberg, 2006. Springer Berlin Heidelberg. ISBN 978-3-540-68141-0. doi: 10.1007/11944874_35. Conference:"International Workshop on Internet and Network Economics". Cited on pp. 40 and 82.

[153] A. Irle. On the best choice problem with random population size. *Z. Oper. Res. Ser. A-B*, 24(5):177–190, 1980. ISSN 0340-9422. doi: 10.1007/bf01919245. Cited on p. 19.

[154] A. Irle. Transitivity in problems of optimal stopping. *Ann. Probab.*, 9(4):642–647, 1981. ISSN 0091-1798. doi: 10.1214/aop/1176994369. MR 624690. Cited on p. 9.

[155] P. Jaillet, E. I. Ronn, and S. Tompaidis. Valuation of commodity-based swing options. *Management Science*, 50(7):909–921, 2004. doi: 10.1287/mnsc.1040.0240. Cited on p. 287.

[156] F. Jamshidian. The duality of optimal exercise and domineering claims: A Doob-Meyer decomposition approach to the Snell envelope. *Stochastics*, 79 (1-2):27–60, 2007. doi: 10.1080/17442500601051914. Cited on p. 287.

[157] A. Jaśkiewicz and A. S. Nowak. Non-Zero-Sum Stochastic Games. In T. Basar and G. Zaccour, editors, *Handbook of Dynamic Game Theory*, pages 1–64. Springer International Publishing, Cham, 2017. ISBN 978-3-319-27335-8. doi: 10.1007/978-3-319-27335-8_33-3. Cited on pp. 78 and 82.

[158] C. Jennison and B. W. Turnbull. *Group Sequential Methods with Applications to Clinical Trials*. Boca Raton, FL: Chapman & Hall/CRC, 2000. ISBN 0-8493-0316-8. Cited on p. 10.

[159] P. Johnson, R. Lagerström, and M. Ekstedt. A Meta Language for Threat Modeling and Attack Simulations. In *Proceedings of the 13th International Conference on Availability, Reliability and Security*, pages 1–8, New York, NY, USA, aug 2018. ACM. ISBN 9781450364485. doi: 10.1145/3230833.3232799. Cited on p. 107.

[160] M. Jones. Prophet inequalities for cost of observation stopping problems. *J. Multivariate Anal.*, 34(2):238–253, 1990. ISSN 0047-259X. doi: 10.1016/0047-259X(90)90038-J. Zbl 0753.60042. Cited on p. 23.

[161] R. E. Kalman. The theory of optimal control and the calculus of variations. In R. Bellman, editor, *Mathematical Optimization Techniques*, pages 309–332. University of California Press, Berkeley and Los Angeles, CA, 1963. ISBN 978-0-520-31987-5. doi: doi:10.1525/9780520319875-018. Zbl 0123.08202. Cited on p. 108.

[162] I. Karatzas. A note on Bayesian detection of change-points with an expected miss criterion. *Stat. Decis.*, 21(1):3–13, 2003. doi: 10.1524/stnd.21.1.3.20317. Cited on p. 122.

[163] A. Karlin and E. Lei. On a Competitive Secretary Problem. *Proceedings of the AAAI Conference on Artificial Intelligence*, 29(1), Feb 2015. ISSN 2374-3468. doi: 10.1609/aaai.v29i1.9312. Cited on p. 82.

[164] G. M. Kaufman. *Statistical Decision and Related Techniques in Oil and Gas Exploration*. Prentice-Hall, Englewood Cliffs, NJ, 1963. Cited on p. 42.

[165] M. Kawai and M. Tamaki. The Gusein-Zade problem under a generalized uniform distribution of the number of applicants. In M. Tamaki, editor, *Mathematics of Decision-making Under Uncertainty*, volume 1306, pages 223–229. Research Institute for Mathematical Sciences, Kyoto University, 2003a. URL `https://www.kurims.kyoto-u.ac.jp/~kyodo/kokyuroku/contents/pdf/1306-28.pdf`. (Japanese) (Kyoto, 2002). MR 1998526. Cited on p. 21.

[166] M. Kawai and M. Tamaki. Choosing either the best or the second best when the number of applicants is random. *Comput. Math. Appl.*, 46(7):1065–1071, 2003b. ISSN 0898-1221,1873-7668. doi: 10.1016/S0898-1221(03)90120-9. Applied stochastic system modeling (Kyoto, 2000). MR 2023150. Cited on p. 21.

[167] D. P. Kennedy and R. P. Kertz. Limit theorems for threshold-stopped random variables with applications to optimal stopping. *Advances in Applied Probability*, 22(2):396–411, 1990. Cited on p. 271.

[168] D. P. Kennedy and R. P. Kertz. The asymptotic behavior of the reward sequence in the optimal stopping of iid random variables. *The Annals of Probability*, 19 (1):329–341, 1991. Cited on p. 271.

[169] D. P. Kennedy and R. P. Kertz. A prophet inequality for independent random variables with finite variances. *J. Appl. Probab.*, 34(4):945–958, 1997. ISSN 0021-9002. doi: 10.2307/3215009. Zbl 0924.60014. Cited on p. 101.

[170] J. Keppo. Pricing of electricity swing options. *Journal of Derivatives*, 11(3): 26–43, 2004. doi: 10.3905/jod.2004.391033. Cited on p. 287.

[171] Y. I. Kifer. Optimal stopped games. *Theory Probab. Appl.*, 16:185–189, 1971. ISSN 0040-585X. doi: 10.1137/1116018. Zbl 0238.90085. Cited on p. 28.

[172] Y. I. Kifer. Optimal stopping in games with continuous time. *Theory Probab. Appl.*, 16:545–550, 1971. ISSN 0040-585X. doi: 10.1137/1116060. Zbl 0251.90066. Cited on p. 78.

[173] Y. I. Kifer. Game options. *Finance Stoch.*, 4(4):443–463, 2000. ISSN 0949-2984. doi: 10.1007/PL00013527. Zbl 1066.91042. Cited on pp. 72 and 73.

[174] Y. I. Kifer. Dynkin's games and Israeli options. *ISRN Probab. Stat.*, 2013: 17, 2013. ISSN 2090-472X. doi: 10.1155/2013/856458. Id/No 856458; Zbl 1271.91022. Cited on pp. 28 and 29.

[175] J. Kleinberg, R. Kleinberg, and S. Oren. Optimal stopping with behaviorally biased agents: the role of loss aversion and changing reference points. *Games Econ. Behav.*, 133:282–299, 2022. ISSN 0899-8256. doi: 10.1016/j.geb.2022.03.007. Zbl 1492.91099. Cited on p. 308.

[176] R. Kleinberg. A multiple-choice secretary algorithm with applications to online auctions. In *Proceedings of the Sixteenth Annual ACM-SIAM Symposium on Discrete Algorithms*, pages 630–631, 01 2005. doi: 10.1145/1070432.1070519. SODA 2005, Vancouver, British Columbia, Canada, January 23-25, 2005. Cited on p. 245.

[177] F. H. Knight. *Risk, Uncertainty, and Profit*. Cosimo Classics. Bibliolife DBA of Bibilio Bazaar II LLC, 2015. ISBN 978-1-293-95925-1. The first edition has 381 pages and was published by Houghton Mifflin Company, The Riberside Press, Boston and New York in 1921. Cited on pp. 300 and 301.

[178] W. Koepf. *Hypergeometric Summation an Algorithmic Approach to Summation and Special Function Identities*. Universitext. Springer London, London, 2nd ed. 2014. ISBN 9781447164647. Cited on p. 232.

[179] M. Kohler, A. Krzyżak, and N. Todorovic. Pricing of high-dimensional american options by neural networks. *Mathematical Finance*, 20(3):383–410, 2010. Cited on p. 263.

[180] A. Kolmogoroff. *Grundbegriffe der Wahrscheinlichkeitsrechnung*. Number 3 in Ergebnisse der Math. und ihrer Grenzgebiete vol. 2. Julius Springer, Berlin, 1933. Cited on pp. 290 and 291.

[181] H. Kösters. A note on multiple stopping rules. *Optimization*, 53(1):69–75, 2004. Cited on p. 51.

[182] A. Krasnosielska. A version of the Elfving problem with random starting time. *Statist. Probab. Lett.*, 79(23):2429–2436, 2009. ISSN 0167-7152. doi: 10.1016/j.spl.2009.08.017. Cited on p. 8.

[183] A. Krasnosielska. A time-dependent best choice problem with costs and random lifetime in organ transplants. *Appl. Math. (Warsaw)*, 37(3):257–274, 2010. ISSN 1233-7234. doi: 10.4064/am37-3-1. Cited on p. 8.

[184] A. Krasnosielska-Kobos. Multiple-stopping problems with random horizon. *Optimization*, 64(7):1625–1645, 2015. ISSN 0233-1934. doi: 10.1080/02331934.2013.869808. Cited on pp. 8, 59, and 190.

[185] U. Krengel and L. Sucheston. Semiamarts and finite values. *Bull. Am. Math. Soc.*, 83:745–747, 1977. ISSN 0002-9904. doi: 10.1090/S0002-9904-1977-14378-4. Zbl 0336.60032. Cited on pp. 22 and 23.

[186] U. Krengel and L. Sucheston. Temps d'arrêt et tactiques pour des processus indexés par un ensemble ordonné. *C. R. Acad. Sci. Paris Sér. A-B*, 290(4): A193–A196, 1980. ISSN 0151-0509. MR 564159. Cited on p. 8.

[187] U. Krengel and L. Sucheston. Stopping rules and tactics for processes indexed by a directed set. *J. Multivariate Anal.*, 11(2):199–229, 1981. ISSN 0047-259X. doi: 10.1016/0047-259X(81)90109-3. MR 618785. Cited on pp. 9 and 22.

[188] A. M. Krieger, M. Pollak, and E. Samuel-Cahn. Select sets: rank and file. *Ann. Appl. Probab.*, 17(1):360–385, 2007. ISSN 1050-5164. doi: 10.1214/105051606000000691. Cited on p. 216.

[189] V. Krishna, editor. *Auction Theory*. Academic Press, San Diego, 2nd edition, 2010. ISBN 978-0-12-374507-1. doi: 10.1016/B978-0-12-374507-1.00024-8. Cited on p. 190.

[190] N. V. Krylov. Control of Markov processes and W-spaces. *Math. USSR-Izv.*, 5:233–266, 1971. Cited on p. 81.

[191] E. M. Kubicka, G. Kubicki, M. Kuchta, and M. Morayne. Secretary problem with hidden information; searching for a high merit candidate. *Adv. Appl. Math.*, 144:35, 2023. ISSN 0196-8858. doi: 10.1016/j.aam.2022.102468. Id/No 102468. Zbl 7637360. Cited on pp. 201 and 202.

[192] M. Kurano, M. Yasuda, and J. Nakagami. Multivariate stopping problem with a majority rule. *J. Oper. Res. Soc. Japan*, 23(3):205–223, 1980. ISSN 0453-4514. doi: 10.15807/jorsj.23.205. MR 594355. Cited on pp. 3, 48, 74, 91, and 287.

[193] T. L. Lai and D. Siegmund. *Herbert Robbins 1915–2001*, page 11pp. National Academy of Sciences, 500 Fifth St., N.W., Washington, D.C. 20001, 2016. URL `http://www.nasonline.org/member-directory/deceased-members/51192.html`. Cited on p. 14.

[194] R. Laraki, J. Renault, and S. Sorin. *Mathematical Foundations of Game Theory*. Universitext. Springer, Cham, 2019. ISBN 978-3-030-26645-5; 978-3-030-26646-2. doi: 10.1007/978-3-030-26646-2. MR 3967750, For the French original see [MR 3135265]. Cited on p. 48.

[195] G. F. Lawler and R. J. Vanderbei. Markov strategies for optimal control problems indexed by a partially ordered set. *Ann. Probab.*, 11(3):642–647, 1983. ISSN 0091-1798. doi: 10.1214/aop/1176993508. Cited on p. 9.

[196] A. Lehtinen. Optimal selection of the four best of a sequence. *Zeitschrift für Operations Research*, 38:309–315, 1993. Cited on pp. 217 and 218.

[197] A. Lehtinen. Optimal selection of the k best of a sequence with k stops. *Mathematical Methods of Operations Research*, 46(2):251–261, 1997. Cited on pp. 217 and 219.

[198] J. P. Lepeltier and M. A. Maingueneau. Le jeu de Dynkin en théorie générale sans l'hypothèse de Mokobodski. *Stochastics*, 13:25–44, 1984. Cited on p. 81.

[199] H. Li. Optimal Multiple Stopping Problems Under g-expectation. *Appl. Math. & Optimization*, 85(2):77–89, 2022. ISSN 1432-0606. doi: 10.1007/s00245-022-09857-0. Id/No 17. Cited on p. 191.

[200] J. Li, S. Zhang, and Y. Li. Modelling and computation of optimal multiple investment timing in multi-stage capacity expansion infrastructure projects. *J. Ind. Manag. Optim.*, 18(1):297–314, 2022. ISSN 1547-5816. doi: 10.3934/jimo.2020154. Zbl 1499.49017. Cited on p. 32.

[201] J. D. Ligon. *The Evolution of Avian Breeding Systems*. Oxford University Press, Oxford, New York, 1999. Cited on p. 44.

[202] Y.-S. Lin, S.-R. Hsiau, and Y.-C. Yao. Optimal selection of the k-th best candidate. *Probab. Eng. Inf. Sci.*, 33(3):327–347, 2019. ISSN 0269-9648. doi: 10.1017/S0269964818000256. Zbl 7629467. Cited on p. 26.

[203] D. V. Lindley. Dynamic programming and decision theory. *Appl. Statist.*, 10: 39–51, 1961. ISSN 0035-9254,1467-9876. doi: 10.2307/2985407. Cited on pp. 15, 16, and 195.

[204] S. A. Lippman and J. J. McCall. Job search in a dynamic economy. *J. Econ. Theory*, 12:365–390, 1976a. ISSN 0022-0531. doi: 10.1016/0022-0531(76)90034-X. Zbl 0346.90019. Cited on p. 42.

[205] S. A. Lippman and J. J. McCall. The economics of job search: a survey. *Economic Inquiry*, 14(2):155–189, 1976b. Cited on p. 42.

[206] S. A. Lippman and J. J. McCall. Search unemployment. In L. Matthiessen and S. Strøm, editors, *Unemployment: Macro and Micro-Economic Explanations*, pages 125–144. Palgrave Macmillan UK, London, 1981. ISBN 978-1-349-05966-9. doi: 10.1007/978-1-349-05966-9_9. Cited on p. 42.

[207] X. Liu, O. Milenkovic, and G. V. Moustakides. Query-based selection of optimal candidates under the Mallows model. *Theor. Comput. Sci.*, 979:24, 2023a. ISSN 0304-3975. doi: 10.1016/j.tcs.2023.114206. Id/No 114206. Zbl 7755523. Cited on p. 18.

[208] X. Liu, O. Milenkovic, and G. V. Moustakides. A combinatorial proof for the secretary problem with multiple choices. Preprint, arXiv:2303.02361 [math.CO] (2023), 2023b. Zbl arXiv:2303.02361. Cited on p. 18.

[209] F. A. Longstaff and E. S. Schwartz. Valuing American options by simulation: A simple least-squares approach. *Review of Financial Studies*, 14(1):113–147, 2001. doi: 10.1093/rfs/14.1.113. Cited on pp. 264, 265, and 287.

[210] W. S. Lovejoy. A survey of algorithmic methods for partially observed Markov decision processes. *Annals of Operations Research*, 28(1):47–65, Dec 1991. ISSN 0254-5330. doi: 10.1007/BF02055574. Cited on p. 106.

[211] J. B. MacQueen and R. G. j. Miller. Optimal persistence policies. *Oper. Res.*, 8(3):362–380, 1960. ISSN 0030-364X. doi: 10.1287/opre.8.3.362. Zbl 0096.12102. Cited on pp. 23 and 73.

[212] A. A. K. Majumdar. Optimal stopping for a two-person sequential game in the discrete case. *Pure and Appl. Math. Sci*, 23:67–75, 1986. Cited on p. 84.

[213] A. A. K. Majumdar. The horse game and the OLA policy. *Indian J. Math.*, 30(3):213–218, 1988. ISSN 0019-5324. MR 971847. Cited on p. 91.

[214] A. A. K. Majumdar. On a generalized secretary problem with three stops under a linear travel cost. *Indian J. Math.*, 33(2):189–202, 1991. ISSN 0019-5324. Zbl 0801.90123. Cited on p. 26.

[215] A. Mandelbaum and R. J. Vanderbei. Optimal stopping and supermartingales over partially ordered sets. *Z. Wahrsch. Verw. Gebiete*, 57(2):253–264, 1981. ISSN 0044-3719. doi: 10.1007/BF00535493. Cited on p. 8.

[216] X. Mao and C. Yuan. *Stochastic Differential Equations with Markovian Switching*. Hackensack, NJ: World Scientific, 2006. ISBN 1-86094-701-8; 978-1-86094-884-8. doi: 10.1142/p473. Zbl 1126.60002. Cited on p. 38.

[217] T. Matsui and K. Ano. A note on a lower bound for the multiplicative odds theorem of optimal stopping. *J. Appl. Probab.*, 51(3):885–889, 2014. ISSN 0021-9002. doi: 10.1239/jap/1409932681. Cited on p. 69.

[218] T. Matsui and K. Ano. Compare the ratio of symmetric polynomials of odds to one and stop. *J. Appl. Probab.*, 54(1):12–22, 2017. ISSN 0021-9002. doi: 10.1017/jpr.2016.83. Cited on p. 69.

[219] V. V. Mazalov. Игровые моменты остановкi. "Nauka" Sibirsk. Otdel., Novosibirsk, 1987. *Igrovye momenty ostanovki*. MR 911102. Cited on p. 91.

[220] V. V. Mazalov. *Mathematical Game Rheory and Applications*. John Wiley & Sons, Ltd., Chichester, 2014. ISBN 978-1-118-89962-5. Cited on p. 91.

[221] V. V. Mazalov and I. A. Falko. The house-selling problem with reward rate criteria and changing costs. *Izv. Ross. Akad. Nauk Teor. Sist. Upr.*, 2:79–88, 2008. ISSN 1029-3620. doi: 10.1134/S106423070802010X. MR 2456169. Cited on p. 42.

[222] V. V. Mazalov and N. V. Peshkov. On asymptotic properties of optimal stopping time. *Theory of Probability & Its Applications*, 48(3):549–555, 2004. Cited on p. 271.

[223] E. J. McShane. Recent developments in the calculus of variations. *Semicent. Adresses Amer. Math. Soc.*, 2:69–97, 1938. Zbl 2516062, JFM 64.0499.01. Cited on p. 6.

[224] S. Mehta and K. S. Kwak. Application of Game Theory to Wireless Networks. In *Convergence and Hybrid Information Technologies*, pages 361–376. InTech, Mar 2010. doi: 10.5772/9642. Cited on p. 76.

[225] Y. Mei. Comments on: "A note on optimal detection of a change in distribution", by Benjamin Yakir. *Ann. Stat.*, 34(3):1570–1076, 2006. ISSN 0090-5364. doi: 10.1214/009053606000000362. Cited on p. 122.

[226] N. Meinshausen and B. Hambly. Monte Carlo methods for the valuation of multiple-exercise options. *Mathematical Finance*, 14(4):557–583, 2004. ISSN 0960-1627. doi: 10.1111/j.0960-1627.2004.00205.x. Cited on pp. 21, 265, and 287.

[227] E. Miehling, M. Rasouli, and D. Teneketzis. Control-Theoretic Approaches to Cyber-Security. In S. Jajodia, G. Cybenko, P. Liu, C. Wang, and M. Wellman, editors, *Adversarial and Uncertain Reasoning for Adaptive Cyber Defense*, volume 11830 of *Lecture Notes in Computer Science*, pages 12–28. Springer International Publishing, Cham, 2019. ISBN 978-3-030-30719-6, 978-3-030-30718-9. doi: 10.1007/978-3-030-30719-6_2. Cited on p. 107.

[228] T. F. Móri. Hitting a small group of middle ranked candidates in the secretary problem. In *Probability Theory and Mathematical Statistics with Applications*

(Visegrád, 1985), pages 155–169, Dordrecht, 1988. Reidel. Proc. 5th Pannonian Symp., Visegrád/Hung. Zbl 0669.60050. Cited on pp. 196 and 215.

[229] H. Morimoto. Non-zero-sum discrete parameter stochastic games with stopping times. *Probab. Theory Relat. Fields*, 72(1):155–160, 1986. ISSN 0178-8051. doi: 10.1007/BF00343901. Cited on p. 80.

[230] L. Moser. On a problem of Cayley. *Scripta Mathematica*, 22:289–292, 1956. Cited on pp. 27, 252, 254, and 271.

[231] H. Moulin. *Game Theory for the Social Sciences*. Studies in Game Theory and Mathematical Economics. New York University Press, New York, 1982. ISBN 0-8147-5386-8/hbk; 0-8147-5387-6/pbk. Cited on p. 92.

[232] G. V. Moustakides. Quickest detection of abrupt changes for a class of random processes. *IEEE Trans. Inf. Theory*, 44(5):1965–1968, 1998. Cited on pp. 38, 122, and 146.

[233] R. B. Myerson. *Game Theory*. Harvard University Press, Cambridge, MA, 1991. ISBN 0-674-34115-5. Analysis of conflict. Cited on p. 48.

[234] T. F. Móri. The random secretary problem with multiple choice. *Ann. Univ. Sci. Budap. Rolando Eötvös, Sect. Comput.*, 5:91–102, 1984. ISSN 0138-9491. MR MR0822604, Zbl 0589.62071. Cited on pp. 3, 20, and 337.

[235] C. H. Nagaraja, L. D. Brown, and L. H. Zhao. An autoregressive approach to house price modeling. *The Annals of Applied Statistics*, 5(1):124–149, 2011. Cited on p. 176.

[236] T. Nakai. Stackelberg solution for a stopping game. *Journal of Information & Optimization Sciences*, 18(3):479–491, 1997. ISSN 0252-2667. doi: 10.1080/02522667.1997.10699354. Cited on pp. 3, 48, 91, and 287.

[237] J. Nash. Non-cooperative games. *Ann. of Math. (2)*, 54:286–295, 1951. ISSN 0003-486X. doi: 10.2307/1969529. Cited on p. 92.

[238] P. Neumann, Z. Porosiński, and K. Szajowski. On two person full-information best choice problem with imperfect observation. In *Game Theory and Applications, II*, volume 2 of *Game Theory Appl.*, pages 47–55. Nova Sci. Publ., Hauppauge, NY, 1996. MR 1428249. Cited on p. 91.

[239] P. Neumann, D. Ramsey, and K. Szajowski. Randomized stopping times in Dynkin games. *ZAMM Z. Angew. Math. Mech.*, 82(11-12):811–819, 2002. ISSN 0044-2267. doi: 10.1002/1521-4001(200211)82:11/12<811::AID-ZAMM811>3.0.CO;2-P. 4th GAMM-Workshop "Stochastic Models and Control Theory" (Lutherstadt Wittenberg, 2001). MR 1944424. Cited on p. 101.

[240] J. Neveu. *Discrete-Parameter Martingales*. North-Holland, Amsterdam, 1975. Cited on pp. 79, 80, and 81.

[241] A. F. Nikolaev. On a formulation of the multiple "disorder" problem. *Theory Probab. Appl.*, 43(2):370–374, 1998. ISSN 0040-361X. doi: 10.1137/S0040585X97976908. MR 1679011. Cited on p. 38.

[242] M. L. Nikolaev. Задача выбора двух объектов с минимальным суммарным рангом. Изв. вузов. Матем., 1976(3(166)):33–42, 1976. ISSN 0021-3446. Zadacha vybora dvukh ob'ektov s minimal'nym summarnym rangom Eng. transl. of the article: „On the selection of two objects with minimal summarized rank". Soviet Math. (Iz. VUZ), 20:3 (1976), 28–36; Zbl 0376.62055. Cited on pp. 19, 202, 228, 231, 235, 246, and 269.

[243] M. L. Nikolaev. On a generalization of the problem of best choice. *Teor. Verojatnost. i Primenen.*, 22(1):191–194, 1977. ISSN 0040-361x. Ob odnom obobshchenii zadachi nailuchshego vybora MR 436309. Cited on pp. 3, 19, 48, 217, 218, 219, 246, and 287.

[244] M. L. Nikolaev. The construction of the value of some sequential game. *Veroyatn. Metody Kibern.*, 14:72–83, 1978. ISSN 0132-2869. Postroenie ceny odnoj posledovatel'noj igry. Zbl 0419.60045. Cited on pp. 59, 61, 63, 65, 67, and 69.

[245] M. L. Nikolaev. Generalized sequential procedures. *Lith. Math. J.*, 19:318–325, 1980. ISSN 0363-1672. doi: 10.1007/BF00969967. Zbl 0436.60037. Cited on pp. 47, 50, 53, and 177.

[246] M. L. Nikolaev. Test for optimality of a generalized sequence of procedures. *Lithuanian Mathematical Journal*, 21(3):253–258, 1981. Cited on pp. 47, 50, and 55.

[247] M. L. Nikolaev. On optimal multiple stopping of Markov sequences. *Theory Probab. Appl.*, 43(2):298–306, 1998. ISSN 0040-585X. doi: 10.1137/S0040585X9797691X. Zbl 0971.60046. Cited on pp. 47, 49, 50, 58, 269, and 287.

[248] M. L. Nikolaev. Optimal multi-stopping rules. *Obozr. Prikl. Prom. Mat.*, 5(2):309–348, 1998. ISSN 0869-8325. Zbl 1075.91511. Cited on pp. 3, 48, 50, 59, 112, 217, and 231.

[249] M. L. Nikolaev and G. Y. Sofronov. A multiple optimal stopping rules for sums of independent random variables. *Discrete Mathematics and Applications*, 17 (5):463–473, 2007. Cited on pp. 3, 48, 49, 177, and 287.

[250] M. L. Nikolaev, G. Y. Sofronov, and T. V. Polushina. A problem of sequential choice of multiple objects with given ranks. *Bulletin of Higher Educational Institutions. The North Caucasus Region. Natural Sciences*, 140(4):11–14, 2007. ISSN 1026-2237. Cited on p. 224.

[251] A. S. Nowak and K. Szajowski. Nonzero-sum stochastic games. In *Stochastic and Differential Games. Theory and Numerical Methods. Dedicated to Prof. A. I. Subbotin*, pages 297–342. Boston: Birkhäuser, 1999. ISBN 0-8176-4029-0. Zbl 0940.91014. Cited on p. 82.

[252] A. Ochman-Gozdek and K. Szajowski. Detection of a random sequence of disorders. In X. He, editor, *Proceedings* 59th *ISI World Statistics Congress 25–30 August 2013, Hong Kong Special Administrative Region, China*, volume CPS018, pages 3795–3800, The Hague, The Netherlands, 2013. Published by the International Statistical Institute. http://2013.isiproceedings.org/Files/CPS018-P8-S.pdf. Cited on pp. 37 and 149.

[253] Y. Ohtsubo. Neveu's martingale conditions and closedness in Dynkin stopping problem with a finite constraint. *Stochastic Process. Appl.*, 22(2):333–342, 1986. ISSN 0304-4149. doi: 10.1016/0304-4149(86)90010-4. MR 860941. Cited on p. 80.

[254] Y. Ohtsubo. Optimal stopping in sequential games with or without a constraint of always terminating. *Math. Oper. Res.*, 11(4):591–607, 1986. ISSN 0364-765X. doi: 10.1287/moor.11.4.591. MR 865554. Cited on p. 80.

[255] Y. Ohtsubo. The Dynkin stopping with a finite constraint and bisequential decision problems. *Mem. Fac. Sci. Kochi Univ. Ser. A Math.*, 7:59–69, 1986. ISSN 0389-0252. Cited on p. 80.

[256] Y. Ohtsubo. A nonzero-sum extension of Dynkin's stopping problem. *Math. Oper. Res.*, 12(2):277–296, 1987. ISSN 0364-765X. doi: 10.1287/moor.12.2.277. MR 888977. Cited on pp. 80 and 81.

[257] G. Owen. *Game Theory*. Emerald Group Publishing Limited, Bingley, Fourth edition, 2013. ISBN 987-1-7819-0507-4. Cited on p. 92.

[258] E. Page. Continuous inspection schemes. *Biometrika*, 41:100–115, 1954. Cited on p. 37.

[259] G. Pagès. *Numerical Probability. An Introduction with Applications to Finance*. Universitext. Cham: Springer, 2018. ISBN 978-3-319-90274-6; 978-3-319-90276-0. doi: 10.1007/978-3-319-90276-0. Zbl 1418.91016. Cited on p. 262.

[260] C. E. Pearce, K. Szajowski, and M. Tamaki. Duration problem with multiple exchanges. *Numerical Algebra, Control and Optimization*, 2(2):333, 2012. Cited on p. 271.

[261] G. Peskir and A. Shiryaev. *Optimal Stopping and Free-boundary Problems*. Lectures in Mathematics, ETH Zürich. Birkhäuser, Basel, 2006. Cited on p. 8.

[262] G. Peskir and A. N. Shiryaev. Solving the Poisson disorder problem. In K. Sandmann and P. J. Schönbucher, editors, *Advances in Finance and Stochastics. Essays in Honour of Dieter Sondermann*, pages 295–312. Springer, Berlin, 2002. Cited on p. 122.

[263] T. E. Pitcher, B. D. Neff, F. H. Rodd, and L. Rowe. Multiple mating and sequential mate choice in guppies: females trade up. *Proc. R. Soc. Lond. B.*, 270:1623–1629, 2003. Cited on p. 44.

[264] J. Pitman. *Combinatorial Stochastic Processes. Ecole d'Eté de Probabilités de Saint-Flour XXXII–2002.*, volume 1875 of *Lect. Notes Math.* Berlin: Springer, 2006. ISBN 3-540-30990-X. doi: 10.1007/b11601500. Cited on p. 216.

[265] E. Platen. About secretary problems. In *Mathematical Statistics*, volume 6 of *Banach Center Publ.*, pages 257–266. PWN, Warsaw, 1980. MR 599392. Cited on p. 246.

[266] S. J. Pocock. *The Historical Development of Clinical Trials*, chapter 2, pages 14–27. John Wiley & Sons, Ltd, 2013. ISBN 9781118793916. doi: 10.1002/9781118793916.ch2. Reprint from 1983. Cited on p. 10.

[267] T. V. Polushina. Estimating optimal stopping rules in the multiple best choice problem with minimal summarized rank via the cross-entropy method. In Y.-P. Chen, editor, *Exploitation of Linkage Learning in Evolutionary Algorithms.*, volume 3 of *Evolutionary Learning and Optimization*, pages 227–241. Springer Berlin Heidelberg, Berlin, Heidelberg, 2010. ISBN 978-3-642-12834-9. doi: 10.1007/978-3-642-12834-9_11. Cited on p. 266.

[268] V. H. Poor and O. Hadjiliadis. *Quickest Detection.* Cambridge: Cambridge University Press, 2009. ISBN 978-0-521-62104-5/hbk. doi: 10.1017/CBO9780511754678. Cited on pp. 38 and 146.

[269] Z. Porosiński. Optimal stopping of a random length sequence of maxima over a random barrier. *Zastos. Mat.*, 20(2):171–184, 1990. ISSN 0044-1899. doi: 10.4064/am-20-2-171-184. MR 1053144. Cited on p. 91.

[270] Z. Porosiński and K. Szajowski. Modified strategies in two person full-information best choice problem with imperfect observation. *Math. Japon.*, 52 (1):103–112, 2000. ISSN 0025-5513. MR 1783184. Cited on p. 91.

[271] J. Preater. *A survey of Secretary Problems with Multiple Appointments.* Technical report, Department of Mathematics, University of Keele, Keele, Staffordshire ST5 5BG, UK, 1988. Research Report 88-12. Cited on p. 24.

[272] J. Preater. *Sequential Multiple Selection Problems.* PhD thesis, University of Keele, Department of Mathematics, University of Keele, Keele, Staffordshire ST5 5BG, UK, 1991. British Library, Shelfmark - 293741. Cited on p. 246.

[273] J. Preater. The senior and junior secretaries problem. *Oper. Res. Lett.*, 14(4): 231–235, 1993. ISSN 0167-6377. doi: 10.1016/0167-6377(93)90074-Q. Zbl 0799.90117. Cited on p. 26.

[274] J. Preater. On multiple choice secretary problems. *Math. Oper. Res.*, 19 (3):597–602, 1994a. ISSN 0364-765X. doi: 10.1287/moor.19.3.597. MR 1288888, Zbl 0813.90080. Cited on pp. 227, 228, and 246.

[275] J. Preater. A multiple stopping problem. *Probability in the Engineering and Informational Sciences*, 8(2):169–177, 1994b. doi: 10.1017/S0269964800003314. Cited on p. 246.

[276] J. Preater. On-line selection of an acceptable pair. *Journal of Applied Probability*, 43(3):729–740, 2006. ISSN 00219002. URL `http://www.jstor.org/stable/27595768`. Cited on p. 246.

[277] E. Presman and I. Sonin. Equilibrium points in a game related to the best choice problem. *Theory Probab. Appl.*, 20:770–781, 1975. ISSN 0040-585X; 1095-7219/e. doi: 10.1137/1120084. Orig. Russian title: Э. Л. Пресман, И. М. Сонин (1975) Точки равновесия в обобщенной игровой задаче наилучшего выбора. 785–796. Zbl 0349.60046. Cited on p. 40.

[278] E. L. Presman and I. M. Sonin. The best choice problem for a random number of objects. *Teor. Veroyatnost. i Primenen.*, 17(4):695–706, 1972. doi: 10.1137/1117078. MR 0314177. Cited on pp. 19 and 90.

[279] M. P. Quine and J. S. Law. Exact results for a secretary problem. *J. Appl. Probab.*, 33(3):630–639, 1996. ISSN 0021-9002. doi: 10.2307/3215345. Zbl 0871.60037. Cited on pp. 196 and 215.

[280] T. Radzik and K. Szajowski. Sequential games with random priority. *Sequential Analysis*, 9(4):361–377, 1990. ISSN 15324176 07474946. doi: 10.1080/07474949008836218. Cited on pp. 84 and 85.

[281] F. P. Ramsey. On a Problem of Formal Logic. *Proc. London Math. Soc. (2)*, 30(4):264–286, 1929. ISSN 0024-6115. doi: 10.1112/plms/s2-30.1.264. MR 1576401. Cited on p. 101.

[282] A. Rapoport, D. A. Seale, and L. Spiliopoulos. Progressive stopping heuristics that excel in individual and competitive sequential search. *Theory and Decision*, 94(1):135–165, 2023. ISSN 0040-5833. doi: 10.1007/s11238-022-09881-0. Cited on p. 261.

[283] M. Rasche. Allgemeine stopprobleme. Unpublished technical report World Bank Policy Research Working paper 4136, Institut für Mathematische Statistik, Universität Münster, 1975. Cited on p. 19.

[284] W. T. Rasmussen. A generalized choice problem. *J. Optim. Theory Appl.*, 15: 311–325, 1975. ISSN 0022-3239. doi: 10.1007/BF00933340. MR 362769. Cited on p. 19.

[285] W. T. Rasmussen and H. Robbins. The candidate problem with unknown population size. *J. Appl. Probability*, 12(4):692–701, 1975. ISSN 0021-9002. doi: 10.2307/3212720. MR 386187. Cited on p. 19.

[286] M. Rasouli, E. Miehling, and D. Teneketzis. A Supervisory Control Approach to Dynamic Cyber-Security. In R. Poovendran and W. Saad, editors, *Decision and Game Theory for Security*, volume 8840 of *Lecture Notes in Computer Science*, pages 99–117. Springer, Cham, 2014. ISBN 978-3-319-12600-5, 978-3-319-12601-2. doi: 10.1007/978-3-319-12601-2_6. Cited on p. 107.

[287] G. Ravindran and K. Szajowski. Non-zero sum game with priority as Dynkin's game. *Math. Japon.*, 37(3):401–413, 1992. Cited on p. 84.

[288] L. Real. Search theory and mate choice. I. Models of single sex-discrimination. *The American Naturalist*, 136(3):376–405, 1990. Cited on p. 44.

[289] L. Real. Search theory and mate choice. II. Mutual interaction, assortative mating, and equilibrium variation in male and female fitness. *The American Naturalist*, 138(4):901–917, 1991. Cited on p. 44.

[290] F. Riedel. Optimal stopping with multiple priors. *Econometrica*, 77(3):857–908, 2009. ISSN 0012-9682. doi: 10.3982/ECTA7594. MR 2531363. Cited on p. 191.

[291] Y. Rinott and E. Samuel-Cahn. Optimal stopping values and prophet inequalities for some dependent random variables. *Lecture Notes-Monograph Series*, 22:343–358, 1992. ISSN 07492170. URL `http://www.jstor.org/stable/4355751`. Cited on p. 101.

[292] H. Robbins. Some aspects of the sequential design of experiments. *Bull. Am. Math. Soc.*, 58:527–535, 1952. ISSN 0002-9904. doi: 10.1090/S0002-9904-1952-09620-8. An address delivered before the Auburn, Alabama, meeting of the Society, November 23, 1951, by invitation of the Committee to Select Hour Speakers for Southeastern Sectional Meetings. Zbl 0049.37009. Cited on p. 302.

[293] H. Robbins. [Who Solved the Secretary Problem?]: Comment. *Statistical Science*, 4(3):291–291, August 1989. ISSN 0883-4237. doi: 10.1214/ss/1177012495. Cited on p. 16.

[294] S. W. Roberts. Properties of control chart zone tests. *The Bell System Technical Journal*, 37(1):83–114, 1958. doi: 10.1002/j.1538-7305.1958.tb03870.x. Cited on p. 144.

[295] S. W. Roberts. A comparison of some control chart procedures. *Technometrics*, 8:411–430, 1966. ISSN 0040-1706,1537-2723. doi: 10.2307/1266688. Cited on p. 144.

[296] L. Rogers. Monte Carlo valuation of American options. *Mathematical Finance*, 12(3):271–286, 2002. doi: 10.1111/1467-9965.02010. Cited on p. 287.

[297] J. Rose. Selection of nonextremal candidates from a sequence. *J. Optimization Theory Appl.*, 38:207–219, 1982. MR 84d:60071. Cited on pp. 26 and 196.

[298] J. S. Rose. Twenty years of secretary problems: a survey of developments in the theory of optimal choice. *Management Studies*, 1:53–64, 1982. Cited on pp. 14, 15, 84, and 195.

[299] J. S. Rose. A problem of optimal choice and assignment. *Oper. Res.*, 30:172–181, 1982. ISSN 0030-364X. doi: 10.1287/opre.30.1.172. Zbl 0481.90049. Cited on p. 41.

[300] D. Rosenberg, E. Solan, and N. Vieille. Stopping games with randomized strategies. *Probab. Theory Related Fields*, 119(3):433–451, 2001. ISSN 0178-8051. doi: 10.1007/PL00008766. MR 1821142. Cited on p. 101.

[301] S. M. Ross. *Applied Probability Models with Optimization Applications*. Holden-Day, San Francisco, Calif.-London-Amsterdam, 1970. MR 0264792. Cited on pp. 18 and 256.

[302] R. Y. Rubinstein. Optimization of computer simulation models with rare events. *Eur. J. Oper. Res.*, 99(1):89–112, 1997. ISSN 0377-2217. doi: 10.1016/S0377-2217(96)00385-2. Zbl 0923.90051. Cited on pp. 265 and 266.

[303] R. Y. Rubinstein and D. P. Kroese. *The Cross-Entropy Method: A Unified Approach to Combinatorial Optimization, Monte-Carlo Simulation and Machine Learning*. Springer-Verlag, New York, 2004. Cited on p. 268.

[304] J. Rukavicka. Secretary problem and two almost the same consecutive applicants. *Math. Appl. (Warsaw)*, 50(2):165–181, 2022. ISSN 1730-2668. doi: 10.14708/ma.v50i2.7143. Cited on p. 26.

[305] Б. А. Березовский, А. В. Гнедин. Задача наилучшего выбора. Издательство “Наука”, Москва, 1984. MR 768372. Cited on pp. 14 and 24.

[306] V. Saario. *Some Techniques and Limiting Properties of the House-selling Problem*. ProQuest LLC, Ann Arbor, MI, 1994. ISBN 978-9514-43536-2. Thesis (Ph.D.)–Tampereen Yliopisto (Finland). Cited on p. 42.

[307] V. Saario and M. Sakaguchi. Some generalized house-selling problems. *Math. Japon.*, 35(5):861–873, 1990. ISSN 0025-5513. MR 1073890. Cited on p. 42.

[308] M. Sakaguchi. Dynamic programming of some sequential sampling design. *J. Math. Anal. Appl.*, 2:446–466, 1961. ISSN 0022-247X. doi: 10.1016/0022-247X(61)90023-3. Cited on p. 42.

[309] M. Sakaguchi. A note on the dowry problem. *Rep. Statist. Appl. Res. Un. Japan. Sci. Engrs.*, 20(1):11–17, 1973. ISSN 0034-4842. MR 329160. Cited on p. 258.

[310] M. Sakaguchi. Optimal stopping in sampling from a bivariate distribution. *J. Operations Res. Soc. Japan*, 16:186–200, 1973. ISSN 0453-4514. MR 0420996; Zbl 0287.62047. Cited on pp. 74 and 92.

[311] M. Sakaguchi. Dowry problems and OLA policies. *Rep. Statist. Appl. Res. Un. Japan. Sci. Engrs.*, 25(3):124–128, 1978. ISSN 0034-4842. MR 0543754; Zbl 0416.62058. Cited on pp. 3, 18, and 21.

[312] M. Sakaguchi. When to stop: Randomly appearing bivariate target values. *J. Oper. Res. Soc. Japan*, 21:45–58, 1978. ISSN 0453-4514. doi: 10.15807/jorsj.21.45. MR 474661, Zbl 0385.62057. Cited on p. 74.

[313] M. Sakaguchi. A bilateral sequential game for sums of bivariate random variables. *J. Oper. Res. Soc. Japan*, 21(4):486–508, 1978. ISSN 0453-4514. doi: 10.15807/jorsj.21.486. MR 517782. Cited on p. 74.

[314] M. Sakaguchi. Multiperson multilateral secretary problems. *Math. Japon.*, 34 (3):459–473, 1989. ISSN 0025-5513. Cited on p. 84.

[315] M. Sakaguchi. Sequential games with priority under expected value maximization. *Math. Japonica*, 36(3):545–562, 1991. Cited on p. 84.

[316] M. Sakaguchi. Optimal stopping games—a review. *Math. Japon.*, 42(2):343–351, 1995. ISSN 0025-5513. MR 1356397. Cited on pp. 3, 40, 48, 82, 246, and 287.

[317] M. Sakaguchi. Correction to: "Optimal stopping games—a review". *Math. Japon.*, 42(3):573, 1995. ISSN 0025-5513. MR 1363848. Cited on pp. 40 and 82.

[318] M. Sakaguchi. A non-zero-sum repeated game – criminal vs. police. *Math. Japon.*, 48(3):427–436, 1998. ISSN 0025-5513. Zbl 0915.90285. Cited on p. 78.

[319] M. Sakaguchi and T. Hamada. A class of best-choice problems on sequences of continuous bivariate random variables. In F. T. Bruss and L. Le Cam, editors, *Game Theory, Optimal Stopping, Probability and Statistics. Papers in Honor of Thomas S. Ferguson*, pages 111–125. Institute of Mathematical Statistics, Beachwood, OH, 2000. ISBN 0-940600-48-X. Cited on p. 249.

[320] M. Sakaguchi and V. Saario. A best-choice problem for bivariate uniform distribution. *Math. Japon.*, 40(3):585–599, 1994. ISSN 0025-5513. Cited on p. 249.

[321] M. Sakaguchi and K. Szajowski. Mixed-type secretary problems on sequences of bivariate random variables. *Math. Japon.*, 51(1):99–111, 2000. ISSN 0025-5513. MR 1739057; Zbl 0942.62094. Cited on pp. 92 and 248.

[322] E. Samuel. *On the Compound Decision Problem in the Nonsequential and the Sequential Case*. PhD thesis, Columbia University, 1961. Cited on p. 6.

[323] S. Samuels. Secretary Problems. In B. Ghosh and P. Sen, editors, *Handbook of Sequential Analysis*, pages 381–405. Marcel Dekker, Inc., New York, Basel, Hong Kong, 1991. Cited on pp. 14, 15, 24, 195, and 249.

[324] S. M. Samuels. MR 0822604 (87e:60074). *MathSciNet*, 1987. ISSN 2167-5163. Mathematical Reviews [234]. Cited on p. 21.

[325] S. M. Samuels. MR 1782448 (2001f:60042). *MathSciNet*, 2000. ISSN 2167-5163. Mathematical Reviews Glickman (2000). Cited on p. 47.

[326] S. M. Samuels. MR 1797879. *MathSciNet*, 2001. ISSN 2167-5163. Mathematical Reviews [52]. Cited on p. 304.

[327] D. A. Sánchez-Salas, J. L. Cuevas-Ruiíz, and M. González-Mendoza. Wireless channel model with Markov chains using MATLAB. In V. N. Katsikis, editor, *MATLAB*, chapter 11. IntechOpen, Rijeka, 2012. doi: 10.5772/46475. Cited on p. 300.

[328] R. W. H. Sargent. Optimal control. *J. Comput. Appl. Math.*, 124(1-2):361–371, 2000. ISSN 0377-0427. doi: 10.1016/S0377-0427(00)00418-0. Zbl 0970.49003. Cited on p. 6.

[329] S. Sarıtaş, E. Shereen, H. Sandberg, and G. Dán. Adversarial Attacks on Continuous Authentication Security: A Dynamic Game Approach. In T. Alpcan, Y. Vorobeychik, J. S. Baras, and G. Dán, editors, *Decision and Game Theory for Security*, pages 439–458, Cham, 2019. Springer International Publishing. ISBN 978-3-030-32430-8. doi: 10.1007/978-3-030-32430-8_26. Cited on p. 107.

[330] W. Sarnowski and K. Szajowski. On-line detection of a part of a sequence with unspecified distribution. *Stat. Probab. Lett.*, 78(15):2511–2516, 2008. doi: 10.1016/j.spl.2008.02.040. Cited on p. 122.

[331] W. Sarnowski and K. Szajowski. Optimal detection of transition probability change in random sequence. *Stochastics*, 83(4-6):569–581, 2011. ISSN 17442508 17442516. doi: 10.1080/17442508.2010.540015. Cited on pp. 37 and 148.

[332] N. Schmitz. *Prophet Theory.* Part I. The general and the independent case. Gesellschaft zur Förderung der Mathematischen Statistik, 2000. ISBN 3-519-02737-2. Skripten zur Mathematische Statistik nr. 34. Cited on p. 22.

[333] J. Schoenmakers. A pure martingale dual for multiple stopping. *Finance Stoch.*, 16(2):319–334, 2012. ISSN 0949-2984. doi: 10.1007/s00780-010-0149-1. MR 2903627. Cited on p. 21.

[334] A. Seierstad. *Stochastic Control in Discrete and Continuous Time.* New York, NY: Springer, 2009. ISBN 978-0-387-76616-4. doi: 10.1007/978-0-387-76617-1. Zbl 1154.93001. Cited on p. 108.

[335] L. S. Shapley. Stochastic games. *Proc. Nat. Acad. Sci. U. S. A.*, 39:1095–1100, 1953. ISSN 0027-8424. doi: 10.1073/pnas.39.10.1953. Cited on p. 78.

[336] W. Shewhart. *Economic Control of Quality of Manufactured Products*. D. Van Nostrand, Yew York, 1931. Cited on pp. 37 and 38.

[337] W. A. Shewhart and W. E. Deming. *Statistical Method from the View-point of Quality Control*. The Graduate School, Department of Agriculture, Washington, D.C., 1939. Cited on p. 37.

[338] A. N. Shiryaev. The detection of spontaneous effects. *Sov. Math., Dokl.*, 2: 740–743, 1961a. ISSN 0197-6788. Ширяев Альберт Николаевич Zbl 0109.12802. Cited on pp. 37, 38, and 122.

[339] A. N. Shiryaev. The problem of the most rapid detection of a disturbance in a stationary process. *Sov. Math., Dokl.*, 2:795–799, 1961b. ISSN 0197-6788. Zbl 0109.11201. Cited on pp. 37 and 38.

[340] A. N. Shiryaev. On the detection of disorder in a manufacturing process. I. *Theory Probab. Appl.*, 8:247–265, 1963a. ISSN 0040-585X. doi: 10.1137/1108029. Zbl 0279.90011. Cited on p. 10.

[341] A. N. Shiryaev. On the detection of disorder in a manufacturing process. II. *Theory Probab. Appl.*, 8:402–413, 1963b. ISSN 0040-585X. doi: 10.1137/1108045. Zbl 0279.90012. Cited on p. 10.

[342] A. N. Shiryaev. *Optimal Stopping Rules*, volume 8 of *Applications of Mathematics*. Springer-Verlag, New York-Heidelberg, New York, 1978. ISBN 0-387-90256-2. Translated from the Russian II edition Статистический последовательный анализ. Оптимальные правила остановки. Moskva: "Nauka". 272 p. (1976) by A. B. Aries. MR 0468067. Cited on pp. 11, 67, 89, 93, 130, 133, and 140.

[343] A. N. Shiryaev. *Essentials of Stochastic Finance: Facts, Models, Theory*, volume 3. World Scientific, 1999. Zbl 0926.62100. Cited on p. 172.

[344] A. N. Shiryaev. From "disorder" to nonlinear filtering and martingale theory. In A. A. Bolibruch, Y. Osipov, Y. Sinai, V. I. Arnold, L. D. Faddeev, Y. I. Manin, V. B. Philippov, V. M. Tikhomirov, and A. M. Vershik, editors, *Mathematical Events of the Twentieth Century*, pages 371–397. Springer, Berlin, 2006. Zbl 1072.01002. Cited on pp. 37, 38, and 122.

[345] A. N. Shiryaev. *Stochastic Disorder Problems*, volume 93 of *Probability Theory and Stochastic Modelling*. Springer, Cham, 2019. ISBN 978-3-030-01525-1; 978-3-030-01526-8. doi: 10.1007/978-3-030-01526-8. With a foreword by H. Vincent Poor. Zbl 1426.93002. Cited on pp. 36 and 37.

[346] E. Shmaya and E. Solan. Two-player nonzero-sum stopping games in discrete time. *Ann. Probab.*, 32(3B):2733–2764, 2004. ISSN 0091-1798. doi: 10.1214/009117904000000162. MR 2078556. Cited on p. 101.

[347] E. Shmaya, E. Solan, and N. Vieille. An application of Ramsey theorem to stopping games. *Games Econom. Behav.*, 42(2):300–306, 2003. ISSN 0899-8256. doi: 10.1016/S0899-8256(02)00539-0. MR 1984248. Cited on p. 101.

[348] D. O. Siegmund. Some problems in the theory of optimal stopping rules. *Ann. Math. Statist.*, 38:1627–1640, 1967. ISSN 0003-4851. doi: 10.1214/aoms/1177698596. MR 221666, Zbl 0183.20706. Cited on pp. 7 and 63.

[349] J. L. Snell. Applications of martingale system theorems. *Trans. Amer. Math. Soc.*, 73:293–312, 1952. ISSN 0002-9947. doi: 10.2307/1990670. MR 50209. Cited on pp. 8, 9, 52, 64, and 302.

[350] L. J. Snell. *Applications of Martingale System Theorems. (Abstract of a thesis)*. PhD thesis, University of Illinois, Urbana-Champaign, 1951. Zbl 0045.22501. Cited on pp. 9 and 302.

[351] G. Sofronov. An optimal sequential procedure for a multiple selling problem with independent observations. *European J. Oper. Res.*, 225(2):332–336, 2013. ISSN 0377-2217. doi: 10.1016/j.ejor.2012.09.042. Cited on pp. 3, 48, 49, 50, 181, 189, and 287.

[352] G. Sofronov, J. M. Keith, and D. P. Kroese. An optimal sequential procedure for a buying-selling problem with independent observations. *J. Appl. Probab.*, 43(2):454–462, 2006. ISSN 0021-9002. doi: 10.1239/jap/1152413734. MR 2248576. Cited on pp. 158, 159, 174, and 175.

[353] G. Y. Sofronov. A multiple optimal stopping rule for a buying-selling problem with a deterministic trend. *Statist. Papers*, 57(4):1107–1119, 2016. ISSN 0932-5026. doi: 10.1007/s00362-016-0776-5. Cited on pp. 3, 48, 163, 170, and 287.

[354] G. Y. Sofronov. An optimal double stopping rule for a buying-selling problem. *Methodology and Computing in Applied Probability*, pages 1–12, 2018. ISSN 1573-7713. doi: 10.1007/s11009-018-9684-6. Cited on pp. 3, 48, 169, 172, 173, and 287.

[355] G. Y. Sofronov and T. V. Polushina. Evaluating optimal stopping rules in the multiple best choice problem using the Cross-Entropy method. In *MIC 2013: Proceedings of the IASTED International Conference on Modelling, Identification and Control: 11-13 February 2013, Innsbruck, Austria*, pages 205–212. ACTA Press Calgary, Canada, 2013. Cited on p. 266.

[356] G. Y. Sofronov, D. P. Kroese, J. M. Keith, and M. L. Nikolaev. Simulations of thresholds in multiple best choice problem. *Obozr. Prikl. Prom. Mat.*, 13(6):975–982, 2006. ISSN 0869-8325. Zbl 1161.60333. Cited on pp. 266 and 268.

[357] J. Song. *Sequential Testing of Multiple Hypotheses*. PhD thesis, University of Southern California, Los Angeles, 2013. The dissertation supervisor: Jay

Bartrof. Copyright - Database copyright ProQuest LLC; ProQuest does not claim copyright in the individual underlying works; Last updated - 2023-03-03. Cited on p. 33.

[358] I. Sonin. The State Reduction and Related Algorithms and Their Applications to the Study of Markov Chains, Graph Theory, and the Optimal Stopping Problem. *Advances in Mathematics*, 145(2):159–188, Aug 1999. ISSN 00018708. doi: 10.1006/aima.1998.1813. Cited on p. 303.

[359] I. M. Sonin. Game problems that are connected with optimal selection. *Kibernetika*, 1976(2):70–75, 1976. ISSN 0023-1274. doi: 10.1007/BF01069894. Orig. Russian title: И. М. Сонин (1976) Игровые задачи, связанные с наилучшим выбором. Eng. transl. in Cybernetics 12 (1976), no. 2, 246–251 (1977). MR 455109; Zbl 0345.90061. Cited on p. 40.

[360] M. Stachowiak. The cross-entropy method and its applications. Technical report, Faculty of Pure and Applied Mathematics, Wrocław University of Science and Technology, Wroclaw, 2019. 33p. Engineering Thesis. Cited on p. 266.

[361] M. K. Stachowiak and K. J. Szajowski. Cross-entropy method in application to the SIRC model. *Algorithms (Basel)*, 13(11):Paper No. 281, 20, 2020. doi: 10.3390/a13110281. MR 4189953. Cited on pp. 266 and 267.

[362] W. Stadje. On multiple stopping rules. *Optimization*, 16(3):401–418, 1985. ISSN 0233-1934. doi: 10.1080/02331938508843030. Cited on pp. 3, 8, 47, 48, 51, and 287.

[363] W. Stadje. An optimal k-stopping problem for the Poisson process. In P. Bauer, F. Konecny, and W. Wertz, editors, *Mathematical statistics and probability theory, Vol. B (Bad Tatzmannsdorf, 1986)*, pages 231–244. Reidel, Dordrecht, 1987. Proceedings of the 6th Pannonian Symposium on Mathematical Statistics, MR 922726. Cited on p. 190.

[364] W. Stadje. A full information pricing problem for the sale of several identical commodities. *Z. Oper. Res.*, 34(3):161–181, 1990. ISSN 0340-9422. doi: 10.1007/BF01415979. Cited on p. 191.

[365] N. Starr. How to win a war if you must: Optimal stopping based on success runs. *The Annals of Mathematical Statistics*, 43(6):1884–1893, 1972. ISSN 00034851. Cited on pp. 112, 115, 116, 117, and 118.

[366] A. Stepanov. On the mathematical theory of records. *Commun. Math.*, 29 (1):151–162, 2021. ISSN 1804-1388. doi: 10.2478/cm-2021-0009. Zbl 1481.60099. Cited on p. 202.

[367] Ł. Stettner. Zero-sum Markov games with stopping and impulsive strategies. *Appl. Math. Optim.*, 9:1–24, 1982. Cited on p. 81.

[368] A. Suchwalko and K. Szajowski. Non standard, no information secretary problems. *Sci. Math. Jpn.*, 56(3):443–456, 2002. ISSN 1346-0862. URL `https://www.jams.jp/scm/contents/Vol-6-5/6-42.pdf`. MR 1937908, Zbl 1034.60045. Cited on pp. 26, 41, and 195.

[369] K. Szajowski. Optimal choice of an object with ath rank. *Mat. Stos. Roczniki Polskiego Towarzystwa Matematycznego. Seria III. Matematyka Stosowana*, 19:51–65, 1982. ISSN 0137-2890. doi: 10.14708/ma.v10i19.1533. MR 664139, Zbl 0539.62094. Cited on pp. 26, 41, 196, 198, 199, 214, 215, 228, and 266.

[370] K. Szajowski. Optimal stopping of a sequence of maxima over an unobservable sequence of maxima. *Zastos. Mat.*, 18(3):359–374, 1984. ISSN 0044-1899. doi: 10.4064/am-18-3-359-374. MR 757886. Cited on p. 91.

[371] K. Szajowski. On non-zero sum game with priority in the secretary problem. *Math. Japonica*, 37(3):415–426, 1992. Cited on p. 84.

[372] K. Szajowski. Optimal on-line detection of outside observation. *J. Stat. Planning and Inference*, 30:413–426, 1992. Cited on pp. 38, 146, and 148.

[373] K. Szajowski. Double stopping by two decision-makers. *Adv. Appl. Probab.*, 25 (2):438–452, 1993. ISSN 0001-8678. doi: 10.2307/1427661. Zbl 0772.60032. Cited on pp. 3, 48, 82, 84, 87, and 287.

[374] K. Szajowski. Markov stopping games with random priority. *Z. Oper. Res.*, 39 (1):69–84, 1994. ISSN 0340-9422; 1432-5217/e. doi: 10.1007/BF01440735. Cited on pp. 85 and 90.

[375] K. Szajowski. Optimal stopping of a discrete Markov process by two decision makers. *SIAM J. Control Optim.*, 33(5):1392–1410, 1995. ISSN 0363-0129. doi: 10.1137/S0363012993246877. MR 1348114, Zbl 0836.90149. Cited on pp. 82, 85, 86, and 91.

[376] K. Szajowski. A two-disorder detection problem. *Appl. Math.*, 24(2):231 241, 1996. Cited on pp. 39, 49, 122, 146, and 147.

[377] K. Szajowski. On a random number of disorders. *Probab. Math. Stat.*, 31(1): 17–45, 2011. ISSN 0208-4147. pms;Zbl 1260.60078. Cited on pp. 37 and 38.

[378] K. Szajowski and M. Skarupski. On multilateral incomplete information decision models. *High Frequency*, 2(3-4):158–168, 2019. ISSN 2470-6981. doi: 10.1002/hf2.10047. Cited on p. 91.

[379] K. Szajowski and M. Yasuda. Voting procedure on stopping games of Markov chain. In A. H. Christer, S. Osaki, and C. Lyn, editors, *UK-Japanese Research Workshop on Stochastic Modelling in Innovative Manufecuring, July 21-22, 1995*, volume 445 of *Lecture Notes in Economics and Mathematical Systems*, pages 68–80. Moller Centre, Churchill College, Univ. Cambridge,

UK, Springer, 1996. doi: 10.1007/978-3-642-59105-1_6. Springer Lecture Notes in Economics and Mathematical Systems, MR 90159; Zbl 0878.90112. Cited on pp. 74, 92, and 94.

[380] M. Tamaki. Recognizing both the maximum and the second maximum of a sequence. *Journal of Applied Probability*, 16(4):803–812, 1979. ISSN 00219002. URL `http://www.jstor.org/stable/3213146`. Cited on pp. 19, 217, 218, 219, 228, and 230.

[381] M. Tamaki. Sum the multiplicative odds to one and stop. *J. Appl. Probab.*, 47(3):761–777, 2010. ISSN 0021-9002,1475-6072. doi: 10.1239/jap/1285335408. Cited on p. 69.

[382] M. Tamaki. *Various Aspects of the Best Choice Problem -the Secretary Problem and its Related Areas*. Part X. Series on Stochastic Models in Informatics and Data Science. Corona Publishing Company, Tokyo, Japan, 2023. ISBN 978-4339028409. URL `https://www.coronasha.co.jp/np/isbn/9784339028409/`. Cited on p. 20.

[383] D.-I. Tang, N. L. Geller, and S. J. Pocock. On the design and analysis of randomized clinical trials with multiple endpoints. *Biometrics*, 49(1):23–30, 1993. ISSN 0006-341X. doi: 10.2307/2532599. Cited on p. 10.

[384] R. S. Targino, G. W. Peters, G. Sofronov, and P. V. Shevchenko. Optimal exercise strategies for operational risk insurance via multiple stopping times. *Methodol. Comput. Appl. Probab.*, 19(2):487–518, 2017. ISSN 1387-5841. doi: 10.1007/s11009-016-9493-8. Cited on pp. 49, 193, and 194.

[385] A. Tartakovsky, I. Nikiforov, and M. Basseville. *Sequential Analysis. Hypothesis Testing and Changepoint Detection*, volume 136 of *Monogr. Stat. Appl. Probab.* Boca Raton, FL: CRC Press, 2015. ISBN 978-1-4398-3820-4; 978-1-4398-3821-1. URL `www.crcnetbase.com/isbn/9781439838211`. Cited on p. 37.

[386] A. G. Tartakovsky, editor. *Special issue: Celebrating the seventy-fifth birthday of Albert N. Shiryaev. Invited papers. Part I. Selected papers based on the presentations at the International Workshop Sequential Methodologies (IWSM), Troyes, France, June 15–17, 2009*, volume 29 of *Sequential Anal.* Taylor & Francis, Philadelphia, PA, 2010. Cited on p. 38.

[387] A. G. Tartakovsky. *Sequential Change Detection and Hypothesis Testing. General non-i.i.d. Stochastic Models and Asymptotically Optimal Rules*, volume 165 of *Monogr. Stat. Appl. Probab.* Boca Raton, FL: CRC Press, 2020. ISBN 978-1-4987-5758-4; 978-0-429-15501-7. doi: 10.1201/9780429155017. Cited on p. 36.

[388] A. G. Tartakovsky, B. L. Rozovskii, R. B. Blažek, and H. Kim. Detection of intrusions in information systems by sequential change-point methods. *Stat.*

Methodol., 3(3):252–293, 2006. ISSN 1572-3127. MR MR2240956. Cited on p. 38.

[389] R Core Team. *R: A Language and Environment for Statistical Computing*. R Foundation for Statistical Computing, Vienna, Austria, 2018. URL `https://www.r-project.org/`. Cited on p. 175.

[390] The MathWorks Inc., Econometrics toolbox™ r2022b, 2022. URL `https://www.mathworks.com`. Econometrics Toolbox™ User's Guide. Cited on p. 300.

[391] J. N. Tsitsiklis and B. Van Roy. Regression methods for pricing complex American-style options. *IEEE Transactions on Neural Networks*, 12(4):694–703, 2001. doi: 10.1109/72.935083. Cited on pp. 264 and 287.

[392] R. J. Vanderbei. The optimal choice of a subset of a population. *Mathematics of Operations Research*, 5(4):481–486, 1980. Cited on pp. 26, 217, and 219.

[393] R. J. Vanderbei. The postdoc variant of the secretary problem. *Math. Appl. Roczniki Polskiego Towarzystwa Matematycznego. Seria III.*, 49(1):3–13, 2021. ISSN 1730-2668. doi: 10.14708/ma.v49i1.7076. Published 2012 on the Author's Web page. Zbl 1499.60133. Cited on p. 41.

[394] N. Vieille. Some recent results on stochastic games. In R. Auman and S. Hart, editors, *Handbook of Game Theory With Economic Applications*, volume 3, chapter 48, pages 1833–1850. Elsevier, 2002. ISBN 5-7997-0412-6. doi: 10.1016/S1574-0005(02)03011-4. Also Ed. Petrosjan, L. A. et al. 10-ый международный симпозиум по динамическим играм и приложениям. , St. Petersburg: International Society of Dynamic Games, St. Petersburg State Univ. 858pp. Zbl 1062.91514. Cited on p. 81.

[395] S. V. Vinnichenko and V. V. Mazalov. Games for a stopping rule of a sequence of observations of fixed length. *Kibernetika (Kiev)*, 1:122–124, 136, 1989. ISSN 0023-1274. MR 997021. Cited on p. 91.

[396] H. von Stackelberg. *Market Structure and Equilibrium*. Berlin: Springer, 2011. ISBN 978-3-642-12585-0; 978-3-642-12586-7. doi: 10.1007/978-3-642-12586-7. Translated from the German by Damian Bazin, Lynn Urch and Rowland Hill. Zbl 6344012. Cited on p. 343.

[397] H. F. von Stackelberg. *Marktform und Gleichgewicht*. Julius Springer, Wien and Berlin, 1934. von Stackelberg (2011). Zbl 1405.91003. Cited on p. 247.

[398] H. F. von Stackelberg. *Grundlagen einer reinen Kostentheorie*. Meilensteine Nationalokonomie. Springer-Verlag Gmbh, Berlin, 2009. Originally published monograph. Reprint of the 1st Ed. Wien, Verlag von Julius Springer, 1932. Read on line, v. [396]. Cited on pp. 91 and 247.

[399] C. Wagner, A. Dulaunoy, G. Wagener, and A. Iklody. Misp: The design and implementation of a collaborative threat intelligence sharing platform. In *Proceedings of the 2016 ACM on Workshop on Information Sharing and Collaborative Security*, pages 49–56, New York, NY, USA, oct 2016b. ACM. ISBN 9781450345651. doi: 10.1145/2994539.2994542. Cited on p. 107.

[400] N. Wagner, C. S. Sahin, M. Winterrose, J. Riordan, J. Pena, D. Hanson, and W. W. Streilein. Towards automated cyber decision support: A case study on network segmentation for security. In *2016 IEEE Symposium Series on Computational Intelligence (SSCI)*, pages 1–10. IEEE, dec 2016a. ISBN 978-1-5090-4240-1. doi: 10.1109/SSCI.2016.7849908. Cited on p. 107.

[401] A. Wald. Sequential tests of statistical hypotheses. *Ann. Math. Stat.*, 16: 117–186, 1945. ISSN 0003-4851. doi: 10.1214/aoms/1177731118. Zbl 0060.30207. Cited on p. 10.

[402] A. Wald. *Sequential Analysis*. John Wiley & Sons, Inc., New York; Chapman & Hall, Ltd., New York, 1947. MR 0020764, Zbl 3045589. Cited on p. 10.

[403] A. Wald and J. Wolfowitz. Bayes solutions of sequential decision problems. *Proc. Natl. Acad. Sci. USA*, 35:99–102, 1949. ISSN 0027-8424. doi: 10.1073/pnas.35.2.99. Zbl 0032.17401. Cited on p. 10.

[404] W. A. Wallis. The statistical research group, 1942-1945. (With comments by F. J. Anscombe and W. H. Kruskal.). *Journal of the American Statistical Association*, 75:320–335, 1980. ISSN 0162-1459. doi: 10.2307/2287451. Cited on p. 10.

[405] J. B. Walsh. Optional increasing paths. In *Two-index Random Processes (Paris, 1980)*, volume 863 of *Lecture Notes in Math.*, pages 172–201. Springer, Berlin, 1981. MR 630313. Cited on p. 9.

[406] W. A. Whitworth. *Choice and Chance with One Thousand Exercises*. Deighton Bell, Cambridge, V edition, 1901. Reprinted by Hafner Publishing Co., New York, 1959. Cited on p. 294.

[407] D. D. Wiegmann and L. M. Angeloni. Mate choice and uncertainty in the decision process. *Journal of Theoretical Biology*, 249(4):654–666, 2007. Cited on p. 44.

[408] J. G. Wilson. Optimal choice and assignment of the best m of n randomly arriving items. *Stochastic Processes Appl.*, 39(2):325–343, 1991. ISSN 0304-4149. doi: 10.1016/0304-4149(91)90086-R. Zbl 0736.60040. Cited on p. 26.

[409] B. Yakir. Optimal detection of a change in distribution when the observations form a Markov chain with a finite state space. In D. S. E. Carlstein, H.-G. Müller, editor, *Change-point Problems. Papers from the AMS-IMS-SIAM Summer Research Conference held at Mt. Holyoke College, South Hadley,*

MA, USA, July 11–16, 1992, volume 23 of *IMS Lecture Notes-Monograph Series*, pages 346–358, Hayward, California, 1994. Institute of Mathematical Statistics. Cited on pp. 38, 122, and 146.

[410] M. Yasuda. Asymptotic results for the best-choice problem with a random number of objects. *Appl, Prob*, 21:521–536, 1984. Cited on p. 271.

[411] M. Yasuda. On a randomized strategy in Neveu's stopping problem. *Stochastic Processes and their Applications*, 21(1):159–166, 1985. Cited on pp. 79, 87, 89, and 101.

[412] M. Yasuda, J. Nakagami, and M. Kurano. Multivariate stopping problems with a monotone rule. *J. Oper. Res. Soc. Japan*, 25(4):334–350, 1982. ISSN 0453-4514. doi: 10.15807/jorsj.25.334. Cited on pp. 3, 48, 74, 93, and 287.

[413] G. F. Yeo. Duration of a secretary problem. *Journal of Applied Probability*, 34(2):556–558, 1997. Cited on p. 271.

[414] M. Yoshida. Probability maximizing approach for a quickest detection problem with complicated Markov chain. *J. Inf. Optim. Sci.*, 4:127–145, 1983. ISSN 0252-2667. doi: 10.1080/02522667.1983.10698755. Zbl 0511.60041. Cited on pp. 38, 49, 122, and 148.

[415] J. Zabczyk. A selection problem associated to a renewal process. In *New Trends in Systems Analysis (Proc. Internat. Sympos., Versailles, 1976)*, Lect. Notes Control Inf. Sci., Vol. 2, pages 508–515. Springer, Berlin-New York, 1977. MR 488270. Cited on p. 20.

[416] J. Zabczyk. Stopping problems in stochastic control. In *Proceedings of the International Congress of Mathematicians, Vol. 1, 2 (Warsaw, 1983)*, pages 1425–1437. PWN, Warsaw, 1984. MR 804789. Cited on p. 191.

[417] W. Zhang and Y. Mei. Bandit change-point detection for real-time monitoring high-dimensional data under sampling control. *Technometrics*, 65(1):33–43, 2023. doi: 10.1080/00401706.2022.2054861. Cited on p. 145.

Acronyms

a.s. almost surely. 50, 51, 345
AIC Akaike's Information Criterion. 106, 345
aSS An aggregated SS. 93, 345

BCP Best Choice Problem. 16, 17, 24, 84, 225, 269, 345
BIC Bayesian Information Criterion. 106, 345

CDF Continuous Distribution Function. 272, 345
CE Cross-Entropy Method. 266–270, 345
CUSUM Cumulative Sums Method. 36, 37, 144, 145, 345

DM Decision-Maker. 71, 345
DRMSP Dual Representations of Multiple-Stopping Problems. 21, 345

FDR False Discovery Rate. 33, 34, 107, 345
FI Full Information. 16, 345
FWER Family-wise Error Rate. 32–34, 345

HPC High Performance Computing. 247, 345
HTC High-Throughput Computing. 247, 345

ISS An individual stopping set. 92–94, 345

MBC Multiple Best Choice. 24–26, 216, 269, 345
MCDM multi-criteria decision-making. 7, 345
MCM Monte Carlo Method. 6, 194, 345
MDP Markov Decision Process. 70, 71, 345
MOS multiple optimal stopping. 7, 8, 345
MS multiple-stopping. 8, 345
MSM multiple-stopping model. 11, 44, 345
MSP multiple-stopping problem. 11, 21, 345
MSR multiple-stopping rule. 345

ODE ordinary differential equation. 8, 345
OLA one step look a head. 18, 345
OMSR optimal multiple-stopping rule. 51, 345
OSP optimal stopping problem. 8, 16, 345

PBS Portable Batch System. 247, 345

PDE partial differential equation. 108, 345
PDF probability density function. 272, 345
PS Player's Strategy. 345

QC Quality Control. 37, 345

SP Secretary Problem. 19, 24, 25, 345
SPRT sequential probability ratio test. 36, 37, 345
SS Stopping Strategy. 93, 345

TSSRP Thompson-Sampling-Shiryaev-Roberts-Pollak. 145, 345

Glossary

AMS Adaptive Multistage Sampling is a method used in survey sampling or experimental design, where the sampling process is adjusted or adapted based on the information gathered at earlier stages. 34, 345

FEA – the finite element analysis. 300, 345

HPP – the homogeneous Poisson process. 7, 345

MCM – a Monte Carlo Methods (**MCM**) is a mathematical framework that involves generating random samples from a stochastic process to estimate its behaviors and the parameters of the underlying distribution. 345

MDP – a Markov Decision Process (**MDP**) is a mathematical framework used to model decision-making problems in situations where outcomes are partially random and partially under the control of a decision-maker. It consists of a set of states, a set of actions that can be taken in each state, and rules governing how the system transitions from one state to another based on the action taken. Importantly, these transitions are stochastic, meaning they are influenced by random factors. 345

OMS – Optimal Multiple-Stopping procedure is sequential decisions about when to stop data collection or experimentation to optimize a specific objective, such as minimizing costs or maximizing utility. 34, 345

i.i.d. – the independent, identically distributed. 345

player's strategy – determine the action which the player will take at any stage of the game. 75, 345

prophet – a person who is believed to have a special power that allows them to say what a god wishes to tell people, especially about things that will happen in the future. 22, 345

relative ranks – in statistics, ranking is the data transformation in which numerical or ordinal values are replaced by their rank when the data are sorted. Relative ranking is a ranking of data with respect to some restriction of data. Usually, the restriction is determined by the source or its time of delivery. 15, 345

sequential – following a particular order. 10, 345

sequential analysis is a statistical technique that allows the study to be carried out in stages, and to discontinue it as soon as one of the stage analyses shows that the results collected so far are sufficient to confirm the research hypothesis. 6, 7, 9, 345

strategy profile – a set of strategies for all players, which fully specifies all actions in a game. 75, 345

Indices

Index of Names

Index of Terms

Printed in the United States
by Baker & Taylor Publisher Services